Verilog HDL
Design Examples

Verilog HDL Design Examples

Joseph Cavanagh

CRC Press is an imprint of the
Taylor & Francis Group, an **informa** business

CRC Press
Taylor & Francis Group
6000 Broken Sound Parkway NW, Suite 300
Boca Raton, FL 33487-2742

Printed on acid-free paper
Version Date: 20170918

International Standard Book Number-13: 978-1-138-09995-1 (Hardback)

Library of Congress Cataloging-in-Publication Data

Names: Cavanagh, Joseph, author.
Title: Verilog HDL design examples / Joseph Cavanagh.
Description: Boca Raton, FL : CRC Press, 2017. | Includes index.
Identifiers: LCCN 2017022734| ISBN 9781138099951 (hardback : acid-free paper)
| ISBN 9781315103846 (ebook)
Subjects: LCSH: Digital electronics--Computer-aided design. | Logic design. |
Verilog (Computer hardware description language)
Classification: LCC TK7868.D5 C3948 2017 | DDC 621.381--dc23
LC record available at https://lccn.loc.gov/2017022734

Visit the Taylor & Francis Web site at
http://www.taylorandfrancis.com

and the CRC Press Web site at
http://www.crcpress.com

To David Dutton
CEO, Silvaco, Inc.
for generously providing the SILOS Simulation Environment software
for all of my books that use Verilog HDL and for his continued support

CONTENTS

PREFACE

The Verilog language provides a means to model a digital system at many levels of abstraction from a logic gate, to a complex digital system, to a mainframe computer. The purpose of this book is to present the Verilog language together with a wide variety of examples so that the reader can gain a firm foundation in the design of digital systems using Verilog HDL. The different modeling constructs supported by Verilog are described in detail.

Numerous examples are designed in each chapter. The examples include logical operations, counters of different moduli, half adders, full adders, a carry lookahead adder, array multipliers, the Booth multiply algorithm, different types of Moore and Mealy machines, including sequence detectors, arithmetic and logic units (ALUs). Also included are synchronous sequential machines and asynchronous sequential machines, including pulse-mode asynchronous sequential machines.

Emphasis is placed on the detailed design of various Verilog projects. The projects include the design module, the test bench module, and the outputs obtained from the simulator that illustrate the complete functional operation of the design. Where applicable, a detailed review of the theory of the topic is presented together with the logic design principles. This includes state diagrams, Karnaugh maps, equations, and the logic diagram.

The book is intended to be tutorial, and as such, is comprehensive and self-contained. All designs are carried through to completion — nothing is left unfinished or partially designed. Each chapter includes numerous problems of varying complexity to be designed by the reader.

Chapter 1 presents an overview of the Verilog HDL language and discusses the different design methodologies used in designing a project. The chapter is intended to introduce the reader to the basic concepts of Verilog modeling techniques, including dataflow modeling, behavioral modeling, and structural modeling. Examples are presented to illustrate the different modeling techniques. There are also sections that incorporate more than one modeling construct in a mixed-design model. The concept of ports and modules is introduced in conjunction with the use of test benches for module design verification.

The chapter introduces gate-level modeling using built-in primitive gates. Verilog has a profuse set of built-in primitive gates that are used to model nets, including **and**, **nand**, **or**, **nor**, **xor**, **xnor**, and **not**, among others. This chapter presents a design methodology that is characterized by a low level of abstraction, in which the logic hardware is described in terms of gates. This is similar to designing logic by drawing logic gate symbols.

The chapter also describes different techniques used to design logic circuits using dataflow modeling. These techniques include the continuous assignment statement, reduction operators, the conditional operator, relational operators, logical operators, bitwise operators, and shift operators.

This chapter also presents behavioral modeling, which describes the *behavior* of a digital system and is not concerned with the direct implementation of logic gates, but more on the architecture of the system. This is an algorithmic approach to hardware implementation and represents a higher level of abstraction than previous modeling methods.

Also included in this chapter is structural modeling, which consists of instantiating one or more of the following design objects into the module:

- Built-in primitives
- User-defined primitives (UDPs)
- Design modules

Instantiation means to use one or more lower-level modules — including logic primitives — that are interconnected in the construction of a higher-level structural module.

Chapter 2 presents combinational logic design using Verilog HDL. Verilog is used to design multiplexers, comparators, programmable logic devices, and a variety of logic equations in this chapter. A combinational logic circuit is one in which the outputs are a function of the present inputs only. This chapter also includes number systems and Boolean algebra. The number systems are binary, octal, decimal, and hexadecimal. Boolean algebra is a systematic treatment of the logic operations AND, OR, NOT, exclusive-OR, and exclusive-NOR. The axioms and theorems of Boolean algebra are also presented. The programmable logic devices include programmable read-only memories, programmable array logic devices, and programmable logic array devices.

Chapter 3 presents the design of sequential logic using Verilog HDL. The examples include both Moore and Mealy sequential machines. Moore machines are synchronous sequential machines in which the output function produces an output vector which is determined by the present state only, and is not a function of the present inputs. This is in contrast to Mealy synchronous sequential machines in which the output function produces an output vector which is determined by both the present input vector and the present state of the machine.

This chapter describes three types of sequential machines: synchronous sequential machines which use a system clock and generally require a state diagram or a state table for its precise description; asynchronous sequential machines in which there is no system clock — state changes occur on the application of input signals only; and pulse-mode asynchronous sequential machines in which state changes occur on the application of input pulses which trigger the storage elements, rather than on a system clock signal.

Chapter 4 presents arithmetic operations for the three primary number representations: fixed-point, binary-coded decimal (BCD), and floating-point. For fixed-point, the radix point is placed to the immediate right of the number for integers or to the immediate left of the number for fractions. For binary-coded decimal, each decimal digit can be encoded into a corresponding binary number; however, only ten decimal digits are valid. For floating-point, the numbers consist of the following three fields: a sign bit, an exponent e, and a fraction f, as shown below for radix r. Addition, subtraction, multiplication, and division will be applied to all three number representations.

$$A = f \times r^e$$

For fixed-point addition,the two operands are the augend and the addend. The addend is added to the augend to produce the sum. Addition of two binary operands treats both signed and unsigned operands the same — there is no distinction between the two types of numbers during the add operation. If the numbers are signed, then the sign bit can be extended to the left indefinitely without changing the value of the number.

For fixed-point subtraction, the two operands are the minuend and the subtrahend. The subtrahend is subtracted from the minuend to produce the difference. Subtraction can be performed in all three number representations: sign magnitude, diminished-radix complement, and radix complement; however, radix complement is the easiest and most widely used method for subtraction in any radix.

For fixed-point multiplication, the two operands are the multiplicand and the multiplier. The n-bit multiplicand is multiplied by the n-bit multiplier to generate the $2n$-bit product. In all methods of multiplication the product is usually $2n$ bits in length. The operands can be either unsigned or signed numbers in 2s complement representation.

For fixed-point division, the two operands are the dividend and the divisor. The $2n$-bit dividend is divided by the n-bit divisor to produce an n-bit quotient and an n-bit remainder, as shown below.

$$2n\text{-bit dividend} = (n\text{-bit divisor} \times n\text{-bit quotient}) + n\text{-bit remainder}$$

For binary-coded decimal addition, and other BCD calculations, the highest-valued decimal digit is 9, which requires four bits in the binary representation (1001). Therefore, each operand is represented by a 4-bit BCD code. Since four binary bits have sixteen combinations (0000 – 1111) and the range for a single decimal digit is 0 – 9, six of the sixteen combinations (1010 – 1111) are invalid for BCD. These invalid BCD digits must be converted to valid digits by adding six to the digit. This is the concept for addition with sum correction. The adder must include correction logic for intermediate sums that are greater than or equal to 1010 in radix 2.

For binary-coded decimal subtraction, the BCD code is not self-complementing as is the radix 2 fixed-point number representation; that is, the $r - 1$ complement cannot be acquired by inverting each bit of the 4-bit BCD digit. Therefore, a 9s complementer must be designed that provides the same function as the diminished-radix complement for the fixed-point number representation. Thus, subtraction in BCD is essentially the same as in fixed-point binary.

For binary-coded decimal multiplication, the algorithms for BCD multiplication are more complex than those for fixed-point multiplication. This is because decimal digits consist of four binary bits and have values in the range of 0 to 9, whereas fixed-point digits have values of 0 or 1. One method that is commonly used is to perform the multiplication in the fixed-point number representation; then convert the product to the BCD number representation. This is accomplished by utilizing a binary-to-decimal converter, which is used to convert a fixed-point multiplication product to the decimal number representation.

For binary-coded decimal division, the division process is first reviewed by using examples of the restoring division method. Then a mixed-design (behavioral/dataflow) module is presented. The dividend is an 8-bit vector, *a[7:0]*; the divisor is a 4-bit vector, *b[3:0]*; and the result is an 8-bit quotient/remainder vector, *rslt[7:0]*.

For floating-point addition, the material presented is based on the Institute of Electrical and Electronics Engineers (IEEE) Standard for Binary Floating-Point Arithmetic IEEE Std 754-1985 (Reaffirmed 1990). Floating-point numbers consist of the following three fields: a sign bit s, an exponent e, and a fraction f. Unbiased and biased exponents are explained. Numerical examples are given that clarify the technique for adding floating-point numbers. The floating-point addition algorithm is given in a step-by-step procedure. A floating-point adder is implemented using behavioral modeling.

For floating-point subtraction, several numerical examples are presented that graphically portray the steps required for true addition and true subtraction for floating-point operands. True addition produces a result that is the sum of the two operands disregarding the signs; true subtraction produces a result that is the difference of the two operands disregarding the signs. A behavioral module is presented that illustrates subtraction operations which yield results that are either true addition or true subtraction.

For floating-point multiplication, numerical examples are presented that illustrate the operation of floating-point multiplication. In floating-point multiplication, the fractions are multiplied and the exponents are added. The fractions are multiplied by any of the methods previously used in fixed-point multiplication. The operands are two normalized floating-point operands. Fraction multiplication and exponent addition are two independent operations and can be done in parallel. Floating-point multiplication is defined as follows:

$$A \times B = (f_A \times f_B) \times r^{(eA + eB)}$$

For floating-point division, the operation is accomplished by dividing the fractions and subtracting the exponents. The fractions are divided by any of the methods presented in the section on fixed-point division and overflow is checked in the same manner. Fraction division and exponent subtraction are two independent operations and can be done in parallel. Floating-point division is defined as follows:

$$A / B = (f_A / f_B) \times r^{(eA - eB)}$$

Appendix A presents a brief discussion on event handling using the event queue. Operations that occur in a Verilog module are typically handled by an event queue.

Appendix B presents a procedure to implement a Verilog project.

Appendix C contains the solutions to selected problems in each chapter.

The material presented in this book represents more than two decades of computer equipment design by the author. The book is not intended as a text on logic design, although this subject is reviewed where applicable. It is assumed that the reader has an adequate background in combinational and sequential logic design. The book presents the Verilog HDL with numerous design examples to help the reader thoroughly understand this popular HDL.

This book is designed for practicing electrical engineers, computer engineers, and computer scientists; for graduate students in electrical engineering, computer engineering, and computer science; and for senior-level undergraduate students.

A special thanks to David Dutton, CEO of Silvaco Incorporated, for allowing use of the SILOS Simulation Environment software for the examples in this book. SILOS is an intuitive, easy-to-use, yet powerful Verilog HDL simulator for logic verification.

I would like to express my appreciation and thanks to the following people who gave generously of their time and expertise to review the manuscript and submit comments: Professor Daniel W. Lewis, Department of Computer Engineering, Santa Clara University who supported me in all my endeavors; Geri Lamble; and Steve Midford. Thanks also to Nora Konopka and the staff at Taylor & Francis for their support.

Joseph Cavanagh

By the Same Author

SEQUENTIAL LOGIC and VERILOG HDL FUNDAMENTALS

X86 ASSEMBLY LANGUAGE and C FUNDAMENTALS

COMPUTER ARITHMETIC and Verilog HDL Fundamentals

DIGITAL DESIGN and Verilog HDL Fundamentals

VERILOG HDL: Digital Design and Modeling

SEQUENTIAL LOGIC: Analysis and Synthesis

DIGITAL COMPUTER ARITHMETIC: Design and Implementation

THE COMPUTER CONSPIRACY
A novel

1

Introduction to Logic Design Using Verilog HDL

This chapter provides an introduction to the design methodologies and modeling constructs of the Verilog hardware description language (HDL). Modules, ports, and test benches will be presented. This chapter introduces Verilog in conjunction with combinational logic and sequential logic. The Verilog simulator used in this book is easy to learn and use, yet powerful enough for any application. It is a logic simulator — called SILOS — developed by Silvaco Incorporated for use in the design and verification of digital systems. The SILOS simulation environment is a method to quickly prototype and debug any application-specific integrated circuit (ASIC), field-programmable gate array (FPGA), or complex programmable logic device (CPLD) design.

Language elements will be described, which consist of comments, logic gates, logic macro functions, parameters, procedural control statements which modify the flow of control in a program, and data types. Also presented will be expressions consisting of operands and operators. Built-in primitives are discussed which are used to describe a net. In addition to built-in primitives, user-defined primitives (UDPs) are presented which are at a higher-level logic function than built-in primitives.

This chapter also presents dataflow modeling which is at a higher level of abstraction than built-in primitives or user-defined primitives. Dataflow modeling corresponds one-to-one with conventional logic design at the gate level. Also introduced is behavioral modeling which describes the behavior of the system and is not concerned with the direct implementation of the logic gates but more on the architecture of the machine. Structural modeling is presented which instantiates one or more lower-level modules into the design. The objects that are instantiated are called *instances*. A module can be a logic gate, an adder, a multiplexer, a counter, or some other logical function. Structural modeling is described by the interconnection of these lower-level logic primitives of modules.

Tasks and functions are also included in this chapter. These constructs allow a behavioral module to be partitioned into smaller segments. Tasks and functions permit modules to execute common code segments that are written once then called when required, thus reducing the amount of code needed.

1.1 Logic Elements

Logic elements are the constituent parts of the Verilog language. They consist of comments, logic gates, parameters, procedural control statements which modify the flow of control in a behavior, and data types.

1.1.1 Comments

Comments can be inserted into a Verilog module to explain the function of a particular block of code or a line of code. There are two types of comments: single line and multiple lines. A single-line comment is indicated by a double forward slash (//) and may be placed on a separate line or at the end of a line of code, as shown below.

```
//This is a single-line comment on a dedicated line
assign z1 = x1 | x2   //This is a comment on a line of code
```

A single-line comment usually explains the function of the following block of code. A comment on a line of code explains the function of that particular line of code. All characters that follow the forward slashes are ignored by the compiler.

A multiple-line comment begins with a forward slash followed by an asterisk (/*) and ends with an asterisk followed by a forward slash (*/), as shown below. Multiple-line comments cannot be nested. All characters within a multiple-line comment are ignored by the compiler.

```
/*This is a multiple-line comment.
   More comments go here.
   More comments. */
```

1.1.2 Logic Gates

Figure 1.1 shows the logic gate distinctive-shape symbols. The polarity symbol "○" indicates an active-low assertion on either an input or an output of a logic symbol.

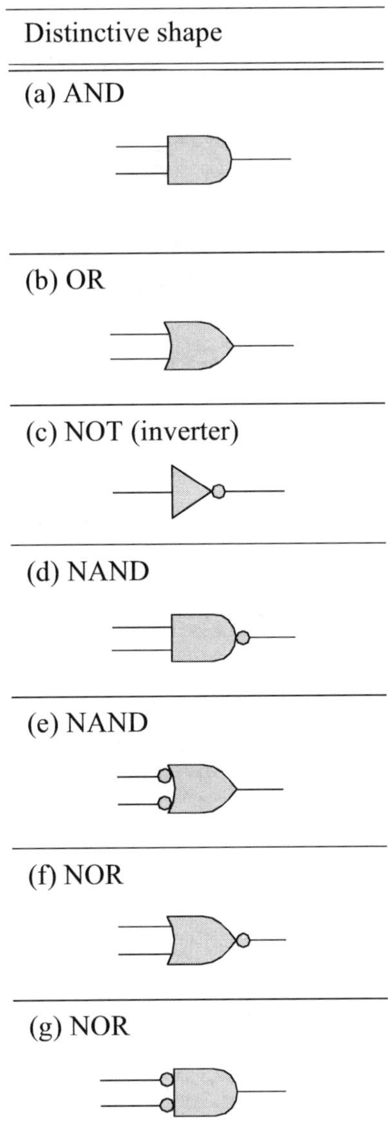

Figure 1.1 Logic gate symbols for logic design: (a) AND gate, (b) OR gate, (c) NOT function (inverter), (d) NAND gate, (e) NAND gate for the OR function, (f) NOR gate, (g) NOR gate for the AND function.

The AND gate can also be used for the OR function, as shown below.

AND gate for the AND function

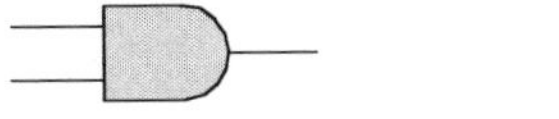

AND gate for the OR function

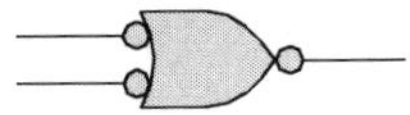

The OR gate can also be used for the AND function, as shown below.

OR gate for the OR function

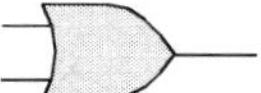

OR gate for the AND function

An exclusive-OR gate is shown below. The output of an exclusive-OR gate is a logical 1 whenever the two inputs are different.

Exclusive-OR gate

An exclusive-NOR gate is shown below. An exclusive-NOR gate is also called an equality function because the output is a logical 1 whenever the two inputs are equal.

Exclusive-NOR gate

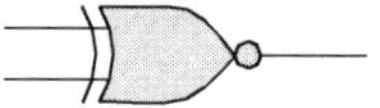

Truth tables for the logic elements are shown in Table 1.1, Table 1.2, Table 1.3, Table 1.4, Table 1.5, and Table 1.6.

Table 1.1 Truth Table for the AND Gate

x_1	x_2	z_1
0	0	0
0	1	0
1	0	0
1	1	1

Table 1.2 Truth Table for the NAND Gate

x_1	x_2	z_1
0	0	1
0	1	1
1	0	1
1	1	0

Table 1.3 Truth Table for the OR Gate

x_1	x_2	z_1
0	0	0
0	1	1
1	0	1
1	1	1

Table 1.4 Truth Table for the NOR Gate

x_1	x_2	z_1
0	0	1
0	1	0
1	0	0
1	1	0

Table 1.5 Truth Table for the Exclusive-OR Function

x_1	x_2	z_1
0	0	0
0	1	1
1	0	1
1	1	0

Table 1.6 Truth Table for the Exclusive-NOR Function

x_1	x_2	z_1
0	0	1
0	1	0
1	0	0
1	1	1

Fan-In Logic gates for the AND and OR functions can be extended to accommodate more than two variables; that is, more than two inputs. The number of inputs available at a logic gate is called the *fan-in*.

Fan-Out The *fan-out* of a logic gate is the maximum number of inputs that the gate can drive and still maintain acceptable voltage and current levels. That is, the fan-out defines the maximum load that the gate can handle.

1.1.3 Logic Macro Functions

Logic macro functions are those circuits that consist of several logic primitives to form larger more complex functions. Combinational logic macros include circuits such as multiplexers, decoders, encoders, comparators, adders, subtractors, array multipliers, array dividers, and error detection and correction circuits. Sequential logic macros include circuits such as: *SR* latches; *D* and *JK* flip-flops; counters of various moduli, including count-up and count-down counters; registers, including shift registers; and sequential multipliers and dividers. This section will present the functional operation of multiplexers, decoders, encoders, priority encoders, and comparators.

Multiplexers A multiplexer is a logic macro device that allows digital information from two or more data inputs to be directed to a single output. Data input selection is controlled by a set of select inputs that determine which data input is gated to the output. The select inputs are labeled $s_0, s_1, s_2, \cdots, s_i, \cdots, s_{n-1}$, where s_0 is the low-order select input with a binary weight of 2^0 and s_{n-1} is the high-order select input with a binary weight of 2^{n-1}. The data inputs are labeled $d_0, d_1, d_2, \cdots, d_j, \cdots, d_{n-1}$. Thus, if a multiplexer has n select inputs, then the number of data inputs will be 2^n and will be labeled d_0 through d_{n-1}. For example, if $n = 2$, then the multiplexer has two select inputs s_0 and s_1 and four data inputs d_0, d_1, d_2, and d_3.

The logic diagram for a 4:1 multiplexer is shown in Figure 1.2. There can also be an *enable* input which gates the selected data input to the output. Each of the four data inputs x_0, x_1, x_2, and x_3 is connected to a separate 3-input AND gate. The select inputs

s_0 and s_1 are decoded to select a particular AND gate. The output of each AND gate is applied to a 4-input OR gate that provides the single output z_1. Input lines that are not selected cannot be transferred to the output and are treated as "don't cares."

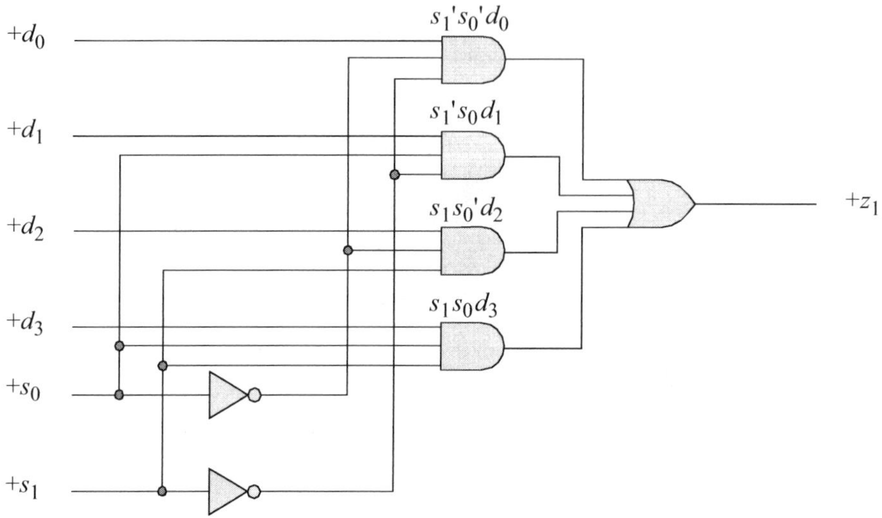

Figure 1.2 Logic diagram for a 4:1 multiplexer.

Figure 1.3 shows a typical multiplexer drawn in the ANSI/IEEE Std. 91-1984 format. Consider the 4:1 multiplexer in Figure 1.3. If $s_1 s_0 = 00$, then data input d_0 is selected and its value is propagated to the multiplexer output z_1. Similarly, if $s_1 s_0 = 01$, then data input d_1 is selected and its value is directed to the multiplexer output.

The equation that represents output z_1 in the 4:1 multiplexer is shown in Equation 1.1. Output z_1 assumes the value of d_0 if $s_1 s_0 = 00$, as indicated by the term $s_1's_0'd_0$. Likewise, z_1 assumes the value of d_1 when $s_1 s_0 = 01$, as indicated by the term $s_1's_0d_1$.

$$z_1 = s_1's_0'd_0 + s_1's_0d_1 + s_1s_0'd_2 + s_1s_0d_3 \tag{1.1}$$

There is a one-to-one correspondence between the data input numbers d_i of a multiplexer and the minterm locations in a Karnaugh map. Equation 1.2 is plotted on the Karnaugh map shown in Figure 1.3(a) using x_3 as a map-entered variable. Minterm location 0 corresponds to data input d_0 of the multiplexer; minterm location 1 corresponds to data input d_1; minterm location 2 corresponds to data input d_2; and minterm location 3 corresponds to data input d_3. The Karnaugh map and the multiplexer implement Equation 1.2, where x_2 is the low-order variable in the Karnaugh map. Figure 1.3(b) shows the implementation using a 4:1 multiplexer.

$$z_1 = x_1 x_2 (x_3') + x_1 x_2' (x_3) + x_1' x_2 \qquad (1.2)$$

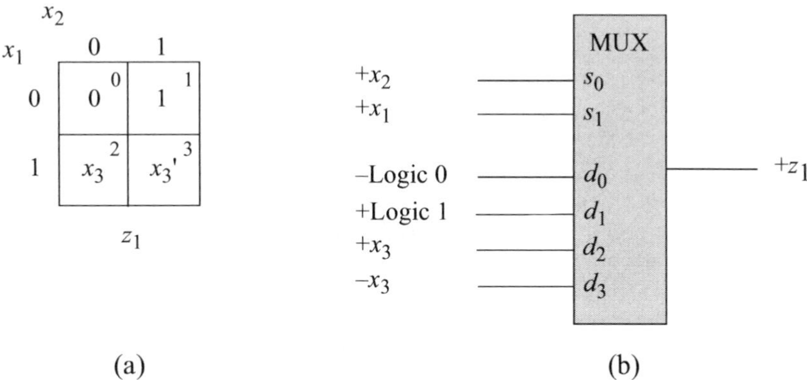

Figure 1.3 Multiplexer using a map-entered variable: (a) Karnaugh map and (b) a 4:1 multiplexer.

Linear-select multiplexers The multiplexer examples described thus far have been classified as *linear-select multiplexers*, because all of the variables of the Karnaugh map coordinates have been utilized as the select inputs for the multiplexer. Since there is a one-to-one correspondence between the minterms of a Karnaugh map and the data inputs of a multiplexer, designing the input logic is relatively straightforward. Simply assign the values of the minterms in the Karnaugh map to the corresponding multiplexer data inputs with the same subscript.

Nonlinear-select multiplexers Although the logic functions correctly according to the equation using a linear-select multiplexer, the design may demonstrate an inefficient use of the 2^p:1 multiplexers. Smaller multiplexers with fewer data inputs could be effectively utilized with a corresponding reduction in machine cost.

For example, the Karnaugh map shown in Figure 1.4 can be implemented with a 4:1 nonlinear-select multiplexer for the function z_1 instead of an 8:1 linear-select multiplexer. Variables x_2 and x_3 will connect to select inputs s_1 and s_0, respectively. When select inputs $s_1 s_0 = x_2 x_3 = 00$, data input d_0 is selected; therefore, $d_0 = 0$. When select inputs $s_1 s_0 = x_2 x_3 = 01$, data input d_1 is selected and d_1 contains the complement of x_1; therefore, $d_1 = x_1'$. When select inputs $s_1 s_0 = x_2 x_3 = 10$, data input d_2 is selected; therefore, $d_2 = 1$. When $s_1 s_0 = x_2 x_3 = 11$, data input d_3 is selected and contains the same value as x_1; therefore, $d_3 = x_1$. The logic diagram is shown in Figure 1.5

The multiplexer of Figure 1.5 can be checked to verify that it operates according to the Karnaugh map of Figure 1.4; that is, for every value of $x_1 x_2 x_3$, output z_1 should generate the same value as in the corresponding minterm location.

x_1 \ x_2x_3	0 0	0 1	1 1	10
0	0 (0)	1 (1)	0 (3)	1 (2)
1	0 (4)	0 (5)	1 (7)	1 (6)
	$d_0 = 0$	$d_1 = x_1'$	$d_3 = x_1$	$d_2 = 1$

z_1

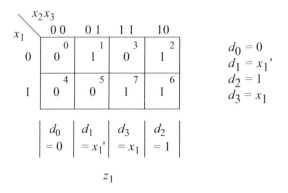

$d_0 = 0$
$d_1 = x_1'$
$d_2 = 1$
$d_3 = x_1$

Figure 1.4 Karnaugh map for an example which will be implemented by a 4:1 nonlinear-select multiplexer.

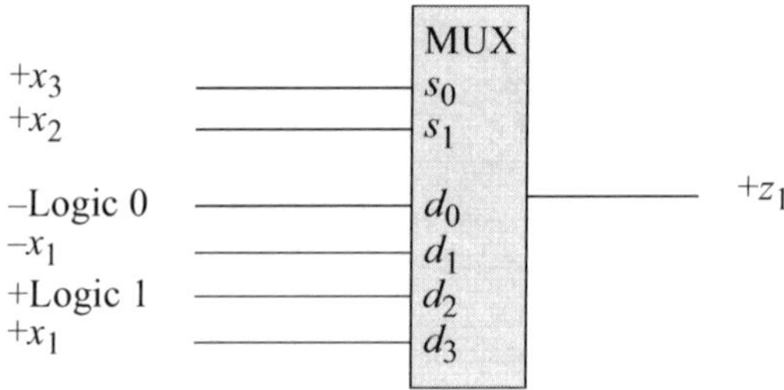

Figure 1.5 A 4:1 nonlinear-select multiplexer to implement the Karnaugh map of Figure 1.4.

Decoders A decoder is a combinational logic macro that is characterized by the following property: For every valid combination of inputs, a unique output is generated. In general, a decoder has n binary inputs and m mutually exclusive outputs, where $2^n \geq m$. An $n{:}m$ (n-to-m) decoder is shown in Figure 1.6, where the label DX specifies a demultiplexer. Each output represents a minterm that corresponds to the binary representation of the input vector. Thus, $z_i = m_i$, where m_i is the ith minterm of the n input variables.

For example, if $n = 3$ and $x_1x_2x_3 = 101$, then output z_5 is asserted. A decoder with n inputs, therefore, has a maximum of 2^n outputs. Because the outputs are mutually exclusive, only one output is active for each different combination of the inputs. The decoder outputs may be asserted high or low. Decoders have many applications in digital engineering, ranging from instruction decoding to memory addressing to code conversion.

Figure 1.7 illustrates the logic symbol for a 2:4 decoder, where x_1 and x_2 are the binary input variables and z_0, z_1, z_2, and z_3 are the output variables. Input x_2 is the low-order variable. Since there are two inputs, each output corresponds to a different minterm of two variables.

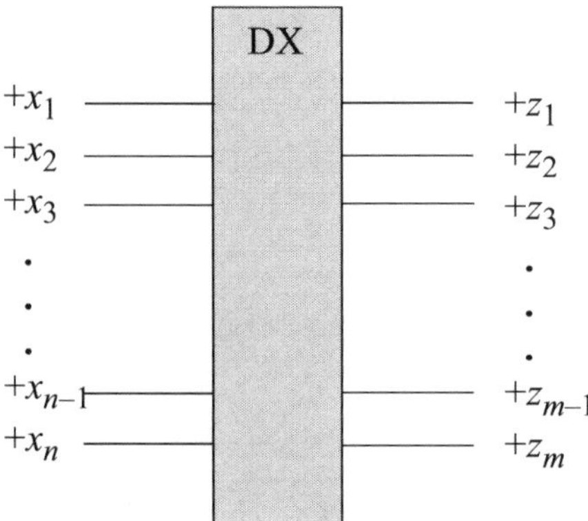

Figure 1.6 An $n{:}m$ decoder.

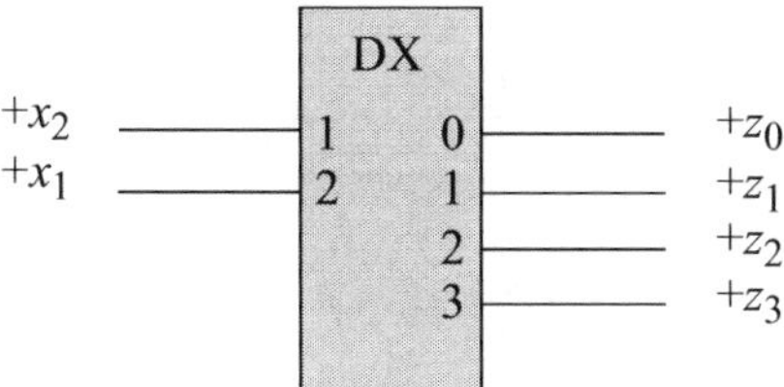

Figure 1.7 Logic symbol for a 2:4 decoder.

A 3:8 decoder is shown in Figure 1.8 which decodes a binary number into the corresponding octal number. The three inputs are x_1, x_2, and x_3 with binary weights of 2^2, 2^1, and 2^0, respectively. The decoder generates an output that corresponds to the decimal value of the binary inputs. For example, if $x_1x_2x_3 = 110$, then output z_6 is asserted high. A decoder may also have an enable function which allows the selected output to be asserted.

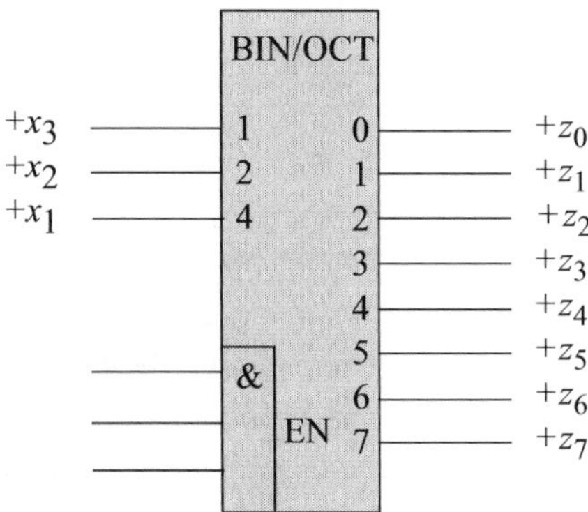

Figure 1.8 A binary-to-octal decoder.

The internal logic for the binary-to-octal decoder of Figure 1.8 is shown in Figure 1.9. The *Enable* gate allows for additional logic functions to control the assertion of the active-high outputs.

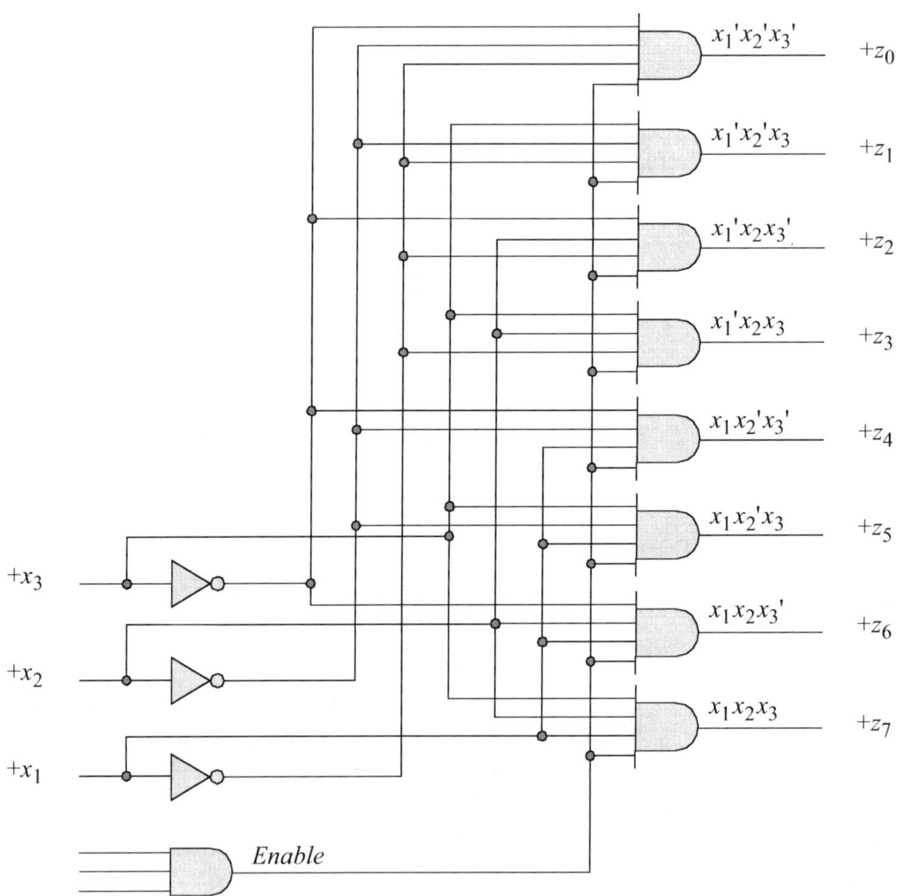

Figure 1.9 Internal logic for the binary-to-octal decoder of Figure 1.8.

Encoders An encoder is a macro logic circuit with n mutually exclusive inputs and m binary outputs, where $n \leq 2^m$. The inputs are mutually exclusive to prevent errors from appearing on the outputs. The outputs generate a binary code that corresponds to the active input value. The function of an encoder can be considered to be the inverse of a decoder; that is, the mutually exclusive inputs are encoded into a corresponding binary number.

A general block diagram for an $n{:}m$ encoder is shown in Figure 1.10. An encoder is also referred to as a code converter. In the label of Figure 1.10, X corresponds to the input code and Y corresponds to the output code. The general qualifying label X/Y is replaced by the input and output codes, respectively, such as, OCT/BIN for an octal-to-binary code converter. Only one input x_i is asserted at a time. The decimal value of x_i is encoded as a binary number which is specified by the m outputs.

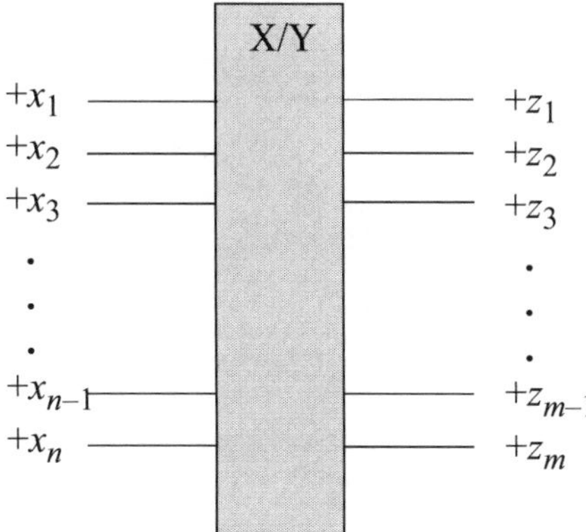

Figure 1.10 An n:m encoder or code converter.

An 8:3 octal-to-binary encoder is shown in Figure 1.11. Although there are 2^8 possible input combinations of eight variables, only eight combinations are valid. The eight inputs each generate a unique octal code word in binary. If the outputs are to be enabled, then the gating can occur at the output gates.

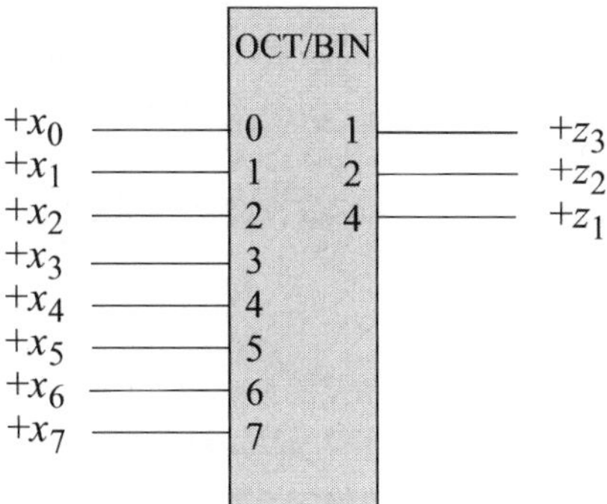

Figure 1.11 An octal-to-binary encoder.

The low-order output z_3 is asserted when one of the following inputs are active: x_1, x_3, x_5, or x_7. Output z_2 is asserted when one of the following inputs are active: x_2, x_3, x_6, or x_7. Output z_1 is asserted when one of the following inputs are active: x_4, x_5, x_6, or x_7. The encoder can be implemented with OR gates whose inputs are established from Equation 1.3 and Figure 1.12.

$$\begin{aligned} z_3 &= x_1 + x_3 + x_5 + x_7 \\ z_2 &= x_2 + x_3 + x_6 + x_7 \\ z_1 &= x_4 + x_5 + x_6 + x_7 \end{aligned} \tag{1.3}$$

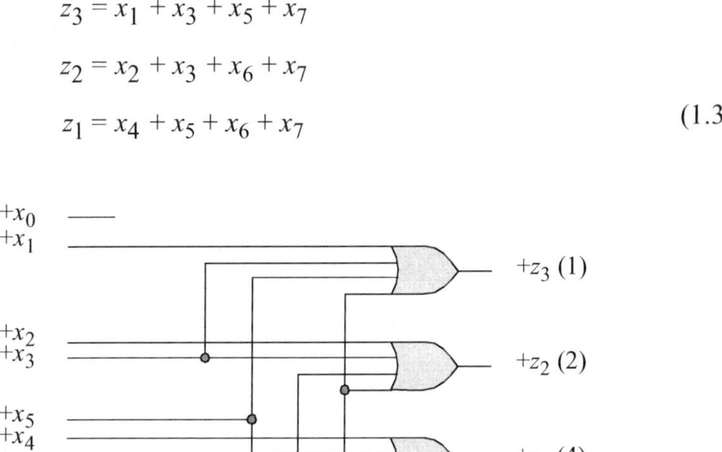

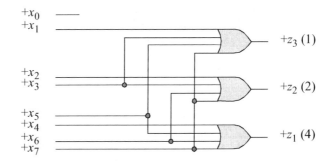

Figure 1.12 Logic diagram for an 8:3 encoder.

Priority encoder It was stated previously that encoder inputs are mutually exclusive. There may be situations, however, where more than one input can be active at a time. Then a priority must be established to select and encode a particular input. This is referred to as a *priority encoder.*

Usually the input with the highest valued subscript is selected as highest priority for encoding. Thus, if x_i and x_j are active simultaneously and $i < j$, then x_j has priority over x_i. The truth table for an octal-to-binary priority encoder is shown in Table 1.7. The outputs $z_1 z_2 z_3$ generate a binary number that is equivalent to the highest priority input. If $x_3 = 1$, the state of x_0, x_1, and x_2 is irrelevant ("don't care") and the output is the binary number 011.

Comparators A comparator is a logic macro circuit that compares the magnitude of two n-bit binary numbers X_1 and X_2. Therefore, there are $2n$ inputs and three outputs that indicate the relative magnitude of the two numbers. The outputs are mutually exclusive, specifying $X_1 < X_2$, $X_1 = X_2$, or $X_1 > X_2$. Figure 1.13 shows a general block diagram of a comparator.

If two or more comparators are connected in cascade, then three additional inputs are required for each comparator. These additional inputs indicate the relative magnitude of the previous lower-order comparator inputs and specify $X_1 < X_2$, $X_1 = X_2$, or $X_1 > X_2$ for the previous stage. Cascading comparators usually apply only to commercially available comparator integrated circuits.

Table 1.7 Octal-to-Binary Priority Encoder

Inputs								Outputs		
x_0	x_1	x_2	x_3	x_4	x_5	x_6	x_7	z_1	z_2	z_3
1	0	0	0	0	0	0	0	0	0	0
–	**1**	0	0	0	0	0	0	0	0	1
–	–	**1**	0	0	0	0	0	0	1	0
–	–	–	**1**	0	0	0	0	0	1	1
–	–	–	–	**1**	0	0	0	1	0	0
–	–	–	–	–	**1**	0	0	1	0	1
–	–	–	–	–	–	**1**	0	1	1	0
–	–	–	–	–	–	–	**1**	1	1	1

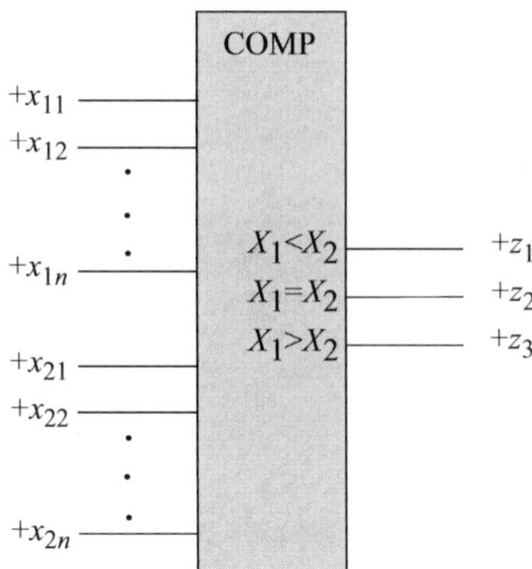

Figure 1.13 General block diagram of a comparator.

Designing the hardware for a comparator is relatively straightforward — it consists of AND gates, OR gates, and exclusive-NOR circuits as shown in Equation 1.4. An alternative approach which may be used to minimize the amount of hardware is to eliminate the equation for $X_1 = X_2$ and replace it with Equation 1.5. That is, if X_1 is neither less nor greater than X_2, then X_1 must equal X_2.

$$(X_1 < X_2) = x_{11}'x_{21} + (x_{11} \oplus x_{21})'x_{12}'x_{22} + (x_{11} \oplus x_{21})'(x_{12} \oplus x_{22})'x_{13}'x_{23}$$

$$(X_1 = X_2) = (x_{11} \oplus x_{21})'(x_{12} \oplus x_{22})'(x_{13} \oplus x_{23})'$$

$$(X_1 > X_2) = x_{11}x_{21}' + (x_{11} \oplus x_{21})'x_{12}x_{22}' + (x_{11} \oplus x_{21})'(x_{12} \oplus x_{22})'x_{13}x_{23}' \quad (1.4)$$

$$(X_1 = X_2) \text{ if } (X_1 < X_2)' \text{ AND } (X_1 > X_2)' \quad (1.5)$$

1.1.4 Procedural Flow Control

Procedural flow control statements modify the flow in a behavior by selecting branch options, repeating certain activities, selecting a parallel activity, or terminating an activity. The activity can occur in sequential blocks or in parallel blocks.

begin . . . end The **begin** . . . **end** keywords are used to group multiple statements into sequential blocks. The statements in a sequential block execute in sequence; that is, a statement does not execute until the preceding statement has executed, except for nonblocking statements. If there is only one procedural statement in the block, then the **begin** . . . **end** keywords may be omitted.

disable The **disable** statement terminates a named block of procedural statements or a task and transfers control to the statement immediately following the block or task. The **disable** statement can also be used to exit a loop.

for The keyword **for** is used to specify a loop. The **for** loop repeats the execution of a procedural statement or a block of procedural statements a specified number of times. The **for** loop is used when there is a specified beginning and end to the loop. The format and function of a **for** loop is similar to the **for** loop used in the C programming language. The parentheses following the keyword **for** contain three expressions separated by semicolons, as shown below.

for (register initialization; test condition; update register control variable)
    procedural statement or block of procedural statements

forever The **forever** loop statement executes the procedural statements continuously. The loop is primarily used for timing control constructs, such as clock pulse generation. The **forever** procedural statement must be contained within an **initial** or an **always** block. In order to exit the loop, the **disable** statement may be used to prematurely terminate the procedural statements. An **always** statement executes at the beginning of simulation; the **forever** statement executes only when it is encountered in a procedural block.

if . . . else These keywords are used as conditional statements to alter the flow of activity through a behavioral module. They permit a choice of alternative paths based upon a Boolean value obtained from a condition. The syntax is shown below.

if (condition)
    {procedural statement 1}
else
    {procedural statement 2}

If the result of the *condition* is true, then procedural statement 1 is executed; otherwise, procedural statement 2 is executed. The procedural statement following the **if** and **else** statements can be a single procedural statement or a block of procedural statements. Two uses for the **if . . . else** statement are to model a multiplexer or decode an instruction register operation code to select alternative paths depending on the instruction. The **if** statement can be nested to provide several alternative paths to execute procedural statements as shown in the syntax below for nested **if** statements.

if (condition 1)
    {procedural statement 1}
else if (condition 2)
    {procedural statement 2}
else if (condition 3)
    {procedural statement 3}
else
    {procedural statement 4)

repeat The **repeat** keyword is used to execute a loop a fixed number of times as specified by a constant contained within parentheses following the **repeat** keyword. The loop can be a single statement or a block of statements contained within **begin . . . end** keywords. The syntax is shown below.

repeat (expression)
    statement or block of statements

When the activity flow reaches the **repeat** construct, the expression in parentheses is evaluated to determine the number of times that the loop is to be executed. The *expression* can be a constant, a variable, or a signal value. If the expression evaluates to **x** or **z**, then the value is treated as 0 and the loop is not executed.

while The **while** statement executes a statement or a block of statements while an expression is true. The syntax is shown below.

while (expression) statement

The expression is evaluated and a Boolean value, either true (a logical 1) or false (a logical 0) is returned. If the expression is true, then the procedural statement or block of statements is executed. The **while** loop executes until the expression becomes false, at which time the loop is exited and the next sequential statement is executed. If the expression is false when the loop is entered, then the procedural statement is not executed. If the value returned is **x** or **z**, then the value is treated as false. An example of the **while** statement is shown below where the initial count = 0.

```
while (count < 16)
begin
        count = count + 1;
end
```

1.1.5 Net Data Types

Verilog defines two data types: nets and registers. These predefined data types are used to connect logical elements and to provide storage. A net is a physical wire or group of wires connecting hardware elements in a module or between modules.

An example of net data types is shown in Figure 1.14, where five internal nets are defined: *net1*, *net2*, *net3*, *net4*, and *net5*. The value of *net1* is determined by the inputs to the *and1* gate represented by the term x_1x_2', where x_2 is active low; the value of *net2* is determined by the inputs to the *and2* gate represented by the term $x_1'x_2$, where x_1 is active low; the value of *net3* is determined by the input to the inverter represented by the term x_3', where x_3 is active low. The equations for outputs z_1 and z_2 are listed in Equation 1.6.

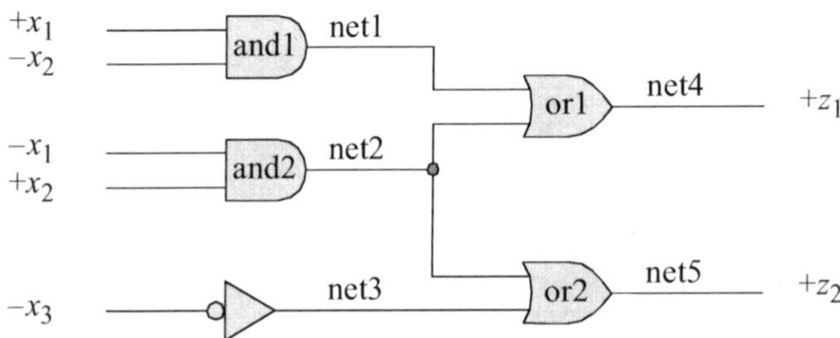

Figure 1.14 A logic diagram showing single-wire nets and one multiple-wire net.

$$z_1 = x_1x_2' + x_1'x_2$$
$$z_2 = x_1'x_2 + x_3 \tag{1.6}$$

1.1.6 Register Data Types

A register data type represents a variable that can retain a value. Verilog registers are similar in function to hardware registers, but are conceptually different. Hardware registers are synthesized with storage elements such as *D* flip-flops, *JK* flip-flops, and *SR* latches. Verilog registers are an abstract representation of hardware registers and are declared as **reg**.

The default size of a register is 1-bit; however, a larger width can be specified in the declaration. The general syntax to declare a width of more than 1-bit is as follows:

reg [most significant bit:least significant bit] register_name.

To declare a one-byte register called data_register is **reg** [7:0] data_register.

Memories Memories can be represented in Verilog by an array of registers and are declared using a **reg** data type as follows:

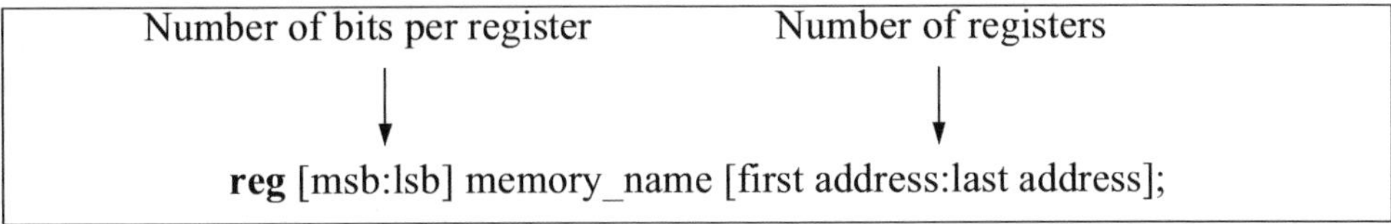

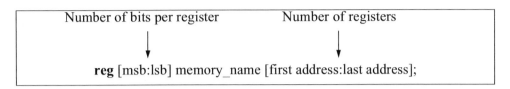

A 32-word register with one byte per word would be declared as follows:

reg [7:0] memory_name [0:31];

An array can have only two dimensions. Memories must be declared as **reg** data types, not as **wire** data types. A register can be assigned a value using one statement, as shown below. Register *buff_reg* is assigned the 16-bit hexadecimal value of 7ab5, which equates to the binary value of $0111\ 1010\ 1011\ 0101_2$.

```
reg [15:0] buff_reg;
buff_reg = 16'h7ab5;
```

Values can also be stored in memories by assigning a value to each word individually, as shown below for an instruction cache of eight registers with eight bits per register.

```
reg [7:0] instr_cache [0:7];
```

```
instr_cache [0] =   8'h08;
instr_cache [1] =   8'h09;
instr_cache [2] =   8'h0a;
instr_cache [3] =   8'h0b;
instr_cache [4] =   8'h0c;
instr_cache [5] =   8'h0d;
instr_cache [6] =   8'h0e;
instr_cache [7] =   8'h0f;
```

1.2 Expressions

Expressions consist of operands and operators, which are the basis of Verilog HDL. The result of a right-hand side expression can be assigned to a left-hand side net variable or register variable using the keyword **assign**. The value of an expression is determined from the combined operations on the operands. An expression can consist of a single operand or two or more operands in conjunction with one or more operators. The result of an expression is represented by one or more bits. Examples of expressions are as follows, where the symbol & indicates an AND operation and the symbol | indicates an OR operation:

assign $z_1 = x_1$ & x_2 & x_3;
assign $z_1 = x_1 \mid x_2 \mid x_3$;
assign cout = (a & cin) | (b & cin) | (a & b);

1.2.1 Operands

Operands can be any of the data types listed in Table 1.8.

Table 1.8 Operands

Operands	Comments
Constant	Signed or unsigned
Parameter	Similar to a constant
Net	Scalar or vector
Register	Scalar or vector
Bit-select	One bit from a vector
Part-select	Contiguous bits of a vector
Memory element	One word of a memory

Constant Constants can be signed or unsigned. A decimal integer is treated as a signed number. An integer that is specified by a base is interpreted as an unsigned number. Examples of both types are shown in Table 1.9.

Table 1.9 Signed and Unsigned Constants

Constant	Comments
127	Signed decimal: Value = 8-bit binary vector: 0111_1111
–1	Signed decimal: Value = 8-bit binary vector: 1111_1111
–128	Signed decimal: Value = 8-bit binary vector: 1000_0000
4'b1110	Binary base: Value = unsigned decimal 14
8'b0011_1010	Binary base: Value = unsigned decimal 58
16'h1A3C	Hexadecimal base: Value = unsigned decimal 6716
16'hBCDE	Hexadecimal base: Value = unsigned decimal 48,350
9'o536	Octal base: Value = unsigned decimal 350
–22	Signed decimal: Value = 8-bit binary vector: 1110_1010
–9'o352	Octal base: Value = 8-bit binary vector: 1110_1010 = unsigned decimal 234

The last two entries in Table 1.9 both evaluate to the same bit configuration, but represent different decimal values. The number -22_{10} is a signed decimal value; the number –9'o352 is treated as an unsigned number with a decimal value of 234_{10}.

Parameter A parameter is similar to a constant and is declared by the keyword **parameter**. Parameter statements assign values to constants; the values cannot be changed during simulation. Examples of parameters are shown in Table 1.10.

Table 1.10 Examples of Parameters

Examples	Comments
parameter width = 8	Defines a bus width of 8 bits
parameter width = 16, depth = 512	Defines a memory with two bytes per word and 512 words
parameter out_port = 8	Defines an output port with an address of 8

Parameters are useful in defining the width of a bus. For example, the adder shown in Figure 1.15 contains two 8-bit vector inputs *a* and *b* and one scalar input *cin*. There is also one 9-bit vector output *sum* comprised of an 8-bit result and a scalar carry-out. The Verilog line of code shown below defines a bus width of eight bits. Wherever *width* appears in the code, it is replaced by the value eight.

```
parameter  width  =  8;
```

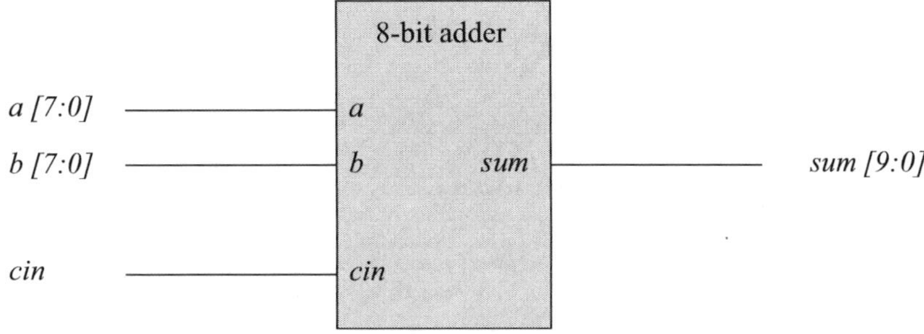

Figure 1.15 Eight-bit adder to illustrate the use of a **parameter** statement.

1.2.2 Operators

Verilog HDL contains a profuse set of operators that perform various operations on different types of data to yield results on nets and registers. Some operators are similar to those used in the C programming language. Table 1.11 lists the categories of operators in order of precedence, from highest to lowest.

Table 1.11 Verilog HDL Operators and Symbols

Operator type	Operator Symbol	Operation	Number of Operands
Arithmetic	+	Add	Two or one
	-	Subtract	Two or one
	*	Multiply	Two
	/	Divide	Two
	%	Modulus	Two
Logical	&&	Logical AND	Two
	||	Logical OR	Two
	!	Logical negation	One
Relational	>	Greater than	Two
	<	Less than	Two
	>=	Greater than or equal	Two
	<=	Less than or equal	Two
Equality	= =	Logical equality	Two
	! =	Logical inequality	Two
	= = =	Case equality	Two
	! = =	Case inequality	Two
Bitwise	&	AND	Two
	|	OR	Two
	~	Negation	One
	^	Exclusive-OR	Two
	^ ~ or ~ ^	Exclusive-NOR	Two
Reduction	&	AND	One
	~ &	NAND	One
	|	OR	One
	~ |	NOR	One
	^	Exclusive-OR	One
	~ ^ or ^ ~	Exclusive-NOR	One
Shift	<<	Left shift	One
	>>	Right shift	One
Conditional	? :	Conditional	Three
Concatenation	{ }	Concatenation	Two or more
Replication	{{ }}	Replication	Two or more

Arithmetic Arithmetic operations are performed on one (unary) operand or two (binary) operands in the following radices: binary, octal, decimal, or hexadecimal. The result of an arithmetic operation is interpreted as an unsigned value or as a signed value in 2s complement representation on both scalar and vector nets and registers. The operands shown in Table 1.12 are used for the operations of addition, subtraction, multiplication, and division.

Table 1.12 Operands Used for Arithmetic Operations

	Addition		Subtraction		Multiplication		Division
	Augend		Minuend		Multiplicand		Dividend
+)	Addend	–)	Subtrahend	×)	Multiplier	÷)	Divisor
	Sum		Difference		Product		Quotient, Remainder

The unary + and – operators change the sign of the operand and have higher precedence than the binary + and – operators. Examples of unary operators are shown below.

$$+45 \text{(Positive } 45_{10}\text{)}$$
$$-72 \text{(Negative } 72_{10}\text{)}$$

Unary operators treat net and register operands as unsigned values, and treat real and integer operands as signed values.

The binary add operator performs unsigned and signed addition on two operands. Register and net operands are treated as unsigned operands; thus, a value of

$$1111\_1111\_1111\_1111_2$$

stored in a register has a value of $65{,}535_{10}$ unsigned, not -1_{10} signed. Real and integer operands are treated as signed operands; thus, a value of

$$1111\_1110\_1010\_0111_2$$

stored in an integer register has a value of -345_{10} signed, not $65{,}191_{10}$ unsigned. The width of the result of an arithmetic operation is determined by the width of the largest operand.

Logical There are three logical operators: the binary logical AND operator (&&), the binary logical OR operator (| |), and the unary logical negation operator (!). Logical operators evaluate to a logical 1 (true), a logical 0 (false), or an **x** (ambiguous). If a logical operation returns a nonzero value, then it is treated as a logical 1 (true); if a bit in an operand is **x** or **z**, then it is ambiguous and is normally treated as a false condition.

Let a and b be two 4-bit operands, where $a = 0110$ and $b = 1100$. Let z_1, z_2, and z_3 be the outputs of the logical operations shown below.

$$z_1 = a \mathbin{\&\&} b$$
$$z_2 = a \mid\mid b$$
$$z_3 = !a$$

Therefore, the operation $z_1 = a \mathbin{\&\&} b$ yields a value of $z_1 = 1$ because both a and b are nonzero. If a vector operand is nonzero, then it treated as a 1 (true). Output z_2 is also equal to 1 for the expression $z_2 = a \mid\mid b$. Output z_3 is equal to 0 because a is true.

Now let $a = 0101$ and $b = 0000$. Thus, $z_1 = a \ \&\& \ b = 1 \ \&\& \ 0 = 0$ because a is true and b is false. Output z_2, however, is equal to 1 because $z_2 = a \ || \ b = 1 \ || \ 0 = 1$. In a similar manner, $z_3 = !a = !1 = 0$, because a is true.

As a final example, let $a = 0000$ and $b = 0000$; that is, both variables are false. Therefore, $z_1 = a \ \&\& \ b = 0 \ \&\& \ 0 = 0$; $z_2 = a \ || \ b = 0 \ || \ 0 = 0$; $z_3 = !a = !0 = 1$. If a bit in either operand is **x**, then the result of a logical operation is **x**. Also, !**x** is **x**.

Relational Relational operators compare operands and return a Boolean result, either 1 (true) or 0 (false) indicating the relationship between the two operands. There are four relational operators as follows: greater than (>), less than (<), greater than or equal (>=), and less than or equal (<=). These operators function the same as identical operators in the C programming language.

If the relationship is true, then the result is 1; if the relationship is false, then the result is 0. Net or register operands are treated as unsigned values; real or integer operands are treated as signed values. An **x** or **z** in any operand returns a result of **x**. When the operands are of unequal size, the smaller operand is zero-extended to the left. Examples are shown below of relational operators, where the identifier *gt* means greater than, *lt* means less than, *gte* means greater than or equal, and *lte* means less than or equal when comparing operand a to operand b.

$$a = 0110, \ b = 1100, \quad gt = 0, \ lt = 1, \ gte = 0, \ lte = 1$$
$$a = 0101, \ b = 0000, \quad gt = 1, \ lt = 0, \ gte = 1, \ lte = 0$$
$$a = 1000, \ b = 1001, \quad gt = 0, \ lt = 1, \ gte = 0, \ lte = 1$$
$$a = 0000, \ b = 0000, \quad gt = 0, \ lt = 0, \ gte = 1, \ lte = 1$$
$$a = 1111, \ b = 1111, \quad gt = 0, \ lt = 0, \ gte = 1, \ lte = 1$$

Equality There are four equality operators: logical equality (==), logical inequality (!=), case equality (===), and case inequality (!==).

Logical equality is used in expressions to determine if two values are identical. The result of the comparison is 1 if the two operands are equal, and 0 if they are not equal. The *logical inequality* operator is used to determine if two operands are unequal. A 1 is returned if the operands are unequal; otherwise a 0 is returned. If the result of the comparison is ambiguous for logical equality or logical inequality, then a value of **x** is returned. An **x** or **z** in either operand will return a value of **x**. If the operands are nets or registers, they are treated as unsigned values; real or integer operands are treated as signed values, but are compared as though they were unsigned operands.

The *case equality* operator compares both operands on a bit-by-bit basis, including **x** and **z**. The result is 1 if both operands are identical in the same bit positions, including those bit positions containing an **x** or a **z**. The *case inequality* operator is used to determine if two operands are unequal by comparing them on a bit-by-bit basis, including those bit positions that contain **x** or **z**.

Examples of the equality operators are shown below, where the 4-bit variables are x_1, x_2, x_3, x_4, and x_5. The outputs are z_1 (logical equality), z_2 (logical inequality), z_3 (case equality), and z_4 (case inequality).

$x_1 = 1000,\ x_2 = 1101,\ x_3 = 01xz,\ x_4 = 01xz,\ x_5 = x1xx$
$z_1 = 0,\ z_2 = 1,\ z_3 = 1,\ z_4 = 1$

$x_1 = 1011,\ x_2 = 1011,\ x_3 = x1xz,\ x_4 = x1xz,\ x_5 = 11xx$
$z_1 = 1,\ z_2 = 1,\ z_3 = 1,\ z_4 = 1$

$x_1 = 1100,\ x_2 = 0101,\ x_3 = x01z,\ x_4 = 11xz,\ x_5 = 11xx$
$z_1 = 0,\ z_2 = 1,\ z_3 = 0,\ z_4 = 1$

Referring to the above outputs for the first set of inputs, the logical equality (z_1) of x_1 and x_2 is false because the operands are unequal. The logical inequality (z_2) of x_2 and x_3 is true. The case equality (z_3) of inputs x_3 and x_4 is 1 because both operands are identical in all bit positions, including the **x** and **z** bits. The case inequality (z_4) of inputs $\boldsymbol{x_4}$ and $\boldsymbol{x_5}$ is also 1 because the operands differ in the high-order and low-order bit positions.

Bitwise The bitwise operators are: AND (&), OR (|), negation (~), exclusive-OR (^), and exclusive-NOR (^~ or ~^). The bitwise operators perform logical operations on the operands on a bit-by-bit basis and produce a vector result. Except for negation, each bit in one operand is associated with the corresponding bit in the other operand. If one operand is shorter, then it is zero-extended to the left to match the length of the longer operand.

The *bitwise AND* operator performs the AND function on two operands on a bit-by-bit basis as shown in the following example:

```
     1  0  1  1  0  1  1  0
 &)  1  1  0  1  0  1  0  1
    ------------------------
     1  0  0  1  0  1  0  0
```

The *bitwise OR* operator performs the OR function on the two operands on a bit-by-bit basis as shown in the following example:

```
     1  0  1  1  0  1  1  0
 |)  1  1  0  1  0  1  0  1
    ------------------------
     1  1  1  1  0  1  1  1
```

The *bitwise negation* operator performs the negation function on one operand on a bit-by-bit basis. Each bit in the operand is inverted as shown in the following example:

```
 ~)  1  1  0  1  0  1  0  1
    ------------------------
     0  0  1  0  1  0  1  0
```

The *bitwise exclusive-OR* operator performs the exclusive-OR function on two operands on a bit-by-bit basis as shown in the following example:

```
     1  0  1  1  0  1  1  0
^)   1  1  0  1  0  1  0  1
    ------------------------
     0  1  1  0  0  0  1  1
```

The *bitwise exclusive-NOR* operator performs the exclusive-NOR function on two operands on a bit-by-bit basis as shown in the following example:

```
     1  0  1  1  0  1  1  0
^~)  1  1  0  1  0  1  0  1
    ------------------------
     1  0  0  1  1  1  0  0
```

Bitwise operators perform operations on operands on a bit-by-bit basis and produce a vector result. This is in contrast to logical operators, which perform operations on operands in such a way that the truth or falsity of the result is determined by the truth or falsity of the operands.

The logical AND operator returns a value of 1 (true) only if both operands are nonzero (true); otherwise, it returns a value of 0 (false). If the result is ambiguous, it returns a value of **x**. The logical OR operator returns a value of 1 (true) if either or both operands are true; otherwise, it returns a value of 0. The logical negation operator returns a value of 1 (true) if the operand has a value of zero and a value of 0 (false) if the operand is nonzero. Examples of the five bitwise operators are shown below. The logical negation operator performs the operation on operand *a*.

a = 11000011, b = 10011001, and_rslt = 10000001, or_rslt = 11011011, neg_rslt = 00111100, xor_rslt = 01011010, xnor_rslt = 10100101	a = 01001111, b = 11011001, and_rslt = 01001001, or_rslt = 11011111, neg_rslt = 10110000, xor_rslt = 10010110, xnor_rslt = 01101001
a = 10010011, b = 11011001, and_rslt = 10010001, or_rslt = 11011011, neg_rslt = 01101100, xor_rslt = 01001010, xnor_rslt =10110101	a = 11001111, b = 11011001, and_rslt = 11001001, or_rslt = 11011111, neg_rslt = 00110000, xor_rslt = 00010110, xnor_rslt = 11101001

Reduction The reduction operators are: AND (&), NAND (~&), OR (|), NOR (~|), exclusive-OR (^), and exclusive-NOR (^~ or ~^). Reduction operators are unary operators; that is, they operate on a single vector and produce a single-bit result. If any bit of the operand is **x** or **z**, the result is **x**. Reduction operators perform their respective operations on a bit-by-bit basis.

For the *reduction AND* operator, if any bit in the operand is 0, then the result is 0; otherwise, the result is 1. For example, let x_1 be the vector shown below.

1	1	1	0	1	0	1	1

The reduction AND ($\& x_1$) operation is equivalent to the following operation:

1 & 1 & 1 & 0 & 1 & 0 & 1 & 1

which returns a result of 1'b0.

For the *reduction NAND* operator, if any bit in the operand is 0, then the result is 1; otherwise, the result is 0. For a vector x_1, the reduction NAND ($\sim\& x_1$) is the inverse of the reduction AND operator.

For the *reduction OR* operator, if any bit in the operand is 1, then the result is 1; otherwise, the result is 0. For example, let x_1 be the vector shown below.

1	1	1	0	1	0	1	1

The reduction OR ($| x_1$) operation is equivalent to the following operation:

1 | 1 | 1 | 0 | 1 | 0 | 1 | 1

which returns a result of 1'b1.

For the *reduction NOR* operator, if any bit in the operand is 1, then the result is 0; otherwise, the result is 1. For a vector x_1, the reduction NOR ($\sim | x_1$) is the inverse of the reduction OR operator.

For the *exclusive-OR* operator, if there are an even number of 1s in the operand, then the result is 0; otherwise, the result is 1. For example, let x_1 be the vector shown below.

1	1	1	0	1	0	1	1

The reduction exclusive-OR ($\wedge x_1$) operation is equivalent to the following operation:

1 ^ 1 ^ 1 ^ 0 ^ 1 ^ 0 ^ 1 ^ 1

which returns a result of 1'b0. The reduction exclusive-OR operator can be used as an even parity generator.

For the *exclusive-NOR* operator, if there are an odd number of 1s in the operand, then the result is 0; otherwise, the result is 1. For a vector x_1, the reduction exclusive-NOR ($\wedge\sim x_1$) is the inverse of the reduction exclusive-OR operator. The reduction exclusive-NOR operator can be used as an odd parity generator.

Shift The shift operators shift a single vector operand left or right a specified number of bit positions. These are logical shift operations, not algebraic; that is, as bits are shifted left or right, zeroes fill in the vacated bit positions. The bits shifted out of the operand are lost; they do not rotate to the high-order or low-order bit positions of the shifted operand. If the shift amount evaluates to **x** or **z**, then the result of the operation is **x**. There are two shift operators, as shown below. The value in parentheses is the number of bits that the operand is shifted.

<< (Left-shift amount)
>> (Right-shift amount)

When an operand is shifted left, this is equivalent to a multiply-by-two operation for each bit position shifted. When an operand is shifted right, this is equivalent to a divide-by-two operation for each bit position shifted. The shift operators are useful to model the sequential add-shift multiplication algorithm and the sequential shift-subtract division algorithm. Examples of shift left and shift right operations are shown below for 8-bit operands. Operand *a_reg* is shifted left three bits with the low-order bits filled with zeroes. Operand *b_reg* is shifted right two bits with the high-order bits filled with zeroes.

```
a_reg = 00000010,  b_reg = 00001000,        //shift a_reg left 3
rslt_a = 00010000,  rslt_b = 00000010       //shift b_reg right 2

a_reg = 00000110,  b_reg = 00011000,        //shift a_reg left 3
rslt_a = 00110000,  rslt_b = 00000110       //shift b_reg right 2

a_reg = 00001111,  b_reg = 00111000,        //shift a_reg left 3
rslt_a = 01111000,  rslt_b = 00001110       //shift b_reg right 2

a_reg = 11100000,  b_reg = 00000011,        //shift a_reg left 3
rslt_a = 00000000,  rslt_b = 00000000       //shift b_reg right 2
```

Conditional The conditional operator (**? :**) has three operands, as shown in the syntax below. The *conditional_expression* is evaluated. If the result is true (1), then the *true_expression* is evaluated; if the result is false (0), then the *false_expression* is evaluated.

conditional_expression **?** true_expression **:** false_expression;

The conditional operator can be used when one of two expressions is to be selected. For example, in the statement below, if x_1 is greater than or equal to x_2, then z_1 is assigned the value of x_3; if x_1 is less than x_2, then z_1 is assigned the value of x_4.

```
z1 = (x1 >= x2) ? x3 : x4;
```

If the operands have different lengths, then the shorter operand is zero-extended on the left. Since the conditional operator selects one of two values, depending on the result of the conditional_expression evaluation, the operator can be used in place of the **if**... **else** construct. The conditional operator is ideally suited to model a 2:1 multiplexer. Conditional operators can be nested; that is, each true_expression and false_expression can be a conditional operation. This is useful for modeling a 4:1 multiplexer.

```
conditional_expression ? (cond_expr1 ? true_expr1 : false_expr1)
                       : (cond_expr2 ? true_expr2 : false_expr2);
```

Concatenation The concatenation operator ({ }) forms a single operand from two or more operands by joining the different operands in sequence separated by commas. The operands to be appended are contained within braces. The size of the operands must be known before concatenation takes place.

The examples below show the concatenation of scalars and vectors of different sizes. Outputs z_1, z_2, z_3, and z_4 are ten bits in length.

```
z1, z2, z3, and z4 are 10 bits in length.

a = 11, b = 001, c = 1100, d = 1

z1 = 0000_11_1100        //z1 = {a, c}
z2 = 00000_001_11        //z2 = {b, a}
z3 = 0_1100_001_11       //z3 = {c, b, a}
z4 = 11_001_1100_1       //z4 = {a, b, c, d}
```

Replication Replication is a means of performing repetitive concatenation. Replication specifies the number of times to duplicate the expressions within the innermost braces. The syntax is shown below together with examples of replication.

```
{number_of_repetitions {expression_1, expression_2, . . . , expression_n}};
```

```
a = 11, b = 010, c = 0011,

z1 = 11_0011_11_0011,                    //z1 = {2{a, c}}
z2 = 010_0011_0111_010_0011_0111         //z2 = {2{b, c, 4'b0111}}
```

1.3 Modules and Ports

A *module* is the basic unit of design in Verilog. It describes the functional operation of some logical entity and can be a stand-alone module or a collection of modules that are instantiated into a structural module. *Instantiation* means to use one or more lower-level modules in the construction of a higher-level structural module. A module can be a logic gate, an adder, a multiplexer, a counter, or some other logical function.

A module consists of declarative text which specifies the function of the module using Verilog constructs; that is, a Verilog module is a software representation of the physical hardware structure and behavior. The declaration of a module is indicated by the keyword **module** and is always terminated by the keyword **endmodule**.

Verilog has predefined logical elements called *primitives*. These built-in primitives are structural elements that can be instantiated into a larger design to form a more complex structure. Examples are: **and**, **or**, **xor**, and **not**. Built-in primitives are discussed in more detail in Section 1.4.

Modules contain *ports* which allow communication with the external environment or other modules. For example, the logic diagram for the full adder of Figure 1.16 has input ports *a*, *b*, and *cin* and output ports *sum* and *cout*. The general structure and syntax of a module is shown in Figure 1.17. An AND gate can be defined as shown in the module of Figure 1.18, where the input ports are x_1 and x_2 and the output port is z_1.

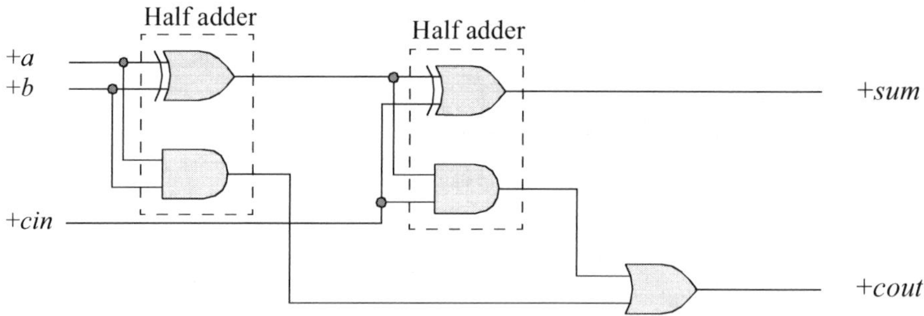

Figure 1.16 Logic diagram for a full adder.

```
module <module name> (port list);
   declarations
      reg, wire, parameter,
      input, output, . . .
      . . .
   <module internals>
      statements
      initial, always, module instantiation, . . .
      . . .
endmodule
```

Figure 1.17 General structure of a Verilog module.

```
//dataflow and gate with two inputs
module and2 (x1, x2, z1);

input x1, x2;
output z1;

wire x1, x2;
wire z1;

assign z1 = x1 & x2;

endmodule
```

Figure 1.18 Verilog module for an AND gate with two inputs.

A Verilog module defines the information that describes the relationship between the inputs and outputs of a logic circuit. A structural module will have one or more instantiations of other modules or logic primitives. In Figure 1.18, the first line is a comment, indicated by (//). In the second line, *and2* is the module name; this is followed by left and right parentheses containing the module ports, which is followed by a semicolon. The inputs and outputs are defined by the keywords **input** and **output**. The ports are declared as **wire** in this dataflow module. Dataflow modeling is covered in detail in Section 1.6. The keyword **assign** describes the behavior of the circuit. Output z_1 is *assigned* the value of x_1 ANDed (&) with x_2.

1.3.1 Designing a Test Bench for Simulation

This section describes the techniques for writing test benches in Verilog HDL. When a Verilog module is finished, it must be tested to ensure that it operates according to the machine specifications. The functionality of the module can be tested by applying stimulus to the inputs and checking the outputs. The test bench will display the inputs and outputs in a radix (binary, octal, hexadecimal, or decimal).

The test bench contains an instantiation of the unit under test and Verilog code to generate input stimulus and to monitor and display the response to the stimulus. Figure 1.19 shows a simple test bench to test the 2-input AND gate of Figure 1.18. Line 1 is a comment indicating that the module is a test bench for a 2-input AND gate. Line 2 contains the keyword **module** followed by the module name, which includes *tb* indicating a test bench module. The name of the module and the name of the module under test are the same for ease of cross-referencing.

Line 4 specifies that the inputs are **reg** type variables; that is, they contain their values until they are assigned new values. Outputs are assigned as type **wire** in test benches. Output nets are driven by the output ports of the module under test. Line 8 contains an **initial** statement, which executes only once.

Verilog provides a means to monitor a signal when its value changes. This is accomplished by the **$monitor** task. The **$monitor** continuously monitors the values of the variables indicated in the parameter list that is enclosed in parentheses. It will display the value of the variables whenever a variable changes state. The quoted string within the task is printed and specifies that the variables are to be shown in binary (%b). The **$monitor** is invoked only once. Line 12 is a second **initial** statement that allows the procedural code between the **begin . . . end** block statements to be executed only once.

```
 1 //and2 test bench
   module and2_tb;

   reg x1, x2;
 5 wire z1;

   //display variables
   initial
   $monitor ("x1 = %b, x2 = %b, z1 = %b", x1, x2, z1);

11 //apply input vectors
   initial
   begin
      #0     x1 = 1'b0;
             x2 = 1'b0;
16
      #10    x1 = 1'b0;
             x2 = 1'b1;

20    #10    x1 = 1'b1;
             x2 = 1'b0;

      #10    x1 = 1'b1;
             x2 = 1'b1;

26    #10    $stop;
   end

   //instantiate the module into the test bench
30 and2 inst1 (
      .x1(x1),
      .x2(x2),
      .z1(z1)
      );

   endmodule
```

Figure 1.19 Test bench for the 2-input AND gate of Figure 1.18.

Lines 14 and 15 specify that at time 0 (#0), inputs x_1 and x_2 are assigned values of 0, where 1 is the width of the value (one bit), ' is a separator, b indicates binary, and 0 is the value. Line 17 specifies that 10 time units later, the inputs change to: $x_1 = 0$ and $x_2 = 1$. This process continues until all possible values of two variables have been applied to the inputs. Simulation stops at 10 time units after the last input vector has been applied (**$stop**). The total time for simulation is 40 time units — the sum of all the time units. The time units can be specified for any duration.

Line 30 begins the instantiation of the module into the test bench. The name of the instantiation must be the same as the module under test, in this case, *and2*. This is followed by an instance name (*inst1*) followed by a left parenthesis. The $.x_1$ variable in line 31 refers to a port in the module that corresponds to a port (x_1) in the test bench. All the ports in the module under test must be listed. The keyword **endmodule** is the last line in the test bench.

The binary outputs for this test bench are shown in Figure 1.20. The output can be presented in binary (b or B), in octal (o or O), in hexadecimal (h or H), or in decimal (d or D).

The Verilog syntax will be covered in greater detail in subsequent sections. It is important at this point to concentrate on how the module under test is simulated and instantiated into the test bench.

```
x1 = 0, x2 = 0, z1 = 0
x1 = 0, x2 = 1, z1 = 0
x1 = 1, x2 = 0, z1 = 0
x1 = 1, x2 = 1, z1 = 1
```

Figure 1.20 Binary outputs for the test bench of Figure 1.19 for a 2-input AND gate.

Several different methods to generate test benches will be shown in subsequent sections. Each design in the book will be tested for correct operation by means of a test bench. Test benches provide clock pulses that are used to control the operation of a synchronous sequential machine. An **initial** statement is an ideal method to generate a waveform at discrete intervals of time for a clock pulse. The Verilog code in Figure 1.21 illustrates the necessary statements to generate clock pulses that have a duty cycle of 20%.

1.4 Built-In Primitives

Logic primitives such as **and**, **nand**, **or**, **nor**, and **not** gates, as well as **xor** (exclusive-OR), and **xnor** (exclusive_NOR) functions are part of the Verilog language and are classified as multiple-input gates. These are built-in primitives that can be instantiated into a module.

```
//generate clock pulses of 20% duty cycle
module clk_gen (clk);
output clk;
reg clk;

initial
begin
   #0    clk = 0;
   #5    clk = 1;
   #5    clk = 0;
   #20   clk = 1;
   #5    clk = 0;
   #20   clk = 1;
   #5    clk = 0;
   #10   $stop;
end
endmodule
```

Figure 1.21 Verilog code to generate clock pulses with a 20% duty cycle.

These are built-in primitive gates used to describe a net and have one or more scalar inputs, but only one scalar output. The output signal is listed first, followed by the inputs in any order. The outputs are declared as **wire;** the inputs can be declared as either **wire** or **reg**. The gates represent a combinational logic function and can be instantiated into a module, as follows, where the instance name is optional:

gate_type inst1 (output, input_1, input_2, . . . , input_*n*);

Two or more instances of the same type of gate can be specified in the same construct, as follows:

gate_type inst1 (output_1, input_11, input_12, . . . , input_1*n*),
inst2 (output_2, input_21, input_22, . . . , input_2*n*),
.
.
.
inst*m* (output_*m*, input_*m*1, input_*m*2, . . . , input_*mn*);

and This is a multiple-input built-in primitive gate that performs the AND function for a multiple-input AND gate. If any input is an **x**, then this represents an unknown logic value. If and entry is a **z**, then this represents a high impedance state, which indicates that the driver of a net is disabled or not connected. AND gates can be represented by two symbols as shown below for the AND function and the OR function.

AND gate for the AND function

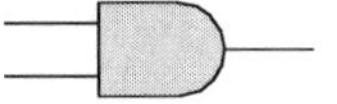

AND gate for the OR function

buf A **buf** gate is a noninverting primitive with one scalar input and one or more scalar outputs. The output terminals are listed first when instantiated; the input is listed last, as shown below. The instance name is optional.

```
buf  inst1 (output, input);                              //one output
buf  inst2 (output_1, output_2, . . . , output_n, input); //multiple outputs
```

nand This is a multiple-input built-in primitive gate that operates as an AND function with a negative output. NAND gates can be represented by two symbols as shown below for the AND function and the OR function.

NAND gate for the AND function

NAND gate for the OR function

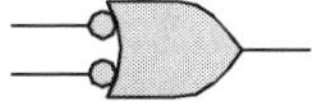

DeMorgan's theorems are associated with NAND and NOR gates and convert the complement of a sum term or a product term into a corresponding product or sum term, respectively. For every $x_1, x_2 \in B$,

(a) $(x_1 \bullet x_2)' = x_1' + x_2'$ Nand gate

(b) $(x_1 + x_2)' = x_1' \bullet x_2'$ NOR gate

DeMorgan's laws can be generalized for any number of variables.

nor This is a multiple-input built-in primitive gate that operates as an OR function with a negative output. NOR gates can be represented by two symbols as shown below for the OR function and the AND function.

NOR gate for the OR function

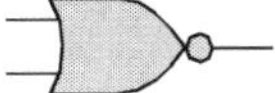

NOR gate for the AND function

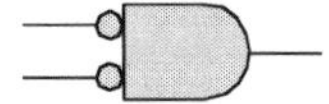

not A **not** gate is an inverting built-in primitive with one scalar input and one or more scalar outputs. The output terminals are listed first when instantiated; the input is listed last, as shown below. The instance name is optional.

```
not inst1 (output, input);                               //one output
not inst2 (output_1, output_2, . . . , output_n, input);  //multiple outputs
```

The NOT function can be represented by two symbols as shown below depending on the assertion levels required. The function of the inverters is identical; the low assertion is placed at the input or output for readability with associated logic.

NOT (inverter) function with low assertion output

NOT (inverter) function with low assertion input

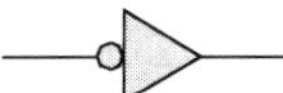

or This is a multiple-input built-in primitive gate that operates as an OR function. OR gates can be represented by two symbols as shown below for the OR function and the AND function.

OR gate for the OR function

OR gate for the AND function

xnor This is a built-in primitive gate that functions as an exclusive-OR gate with a negative output. Exclusive-NOR gates can be represented by the symbol shown below. An exclusive-NOR gate is also called an equality function because the output is a logical 1 whenever the two inputs are equal.

Exclusive-NOR gate

$+x_1$
$+x_2$
$+z_1$

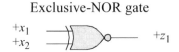

The equation for the exclusive-NOR gate shown above is

$$z_1 = (x_1 x_2) + (x_1' x_2')$$

xor This is a built-in primitive gate that functions as an exclusive-OR circuit. Exclusive-OR gates can be represented by the symbol shown below. The output of an exclusive-OR gate is a logical 1 whenever the two inputs are different.

Exclusive-OR gate

$+x_1$
$+x_2$
$+z_1$

The equation for the exclusive-OR gate shown above is

$$z_1 = (x_1 x_2') + (x_1' x_2)$$

1.4.1 Built-In Primitive Design Examples

The best way to learn design methodologies using built-in primitives is by examples. Therefore, examples will be presented ranging from very simple to moderately complex. When necessary, the theory for the examples will be presented prior to the Verilog design. All examples are carried through to completion at the gate level. Nothing is left unfinished or partially designed.

Example 1.1 The Karnaugh map of Figure 1.22 will be implemented using only NOR gates in a product-of-sums format. Equation 1.7 shown the product-of-sums expression obtained from the Karnaugh map. The logic diagram is shown in Figure 1.23 which indicates the instantiation names and net names.

$x_1 x_2$ \ $x_3 x_4$	0 0	0 1	1 1	1 0
0 0	0 (0)	1 (1)	1 (3)	0 (2)
0 1	1 (4)	1 (5)	0 (7)	1 (6)
1 1	1 (12)	1 (13)	0 (15)	1 (14)
1 0	1 (8)	1 (9)	1 (11)	0 (10)

z_1

Figure 1.22 Karnaugh map for Example 1.1.

$$z_1 = (x_1 + x_2 + x_4)(x_2 + x_3' + x_4)(x_2' + x_3' + x_4') \tag{1.7}$$

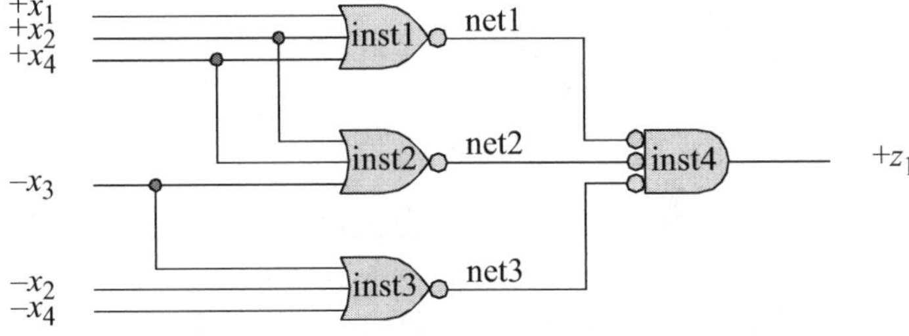

Figure 1.23 Logic diagram for Example 1.1.

The design module is shown in Figure 1.24 using NOR gate built-in primitives. The test bench is shown in Figure 1.25 using a different approach to generate all 16 combinations of the four inputs. Several new modeling constructs are shown in the test bench. Since there are four inputs to the circuit, all 16 combinations of four variables must be applied to the circuit. This is accomplished by a **for** loop statement, which is similar in construction to a **for** loop in the C programming language.

```
//logic diagram using built-in primitives
module log_eqn_pos5 (x1, x2, x3, x4, z1);

input x1, x2, x3, x4;
output z1;

//instantiate the nor built-in primitives
nor   inst1 (net1, x1, x2, x4);
nor   inst2 (net2, x2, x4, ~x3);
nor   inst3 (net3, ~x3, ~x2, ~x4);
nor   inst4 (z1, net1, net2, net3);

endmodule
```

Figure 1.24 Module for the product-of-sums logic diagram of Figure 1.23.

```
//test bench for log_eqn_pos5
module log_eqn_pos5_tb;

reg x1, x2, x3, x4;
wire z1;

//apply input vectors
initial
begin: apply_stimulus
   reg [4:0] invect;     //invect[4] terminates the loop
   for (invect = 0; invect < 16; invect = invect + 1)
      begin
         {x1, x2, x3, x4} = invect[4:0];
         #10 $display ("x1x2x3x4 = %b, z1 = %b",
                           {x1, x2, x3, x4}, z1);
      end
end
//continued on next page
```

Figure 1.25 Test bench for the design module of Figure 1.24.

```
//instantiate the module into the test bench
log_eqn_pos5 inst1 (
    .x1(x1),
    .x2(x2),
    .x3(x3),
    .x4(x4),
    .z1(z1)
    );
endmodule
```

Figure 1.25 (Continued)

Referring to the test bench of Figure 1.25, following the keyword **begin** is the name of the block: *apply_stimulus*. In this block, a 5-bit **reg** variable is declared called *invect*. This guarantees that all combinations of the four inputs will be tested by the **for** loop, which applies input vectors of $x_1x_2x_3x_4$ = 0000, 0001, 0010, 0011 . . . 1111 to the circuit. The **for** loop stops when the pattern 10000 is detected by the test segment (*invect* < 16). If only a 4-bit vector were applied, then the expression (*invect* < 16) would always be true and the loop would never terminate. The increment segment of the **for** loop does not support an increment designated as *invect*++; therefore, the long notation must be used: *invect* = *invect* + *1*.

The target of the first assignment within the **for** loop ($\{x_1, x_2, x_3, x_4\}$ = *invect [4:0]*) represents a concatenated target. The concatenation of inputs x_1, x_2, x_3, and x_4 is performed by positioning them within braces: $\{x_1, x_2, x_3, x_4\}$. A vector of five bits ([4:0]) is then assigned to the inputs. This will apply inputs of 0000, 0001, 0010, 0011, . . . 1111 and stop when the vector is 10000.

The **initial** statement also contains a system task (**$display**) which prints the argument values — within the quotation marks — in binary. The concatenated variables x_1, x_2, x_3, and x_4 are listed first; therefore, their values are obtained from the first argument to the right of the quotation marks: $\{x_1, x_2, x_3, x_4\}$. The value for the second variable z_1 is obtained from the second argument to the right of the quotation marks. The variables to the right of the quotation marks are listed in the same order as the variables within the quotation marks.

The delay time (#10) in the system task specifies that the task is to be executed after 10 time units; that is, the delay between the application of a vector and the response of the module. This delay represents the propagation delay of the logic. The simulation results are shown in binary format in Figure 1.26.

```
x1x2x3x4 = 0000, z1 = 0
x1x2x3x4 = 0001, z1 = 1
x1x2x3x4 = 0010, z1 = 0
x1x2x3x4 = 0011, z1 = 1              //continued on next page
```

Figure 1.26 Outputs generated by the test bench of Figure 1.25.

```
x1x2x3x4 = 0100, z1 = 1
x1x2x3x4 = 0101, z1 = 1
x1x2x3x4 = 0110, z1 = 1
x1x2x3x4 = 0111, z1 = 0
x1x2x3x4 = 1000, z1 = 1
x1x2x3x4 = 1001, z1 = 1
x1x2x3x4 = 1010, z1 = 0
x1x2x3x4 = 1011, z1 = 1
x1x2x3x4 = 1100, z1 = 1
x1x2x3x4 = 1101, z1 = 1
x1x2x3x4 = 1110, z1 = 1
x1x2x3x4 = 1111, z1 = 0
```

Figure 1.26 (Continued)

Example 1.2 Equation 1.8 will be minimized as a sum-of-products form and then implemented using built-in primitives of AND and OR with x_4 and x_5 as map-entered variables. Variables may be entered in a Karnaugh map as map-entered variables, together with 1s and 0s. A map of this type is more compact than a standard Karnaugh map, but contains the same information. A map containing map-entered variables is particularly useful in analyzing and synthesizing synchronous sequential machines. When variables are entered in a Karnaugh map, two or more squares can be combined only if the squares are adjacent and contain the same variable(s).

$$z_1 = x_1'x_2'x_3'x_4x_5' + x_1'x_2 + x_1'x_2'x_3'x_4x_5 + x_1x_2'x_3'x_4x_5$$
$$+ x_1x_2'x_3 + x_1x_2'x_3'x_4' + x_1x_2'x_3'x_5' \quad (1.8)$$

The Karnaugh map is shown in Figure 1.27 in which the following minterm locations combine:

Minterm location 0 $= x_4x_5' + x_4x_5 = x_4$
Minterm location 2 $= 1 + x_4$
Combine minterm locations 0 and 2 to yield the sum term $x_1'x_3'x_4$

Combine minterm locations 2 and 3 to yield $x_1'x_2$

Minterm location 4 $= x_4x_5 + x_4' + x_5' = 1$
Minterm location 5 $= 1$
Combine minterm locations 4 and 5 to yield x_1x_2'

$x_1 \backslash x_2x_3$	0 0	0 1	1 1	1 0
0	[0] $x_4 x_5' + x_4 x_5$	[1] 0	[3] 1	[2] 1
1	[4] $x_4 x_5 + x_4' + x_5'$	[5] 1	[7] 0	[6] 0

z_1

Figure 1.27 Karnaugh map for Example 1.2.

The minimized sum-of-products equation from the Karnaugh map is shown in Equation 1.9. The logic diagram is shown in Figure 1.28. The design module is shown in Figure 1.29 and the test bench is shown in Figure 1.30. Figure 1.31 lists the outputs obtained from the test bench.

$$z_1 = x_1'x_3'x_4 + x_1'x_2 + x_1x_2' \tag{1.9}$$

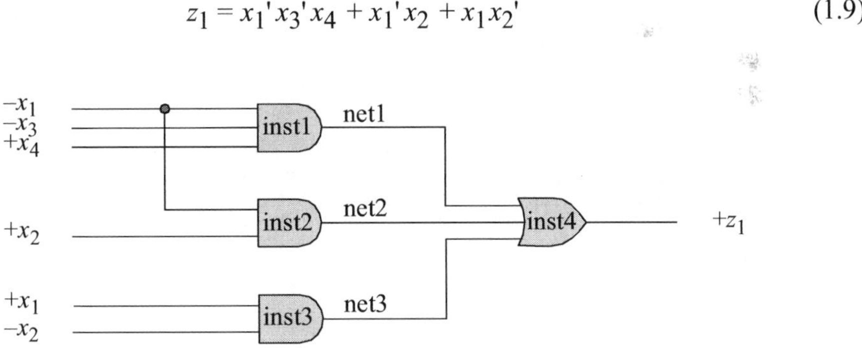

Figure 1.28 Logic diagram for Equation 1.9.

```
//logic equation using map-entered variables
module mev (x1, x2, x3, x4, z1);

input x1, x2, x3, x4;
output z1;

and   inst1 (net1, ~x1, ~x3, x4);
and   inst2 (net2, ~x1, x2);
and   inst3 (net3, x1, ~x2);
or    inst4 (z1, net1, net2, net3);
endmodule
```

Figure 1.29 Design module to implement Equation 1.9 using built-in primitives.

```
//test bench for logic equation using map-entered variables
module mev_tb;

reg x1, x2, x3, x4;
wire z1;

initial         //apply input vectors
begin: apply_stimulus
   reg [4:0] invect;
   for (invect=0; invect<16; invect=invect+1)
      begin
         {x1, x2, x3, x4} = invect [4:0];
         #10 $display ("x1x2x3x4 = %b, z1 = %b",
                        {x1, x2, x3, x4}, z1);
   end
end

//instantiate the module into the test bench
mev inst1 (
   .x1(x1),
   .x2(x2),
   .x3(x3),
   .x4(x4),
   .z1(z1)
   );
endmodule
```

Figure 1.30 Test bench for the design module of Figure 1.29.

```
x1x2x3x4 = 0000, z1 = 0
x1x2x3x4 = 0001, z1 = 1
x1x2x3x4 = 0010, z1 = 0
x1x2x3x4 = 0011, z1 = 0
x1x2x3x4 = 0100, z1 = 1
x1x2x3x4 = 0101, z1 = 1
x1x2x3x4 = 0110, z1 = 1
x1x2x3x4 = 0111, z1 = 1
x1x2x3x4 = 1000, z1 = 1
x1x2x3x4 = 1001, z1 = 1
x1x2x3x4 = 1010, z1 = 1
x1x2x3x4 = 1011, z1 = 1
x1x2x3x4 = 1100, z1 = 0
x1x2x3x4 = 1101, z1 = 0
x1x2x3x4 = 1110, z1 = 0
x1x2x3x4 = 1111, z1 = 0
```

Figure 1.31 Outputs for the test bench of Figure 1.30.

Example 1.3 A 4:1 multiplexer will be designed using built-in logic primitives. The 4:1 multiplexer of Figure 1.32 will be designed using built-in primitives of AND, OR, and NOT. The design is simpler and takes less code if a continuous assignment statement is used, but this section presents gate-level modeling only — continuous assignment statements are used in dataflow modeling.

The multiplexer has four data inputs: d_3, d_2, d_1, and d_0, which are specified as a 4-bit vector *d[3:0]*, two select inputs: s_1 and s_0, specified as a 2-bit vector *s[1:0]*, one scalar input *Enable*, and one scalar output z_1, as shown in the logic diagram of Figure 1.32.. Also, the system function **$time** will be used in the test bench to return the current simulation time measured in nanoseconds (ns). The design module is shown in Figure 1.33, the test bench in Figure 1.34, and the outputs in Figure 1.35.

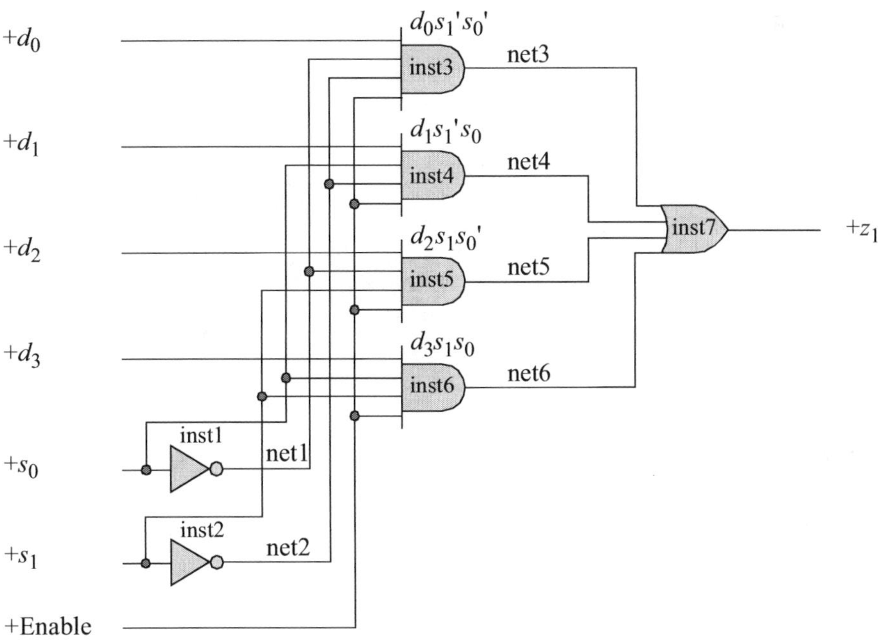

Figure 1.32 Logic diagram of a 4:1 multiplexer to be designed using built-in primitives.

```
//a 4:1 multiplexer using built-in primitives
module mux_4to1 (d, s, enbl, z1);

input [3:0] d;
input [1:0] s;
input enbl;
output z1;
                                   //continued on next page
```

Figure 1.33 Module for a 4:1 multiplexer with *Enable* using built-in primitives.

```
not    inst1 (net1, s[0]),
       inst2 (net2, s[1]);

and    inst3 (net3, d[0], net1, net2, enbl),
       inst4 (net4, d[1], s[0], net2, enbl),
       inst5 (net5, d[2], net1, s[1], enbl),
       inst6 (net6, d[3], s[0], s[1], enbl);

or     inst7 (z1, net3, net4, net5, net6);

endmodule
```

Figure 1.33 (Continued)

```
//test bench for 4:1 multiplexer
module mux_4to1_tb;

reg [3:0] d;
reg [1:0] s;
reg enbl;
wire z1;

initial
$monitor ($time,"ns, select:s=%b, inputs:d=%b, output:z1=%b",
         s, d, z1);
initial
begin
   #0    s[0]=1'b0;  s[1]=1'b0;
         d[0]=1'b0;  d[1]=1'b1;  d[2]=1'b0;  d[3]=1'b1;
         enbl=1'b1;  //d[0]=0; z1=0

   #10   s[0]=1'b0;  s[1]=1'b0;
         d[0]=1'b1;  d[1]=1'b1;  d[2]=1'b0;  d[3]=1'b1;
         enbl=1'b1;  //d[0]=1; z1=1

   #10   s[0]=1'b1;  s[1]=1'b0;
         d[0]=1'b1;  d[1]=1'b1;  d[2]=1'b0;  d[3]=1'b1;
         enbl=1'b1;  //d[1]=1; z1=1

   #10   s[0]=1'b0;  s[1]=1'b1;
         d[0]=1'b1;  d[1]=1'b1;  d[2]=1'b0;  d[3]=1'b1;
         enbl=1'b1;  //d[2]=0; z1=0
                                    //continued on next page
```

Figure 1.34 Test bench for the 4:1 multiplexer of Figure 1.33.

```
   #10    s[0]=1'b1;  s[1]=1'b0;
          d[0]=1'b1;  d[1]=1'b0;  d[2]=1'b0;  d[3]=1'b1;
          enbl=1'b1;  //d[1]=1; z1=0

   #10    s[0]=1'b1;  s[1]=1'b1;
          d[0]=1'b1;  d[1]=1'b1;  d[2]=1'b0;  d[3]=1'b1;
          enbl=1'b1;  //d[3]=1; z1=1

   #10    s[0]=1'b1;  s[1]=1'b1;
          d[0]=1'b1;  d[1]=1'b1;  d[2]=1'b0;  d[3]=1'b0;
          enbl=1'b1;  //d[3]=0; z1=0

   #10    s[0]=1'b1;  s[1]=1'b1;
          d[0]=1'b1;  d[1]=1'b1;  d[2]=1'b0;  d[3]=1'b0;
          enbl=1'b0;  //d[3]=0; z1=0

   #10    $stop;
end

//instantiate the module into the test bench
mux_4to1 inst1 (
   .d(d),
   .s(s),
   .z1(z1),
   .enbl(enbl)
   );
endmodule
```

Figure 1.34 (Continued)

```
0  ns, select:s=00, inputs:d=1010, output:z1=0
10 ns, select:s=00, inputs:d=1011, output:z1=1
20 ns, select:s=01, inputs:d=1011, output:z1=1
30 ns, select:s=10, inputs:d=1011, output:z1=0
40 ns, select:s=01, inputs:d=1001, output:z1=0
50 ns, select:s=11, inputs:d=1011, output:z1=1
60 ns, select:s=11, inputs:d=0011, output:z1=0
```

Figure 1.35 Outputs for the 4:1 multiplexer test bench of Figure 1.34.

Example 1.4 This example illustrates the design of a majority circuit using built-in primitives. The output of a majority circuit is a logic 1 if the majority of the inputs is a logic 1; otherwise, the output is a logic 0. Therefore, a majority circuit must have an odd number of inputs in order to have a majority of the inputs at the same logic level.

A 5-input majority circuit will be designed using the Karnaugh map of Figure 1.36, where a 1 entry indicates that the majority of the inputs is a logic 1.

$x_5 = 0$

x_1x_2 \ x_3x_4	0 0	0 1	1 1	1 0
0 0	0 (0)	0 (2)	0 (6)	0 (4)
0 1	0 (8)	0 (10)	1 (14)	0 (12)
1 1	0 (24)	1 (26)	1 (30)	1 (28)
1 0	0 (16)	0 (18)	1 (22)	0 (20)

$x_5 = 1$

x_1x_2 \ x_3x_4	0 0	0 1	1 1	1 0
0 0	0 (1)	0 (3)	1 (7)	0 (5)
0 1	0 (9)	1 (11)	1 (15)	1 (13)
1 1	1 (25)	1 (27)	1 (31)	1 (29)
1 0	0 (17)	1 (19)	1 (23)	1 (21)

z_1

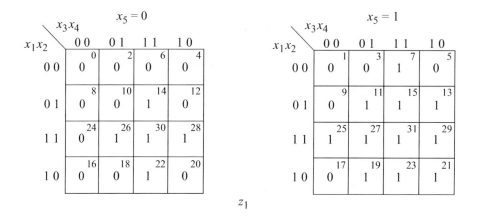

Figure 1.36 Karnaugh map for the majority circuit of Example 1.4.

Equation 1.10 represents the logic for output z_1 in a sum-of-products form. The design module is shown in Figure 1.37, which is designed directly from Equation 1.10 without the use of a logic diagram. The test bench is shown in Figure 1.38, and the outputs are shown in Figure 1.39.

$$z_1 = x_3x_4x_5 + x_2x_3x_5 + x_1x_3x_5 + x_2x_4x_5 + x_1x_4x_5 + x_1x_2x_5 + x_1x_2x_4 + x_2x_3x_4 + x_1x_3x_4 \qquad (1.10)$$

```
//5-input majority circuit
module majority (x1, x2, x3, x4, x5, z1);
input x1, x2, x3, x4, x5;
output z1;

and   inst1  (net1, x3, x4, x5),
      inst2  (net2, x2, x3, x5),
      inst3  (net3, x1, x3, x5),
      inst4  (net4, x2, x4, x5),    //continued on next page
```

Figure 1.37 Design module for the majority circuit of Figure 1.36.

```
        inst5  (net5, x1, x4, x5),
        inst6  (net6, x1, x2, x5),
        inst7  (net7, x1, x2, x4),
        inst8  (net8, x2, x3, x4),
        inst9  (net9, x1, x3, x4);
or      inst10 (z1, net1, net2, net3, net4, net5,
                    net6, net7, net8, net9);
endmodule
```

Figure 1.37 (Continued)

```
//test bench for 5-input majority circuit
module majority_tb;
reg x1, x2, x3, x4, x5;
wire z1;

//apply input vectors
initial
begin: apply_stimulus
      reg [6:0] invect;
      for (invect=0; invect<32; invect=invect+1)
            begin
                  {x1, x2, x3, x4, x5} = invect [6:0];
                  #10 $display ("x1x2x3x4x5 = %b, z1 = %b",
                              {x1, x2, x3, x4, x5}, z1);
            end
end

//instantiate the module into the test bench
majority inst1 (
      .x1(x1),
      .x2(x2),
      .x3(x3),
      .x4(x4),
      .x5(x5),
      .z1(z1)
      );

endmodule
```

Figure 1.38 Test bench for the majority circuit module of Figure 1.37.

```
x1x2x3x4x5 = 00000, z1 = 0
x1x2x3x4x5 = 00001, z1 = 0
x1x2x3x4x5 = 00010, z1 = 0
x1x2x3x4x5 = 00011, z1 = 0
x1x2x3x4x5 = 00100, z1 = 0
x1x2x3x4x5 = 00101, z1 = 0
x1x2x3x4x5 = 00110, z1 = 0
x1x2x3x4x5 = 00111, z1 = 1
x1x2x3x4x5 = 01000, z1 = 0
x1x2x3x4x5 = 01001, z1 = 0
x1x2x3x4x5 = 01010, z1 = 0
x1x2x3x4x5 = 01011, z1 = 1
x1x2x3x4x5 = 01100, z1 = 0
x1x2x3x4x5 = 01101, z1 = 1
x1x2x3x4x5 = 01110, z1 = 1
x1x2x3x4x5 = 01111, z1 = 1
x1x2x3x4x5 = 10000, z1 = 0
x1x2x3x4x5 = 10001, z1 = 0
x1x2x3x4x5 = 10010, z1 = 0
x1x2x3x4x5 = 10011, z1 = 1
x1x2x3x4x5 = 10100, z1 = 0
x1x2x3x4x5 = 10101, z1 = 1
x1x2x3x4x5 = 10110, z1 = 1
x1x2x3x4x5 = 10111, z1 = 1
x1x2x3x4x5 = 11000, z1 = 0
x1x2x3x4x5 = 11001, z1 = 1
x1x2x3x4x5 = 11010, z1 = 1
x1x2x3x4x5 = 11011, z1 = 1
x1x2x3x4x5 = 11100, z1 = 1
x1x2x3x4x5 = 11101, z1 = 1
x1x2x3x4x5 = 11110, z1 = 1
x1x2x3x4x5 = 11111, z1 = 1
```

Figure 1.39 Outputs for the majority circuit of Figure 1.37.

Example 1.5 A code converter will be designed to convert a 4-bit binary number to the corresponding Gray code number. The inputs of the binary number $x_1x_2x_3x_4$ are available in both high and low assertion, where x_4 is the low-order bit. The outputs for the Gray code $z_1z_2z_3z_4$ are asserted high, where z_4 is the low-order bit. The binary-to-Gray code conversion table is shown in Table 1.13.

Table 1.13 Binary-to-Gray Code Conversion

Binary Code				Gray Code			
x_1	x_2	x_3	x_4	z_1	z_2	z_3	z_4
0	0	0	0	0	0	0	0
0	0	0	1	0	0	0	1
0	0	1	0	0	0	1	1
0	0	1	1	0	0	1	0
0	1	0	0	0	1	1	0
0	1	0	1	0	1	1	1
0	1	1	0	0	1	0	1
0	1	1	1	0	1	0	0
1	0	0	0	1	1	0	0
1	0	0	1	1	1	0	1
1	0	1	0	1	1	1	1
1	0	1	1	1	1	1	0

//continued on next page

Table 1.13 Binary-to-Gray Code Conversion

Binary Code				Gray Code			
x_1	x_2	x_3	x_4	z_1	z_2	z_3	z_4
1	1	0	0	1	0	1	0
1	1	0	1	1	0	1	1
1	1	1	0	1	0	0	1
1	1	1	1	1	0	0	0

There are four Karnaugh maps shown in Figure 1.40, one map for each of the Gray code outputs. The equations obtained from the Karnaugh maps are shown in Equation 1.11. The logic diagram is shown in Figure 1.41. The design module is shown in Figure 1.42, the test bench module is shown in Figure 1.43, and the outputs are shown in Figure 1.44.

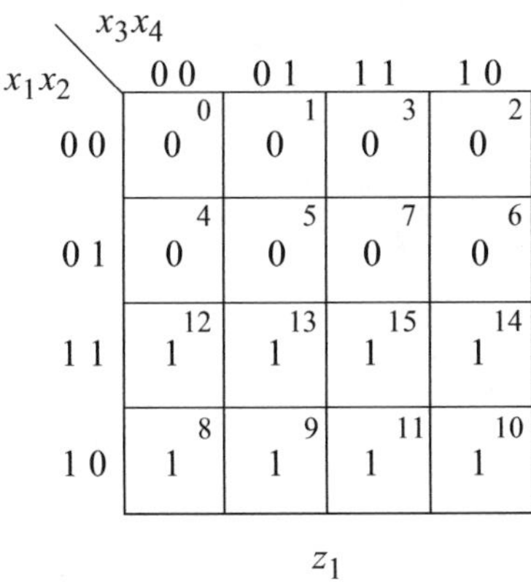

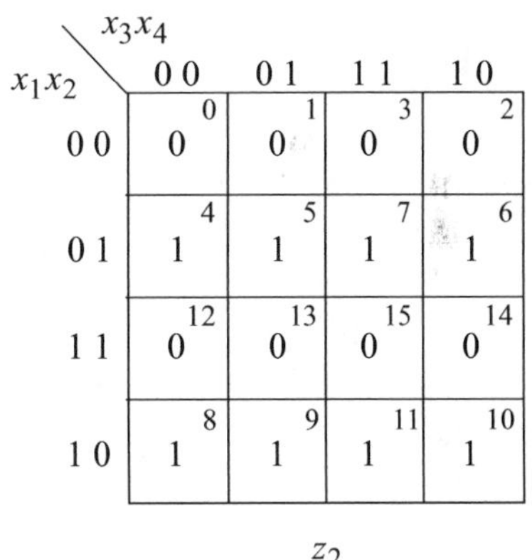

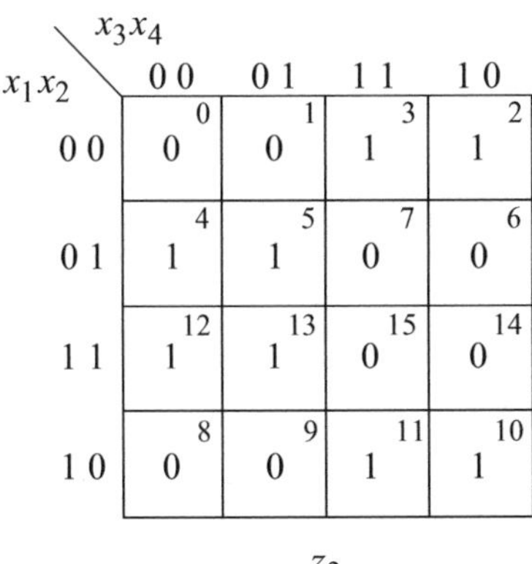

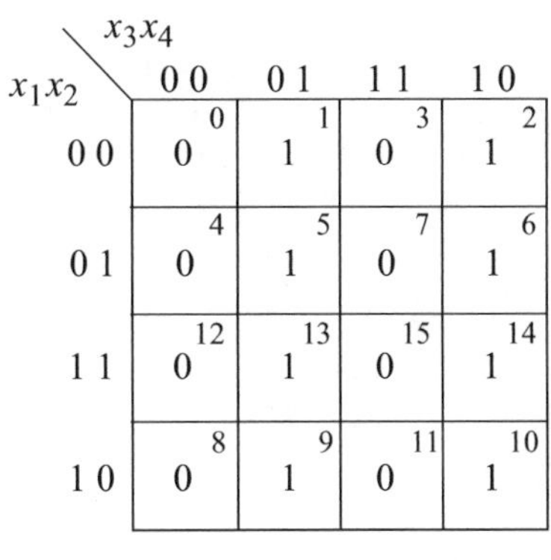

Figure 1.40 Karnaugh maps for the binary-to-Gray code converter.

$$
\begin{aligned}
z_1 &= x_1 \\
z_2 &= x_1'x_2 + x_1x_2' = x_1 \oplus x_2 \\
z_3 &= x_2x_3' + x_2'x_3 = x_2 \oplus x_3 \\
z_4 &= x_3'x_4 + x_3x_4' = x_3 \oplus x_4
\end{aligned} \tag{1.11}
$$

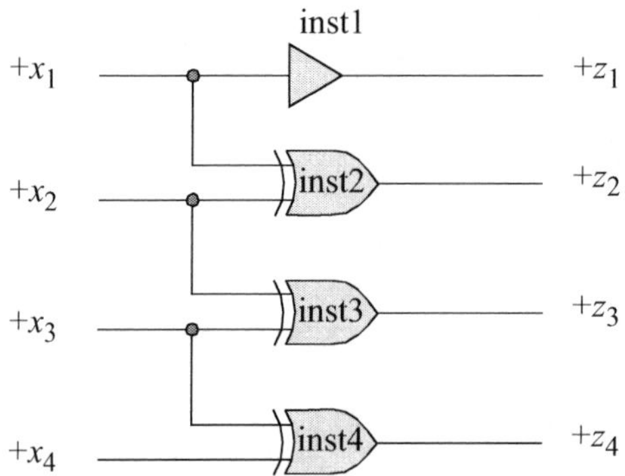

Figure 1.41 Logic diagram for the binary-to-Gray code converter.

```
//binary-to-gray code converter
module bin_to_gray (x1, x2, x3, x4, z1, z2, z3, z4);

input x1, x2, x3, x4;
output z1, z2, z3, z4;

buf inst1 (z1, x1);
xor inst2 (z2, x1, x2);
xor inst3 (z3, x2, x3);
xor inst4 (z4, x3, x4);
endmodule
```

Figure 1.42 Design module for the binary-to-Gray code converter.

```
//test bench for binary-to-gray code converter
module bin_to_gray_tb;

reg x1, x2, x3, x4;
wire z1, z2, z3, z4;
                                        //continued on next page
```

Figure 1.43 Test bench for the binary-to-Gray code converter.

```
//apply input vectors
initial
begin: apply_stimulus
   reg [4:0] invect;
   for (invect=0; invect<16; invect=invect+1)
      begin
         {x1, x2, x3, x4} = invect [4:0];
         #10 $display ("{x1x2x3x4}=%b, {z1z2z3z4}=%b",
                           {x1, x2, x3, x4}, {z1, z2, z3, z4});
      end
end

//instantiate the module into the test bench
bin_to_gray inst1 (
   .x1(x1),
   .x2(x2),
   .x3(x3),
   .x4(x4),
   .z1(z1),
   .z2(z2),
   .z3(z3),
   .z4(z4)
   );

endmodule
```

Figure 1.43 (Continued)

```
{x1x2x3x4}=0000, {z1z2z3z4}=0000
{x1x2x3x4}=0001, {z1z2z3z4}=0001
{x1x2x3x4}=0010, {z1z2z3z4}=0011
{x1x2x3x4}=0011, {z1z2z3z4}=0010
{x1x2x3x4}=0100, {z1z2z3z4}=0110
{x1x2x3x4}=0101, {z1z2z3z4}=0111
{x1x2x3x4}=0110, {z1z2z3z4}=0101
{x1x2x3x4}=0111, {z1z2z3z4}=0100
{x1x2x3x4}=1000, {z1z2z3z4}=1100
{x1x2x3x4}=1001, {z1z2z3z4}=1101
{x1x2x3x4}=1010, {z1z2z3z4}=1111
{x1x2x3x4}=1011, {z1z2z3z4}=1110
{x1x2x3x4}=1100, {z1z2z3z4}=1010
{x1x2x3x4}=1101, {z1z2z3z4}=1011
{x1x2x3x4}=1110, {z1z2z3z4}=1001
{x1x2x3x4}=1111, {z1z2z3z4}=1000
```

Figure 1.44 Outputs for the binary-to-Gray code converter.

Example 1.6 A *full adder* is a combinational circuit that adds two operand bits: *a* and *b* plus a carry-in bit *cin*. The carry-in bit represents the carry-out of the previous lower-order stage. A full adder produces two outputs: a sum bit *sum* and carry-out bit *cout*. This example will use built-in primitives to design a full adder consisting of two half adders plus additional logic as shown in Figure 1.45.

The design module is shown in Figure 1.46, test bench module is shown in Figure 1.47 and the outputs are shown in Figure 1.48.

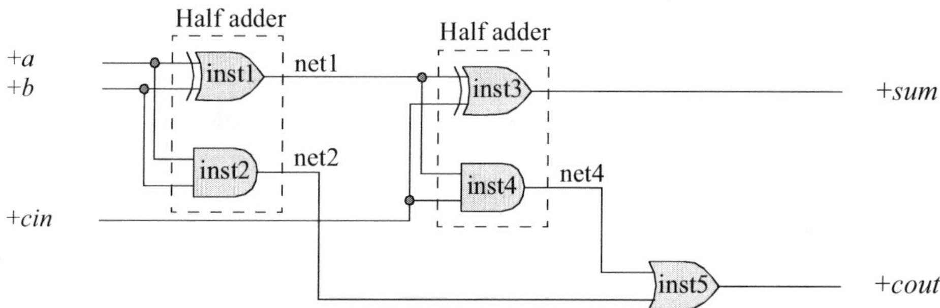

Figure 1.45 Full adder to be designed with built-in primitives.

```
//full adder using built-in primitives
module full_adder_bip (a, b, cin, sum, cout);

input a, b, cin;
output sum, cout;

xor   inst1 (net1, a, b);
and   inst2 (net2, a, b);
xor   inst3 (sum, net1, cin);
and   inst4 (net4, net1, cin);
or    inst5 (cout, net4, net2);
endmodule
```

Figure 1.46 Module for a full adder using built-in primitives.

```
//test bench for full adder using built-in primitives
module full_adder_bip_tb;

reg a, b, cin;
wire sum, cout;                //continued on next page
```

Figure 1.47 Test bench for the full adder of Figure 1.46.

```
//apply input vectors
initial
begin: apply_stimulus
   reg[3:0] invect;      //invect[3] terminates the for loop
   for (invect = 0; invect < 8; invect = invect + 1)
      begin
         {a, b, cin} = invect [3:0];
         #10 $display ("abcin = %b, cout = %b, sum = %b",
                        {a, b, cin}, cout, sum);
      end
end

//instantiate the module into the test bench
full_adder_bip inst1 (
   .a(a),
   .b(b),
   .cin(cin),
   .sum(sum),
   .cout(cout)
   );
endmodule
```

Figure 1.47 (Continued)

```
abcin = 000, cout = 0, sum = 0
abcin = 001, cout = 0, sum = 1
abcin = 010, cout = 0, sum = 1
abcin = 011, cout = 1, sum = 0
abcin = 100, cout = 0, sum = 1
abcin = 101, cout = 1, sum = 0
abcin = 110, cout = 1, sum = 0
abcin = 111, cout = 1, sum = 1
```

Figure 1.48 Outputs for the full adder of Figure 1.46.

1.5 User-Defined Primitives

Verilog also provides the ability to design primitives according to user specifications. These are called *user-defined primitives* (*UDPs*) and are usually at a higher-level logic function than built-in primitives. They are independent primitives and do not instantiate other primitives or modules. UDPs are instantiated into a module the same way as built-in primitives; that is, the syntax for a UDP instantiation is the same as that for a built-in primitive instantiation. A UDP is defined outside the module into which it is instantiated. There are two types of UDPs: combinational and sequential. Sequential primitives include level-sensitive and edge-sensitive circuits.

1.5.1 Defining a User-Defined Primitive

The syntax for a UDP is similar to that for declaring a module. The definition begins with the keyword **primitive** and ends with the keyword **endprimitive**. The UDP contains a name and a list of ports, which are declared as **input** or **output**. For a sequential UDP, the output port is declared as **reg**. UDPs can have one or more scalar inputs, but only one scalar output. The output port is listed first in the terminal list followed by the input ports, in the same way that the terminal list appears in built-in primitives. UDPs do not support **inout** ports.

The UDP table is an essential part of the internal structure and defines the functionality of the circuit. It is a lookup table similar in concept to a truth table. The table begins with the keyword **table** and ends with the keyword **endtable**. The contents of the table define the value of the output with respect to the inputs. The syntax for a UDP is shown below.

```
primitive udp_name (output, input_1, input_2, . . . , input_n);
    output output;
    input input_1, input_2, . . . , input_n;
    reg sequential_output;          //for sequential UDPs

    initial                         //for sequential UDPs

    table
        state table entries
    endtable
endprimitive
```

1.5.2 Combinational User-Defined Primitives

To illustrate the method for defining and using combinational UDPs, examples will be presented ranging from simple designs to more complex designs. UDPs are not compiled separately. They are saved in the same project as the module with a .v extension; for example, *udp_and.v*.

Example 1.7 A 2-input OR gate *udp_or2* will be designed using a UDP. The module is shown in Figure 1.49. The inputs in the state table must be in the same order as in the input list. The table heading is a comment for readability. The inputs and output are separated by a colon and the table entry is terminated by a semicolon. All combinations of the inputs must be entered in the table in order to obtain a correct output; otherwise, the output will be designated as **x** (unknown). To completely specify all combinations of the inputs, a value of **x** should be included in the input values where appropriate.

```
//used-defined primitive for a 2-input OR gate

primitive udp_or2 (z1, x1, x2);//list output first

//input/output declarations
input x1, x2;
output z1;              //must be output (not reg)
                        //...for combinational logic

//state table definition
table
//inputs are in same order as input list
// x1 x2 :  z1;    comment is for readability
   0  0  :  0;
   0  1  :  1;
   1  0  :  1;
   1  1  :  1;
   x  1  :  1;
   1  x  :  1;
endtable

endprimitive
```

Figure 1.49 A user-defined primitive for a 2-input OR gate.

Example 1.8 This example will use a combination of built-in primitives and UDPs to design a full adder from two half adders. The truth tables for a half adder and full adder are shown in Table 1.14 and Table 1.15, respectively. A *half adder* is a combinational circuit that performs the addition of two operand bits and produces two outputs: a sum bit and a carry-out bit. The half adder does not accommodate a carry-in bit. A *full adder* is a combinational circuit that performs the addition of two operand bits plus a carry-in bit. The carry-in represents the carry-out of the previous lower-order stage. The full adder produces two outputs: a sum bit and a carry-out bit.

The sum and carry-out equations for the half adder are shown in Equation 1.12. The sum and carry-out equations for the full adder are shown in Equation 1.13. The logic diagram for a full adder obtained from two half adders using Equation 1.13 is shown in Figure 1.50.

Table 1.14 Truth Table for a Half Adder

a	*b*	*sum*	*carry-out*
0	0	0	0
0	1	1	0
1	0	1	0
1	1	0	1

Table 1.15 Truth Table for a Full Adder

a	*b*	*cin*	*sum*	*carry-out*
0	0	0	0	0
0	0	1	1	0
0	1	0	1	0
0	1	1	0	1
1	0	0	1	0
1	0	1	0	1
1	1	0	0	1
1	1	1	1	1

$$\begin{aligned} sum &= a'b + ab' \\ &= a \oplus b \\ carry\text{-}out &= ab \end{aligned} \tag{1.12}$$

$$\begin{aligned} sum &= a'b'cin + a'bcin' + ab'cin' + abcin \\ &= a \oplus b \oplus cin \\ carry\text{-}out &= a'bcin + ab'cin + abcin' + abcin \\ &= cin(a \oplus b) + ab \\ &= ab + acin + bcin \qquad \text{(see Figure 1.51)} \end{aligned} \tag{1.13}$$

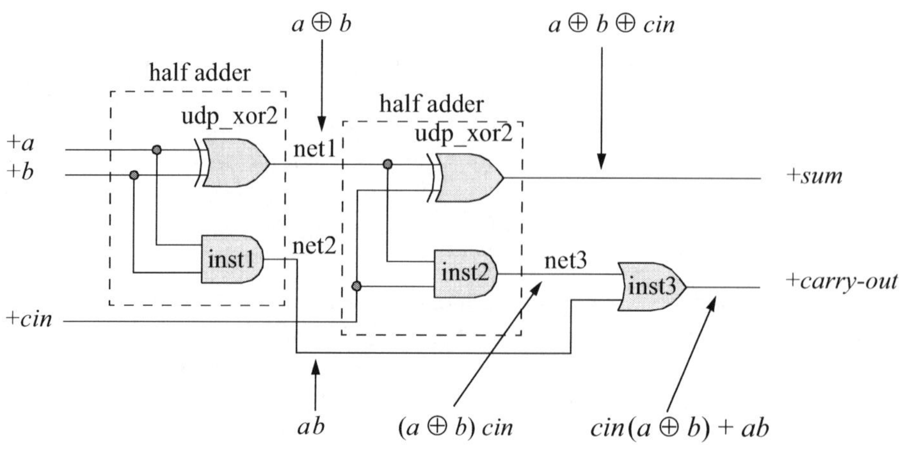

Figure 1.50 Full adder designed from two half adders.

The equation for *carry-out* can also be obtained by plotting Table 1.15 on a Karnaugh map, as shown in Figure 1.51. The equation is then easily obtained in a sum-of-products notation as: $ab + acin + bcin$.

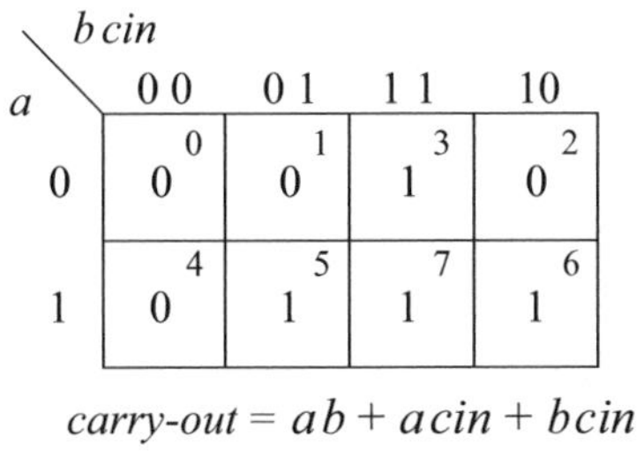

a \ $bcin$	0 0	0 1	1 1	10
0	0 (0)	0 (1)	1 (3)	0 (2)
1	0 (4)	1 (5)	1 (7)	1 (6)

$$carry\text{-}out = ab + acin + bcin$$

Figure 1.51 Karnaugh map for the carry-out of a full adder.

The full adder will be designed by means of a UDP for the exclusive-OR gates and built-in primitives for the AND gates and OR gate, all of which will be instantiated into the project *full_adder_udp*. The module for the *udp_xor2* is shown in Figure 1.52. The *full_adder_udp* module is shown in Figure 1.53, the test bench is shown in Figure 1.54, and the outputs are shown in Figure 1.55.

```
//UDP for a 2-input exclusive-OR
primitive udp_xor2 (z1, x1, x2);

input x1, x2;
output z1;

//define state table
table
//inputs are in the same order as the input list
// x1 x2  :  z1;      comment is for readability
   0  0   :  0;
   0  1   :  1;
   1  0   :  1;
   1  1   :  0;
endtable

endprimitive
```

Figure 1.52 Module for the *udp_xor2* to be instantiated into the full adder module *full_adder_udp*.

```
//full adder using a UDP and built-in primitives
module full_adder_udp (a, b, cin, sum, cout);

input a, b, cin;
output sum, cout;

wire net1, net2, net3;      //define internal nets

//instantiate the udps and built-in primitive
udp_xor2 (net1, a, b);
and inst1 (net2, a, b);


udp_xor2 (sum, net1, cin);
and inst2 (net3, net1, cin);

or inst3 (cout, net3, net2);
endmodule
```

Figure 1.53 Module for a full adder using a UDP and built-in primitives.

```
//test bench for full adder
module full_adder_udp_tb;

reg a, b, cin;
wire sum, cout;

initial         //apply input vectors
begin: apply_stimulus
   reg [3:0] invect;
   for (invect=0; invect<8; invect=invect+1)
      begin
         {a, b, cin} = invect [3:0];
         #10 $display ("a b cin = %b, sum cout = %b",
                        {a, b, cin}, {sum, cout});
      end
end

//instantiate the module into the test bench
full_adder_udp inst1 (
   .a(a),
   .b(b),
   .cin(cin),
   .sum(sum),
   .cout(cout)
   );
endmodule
```

Figure 1.54 Test bench for the full adder of Figure 1.53.

```
a b cin = 000, sum cout = 00
a b cin = 001, sum cout = 10
a b cin = 010, sum cout = 10
a b cin = 011, sum cout = 01
a b cin = 100, sum cout = 10
a b cin = 101, sum cout = 01
a b cin = 110, sum cout = 01
a b cin = 111, sum cout = 11
```

Figure 1.55 Outputs for the full adder of Figure 1.53.

Example 1.9 This example will design a 4:1 multiplexer as a UDP. The multiplexer will then be checked for correct functional operation by means of a test bench which will generate the outputs. A block diagram of the multiplexer is shown in Figure 1.56 together with a table defining the output as a function of the two select inputs s_1 and s_0 and the four data inputs d_0, d_1, d_2, and d_3. The equation for the output can be written directly from the table as shown in Equation 1.14. An *Enable* input may also be associated with a multiplexer to enable the output.

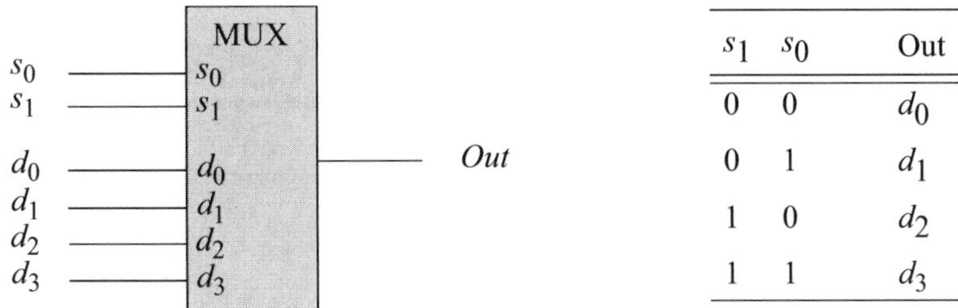

s_1	s_0	Out
0	0	d_0
0	1	d_1
1	0	d_2
1	1	d_3

Figure 1.56 A 4:1 multiplexer to be designed as a UDP.

$$Out = s_1's_0'd_0 + s_1's_0d_1 + s_1s_0'd_2 + s_1s_0d_3 \tag{1.14}$$

The 4:1 multiplexer UDP is shown in Figure 1.57. Note the entries in the table that contain the symbol (?), which indicates a "don't care" condition. Referring to the first line in the table, if $s_1s_0 = 00$, then it does not matter what the values are for inputs $d_1d_2d_3$ because only input d_0 is selected.

The test bench for the 4:1 multiplexer is shown in Figure 1.58. The input lines are set to known values such that $d_0d_1d_2d_3 = 1010$. The input values are then displayed using the **$display** system task. The backslash (\) character is used to escape certain special characters such as \n, which is a newline character.

```
//4:1 multiplexer as a UDP
primitive udp_mux4 (out, s1, s0, d0, d1, d2, d3);

input s1, s0, d0, d1, d2, d3;
output out;

table    //define state table
//inputs are in the same order as the input list
// s1 s0 d0 d1 d2 d3 :  out     comment is for readability
   0  0  1  ?  ?  ?  :  1;      //? is "don't care"
   0  0  0  ?  ?  ?  :  0;

   0  1  ?  1  ?  ?  :  1;
   0  1  ?  0  ?  ?  :  0;

   1  0  ?  ?  1  ?  :  1;
   1  0  ?  ?  0  ?  :  0;

   1  1  ?  ?  ?  1  :  1;
   1  1  ?  ?  ?  0  :  0;

   ?  ?  0  0  0  0  :  0;
   ?  ?  1  1  1  1  :  1;
endtable
endprimitive
```

Figure 1.57 A UDP for a 4:1 multiplexer.

```
//test bench for the 4:1 multiplexer udp
module udp_mux4_tb;

reg s1, s0, d0, d1, d2, d3;
wire out;

initial
begin
//set the input lines to known values
   d0 = 1; d1 = 0; d2 = 1; d3 = 0;

//display the input values
   #10 $display ("d0=%b, d1=%b, d2=%b, d3=%b \n",
                  d0, d1, d2, d3);         // \n is new line
//continued on next page
```

Figure 1.58 Test bench for the UDP 4:1 multiplexer.

```
//select d0 = 1
   s1 = 0; s0 = 0;
   #10 $display ("s1=%b, s0=%b, output=%b \n",
                  s1, s0, out);

//select d1 = 0
   s1 = 0; s0 = 1;
   #10 $display ("s1=%b, s0=%b, output=%b \n",
                  s1, s0, out);

//select d2 = 1
   s1 = 1; s0 = 0;
   #10 $display ("s1=%b, s0=%b, output=%b \n",
                  s1, s0, out);

//select d3 = 0
   s1 = 1; s0 = 1;
   #10 $display ("s1=%b, s0=%b, output=%b \n",
                  s1, s0, out);

   #10 $stop;
end

//instantiate the module into the test bench.
//if instantiating only the primitive of USB with no module,
//then instantiation must be done using positional notation

udp_mux4 inst1 (out, s1, s0, d0, d1, d2, d3);
endmodule
```

Figure 1.58 (Continued)

Beginning at 10 time units, the select lines are rotated through all four combinations of the two variables, which in turn transmit the input values to the output. For example, if $s_1 s_0 = 11$, then the value of input line d_3 is transmitted to the output. When instantiating a UDP module into a test bench — when there is no design module — the ports must be instantiated by position. The outputs are shown in Figure 1.59.

```
d0=1, d1=0, d2=1, d3=0

s1=0, s0=0, output=1
s1=0, s0=1, output=0
s1=1, s0=0, output=1
s1=1, s0=1, output=0
```

Figure 1.59 Outputs for the UDP 4:1 multiplexer.

Example 1.10 Variables may also be entered in a Karnaugh map as *map-entered variables*, together with 1s and 0s. A map of this type is more compact than a standard Karnaugh map, but contains the same information. A map containing map-entered variables is particularly useful in analyzing and designing synchronous sequential machines. When variables are entered in a Karnaugh map, two or more squares can be combined only if the squares are adjacent and contain the same variable(s).

The Karnaugh map of Figure 1.60 will be implemented using a 4:1 multiplexer and any additional logic. First, the equations for the multiplexer data inputs, d_0, d_1, d_2, and d_3 will be obtained using E as a map-entered variable, where the multiplexer select inputs are $s_1 s_0 = x_1 x_2$. Then the circuit will be designed using UDPs for the multiplexer and associated logic gates.

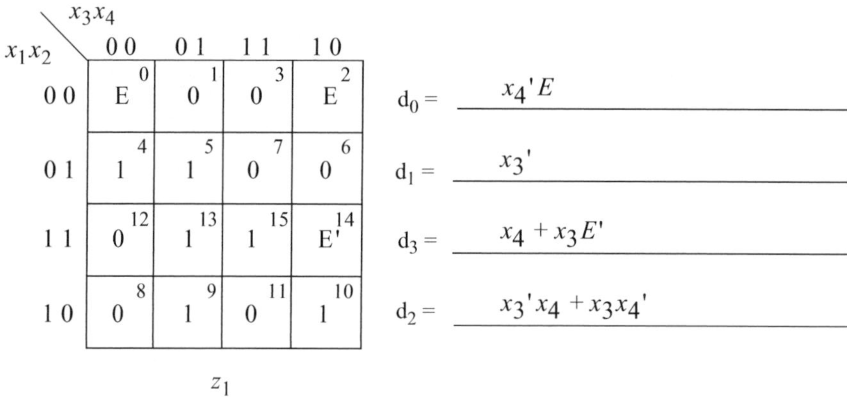

Figure 1.60 Karnaugh map for Example 1.10 using E as a map-entered variable.

To obtain the equation for data input d_0, where $s_1 s_0 = x_1 x_2 = 00$, minterm locations 0 and 2 are adjacent and contain the same variable E; therefore, the term is $x_4' E$. Data input d_1, where $s_1 s_0 = x_1 x_2 = 01$, contains 1s in minterm locations 4 and 5; therefore, $d_1 = x_3'$. To obtain the equation for d_3, where $s_1 s_0 = x_1 x_2 = 11$, minterm locations 13 and 15 combine to yield x_4. Minterm location 15 is equivalent to $1 + E'$; therefore, minterm locations 14 and 15 combine to yield the product term $x_3 E'$. The equation for d_3 is $x_4 + x_3 E'$. Data input d_2 is obtained in a similar manner.

The logic diagram is shown in Figure 1.61 using a 4:1 multiplexer (*udp_mux4*), a 2-input AND gate (*udp_and2*), a 2-input exclusive-OR function (*udp_xor2*) previously designed, and a 2-input OR gate (*udp_or2*) previously designed.

The module for the logic diagram is shown in Figure 1.62 and the test bench is shown in Figure 1.63. The Karnaugh map of Figure 1.60 is expanded to the 5-variable map of Figure 1.64 to better visualize the minterm entries when comparing them with the outputs of Figure 1.65.

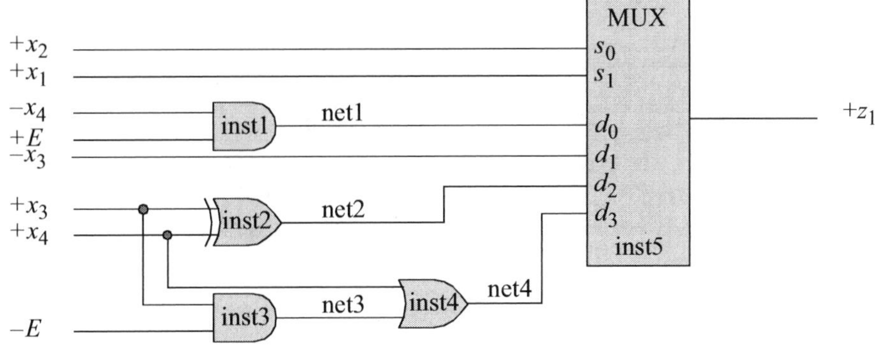

Figure 1.61 Logic diagram for the Karnaugh map of Figure 1.60.

```
//logic circuit using a multiplexer udp
//together with other logic gate udps
module mux4_mev (x1, x2, x3, x4, E, z1);

input x1, x2, x3, x4, E;
output z1;

//instantiate the udps
udp_and2 inst1 (net1, ~x4, E);
udp_xor2 inst2 (net2, x3, x4);
udp_and2 inst3 (net3, x3, ~E);
udp_or2  inst4 (net4, x4, net3);

//the mux inputs are: s1, s0, d0, d1, d2, d3
udp_mux4 inst5 (z1, x1, x2, net1, ~x3, net2, net4);

endmodule
```

Figure 1.62 Module for the logic diagram of Figure 1.61.

```
//test bench for mux4_mev

module mux4_mev_tb;

reg x1, x2, x3, x4, E;
wire z1;
                              //continued on next page
```

Figure 1.63 Test bench for Figure 1.62 for the logic diagram of Figure 1.61.

```
//apply input vectors
initial
begin: apply_stimulus
   reg [5:0] invect;
   for (invect=0; invect<32; invect=invect+1)
      begin
         {x1, x2, x3, x4, E} = invect [5:0];
         #10 $display ("x1x2x3x4E = %b, z1 = %b",
                  {x1, x2, x3, x4, E}, z1);
      end
end


//instantiate the module into the test bench
mux4_mev inst1 (
   .x1(x1),
   .x2(x2),
   .x3(x3),
   .x4(x4),
   .E(E),
   .z1(z1)
   );

endmodule
```

Figure 1.63 (Continued)

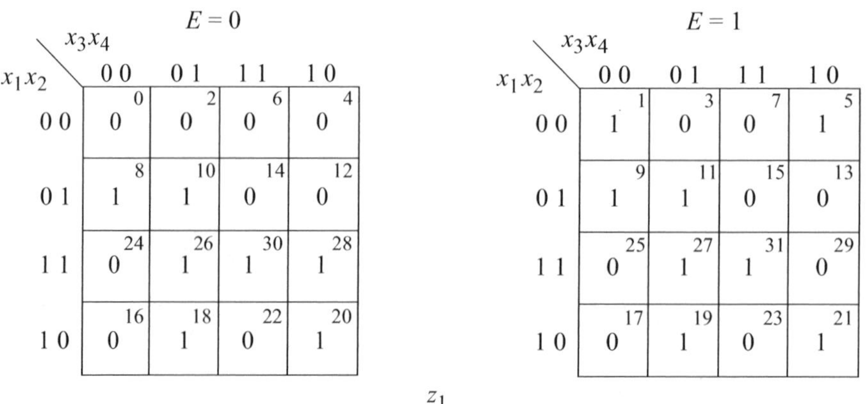

Figure 1.64 Five-variable Karnaugh map equivalent to the 4-variable map of Figure 1.60.

```
x1x2x3x4E = 00000, z1 = 0
x1x2x3x4E = 00001, z1 = 1
x1x2x3x4E = 00010, z1 = 0
x1x2x3x4E = 00011, z1 = 0
x1x2x3x4E = 00100, z1 = 0
x1x2x3x4E = 00101, z1 = 1
x1x2x3x4E = 00110, z1 = 0
x1x2x3x4E = 00111, z1 = 0

x1x2x3x4E = 01000, z1 = 1
x1x2x3x4E = 01001, z1 = 1
x1x2x3x4E = 01010, z1 = 1
x1x2x3x4E = 01011, z1 = 1
x1x2x3x4E = 01100, z1 = 0
x1x2x3x4E = 01101, z1 = 0
x1x2x3x4E = 01110, z1 = 0
x1x2x3x4E = 01111, z1 = 0

x1x2x3x4E = 10000, z1 = 0
x1x2x3x4E = 10001, z1 = 0
x1x2x3x4E = 10010, z1 = 1
x1x2x3x4E = 10011, z1 = 1
x1x2x3x4E = 10100, z1 = 1
x1x2x3x4E = 10101, z1 = 1
x1x2x3x4E = 10110, z1 = 0
x1x2x3x4E = 10111, z1 = 0

x1x2x3x4E = 11000, z1 = 0
x1x2x3x4E = 11001, z1 = 0
x1x2x3x4E = 11010, z1 = 1
x1x2x3x4E = 11011, z1 = 1
x1x2x3x4E = 11100, z1 = 1
x1x2x3x4E = 11101, z1 = 0
x1x2x3x4E = 11110, z1 = 1
x1x2x3x4E = 11111, z1 = 1
```

Figure 1.65 Outputs obtained from the test bench of Figure 1.63 for the module of Figure 1.62.

1.5.3 Sequential User-Defined Primitives

Verilog provides a means to model sequential UDPs in much the same way as built-in primitives are modeled. Sequential UDPs can be used to model both level-sensitive and edge-sensitive sequential circuits. Level-sensitive behavior is controlled by the value of an input signal; edge-sensitive behavior is controlled by the edge of an input signal. The inputs are implied to be of type **wire**. Sequential devices have an internal state that is a 1-bit register and must be modeled as a type **reg** variable, which is the output of the device and specifies the present state. One **initial** statement can be used to initialize the output of a sequential UDP.

Level-sensitive user-defined primitives The state — and thus the output — of a level-sensitive device is a function of the input levels only, not on a low-to-high or a high-to-low transition. A latch is an example of a level-sensitive UDP.

Example 1.11 The logic diagram of a latch is shown in Figure 1.66. The UDP module is shown in Figure 1.67, the test bench module is shown in Figure 1.68, and the outputs are shown in Figure 1.69.

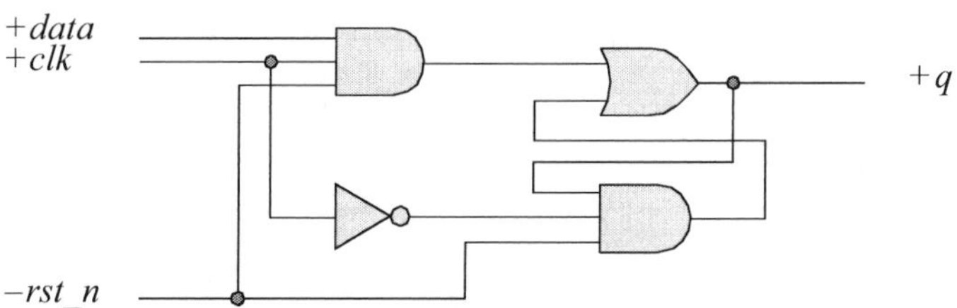

Figure 1.66 Logic diagram for a gated latch to be modeled as a sequential level-sensitive UDP.

```
//a gated latch as a level-sensitive udp
primitive udp_latch_level (q, data, clk, rst_n);
input data, clk, rst_n;
output q;
reg q;       //q is internal storage

initial
   q = 0;    //initialize output q to 0

//define state table
table
//inputs are in the same order as the input list
// data  clk   rst_n :  q  :  q+;    q+ is next state
   ?     ?     0     :  ?  :  0;     //latch is reset
   0     0     1     :  ?  :  -;     //- means no change
   0     1     1     :  ?  :  0;     //data=0; clk=1; q+=0
   1     0     1     :  ?  :  -;
   1     1     1     :  ?  :  1;     //data=1; clk=1; q+=1
   ?     0     1     :  ?  :  -;
endtable
endprimitive
```

Figure 1.67 Design module for a level-sensitive gated latch UDP.

```
//test bench for level-sensitive latch
module udp_latch_level_tb;

reg data, clk, rst_n;
wire q;

//display variables
initial
$monitor ("rst_n=%b, data=%b, clk=%b, q=%b",
            rst_n, data, clk, q);

//apply input vectors
initial
begin
   #0     rst_n=1'b0;  data=1'b0;  clk=1'b0;
   #10    rst_n=1'b1;  data=1'b1;  clk=1'b1;
   #10    rst_n=1'b1;  data=1'b1;  clk=1'b0;
   #10    rst_n=1'b1;  data=1'b0;  clk=1'b1;
   #10    rst_n=1'b1;  data=1'b1;  clk=1'b1;
end

//instantiation must be done by position, not by name
udp_latch_level inst1 (q, data, clk, rst_n);
endmodule
```

Figure 1.68 Test bench for the level-sensitive gated latch of Figure 1.67.

```
rst_n=0, data=0, clk=0, q=0
rst_n=1, data=1, clk=1, q=1
rst_n=1, data=1, clk=0, q=1
rst_n=1, data=0, clk=1, q=0
rst_n=1, data=1, clk=1, q=1
```

Figure 1.69 Outputs for the level-sensitive gated latch of Figure 1.67.

Edge-sensitive user-defined primitives Edge-sensitive UDPs can model behavior that is triggered by either a positive edge or a negative edge. The table entries in edge-sensitive circuits are similar to those in level-sensitive circuits. The difference is that a rising or falling edge must be specified on the clock input (or any other input that triggers the circuit).

Most counters count in either a count-up or count-down sequence. Still other counters can be designed for a unique application in which the counting sequence is neither entirely up nor entirely down. These have a nonsequential counting sequence that is prescribed by external requirements. Such a counter has a counting sequence as follows: $y_1y_2y_3y_4$ = 0000, 1000, 1100, 1110, 1111, 0111, 0011, 0001, 0000, and is classified as a *Johnson counter*. The counter is reset initially to $y_1y_2y_3y_4 = 0000$. The unspecified states can be regarded as "don't care" states in order to minimize the δ next-state logic. The inverted output of the last flip-flop is fed back to the *D* input of the first flip-flop.

The logic diagram for a 4-bit Johnson counter is shown in Figure 1.70 using positive-edge-triggered *D* flip-flops. The *D* flip-flop will be designed as a user-defined primitive, then instantiated four times into the design module of the Johnson counter. The *D* flip-flop is shown in Figure 1.71 as a UDP.

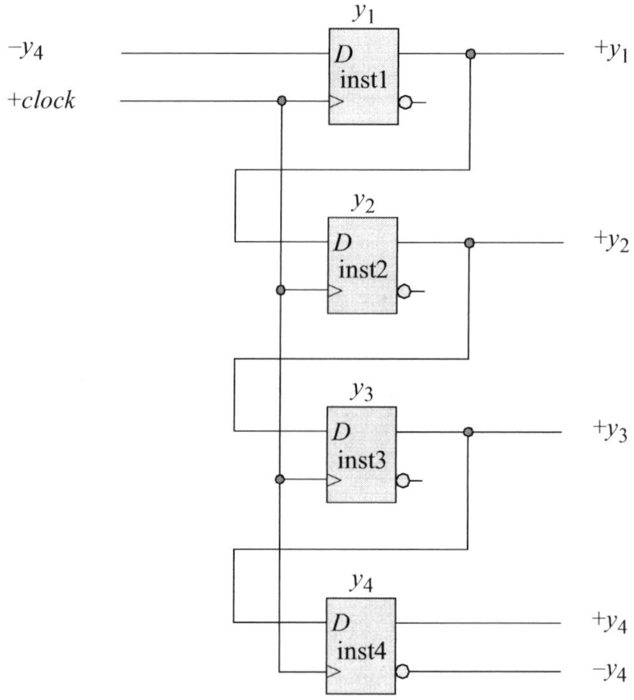

Figure 1.70 Logic diagram for a 4-bit Johnson counter.

```
//a positive-edge-sensitive D flip-flop
primitive udp_dff_edge1 (q, d, clk, rst_n);

input d, clk, rst_n;
output q;

reg q;       //q is internal storage

//initialize q to 0
initial
   q = 0;                        //continued on next page
```

Figure 1.71 A user-defined primitive for a *D* flip-flop.

```
//define state table
table
//inputs are in the same order as the input list
// d      clk   rst_n :  q  :  q+;    q+ is the next state
   0      (01)  1     :  ?  :  0;     //(01) is rising edge
   1      (01)  1     :  ?  :  1;     //rst_n = 1 means no rst
   1      (0x)  1     :  1  :  1;     //(0x) is no change
   0      (0x)  1     :  0  :  0;
   ?      (?0)  1     :  ?  :  -;     //ignore negative edge
//reset case when rst_n is 0 and clk has any transition
   ?      (??)  0     :  ?  :  0;     //rst_n = 0 means reset
//reset case when rst_n is 0.  d & clk can be anything, q+=0
   ?      ?     0     :  ?  :  0;
//reset case when 0 --> 1 transition on rst_n.  Hold q+ state
   ?      ?      (01) :  ?  :  -;
//non-reset case when d has any trans, but clk has no trans
   (??)   ?     1     :  ?  :  -;     //clk = ?, means no edge
endtable
endprimitive
```

Figure 1.71 (Continued)

The design module for the Johnson counter is shown in Figure 1.72 which instantiates the user-defined primitive *udp_dff_edge1* four times to implement the Johnson counter. The test bench is shown in Figure 1.73. The outputs are shown in Figure 1.74.

```
//udp for a 4-bit johnson counter
module ctr_johnson4 (rst_n, clk, y1, y2, y3, y4);
input rst_n, clk;
output y1, y2, y3, y4;

//instantiate D flip-flop for y1
udp_dff_edge1 inst1 (y1, ~y4, clk, rst_n);

//instantiate D flip-flop for y2
udp_dff_edge1 inst2 (y2, y1, clk, rst_n);

//instantiate D flip-flop for y3
udp_dff_edge1 inst3 (y3, y2, clk, rst_n);

//instantiate D flip-flop for y4
udp_dff_edge1 inst4 (y4, y3, clk, rst_n);
endmodule
```

Figure 1.72 A Johnson counter designed using a UDP for a *D* flip-flop.

```
//test bench for the 4-bit johnson counter
module ctr_johnson4_tb;

reg clk, rst_n;             //inputs are reg for tb
wire y1, y2, y3, y4;        //outputs are wire for tb

initial
$monitor ("count = %b", {y1, y2, y3, y4});

initial                     //define clk
begin
   clk = 1'b0;
   forever
      #10   clk = ~clk;
end

initial                     //define reset
begin
   #0 rst_n = 1'b0;
   #5 rst_n = 1'b1;
   #200  $stop;
end

ctr_johnson4 inst1 (        //instantiate the module
   .rst_n(rst_n),
   .clk(clk),
   .y1(y1),
   .y2(y2),
   .y3(y3),
   .y4(y4)
   );
endmodule
```

Figure 1.73 Test bench for the 4-bit UDP Johnson counter.

```
count = 0000
count = 1000
count = 1100
count = 1110
count = 1111
count = 0111
count = 0011
count = 0001
count = 0000
count = 1000
```

Figure 1.74 Outputs for the 4-bit UDP Johnson counter.

1.6 Dataflow Modeling

Gate-level modeling using built-in primitives is an intuitive approach to digital design because it corresponds one-to-one with traditional digital logic design at the gate level. Dataflow modeling, however, is at a higher level of abstraction than gate-level modeling. Design automation tools are used to create gate-level logic from dataflow modeling by a process called *logic synthesis*. Register transfer level (RTL) is a combination of dataflow modeling and behavioral modeling and characterizes the flow of data through logic circuits. The following sections describe different techniques used to design logic circuits using dataflow modeling. These techniques include the continuous assignment statement, reduction operators, the conditional operator, relational operators, logical operators, bitwise operators, and shift operators.

1.6.1 Continuous Assignment

The *continuous assignment* statement models dataflow behavior and is used to design combinational logic without using gates and interconnecting nets. Continuous assignment statements provide a Boolean correspondence between the right-hand side expression and the left-hand side target. The continuous assignment statement uses the keyword **assign** and has the following syntax with optional drive strength and delay:

assign [drive_strength] [delay] left-hand side target = right-hand side expression

The continuous assignment statement assigns a value to a net (**wire**) that has been previously declared — it cannot be used to assign a value to a register. Therefore, the left-hand target must be a scalar or vector net or a concatenation of scalar and vector nets. The operands on the right-hand side can be registers, nets, or function calls. The registers and nets can be declared as either scalars or vectors. The following are examples of continuous assignment statements for scalar nets:

assign $z_1 = x_1 \ \& \ x_2 \ \& \ x_3$;
assign $z_1 = x_1 \ \hat{} \ x_2$;
assign $z_1 = (x_1 \ \& \ x_2) \mid x_3$;

where the symbol "&" is the AND operation, the symbol "^" is the exclusive-OR operation, and the symbol "|" is the OR operation.

The following are examples of continuous assignment statements for vector and scalar nets, where *sum* is a 9-bit vector to accommodate the *sum* and carry-out, *a* and *b* are 8-bit vectors, and *cin* is a scalar:

assign sum = a + b + cin
assign sum = a ^ b ^ cin

where the symbol "+" is the add operation.

Example 1.12 Figure 1.75 is an example of a continuous assignment statement utilized in the design of an exclusive-NOR circuit where x_1 and x_2 are the inputs and z_1 is the output. Figure 1.76 shows the test bench and Figure 1.77 shows the outputs. Recall that an exclusive-NOR circuit is defined as:

x_1 x_2	z_1
0 0	1
0 1	0
1 0	0
1 1	1

```
//dataflow 2-input exclusive-nor
module xnor2_df (x1, x2, z1);

input x1, x2;        //list all inputs and outputs
output z1;

wire x1, x2;         //all signals are wire
wire z1;

assign z1 = ~(x1 ^ x2);       //continuous assign used for
endmodule                     //...dataflow modeling
```

Figure 1.75 Continuous assignment statement used to design an exclusive-NOR circuit.

```
//dataflow xnor2_df test bench
module xnor2_tb;

reg x1, x2;       //inputs are reg for test bench
wire z1;          //outputs are wire for test bench

initial           //apply input vectors and display variables
begin: apply_stimulus
   reg [2:0] invect;
   for (invect = 0; invect < 4; invect = invect + 1)
      begin
         {x1, x2} = invect [2:0];
         #10 $display ("x1 x2 = %b, z1 = %b", {x1, x2}, z1);
      end
end
//instantiate the module into the test bench as a  single line
xnor2_df inst1 (x1, x2, z2);
endmodule
```

Figure 1.76 Test bench for the exclusive-NOR circuit.

```
x1 x2 = 00, z1 = 1
x1 x2 = 01, z1 = 0
x1 x2 = 10, z1 = 0
x1 x2 = 11, z1 = 1
```

Figure 1.77 Outputs for the exclusive-NOR circuit.

1.6.2 Reduction Operators

The reduction operators are: AND (&), NAND (~&), OR (|), NOR (~ |), exclusive-OR (^), and exclusive-NOR (^ ~ or ~ ^). Reduction operators are unary operators; that is, they operate on a single vector and produce a single-bit result. Reduction operators perform their respective operations on a bit-by-bit basis from right to left. If any bit in the operand is an **x** or a **z**, then the result of the operation is an **x**. The reduction operators are defined as follows:

Reduction Operator	Description
& (Reduction AND)	If any bit is a 0, then the result is 0, otherwise the result is 1.
~& (Reduction NAND)	This is the complement of the reduction AND operation.
| (Reduction OR)	If any bit is a 1, then the result is 1, otherwise the result is 0.
~ | (Reduction NOR)	This is the complement of the reduction OR operation.
^ (Reduction exclusive-OR)	If there are an even number of 1s in the operand, then the result is 0, otherwise the result is 1.
~ ^ (Reduction exclusive-NOR)	This is the complement of the reduction exclusive-OR operation.

Example 1.13 This example illustrates the continuous assignment statement to demonstrate the reduction operators. Figure 1.78 contains the design module to illustrate the operation of the six reduction operators using a 4-bit operand *a[3:0]*. If no delays are specified for the continuous assignment statement, then only one **assign** keyword is required. Only the final statement is terminated by a semicolon; all other statements are terminated by a comma. The test bench and outputs are shown in Figure 1.79 and Figure 1.80, respectively.

```
//module to illustrate the use of reduction operators
module reduction3 (a, red_and, red_nand, red_or, red_nor,
                    red_xor, red_xnor);

//list inputs and outputs
input [3:0] a;
output red_and, red_nand, red_or, red_nor, red_xor, red_xnor;

//define signals
wire [3:0] a;
wire red_and, red_nand, red_or, red_nor, red_xor, red_xnor;

assign   red_and   = &a,   //reduction AND
         red_nand  = ~&a,  //reduction NAND
         red_or    = |a,   //reduction OR
         red_nor   = ~|a,  //reduction NOR
         red_xor   = ^a,   //reduction exclusive-OR
         red_xnor  = ^~a;  //reduction exclusive-NOR

endmodule
```

Figure 1.78 Design module for reduction operators.

```
//test bench for reduction2 module
module reduction3_tb;

reg [3:0] a;  //inputs are reg for test bench; outputs are wire
wire red_and, red_nand, red_or, red_nor, red_xor, red_xnor;

initial
$monitor ("a=%b, red_and=%b, red_nand=%b, red_or=%b,
            red_nor=%b, red_xor=%b, red_xnor=%b",
         a, red_and, red_nand, red_or, red_nor, red_xor,
            red_xnor);

//apply input vectors
initial
begin
   #0    a = 4'b0001;
   #10   a = 4'b0010;
   #10   a = 4'b0011;
   #10   a = 4'b0100;
   #10   a = 4'b0101;
   #10   a = 4'b0110;              //continued on next page
```

Figure 1.79 Test bench module for reduction operators.

```
   #10   a = 4'b0111;
   #10   a = 4'b1000;
   #10   a = 4'b1001;
   #10   a = 4'b1010;
   #10   a = 4'b1011;
   #10   a = 4'b1100;
   #10   a = 4'b1101;
   #10   a = 4'b1110;
   #10   a = 4'b1111;

   #10   $stop;
end

//instantiate the module into the test bench as a single line
reduction3 inst1 (a, red_and, red_nand, red_or, red_nor,
                     red_xor, red_xnor);
endmodule
```

Figure 1.79 (Continued)

```
a=0001,
red_and=0, red_nand=1, red_or=1,
red_nor=0, red_xor=1, red_xnor=0

a=0010,
red_and=0, red_nand=1, red_or=1,
red_nor=0, red_xor=1, red_xnor=0

a=0011,
red_and=0, red_nand=1, red_or=1,
red_nor=0, red_xor=0, red_xnor=1

a=0100,
red_and=0, red_nand=1, red_or=1,
red_nor=0, red_xor=1, red_xnor=0

a=0101,
red_and=0, red_nand=1, red_or=1,
red_nor=0, red_xor=0, red_xnor=1

a=0110,
red_and=0, red_nand=1, red_or=1,
red_nor=0, red_xor=0, red_xnor=1
                              //continued on next  page
```

Figure 1.80 Outputs for reduction operators.

```
a=0111,
red_and=0, red_nand=1, red_or=1,
red_nor=0, red_xor=1, red_xnor=0

a=1000,
red_and=0, red_nand=1, red_or=1,
red_nor=0, red_xor=1, red_xnor=0

a=1001,
red_and=0, red_nand=1, red_or=1,
red_nor=0, red_xor=0, red_xnor=1

a=1010,
red_and=0, red_nand=1, red_or=1,
red_nor=0, red_xor=0, red_xnor=1

a=1011,
red_and=0, red_nand=1, red_or=1,
red_nor=0, red_xor=1, red_xnor=0

a=1100,
red_and=0, red_nand=1, red_or=1,
red_nor=0, red_xor=0, red_xnor=1

a=1101,
red_and=0, red_nand=1, red_or=1,
red_nor=0, red_xor=1, red_xnor=0

a=1110,
red_and=0, red_nand=1, red_or=1,
red_nor=0, red_xor=1, red_xnor=0

a=1111,
red_and=1, red_nand=0, red_or=1,
red_nor=0, red_xor=0, red_xnor=1
```

Figure 1.80 (Continued)

1.6.3 Conditional Operator

The conditional operator (**? :**) has three operands, as shown in the syntax below. The *conditional_expression* is evaluated. If the result is true (1), then the *true_expression* is evaluated; if the result is false (0), then the *false_expression* is evaluated.

conditional_expression **?** true_expression **:** false_expression;

The conditional operator can be used when one of two expressions is to be selected. For example, in Equation 1.15 shown below, if x_1 is greater than or equal to x_2, then z_1 is assigned the value of x_3; if x_1 is less than x_2, then z_1 is assigned the value of x_4.

$$z_1 = (x_1 >= x_2) \,?\, x_3 : x_4; \tag{1.15}$$

Conditional operators can be nested; that is, each true_expression and false_expression can be a conditional operation, as shown below. This is useful for modeling a 4:1 multiplexer.

conditional_expression ? (cond_expr1 ? true_expr1 : false_expr1)
: (cond_expr2 ? true_expr2 : false_expr2);

Example 1.14 Equation 1.15 will be implemented using the conditional operator. If x_1 is greater than or equal to x_2, then output z_1 will be assigned the value of x_3, otherwise z_1 will be assigned the value of x_4. The design module is shown in Figure 1.81. The test bench module is shown in Figure 1.82 and the outputs are shown in Figure 1.83.

```
//conditional operator for the following equation:
//z1 = (x1 >= x2) ? x3 : x4;

module conditional_op (x1, x2, x3, x4, z1);

//define inputs and outputs
input x1, x2, x3, x4;
output z1;

assign z1 = (x1 >= x2) ? x3 : x4;

endmodule
```

Figure 1.81 Design module for the conditional operator of Equation 1.15.

```
//test bench for conditional_op

module conditional_op_tb;
//inputs are reg for test bench; outputs are wire
reg x1, x2, x3, x4;
wire z1;                              //continued on next page
```

Figure 1.82 Test bench module for Figure 1.81.

```
//display variables
initial
$monitor ("x1 = %b, x2 = %b, x3 = %b, x4 = %b, z1 = %b",
            x1, x2, x3, x4, z1);

//apply input vectors
initial
begin
   #0    x1=1'b0;    x2=1'b0;    x3=1'b0;    x4=1'b0;
   #10   x1=1'b0;    x2=1'b0;    x3=1'b0;    x4=1'b1;
   #10   x1=1'b0;    x2=1'b0;    x3=1'b1;    x4=1'b0;
   #10   x1=1'b0;    x2=1'b0;    x3=1'b1;    x4=1'b1;
   #10   x1=1'b0;    x2=1'b1;    x3=1'b0;    x4=1'b0;
   #10   x1=1'b0;    x2=1'b1;    x3=1'b0;    x4=1'b1;
   #10   x1=1'b0;    x2=1'b1;    x3=1'b1;    x4=1'b0;
   #10   x1=1'b0;    x2=1'b1;    x3=1'b1;    x4=1'b1;

   #10   x1=1'b1;    x2=1'b0;    x3=1'b0;    x4=1'b0;
   #10   x1=1'b1;    x2=1'b0;    x3=1'b0;    x4=1'b1;
   #10   x1=1'b1;    x2=1'b0;    x3=1'b1;    x4=1'b0;
   #10   x1=1'b1;    x2=1'b0;    x3=1'b1;    x4=1'b1;
   #10   x1=1'b1;    x2=1'b1;    x3=1'b0;    x4=1'b0;
   #10   x1=1'b1;    x2=1'b1;    x3=1'b0;    x4=1'b1;
   #10   x1=1'b1;    x2=1'b1;    x3=1'b1;    x4=1'b0;
   #10   x1=1'b1;    x2=1'b1;    x3=1'b1;    x4=1'b1;

   #10   $stop;
end

//instantiate the module into the test bench
conditional_op inst1 (x1, x2, x3, x4, z1);

endmodule
```

Figure 1.82 (Continued)

```
x1 = 0, x2 = 0, x3 = 0, x4 = 0, z1 = 0
x1 = 0, x2 = 0, x3 = 0, x4 = 1, z1 = 0
x1 = 0, x2 = 0, x3 = 1, x4 = 0, z1 = 1
x1 = 0, x2 = 0, x3 = 1, x4 = 1, z1 = 1
x1 = 0, x2 = 1, x3 = 0, x4 = 0, z1 = 0
x1 = 0, x2 = 1, x3 = 0, x4 = 1, z1 = 1
x1 = 0, x2 = 1, x3 = 1, x4 = 0, z1 = 0
x1 = 0, x2 = 1, x3 = 1, x4 = 1, z1 = 1     //continued next pg
```

Figure 1.83 Outputs for the conditional operator of Figure 1.81.

```
x1 = 1, x2 = 0, x3 = 0, x4 = 0,      z1 = 0
x1 = 1, x2 = 0, x3 = 0, x4 = 1,      z1 = 0
x1 = 1, x2 = 0, x3 = 1, x4 = 0,      z1 = 1
x1 = 1, x2 = 0, x3 = 1, x4 = 1,      z1 = 1
x1 = 1, x2 = 1, x3 = 0, x4 = 0,      z1 = 0
x1 = 1, x2 = 1, x3 = 0, x4 = 1,      z1 = 0
x1 = 1, x2 = 1, x3 = 1, x4 = 0,      z1 = 1
x1 = 1, x2 = 1, x3 = 1, x4 = 1,      z1 = 1
```

Figure 1.83 (Continued)

1.6.4 Relational Operators

Relational operators compare operands and return a Boolean result, either 1 (true) or 0 (false) indicating the relationship between the two operands. There are four relational operators: greater than (>), less than (<), greater than or equal (>=), and less than or equal (<=). These operators function the same as identical operators in the C programming language.

If the relationship is true, then the result is 1; if the relationship is false, then the result is 0. Net or register operands are treated as unsigned values; real or integer operands are treated as signed values. An **x** or **z** in any operand returns a result of **x**. When the operands are of unequal size, the smaller operand is zero-extended to the left.

Example 1.15 Figure 1.84 shows examples of relational operators using dataflow modeling, where the identifier *gt* means greater than, *lt* means less than, *gte* means greater than or equal, and *lte* means less than or equal. The test bench, which applies several different values to the two operands, is shown in Figure 1.85. The outputs are shown in Figure 1.86.

```
//example of relational operands
module relational_opnds (x1, x2, gt, lt, gte, lte);

//define inputs and outputs
input [1:4] x1, x2;
output gt, lt, gte, lte;

//define outputs
assign   gt = x1 > x2,
         lt = x1 < x2,
         gte = x1 >= x2,
         lte = x1 <= x2;

endmodule
```

Figure 1.84 Design module for relational operators.

```
//test bench for relational_opnds
module relational_opnds_tb;

reg [1:4] x1, x2;           //inputs are reg for test bench
wire gt, lt, gte, lte;      //outputs are wire for test bench

initial         //display variables
$monitor ("x1 = %b, x2 = %b, gt = %b, lt = %b,
            gte = %b, lte = %b",
               x1, x2, gt, lt, gte, lte);

//apply input vectors
initial
begin
   #0     x1 = 4'b0000;
          x2 = 4'b0000;

   #10    x1 = 4'b0001;
          x2 = 4'b0010;

   #10    x1 = 4'b0011;
          x2 = 4'b0010;

   #10    x1 = 4'b0101;
          x2 = 4'b0101;

   #10    x1 = 4'b1000;
          x2 = 4'b0110;

   #10    x1 = 4'b1100;
          x2 = 4'b1110;

   #10    x1 = 4'b0111;
          x2 = 4'b0111;

   #10    x1 = 4'b0100;
          x2 = 4'b0010;

   #10    x1 = 4'b0010;
          x2 = 4'b0011;
   #10    $stop;
end

//instantiate the module into the test bench as a single line
relational_opnds inst1 (x1, x2, gt, lt, gte, lte);

endmodule
```

Figure 1.85 Test bench module for relational operators.

```
x1 = 0000, x2 = 0000, gt = 0, lt = 0, gte = 1, lte = 1
x1 = 0001, x2 = 0010, gt = 0, lt = 1, gte = 0, lte = 1
x1 = 0011, x2 = 0010, gt = 1, lt = 0, gte = 1, lte = 0
x1 = 0101, x2 = 0101, gt = 0, lt = 0, gte = 1, lte = 1
x1 = 1000, x2 = 0110, gt = 1, lt = 0, gte = 1, lte = 0
x1 = 1100, x2 = 1110, gt = 0, lt = 1, gte = 0, lte = 1
x1 = 0111, x2 = 0111, gt = 0, lt = 0, gte = 1, lte = 1
x1 = 0100, x2 = 0010, gt = 1, lt = 0, gte = 1, lte = 0
x1 = 0010, x2 = 0011, gt = 0, lt = 1, gte = 0, lte = 1
```

Figure 1.86 Outputs for relational operators.

1.6.5 Logical Operators

There are three logical operators: the binary logical AND operator (&&), the binary logical OR operator (||), and the unary logical negation operator (!). Logical operators evaluate to a logical 1 (true), a logical 0 (false), or an **x** (ambiguous). If a logical operation returns a nonzero value, then it is treated as a logical 1 (true); if a bit in an operand is **x** or **z**, then it is ambiguous and is normally treated as a false condition. For vector operands, a nonzero vector is treated as a 1.

Example 1.16 Figure 1.87 shows examples of the logical operators using dataflow modeling. Figure 1.88 and Figure 1.89 show the test bench and outputs, respectively.

```
//dataflow for logical operators
module logical_operators (x1, x2, x3, z1, z2, z3, z4);

//define inputs and outputs
input [1:4] x1, x2, x3;
output z1, z2, z3, z4;

//perform the logical operations
assign    z1 = (x1 && x2) && x3,
          z2 = (x1 || x2) && x3,
          z3 = (x1 && x3) || x2,
          z4 = !(x1 || x3);

endmodule
```

Figure 1.87 Design module for examples of logical operators.

```
//test bench for logical operators
module logical_operators_tb;

reg [1:4] x1, x2, x3;    //inputs are reg for test bench
wire z1, z2, z3, z4;     //outputs are wire for test bench

initial      //display variables
$monitor ("x1=%b, x2=%b, x3=%b, z1=%b, z2=%b, z3=%b, z4=%b",
            x1, x2, x3, z1, z2, z3, z4);

initial      //apply input vectors
begin
   #0     x1 = 4'b0001;  x2 = 4'b0001;  x3 = 4'b0001;
   #10    x1 = 4'b0011;  x2 = 4'b0011;  x3 = 4'b0011;
   #10    x1 = 4'b1111;  x2 = 4'b0000;  x3 = 4'b1000;
   #10    x1 = 4'b0000;  x2 = 4'b1000;  x3 = 4'b0000;

   #10    x1 = 4'b0100;  x2 = 4'b0110;  x3 = 4'b0111;
   #10    x1 = 4'b0111;  x2 = 4'b0000;  x3 = 4'b1000;
   #10    x1 = 4'b0000;  x2 = 4'b0000;  x3 = 4'b0000;
   #10    x1 = 4'b1111;  x2 = 4'b1111;  x3 = 4'b1111;

   #10    x1 = 4'b0110;  x2 = 4'b0110;  x3 = 4'b0111;
   #10    x1 = 4'b1011;  x2 = 4'b0000;  x3 = 4'b1011;
   #10    x1 = 4'b0000;  x2 = 4'b0000;  x3 = 4'b0000;
   #10    x1 = 4'b1110;  x2 = 4'b1011;  x3 = 4'b1101;

   #10    $stop;
end

//instantiate the module into the test bench
logical_operators inst1 (x1, x2, x3, z1, z2, z3, z4);

endmodule
```

Figure 1.88 Test bench module for examples of logical operators.

```
z1 = (x1 && x2) && x3,        z2 = (x1 || x2) && x3,
z3 = (x1 && x3) || x2,        z4 = !(x1 || x3);

x1=0001, x2=0001, x3=0001, | z1=1, z2=1, z3=1, z4=0
x1=0011, x2=0011, x3=0011, | z1=1, z2=1, z3=1, z4=0
x1=1111, x2=0000, x3=1000, | z1=0, z2=1, z3=1, z4=0
x1=0000, x2=1000, x3=0000, | z1=0, z2=0, z3=1, z4=1
                                 //continued on next page
```

Figure 1.89 Outputs for examples of logical operators.

```
z1 = (x1 && x2) && x3,          z2 = (x1 || x2) && x3,
z3 = (x1 && x3) || x2,          z4 = !(x1 || x3);

x1=0100, x2=0110, x3=0111,      z1=1, z2=1, z3=1, z4=0
x1=0111, x2=0000, x3=1000,      z1=0, z2=1, z3=1, z4=0
x1=0000, x2=0000, x3=0000,      z1=0, z2=0, z3=0, z4=1
x1=1111, x2=1111, x3=1111,      z1=1, z2=1, z3=1, z4=0

x1=0110, x2=0110, x3=0111,      z1=1, z2=1, z3=1, z4=0
x1=1011, x2=0000, x3=1011,      z1=0, z2=1, z3=1, z4=0
x1=0000, x2=0000, x3=0000,      z1=0, z2=0, z3=0, z4=1
x1=1110, x2=1011, x3=1101,      z1=1, z2=1, z3=1, z4=0
```

Figure 1.89 (Continued)

1.6.6 Bitwise Operators

The bitwise operators are: AND (&), OR (|), negation (~), exclusive-OR (^), and exclusive-NOR (^~ or ~^). The bitwise operators perform logical operations on the operands on a bit-by-bit basis and produce a vector result. Except for negation, each bit in one operand is associated with the corresponding bit in the other operand. If one operand is shorter, then it is zero-extended to the left to match the length of the longer operand.

The *bitwise AND* operator performs the AND function on two operands on a bit-by-bit basis. An example of the bitwise AND operator is shown below.

```
     1 1 0 0 0 1 0 1
&)   1 1 0 1 1 1 0 0
     ---------------
     1 1 0 0 0 1 0 0
```

The *bitwise OR* operator performs the OR function on the two operands on a bit-by-bit basis. An example of the bitwise OR operator is shown below.

```
     0 1 0 1 0 0 0 1
|)   0 1 0 0 0 1 0 1
     ---------------
     0 1 0 1 0 1 0 1
```

The *bitwise negation* operator performs the negation function on one operand on a bit-by-bit basis. Each bit in the operand is inverted. An example of the bitwise negation operator is shown below.

```
~) 1  1  1  0  0  0  1  0
   ----------------------
   0  0  0  1  1  1  0  1
```

The *bitwise exclusive-OR* operator performs the exclusive-OR function on two operands on a bit-by-bit basis. An example of the bitwise exclusive-OR operator is shown below.

```
   1  0  0  1  1  0  1  0
^) 1  1  0  1  0  1  0  0
   ----------------------
   0  1  0  0  1  1  1  0
```

The *bitwise exclusive-NOR* operator performs the exclusive-NOR function on two operands on a bit-by-bit basis. An example of the bitwise exclusive-NOR operator is shown below.

```
    0  1  0  1  0  1  0  0
^~) 0  1  1  0  0  1  0  1
    ----------------------
    1  1  0  0  1  1  1  0
```

Bitwise operators perform operations on operands on a bit-by-bit basis and produce a vector result. This is in contrast to logical operators, which perform operations on operands in such a way that the truth or falsity of the result is determined by the truth or falsity of the operands. That is, the logical AND operator returns a value of 1 (true) only if both operands are nonzero (true); otherwise, it returns a value of 0 (false). If the result is ambiguous, it returns a value of **x**.

Example 1.17 Figure 1.90 shows a coding example to illustrate the use of the five bitwise operators. The test bench and outputs are in Figure 1.91 and Figure 1.92, respectively.

```
//dataflow bitwise operators
module bitwise4 (a, b, c, z1, z2, z3, z4);

input [3:0] a, b, c;      //define inputs and outputs
output [3:0] z1, z2, z3, z4;

assign   z1 = (a & b) | c,
         z2 = (a ^ b) & c,
         z3 = (a | c) ^ b,
         z4 = (a ^~ c);
endmodule
```

Figure 1.90 Design module for the bitwise operators.

```
//test bench for bitwise operators
module bitwise4_tb;

reg [3:0] a, b, c;          //inputs are reg for test bench
wire [3:0] z1, z2, z3, z4; //outputs are wire for test bench

initial     //display variables
$monitor ("a=%b, b=%b, c=%b, z1=%b, z2=%b, z3=%b, z4=%b",
            a, b, c, z1, z2, z3, z4);

initial     //apply input vectors
begin
   #0    a = 4'b0001;   b = 4'b0001;   c = 4'b0001;
   #10   a = 4'b0011;   b = 4'b0011;   c = 4'b0011;
   #10   a = 4'b1111;   b = 4'b0000;   c = 4'b1000;
   #10   a = 4'b0000;   b = 4'b1000;   c = 4'b0000;

   #10   a = 4'b0100;   b = 4'b0110;   c = 4'b0111;
   #10   a = 4'b0111;   b = 4'b0000;   c = 4'b1000;
   #10   a = 4'b0000;   b = 4'b0000;   c = 4'b0000;
   #10   a = 4'b1111;   b = 4'b1111;   c = 4'b1111;

   #10   a = 4'b0000;   b = 4'b0001;   c = 4'b0010;
   #10   a = 4'b0011;   b = 4'b0100;   c = 4'b0101;
   #10   a = 4'b0110;   b = 4'b0111;   c = 4'b1000;
   #10   a = 4'b1001;   b = 4'b1010;   c = 4'b1011;

   #10   a = 4'b1100;   b = 4'b1101;   c = 4'b1110;
   #10   $stop;
end

//instantiate the module into the test bench
bitwise4 inst1 (a, b, c, z1, z2, z3, z4);
endmodule
```

Figure 1.91 Test bench module for bitwise operators.

```
z1 = (a & b) | c,      z2 = (a ^ b) & c,
z3 = (a | c) ^ b,      z4 = (a ^~ c);

a=0001, b=0001, c=0001, z1=0001, z2=0000, z3=0000, z4=1111
a=0011, b=0011, c=0011, z1=0011, z2=0000, z3=0000, z4=1111
a=1111, b=0000, c=1000, z1=1000, z2=1000, z3=1111, z4=1000
a=0000, b=1000, c=0000, z1=0000, z2=0000, z3=1000, z4=1111
                              //continued on next page
```

Figure 1.92 Outputs for bitwise operators.

```
z1 = (a & b) | c,        z2 = (a ^ b) & c,
z3 = (a | c) ^ b,        z4 = (a ^~ c);

a=0100, b=0110, c=0111, | z1=0111, z2=0010, z3=0001, z4=1100
a=0111, b=0000, c=1000, | z1=1000, z2=0000, z3=1111, z4=0000
a=0000, b=0000, c=0000, | z1=0000, z2=0000, z3=0000, z4=1111
a=1111, b=1111, c=1111, | z1=1111, z2=0000, z3=0000, z4=1111

a=0000, b=0001, c=0010, | z1=0010, z2=0000, z3=0011, z4=1101
a=0011, b=0100, c=0101, | z1=0101, z2=0101, z3=0011, z4=1001
a=0110, b=0111, c=1000, | z1=1110, z2=0000, z3=1001, z4=0001
a=1001, b=1010, c=1011, | z1=1011, z2=0011, z3=0001, z4=1101

a=1100, b=1101, c=1110, | z1=1110, z2=0000, z3=0011, z4=1101
```

Figure 1.92 (Continued)

1.6.7 Shift Operators

The shift operators shift a single vector operand left or right a specified number of bit positions. These are logical shift operations, not algebraic; that is, as bits are shifted left or right, zeroes fill in the vacated bit positions. The bits shifted out of the operand are lost; they do not rotate to the high-order or low-order bit positions of the shifted operand. If the shift amount evaluates to **x** or **z**, then the result of the operation is **x**. Algebraic shifters are presented in behavioral modeling, which is described in Section 1.7. There are two logical shift operators, as shown below. The value in parentheses is the number of bits that the operand is shifted.

<< (Left-shift amount)
>> (Right-shift amount)

When an operand is shifted left, this is equivalent to a multiply-by-two operation for each bit position shifted. When an operand is shifted right, this is equivalent to a divide-by-two operation for each bit position shifted. The shift operators are useful to model the sequential add-shift multiplication algorithm and the sequential shift-subtract division algorithm.

Example 1.18 Figure 1.93 shows examples of the shift-left and shift-right operators using dataflow modeling. The test bench is shown in Figure 1.94 and the outputs are shown in Figure 1.95.

```
//dataflow for shift left and shift right
module shift3 (a1, a2, a3, b1, b2, b3, a1_rslt, a2_rslt,
         a3_rslt, b1_rslt, b2_rslt, b3_rslt);

//define inputs and outputs
input [7:0] a1, a2, a3, b1, b2, b3;
output [7:0] a1_rslt, a2_rslt, a3_rslt,
            b1_rslt, b2_rslt, b3_rslt;

//define outputs
assign   a1_rslt = a1 << 2,    //multiply by 4
         a2_rslt = a2 << 3,    //multiply by 8
         a3_rslt = a3 << 4,    //multiply by 16

         b1_rslt = b1 >> 1,    //divide by 2
         b2_rslt = b2 >> 2,    //divide by 4
         b3_rslt = b3 >> 3;    //divide by 8
endmodule
```

Figure 1.93 Design module for examples of the shift-left and shift-right operators

```
//test bench for shift operators
module shift3_tb;

//inputs are reg for test bench
reg [7:0] a1, a2, a3, b1, b2, b3;

//outputs are wire for test bench
wire [7:0] a1_rslt, a2_rslt, a3_rslt,
            b1_rslt, b2_rslt, b3_rslt;

initial     //display variables
$monitor ("a1=%b, a2=%b, a3=%b, b1=%b, b2=%b, b3=%b,
            a1_rslt=%b, a2_rslt=%b, a3_rslt=%b, b1_rslt=%b,
            b2_rslt=%b, b3_rslt=%b",
               a1, a2, a3, b1, b2, b3, a1_rslt, a2_rslt,
               a3_rslt, b1_rslt, b2_rslt, b3_rslt);

//apply input vectors
initial
begin
   #0    a1 = 8'b0000_0011;    //multiply by 4
         a2 = 8'b0000_1000;    //multiply by 8
         a3 = 8'b0000_0011;    //multiply by 16
                                  //continued on next page
```

Figure 1.94 Test bench module for the logical shift operators.

```
          b1 = 8'b0011_0000;    //divide by 2
          b2 = 8'b0001_0000;    //divide by 4
          b3 = 8'b0011_0000;    //divide by 8

   #10    a1 = 8'b0000_0010;    //multiply by 4
          a2 = 8'b0000_0111;    //multiply by 8
          a3 = 8'b0000_0010;    //multiply by 16

          b1 = 8'b0100_0000;    //divide by 2
          b2 = 8'b0010_0000;    //divide by 4
          b3 = 8'b0000_1000;    //divide by 8

   #10    $stop;
end

//instantiate the module into the test bench
shift3 inst1 (a1, a2, a3, b1, b2, b3, a1_rslt, a2_rslt,
              a3_rslt, b1_rslt, b2_rslt, b3_rslt);

endmodule
```

Figure 1.94 (Continued)

```
a1 = multiply by 4; a2= multiply by 8; a3 = multiply by 16
a1=00000011, a2=00001000, a3=00000011,
a1_rslt=00001100, a2_rslt=01000000, a3_rslt=00110000,

b1 = divide by 2; b2 = divide by 4; b3 = divide by 8
b1=00110000, b2=00010000, b3=00110000,
b1_rslt=00011000, b2_rslt=00000100, b3_rslt=00000110

-----------------------------------------------------------

a1 = multiply by 4; a2= multiply by 8; a3 = multiply by 16
a1=00000010, a2=00000111, a3=00000010,
a1_rslt=00001000, a2_rslt=00111000, a3_rslt=00100000,

b1 = divide by 2; b2 = divide by 4; b3 = divide by 8
b1=01000000, b2=00100000, b3=00001000,
b1_rslt=00100000, b2_rslt=00001000, b3_rslt=00000001
```

Figure 1.95 Outputs for the logical shift operators.

1.7 Behavioral Modeling

This section describes the *behavior* of a digital system and is not concerned with the direct implementation of logic gates, but more on the architecture of the system. This is an algorithmic approach to hardware implementation and represents a higher level of abstraction than previous modeling methods. A Verilog module may contain a mixture of built-in primitives, UDPs, dataflow constructs, and behavioral constructs. The constructs in behavioral modeling closely resemble those used in the C programming language.

Describing a module in *behavioral* modeling is an abstraction of the functional operation of the design. It does not describe the implementation of the design at the gate level. The outputs of the module are characterized by their relationship to the inputs. The behavior of the design is described using procedural constructs. These constructs are the **initial** statement and the **always** statement.

A *procedure* is series of operations taken to design a module. A Verilog module that is designed using behavioral modeling contains no internal structural details, it simply defines the behavior of the hardware in an abstract, algorithmic description. Verilog contains two structured procedure statements or behaviors: **initial** and **always**. A behavior may consist of a single statement or a block of statements delimited by the keywords **begin** . . . **end**. A module may contain multiple **initial** and **always** statements. These statements are the basic statements used in behavioral modeling and execute concurrently starting at time zero in which the order of execution is not important. All other behavioral statements are contained inside these structured procedure statements.

1.7.1 Initial Statement

All statements within an **initial** statement comprise an **initial** block. An **initial** statement executes only once beginning at time zero, then suspends execution. An **initial** statement provides a method to initialize and monitor variables before the variables are used in a module; it is also used to generate waveforms. For a given time unit, all statements within the **initial** block execute sequentially.

Execution or assignment is controlled by the time symbol #. Examples of the time symbol are shown below. At time zero (#0), variable x_1 is set to a one-bit (1') binary (b) value of 0. Ten time units later x_1 and x_2 are set to a value of 1. Ten time units later x_1 is set to a value of 0 and ten time units later (at 30 time units) x_2 is set to a value of 0.

```
#0    x1 = 1'b0;
#10   x1 = 1'b1;   x2 = 1'b1;
#10   x1 = 1'b0;
#10   x2 = 1'b0;
```

The syntax for an **initial** statement is as follows:

initial [optional timing control] procedural statement or
block of procedural statements

1.7.2 Always Statement

The **always** statement executes the behavioral statements within the **always** block repeatedly in a looping manner and begins execution at time zero. Execution of the statements continues indefinitely until the simulation is terminated. The syntax for the **always** statement is shown below.

always [optional timing control] procedural statement or
block of procedural statements

An **always** statement is often used with an *event control list* — or *sensitivity list* — to execute a sequential block. When a change occurs to a variable in the sensitivity list, the statement or block of statements in the **always** block is executed. The keyword **or** is used to indicate multiple events. When one or more inputs change state, the statement in the **always** block is executed. The **begin** . . . **end** keywords are necessary only when there is more than one behavioral statement. Target variables used in an **always** statement are declared as type **reg**.

Example 1.19 Figure 1.96 shows a 3-input OR gate, which will be designed using behavioral modeling. The behavioral module is shown in Figure 1.97 using an **always** statement. The expression within the parentheses is called an *event control* or *sensitivity list*. Whenever a variable in the event control list changes value, the statements in the **begin** . . . **end** block will be executed; that is, if either x_1 or x_2 or x_3 changes value, the following statement will be executed: $z_1 = x_1 \mid x_2 \mid x_3$; where the symbol (|) signifies the logical OR operation.

If only a single statement appears after the **always** statement, then the keywords **begin** and **end** are not required. The **always** statement has a sequential block (**begin** . . . **end**) associated with an event control. The statements within a **begin** . . . **end** block execute sequentially and execution suspends when the last statement has been executed. When the sequential block completes execution, the **always** statement checks for another change of variables in the event control list.

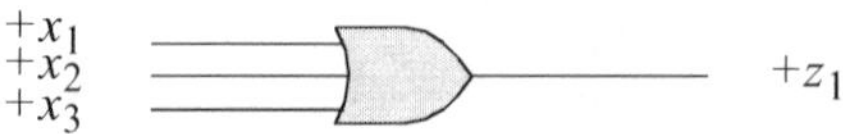

Figure 1.96 Three-input OR gate to be implemented using behavioral modeling.

```
//behavioral 3-input or gate
module or3a (x1, x2, x3, z1);

input x1, x2, x3;     //define inputs and output
output z1;                        //continued on next page
```

Figure 1.97 Design module for the three-input OR gate.

```
//define signals
wire x1, x2, x3;   //alternatively do not declare wires
                   //because inputs are wire by default

reg z1;         //outputs are reg for behavioral
                //z1 is used in the always statement
                //and must be declared as type reg

always @ (x1 or x2 or x3)  //sensitivity list is x1, x2, x3
begin
   z1 = x1 | x2 | x3;
end

endmodule
```

Figure 1.97 (Continued)

The test bench for the OR gate module is shown in Figure 1.98 using the **initial** statement. The inputs for a test bench are of type **reg** because they retain their value until changed, and the outputs are of type **wire**. All eight combinations of the inputs are tested. The inputs are applied in sequence, $x_1x_2x_3 = 000$ through 111. The binary outputs of the simulator are shown in Figure 1.99 listing the output value for z_1 for all combinations of inputs.

```
//test bench for three-input or gate
module or3a_tb;

//inputs are reg for test bench
reg x1, x2, x3;

//outputs are wire for test bench
wire z1;

//display variables
initial
$monitor ("x1 = %b, x2 = %b, x3 = %b, z1 = %b",
                x1, x2, x3, z1);

initial       //apply input vectors
begin
   #0    x1 = 1'b0;  x2 = 1'b0;  x3 = 1'b0;
   #10   x1 = 1'b0;  x2 = 1'b0;  x3 = 1'b1;
   #10   x1 = 1'b0;  x2 = 1'b1;  x3 = 1'b0;
   #10   x1 = 1'b0;  x2 = 1'b1;  x3 = 1'b1;      //next page
```

Figure 1.98 Test bench module for the three-input OR gate.

```
    #10   x1 = 1'b1;   x2 = 1'b0;   x3 = 1'b0;
    #10   x1 = 1'b1;   x2 = 1'b0;   x3 = 1'b1;
    #10   x1 = 1'b1;   x2 = 1'b1;   x3 = 1'b0;
    #10   x1 = 1'b1;   x2 = 1'b1;   x3 = 1'b1;

    #10   $stop;
end

//instantiate the module into the test bench
or3a inst1 (x1, x2, x3, z1);

endmodule
```

Figure 1.98 (Continued)

```
x1 = 0, x2 = 0, x3 = 0,      z1 = 0
x1 = 0, x2 = 0, x3 = 1,      z1 = 1
x1 = 0, x2 = 1, x3 = 0,      z1 = 1
x1 = 0, x2 = 1, x3 = 1,      z1 = 1
x1 = 1, x2 = 0, x3 = 0,      z1 = 1
x1 = 1, x2 = 0, x3 = 1,      z1 = 1
x1 = 1, x2 = 1, x3 = 0,      z1 = 1
x1 = 1, x2 = 1, x3 = 1,      z1 = 1
```

Figure 1.99 Outputs for the three-input OR gate.

1.7.3 Intrastatement Delay

An *intrastatement* delay is a delay on the right-hand side of the statement and indicates that the right-hand side is to be evaluated, wait the specified number of time units, and then assign the value to the left-hand side. This can be used to simulate logic gate delays. Equation 1.16 is an example of an intrastatement delay.

$$z_1 = \#5\ x_1\ \&\ x_2 \tag{1.16}$$

The statement evaluates the logical function x_1 AND x_2, waits five time units, then assigns the result to z_1. If no delay is specified in a procedural assignment, then zero delay is the default delay and the assignment occurs instantaneously.

1.7.4 Interstatement Delay

An *interstatement* delay is the delay by which a statement's execution is delayed; that is, it is the delay between statements. The code segment of Equation 1.17 is an example of an interstatement delay.

$$
\begin{array}{l}
z_1 = x_1 \mid x_2 \\
\#5\ z_2 = x_1\ \&\ x_2
\end{array}
\tag{1.17}
$$

When the first statement has completed execution, a delay of five time units will be taken before the second statement is executed. If no delays are specified in a procedural assignment, then there is zero delay in the assignment.

1.7.5 Blocking Assignments

A blocking procedural assignment completes execution before the next statement executes. The assignment operator (=) is used for blocking assignments. The right-hand expression is evaluated, then the assignment is placed in an internal temporary register called the *event queue* and scheduled for assignment. If no time units are specified, then the scheduling takes place immediately. The event queue is covered in Appendix A.

In the code segment below, an interstatement delay of two time units is specified for the assignment to z_2. The evaluation of z_2 is delayed by the timing control; that is, the expression for z_2 will not be evaluated until the expression for z_1 has been executed, plus two time units. The execution of any following statements is blocked until the assignment occurs.

```
initial
   begin
         z1 = x1 & x2;
      #2 z2 = x2 | x3;
   end
```

1.7.6 Nonblocking Assignments

The assignment symbol (<=) is used to represent a nonblocking procedural assignment. Nonblocking assignments allow the scheduling of assignments without blocking execution of the following statements in a sequential procedural block. A nonblocking assignment is used to synchronize assignment statements so that they appear to execute at the same time. In the code segment shown below using blocking assignments, the result is indeterminate because both **always** blocks execute concurrently resulting in a race condition. Depending on the simulator implementation, either $x_1 = x_2$ would be executed before $x_2 = x_3$ or vice versa.

```
always @ (posedge clk)
    x1 = x2;

always @ (posedge clk)
    x2 = x3;
```

The race condition is solved by using nonblocking assignments as shown below.

```
always @ (posedge clk)
    x1 <= x2;

always @ (posedge clk)
    x2 <= x3;
```

The Verilog simulator schedules a nonblocking assignment statement to execute, then proceeds to the next statement in the block without waiting for the previous nonblocking statement to complete execution. That is, the right-hand expression is evaluated and the value is stored in the event queue and is *scheduled* to be assigned to the left-hand target. The assignment is made at the end of the current time step if there are no intrastatement delays specified.

Nonblocking assignments are typically used to model several concurrent assignments that are caused by a common event such as @ **posedge** clk. The order of the assignments is irrelevant because the right-hand side evaluations are stored in the event queue before any assignments are made.

1.7.7 Conditional Statements

Conditional statements alter the flow within a behavior based upon certain conditions. The choice among alternative statements depends on the Boolean value of an expression. The alternative statements can be a single statement or a block of statements delimited by the keywords **begin** . . . **end**. The keywords **if** and **else** are used in conditional statements. There are three categories of the conditional statement as shown below. A true value is 1 or any nonzero value; a false value is 0, **x**, or **z**. If the evaluation is false, then the next expression in the activity flow is evaluated.

```
//no else statement
if (expression) statement1;      //if expression is true, then statement1 is executed.

//one else statement             //choice of two statements. Only one is executed.
if (expression) statement1;      //if expression is true, then statement1 is executed.
else statement2;                 //if expression is false, then statement2 is executed.
```

//nested **if-else if** //choice of multiple statements. One is executed.
if (expression1) statement1; //if expression1 is true, then statement1 is executed.
else if (expression2) statement2; //if expression2 is true, then statement2 is executed.
else if (expression3) statement3; //if expression3 is true, then statement3 is executed.
else default statement;

Examples of the three categories are shown below.

//no **else** statement
if (x_1 & x_2) z_1 = 1;

//one **else** statement
if (rst_n = = 0)
ctr = 3'b000;
else ctr = next_count;

//nested **if-else if**
if (opcode = = 00)
$z_1 = x_1 + x_2$;
else if (opcode = = 01)
$z_1 = x_1 - x_2$;
else if (opcode = = 10)
$z_1 = x_1 * x_2$;
else
$z_1 = x_1 / x_2$;

Example 1.20 This example uses scalar variables $x_1x_2x_3$ to illustrates the use of conditional statements to implement the expression: $z_1 = (x_1 \ \&\& \ x_2) \,||\, (x_1 \ \&\& \ x_3) \,||\, (x_2 \ \&\& \ x_3))$, where the symbol | | represents the logical OR operation and the symbol && represents the logical AND operation. Recall that the logical OR and logical AND operators are binary operations that evaluate to a logical 1 (true), a logical 0 (false), or an **x** (ambiguous). If a logical operation returns a nonzero value, then it is treated as a logical 1 (true); if a bit in an operand is **x** or **z**, then it is ambiguous and is normally treated as a false condition.

The design module and test bench module are shown in Figure 1.100 and Figure 1.101, respectively. The outputs are illustrated in Figure 1.102 which display the correct value for output z_1 for all combinations of the input values.

```
//behavioral conditional if ... else if
module cond_if_else (x1, x2, x3, z1);

input x1, x2, x3;      //define inputs and output
output z1;                         //continued on next page
```

Figure 1.100 Design module to illustrate the conditional statement **if . . . else**.

```
//define signals
reg z1;      //outputs are declared as reg for behavioral
             //z1 is used as target in always statement

always @ (x1 or x2 or x3)  //sensitivity list
begin
   if ((x1 && x2) || (x1 && x3) || (x2 && x3))
      z1= 1;
   else
      z1 = 0;
end
endmodule
```

Figure 1.100 (Continued)

```
//test bench for cond_if_else module

module cond_if_else_tb;

reg x1, x2, x3;    //inputs are reg for test bench
wire z1;           //outputs are wire for test bench

initial            //display variables
$monitor ("x1 = %b, x2 = %b, x3 = %b, z1 = %b",
            x1, x2, x3, z1);

initial            //apply input vectors
begin
   #0    x1 = 1'b0;  x2 = 1'b0;x3 = 1'b0;
   #10   x1 = 1'b0;  x2 = 1'b0;x3 = 1'b1;
   #10   x1 = 1'b0;  x2 = 1'b1;x3 = 1'b0;
   #10   x1 = 1'b0;  x2 = 1'b1;x3 = 1'b1;

   #10   x1 = 1'b1;  x2 = 1'b0;x3 = 1'b0;
   #10   x1 = 1'b1;  x2 = 1'b0;x3 = 1'b1;
   #10   x1 = 1'b1;  x2 = 1'b1;x3 = 1'b0;
   #10   x1 = 1'b1;  x2 = 1'b1;x3 = 1'b1;

   #10   $stop;
end

//instantiate the module into the test bench
cond_if_else inst1 (x1, x2, x3, z1);

endmodule
```

Figure 1.101 Test bench module for the conditional statement **if . . . else**.

```
x1 = 0, x2 = 0, x3 = 0,      z1 = 0
x1 = 0, x2 = 0, x3 = 1,      z1 = 0
x1 = 0, x2 = 1, x3 = 0,      z1 = 0
x1 = 0, x2 = 1, x3 = 1,      z1 = 1
x1 = 1, x2 = 0, x3 = 0,      z1 = 0
x1 = 1, x2 = 0, x3 = 1,      z1 = 1
x1 = 1, x2 = 1, x3 = 0,      z1 = 1
x1 = 1, x2 = 1, x3 = 1,      z1 = 1
```

Figure 1.102 Outputs for the conditional statement **if . . . else**.

1.7.8 Case Statement

The **case** statement is an alternative to the **if . . . else if** construct and may simplify the readability of the Verilog code. The **case** statement is a multiple-way conditional branch. It executes one of several different procedural statements depending on the comparison of an expression with a case item. The expression and the case item are compared bit-by-bit and must match exactly. The statement that is associated with a case item may be a single procedural statement or a block of statements delimited by the keywords **begin** . . . **end.** The **case** statement has the following syntax:

```
case (expression)
    case_item1 : procedural_statement1;
    case_item2 : procedural_statement2;
    case_item3 : procedural_statement3;
                 .
                 .
                 .
    case_itemn : procedural_statementn;
    default : default_statement;
endcase
```

The case expression may be an expression or a constant. The case items are evaluated in the order in which they are listed. If a match occurs between the case expression and the case item, then the corresponding procedural statement, or block of statements, is executed. If no match occurs, then the optional default statement is executed.

Example 1.21 An 8-bit Johnson counter will be designed that counts in the following sequence: 00000000, 10000000, 11000000, 11100000, 11110000, 11111000, 11111100, 11111110, 11111111, 01111111, 00111111, 00011111, 00001111, 00000111, 00000011, 00000001, 00000000, . . . The **case** statement will be used to determine the next count from any current count. For example, if the current count is 00000000, then the expression *count* is compared with the *case item* 00000000 yielding a next count of 10000000. The flow then exits the **case** statement and continues with the next statement in the module.

The design module is shown in Figure 1.103 using behavioral modeling. The expression *count* in the **always** statement for the **case** statement represents the event control or sensitivity list. Whenever a change occurs to *count*, the code in the **begin** . . . **end** block executes. Each count is then compared to the value of the expression *count*. The test bench is shown in Figure 1.104. The outputs are shown in Figure 1.105.

```
//8-bit johnson counter
module johnson_ctr (clk, rst_n, count);

//define inputs and outputs
input clk, rst_n;
output [7:0] count;

//define signals
wire clk, rst_n;          //inputs are wire
reg [7:0] count;          //outputs are reg used in always
reg [7:0] next_count;     //define internal reg used in always

//set next count -----------------------------
always @ (posedge clk or negedge rst_n)
begin
   if (rst_n == 1'b0)
      count <= 8'b0000_0000;
   else
      count <= next_count;
end

//determine next count -----------------------
always @ (count)
begin
   case (count)    //case item is 8'b00000000
      8'b00000000 : next_count = 8'b10000000;
      8'b10000000 : next_count = 8'b11000000;
      8'b11000000 : next_count = 8'b11100000;
      8'b11100000 : next_count = 8'b11110000;
      8'b11110000 : next_count = 8'b11111000;
      8'b11111000 : next_count = 8'b11111100;
      8'b11111100 : next_count = 8'b11111110;
      8'b11111110 : next_count = 8'b11111111;
      8'b11111111 : next_count = 8'b01111111;
      8'b01111111 : next_count = 8'b00111111;
      8'b00111111 : next_count = 8'b00011111;
      8'b00011111 : next_count = 8'b00001111;
                                 //continued on next page
```

Figure 1.103 Design module for the 8-bit Johnson counter.

```
        8'b00001111 : next_count = 8'b00000111;
        8'b00000111 : next_count = 8'b00000011;
        8'b00000011 : next_count = 8'b00000001;
        8'b00000001 : next_count = 8'b00000000;
        default     : next_count = 8'b00000000;
    endcase
end

endmodule
```

Figure 1.103 (Continued)

```
//test bench for Johnson counter

module johnson_ctr_tb;

reg clk, rst_n;         //inputs are reg for test bench
wire [7:0] count;       //outputs are wire for test bench

//display variables
initial
$monitor ("count = %b", count);

//define reset
initial
begin
   #0 rst_n = 1'b0;
   #5 rst_n = 1'b1;
   #320  $stop;         //establish length of simulation
end

//define clk
initial
begin
   clk = 1'b0;
   forever
      #10clk = ~clk;
end

//instantiate the module into the test bench
johnson_ctr inst1 (clk, rst_n, count);

endmodule
```

Figure 1.104 Test bench module for the 8-bit Johnson counter.

```
count = 00000000
count = 10000000
count = 11000000
count = 11100000
count = 11110000
count = 11111000
count = 11111100
count = 11111110
count = 11111111
count = 01111111
count = 00111111
count = 00011111
count = 00001111
count = 00000111
count = 00000011
count = 00000001
count = 00000000
```

Figure 1.105 Outputs for the 8-bit Johnson counter.

1.7.9 Loop Statements

There are four types of loop statements in Verilog: **for**, **while**, **repeat**, and **forever**. Loop statements must be placed within an **initial** or an **always** block and may contain delay controls. The loop constructs allow for repeated execution of procedural statements within an **initial** or an **always** block.

For loop The **for** loop contains three parts:

1. An *initial* condition to assign a value to a register control variable. This is executed once at the beginning of the loop to initialize a register variable that controls the loop.

2. A *test* condition to determine when the loop terminates. This is an expression that is executed before the procedural statements of the loop to determine if the loop should execute. The loop is repeated as long as the expression is true. If the expression is false, the loop terminates and the activity flow proceeds to the next statement in the module.

3. An *assignment* to modify the control variable, usually an increment or a decrement. This assignment is executed after each execution of the loop and before the next test to terminate the loop.

The syntax of a **for** loop is shown below. The body of the loop can be a single procedural statement or a block of procedural statements.

for (initial control variable assignment; test expression; control variable assignment)
procedural statement or block of procedural statements

The **for** loop is generally used when there is a known beginning and an end to a loop. The **for** loop is similar in function to the **for** loop in the C programming language and is used in a test bench. Example 1.20 will be used in the following example to illustrate the use of the **for** loop.

Example 1.22 This example uses scalar variables $x_1x_2x_3$ to illustrates the use of conditional statements to implement the expression: $z_1 = (x_1 \ \&\& \ x_2) \,||\, (x_1 \ \&\& \ x_3) \,||\, (x_2 \ \&\& \ x_3))$, where the symbol || represents the logical OR operation and the symbol && represents the logical AND operation. The design module is shown in Figure 1.106. The test bench module is shown in Figure 1.107 and the outputs are shown in Figure 1.108.

Since there are three inputs to the sum-of-products expression, all eight combinations of three variables must be applied to the circuit. This is accomplished by a **for** loop statement.

Following the keyword **begin** is the name of the block: *apply_stimulus*. In this block, a 4-bit **reg** variable is declared called *invect*. This guarantees that all eight combinations of the four inputs will be tested by the **for** loop, which applies input vectors of $x_1x_2x_3 = 000$ through 111 to the circuit. The **for** loop stops when the pattern 1000 is detected by the test segment (*invect* < 8). If only a 3-bit vector were applied, then the expression (*invect* < 8) would always be true and the loop would never terminate. The increment segment of the **for** loop does not support an increment designated as *invect++*; therefore, the long notation must be used: *invect = invect + 1*.

```
//behavioral conditional if ... else if
module cond_if_else2 (x1, x2, x3, z1);

input x1, x2, x3;       //define inputs and output
output z1;

//define signals
reg z1;        //outputs are declared as reg for behavioral
               //z1 is used as target in always statement

always @ (x1 or x2 or x3)       //sensitivity list
begin
   if ((x1 && x2) || (x1 && x3) || (x2 && x3))
      z1= 1;
   else
      z1 = 0;
end

endmodule
```

Figure 1.106 Design module for the **for** loop statement.

```
//test bench for cond_if_else2

module cond_if_else2_tb;

reg x1, x2, x3;      //inputs are reg for test bench
wire z1;             //outputs are wire for test bench

//apply input vectors and display variables
initial
begin: apply_stimulus
   reg [3:0] invect;
   for (invect = 0; invect < 8; invect = invect + 1)
      begin
         {x1, x2, x3} = invect [3:0];
         #10 $display ("x1 x2 x3 = %b, z1 = %b",
                           {x1, x2, x3}, z1);
      end
end

//instantiate the module into the test bench
cond_if_else2 inst1 (x1, x2, x3, z1);

endmodule
```

Figure 1.107 Test bench module for the **for** loop statement.

```
x1 x2 x3 = 000, z1 = 0
x1 x2 x3 = 001, z1 = 0
x1 x2 x3 = 010, z1 = 0
x1 x2 x3 = 011, z1 = 1
x1 x2 x3 = 100, z1 = 0
x1 x2 x3 = 101, z1 = 1
x1 x2 x3 = 110, z1 = 1
x1 x2 x3 = 111, z1 = 1
```

Figure 1.108 Outputs for the **for** loop statement.

While loop The **while** loop executes a procedural statement or a block of procedural statements as long as a Boolean expression returns a value of true. When the procedural statements are executed, the Boolean expression is reevaluated. The loop is executed until the expression returns a value of false. If the evaluation of the expression is false, then the **while** loop is terminated and control is passed to the next statement in the module. If the expression is false before the loop is initially entered, then the **while** loop is never executed.

The Boolean expression may contain any of the following types: arithmetic, logical, relational, equality, bitwise, reduction, shift, concatenation, replication, or conditional. If the **while** loop contains multiple procedural statements, then they are contained within the **begin** . . . **end** keywords. The syntax for a **while** statement is as follows:

> **while** (expression)
> procedural statement or block of procedural statements

Example 1.23 This example demonstrates the use of the **while** construct to count the number of 1s in a 16-bit register *reg_a*. The design module is shown in Figure 1.109. The variable *count* is declared as type **integer** and is used to obtain the cumulative count of the number of 1s. The first **begin** keyword must have a name associated with the keyword because this declaration is allowed only with named blocks.

The register is initialized to contain twelve 1s (*16'h75fd*). Alternatively, the register can be loaded from any other register. If *reg_a* contains a 1 bit in any bit position, then the **while** loop is executed. If *reg_a* contains all zeroes, then the **while** loop is terminated.

The low-order bit position (*reg_a[0]*) is tested for a 1 bit. If a value of 1 (true) is returned, *count* is incremented by one and the register is shifted right one bit position. There is only one procedural statement following the **if** statement; therefore, if a value of 0 (false) is returned, then *count* is not incremented and the register is shifted right one bit position.

The **$display** system task then displays the number of 1s that were contained in the register *reg_a* as shown in Figure 1.109. Notice that the count changes value only when there is a 1 bit in the low-order bit position of *reg_a*. If *reg_a[0]* = 0, then the count is not incremented, but the total count is still displayed.

```
//example of a while loop
//count the number of 1s in a 16-bit register
module while_loop3;

integer count;

initial
begin: number_of_1s
   reg [16:0] x;

   count = 0;

   x = 16'h75fd;        //set x to a known hex value (twelve 1s)

                                      //continued on next page
```

Figure 1.109 Design module to illustrate the use of the **while** loop.

```
    while (x)               //execute while loop if x contains 1s
        begin
            if (x[0])       //check low-order bit position
                count = count + 1;     //if true, add one to count
            x = x >> 1;     //shift right x one bit position
        end

//shows final count
        $display ("final count = %d", count);
end

endmodule

-----------------------------------------------

final count = 12
```

Figure 1.109 (Continued)

Repeat loop The **repeat** loop executes a procedural statement or a block of procedural statements a specified number of times. The **repeat** construct can contain a constant, an expression, a variable, or a signed value. The syntax for the **repeat** loop is as follows:

> **repeat** (loop count expression)
> procedural statement or block of procedural statements

If the loop count is **x** (unknown value) or **z** (high impedance), then the loop count is treated as zero. The value of the loop count expression is evaluated once at the beginning of the loop.

Example 1.24 An example of the **repeat** loop is shown Figure 1.110, in which two 8-bit registers are added to yield a sum of eight bits. Register *reg_a* is initialized to a value of twenty-four; *reg_b* is initialized to a value of two. The add operation is repeated eight times and register *b* is incremented by one after each add operation. The outputs are shown in Figure 1.111.

```
//example of the repeat keyword
module add_regs_repeat;

reg [7:0] reg_a, reg_b, sum;
                                    //continued on next page
```

Figure 1.110 Design module for the **repeat** loop.

```
initial
begin
   reg_a = 8'b0001_1000;
   reg_b = 8'b0000_0010;

   repeat (8)
   begin
      sum = reg_a + reg_b;
      $display ("reg_a=%b, reg_b=%b, sum=%b",
                  reg_a, reg_b, sum);
      reg_b = reg_b + 1;
   end
end

endmodule
```

Figure 1.110 (Continued)

```
reg_a=00011000, reg_b=00000010, sum=00011010
reg_a=00011000, reg_b=00000011, sum=00011011
reg_a=00011000, reg_b=00000100, sum=00011100
reg_a=00011000, reg_b=00000101, sum=00011101
reg_a=00011000, reg_b=00000110, sum=00011110
reg_a=00011000, reg_b=00000111, sum=00011111
reg_a=00011000, reg_b=00001000, sum=00100000
reg_a=00011000, reg_b=00001001, sum=00100001
```

Figure 1.111 Outputs for the **repeat** loop.

Forever loop The **forever** loop executes the procedural statement continuously until the system tasks **$finish** or **$stop** are encountered. It can also be terminated by the **disable** statement. The **disable** statement is a procedural statement; therefore, it must be used within an **initial** or an **always** block. It is used to prematurely terminate a block of procedural statements or a system task. When a **disable** statement is executed, control is transferred to the statement immediately following the procedural block or task.

The **forever** loop is similar to a **while** loop in which the expression always evaluates to true (1). A timing control must be used with the **forever** loop; otherwise, the simulator would execute the procedural statement continuously without advancing the simulation time. The syntax of the **forever** loop is as follows:

forever
 procedural statement

The **forever** statement is typically used for clock generation as shown in Figure 1.112 together with the system task **$finish**. The variable *clk* will toggle every 10 time units for a period of 20 time units. The length of simulation is 100 time units.

```
//define clock
initial
begin
   clk = 1'b0;
   forever
      #10  clk = ~clk;
end

//define length of simulation
initial
   #100  $finish;
```

Figure 1.112 Clock generation using the **forever** statement.

1.7.10 Logical, Algebraic, and Rotate Shift Operations

Shift registers that perform the operations of shift left logical (SLL), shift left algebraic (SLA), shift right logical (SRL), shift right algebraic (SRA), rotate left (ROL), and rotate right (ROR) will be presented in this section.

Shift left logical (SLL) The logical shift operations are much simpler to implement than the arithmetic (algebraic) shift operations. For SLL, the high-order bit of the unsigned operand is shifted out of the left end of the shifter for each shift cycle. Zeroes are entered from the right and fill the vacated low-order bit positions.

Shift left algebraic (SLA) SLA operates on signed operands in 2s complement representation for radix 2. The numeric part of the operand is shifted left the number of bit positions specified in the shift count field. The sign remains unchanged and does not participate in the shift operation. All remaining bits participate in the left shift. Bits are shifted out of the high-order numeric position. Zeroes are entered from the right and fill the vacated low-order bit positions.

Shift right logical (SRL) For SRL, the low-order bit of the unsigned operand is shifted out of the right end of the shifter for each shift cycle. Zeroes are entered from the left and fill the vacated high-order bit positions.

Shift right algebraic (SRA) The numeric part of the signed operand is shifted right the number of bits specified by the shift count. The sign of the operand remains unchanged. All numeric bits participate in the right shift. The sign bit propagates right to fill in the vacated high-order numeric bit positions.

Rotate left (ROL) Rotate operations execute on unsigned operands. The ROL operation shifts the operand left one bit position and the high-order bit is then rotated into the low-order bit position.

Rotate right (ROR) The ROR operation shifts the operand right one bit position and the low-order bit is then rotated into the high-order bit position.

Example 1.25 Figure 1.113 shows the design module to illustrate utilizing the six shift and rotate operations described above. Behavioral modeling is used in conjunction with the **case** statement. Figure 1.114 shows the test bench module and Figure 1.115 shows the outputs.

```
//behavioral shift rotate
module shift_rotate (a, opcode, result);

//list inputs and outputs
input [7:0] a;
input [2:0] opcode;

output [7:0] result;

//specify wire for input and reg for output
wire [7:0] a;
wire [2:0] opcode;
reg [7:0] result;

//define the opcodes
parameter    sra_op = 3'b000,
             srl_op = 3'b001,
             sla_op = 3'b010,
             sll_op = 3'b011,
             ror_op = 3'b100,
             rol_op = 3'b101;

//execute the operations
always @ (a or opcode)
begin
   case (opcode)
      sra_op : result = {a[7], a[7], a[6], a[5],
                         a[4], a[3], a[2], a[1]};
      srl_op : result = a >> 1;
      sla_op : result = {a[7], a[5], a[4], a[3],
                         a[2], a[1], a[0], 1'b0};

                                 //continued on next page
```

Figure 1.113 Design module for SLL, SLA, SRL, SRA, ROL, and ROR.

```
      sll_op : result = a << 1;
      ror_op : result = {a[0], a[7], a[6], a[5],
                         a[4], a[3], a[2], a[1]};
      rol_op : result = {a[6], a[5], a[4], a[3],
                         a[2], a[1], a[0], a[7]};
      default : result = 0;
   endcase

end

endmodule
```

Figure 1.113 (Continued)

```
//test bench for shift rotate module
module shift_rotate_tb;

reg [7:0] a;          //inputs are reg for test bench
reg [2:0] opcode;
wire [7:0] result;    //outputs are wire for test bench

initial               //display variables
$monitor ("a=%b, opcode=%b, rslt=%b",
            a, opcode, result);

//apply input vectors
initial
begin
//sra op
   #10   a = 8'b1000_1110;    opcode = 3'b000;
                              //result = 11000111
   #10   a = 8'b0110_1111;    opcode = 3'b000;
                              //result = 00110111
   #10   a = 8'b1111_1110;    opcode = 3'b000;
                              //result = 11111111

//srl op
   #10   a = 8'b1111_0011;    opcode = 3'b001;
                              //result = 01111001
   #10   a = 8'b0111_1011;    opcode = 3'b001;
                              //result = 00111101
   #10   a = 8'b1000_1110;    opcode = 3'b001;
                              //result = 01000111
                                 //continued on next  page
```

Figure 1.114 Test bench module for SLL, SLA, SRL, SRA, ROL, and ROR.

```
//sla op
   #10   a = 8'b1000_1111;     opcode = 3'b010;
                               //result = 10011110
   #10   a = 8'b1110_1111;     opcode = 3'b010;
                               //result = 11011110
   #10   a = 8'b1111_0000;     opcode = 3'b010;
                               //result = 11100000

//sll op
   #10   a = 8'b0111_0111;     opcode = 3'b011;
                               //result = 11101110
   #10   a = 8'b0110_0011;     opcode = 3'b011;
                               //result = 11000110
   #10   a = 8'b1111_1111;     opcode = 3'b011;
                               //result = 11111110

//ror op
   #10   a = 8'b0101_0101;     opcode = 3'b100;
                               //result = 10101010
   #10   a = 8'b0101_1101;     opcode = 3'b100;
                               //result = 10101110
   #10   a = 8'b1111_1110;     opcode = 3'b100;
                               //result = 01111111

//rol op
   #10   a = 8'b0101_0101;     opcode = 3'b101;
                               //result = 10101010
   #10   a = 8'b0101_0111;     opcode = 3'b101;
                               //result = 10101110
   #10   a = 8'b0000_0001;     opcode = 3'b101;
                               //result = 00000010

   #10   $stop;

end

//instantiate the module into the test bench
shift_rotate inst1 (
   .a(a),
   .opcode(opcode),
   .result(result)
   );

endmodule
```

Figure 1.114 (Continued)

```
a=10001110, opcode=000, rslt=11000111  //sra
a=01101111, opcode=000, rslt=00110111
a=11111110, opcode=000, rslt=11111111

a=11110011, opcode=001, rslt=01111001  //srl
a=01111011, opcode=001, rslt=00111101
a=10001110, opcode=001, rslt=01000111

a=10001111, opcode=010, rslt=10011110  //sla
a=11101111, opcode=010, rslt=11011110
a=11110000, opcode=010, rslt=11100000

a=01110111, opcode=011, rslt=11101110  //sll
a=01100011, opcode=011, rslt=11000110
a=11111111, opcode=011, rslt=11111110

a=01010101, opcode=100, rslt=10101010  //ror
a=01011101, opcode=100, rslt=10101110
a=11111110, opcode=100, rslt=01111111

a=01010101, opcode=101, rslt=10101010  //rol
a=01010111, opcode=101, rslt=10101110
a=00000001, opcode=101, rslt=00000010
```

Figure 1.115 Outputs for SLL, SLA, SRL, SRA, ROL, and ROR.

1.8 Structural Modeling

Structural modeling consists of instantiating one or more of the following design objects:

- Built-in primitives
- User-defined primitives (UDPs)
- Design modules

Instantiation means to use one or more lower-level modules — including logic primitives — that are interconnected in the construction of a higher-level structural module. A module can be a logic gate, an adder, a multiplexer, a counter, or some other logical function. The objects that are instantiated are called *instances*. Structural modeling is described by the interconnection of these lower-level logic primitives or modules. The interconnections are made by wires that connect primitive terminals or module ports.

1.8.1 Module Instantiation

Design modules are instantiated into test bench modules. The ports of the design module are instantiated by name and connected to the corresponding net names of the test bench. Each named instantiation can be of the following form:

```
.design_module_port_name (test_bench_module_net_name)
```

An example is shown below from Figure 1.114 of the previous section for each individual port:

```
shift_rotate inst1 (
        .a(a),
        .opcode(opcode),
        .result(result)
        );
```

The instantiation can also be of the following form, which instantiates all the ports in a single line, as shown below.

```
shift_rotate inst1 (a, opcode, result);
```

Design module ports can be instantiated by name explicitly or by position. Modules cannot be nested, but they can be instantiated into other modules. Structural modeling is analogous to placing the instances on a logic diagram and then connecting them by wires. When instantiating built-in primitives, an instance name is optional; however, when instantiating a module, an instance name must be used. Instances that are instantiated into a structural module are connected by nets of type **wire**.

A structural module may contain behavioral statements (**always**), continuous assignment statements (**assign**), built-in primitives (**and**, **or**, **nand**, **nor**, etc.), UDPs (*mux4*, *half_adder*, *adder4*, etc.), design modules, or any combination of these objects. Design modules can be instantiated into a higher-level structural module in order to achieve a hierarchical design.

Each module in Verilog is either a top-level (higher-level) module or an instantiated module. There is only one top-level module and it is not instantiated anywhere else in the design project. Instantiated primitives or modules, however, can be instantiated many times into a top-level module and each instance of a module is unique.

1.8.2 Ports

Ports provide a means for the module to communicate with its external environment. Ports, also referred to as terminals, can be declared as **input**, **output**, or **inout**. A port is a net by default; however, it can be declared explicitly as a net. A module contains an optional list of ports, as shown below for a full adder.

```
module full_adder (a, b, cin, sum, cout);
```

Ports *a*, *b*, and *cin* are input ports; ports *sum* and *cout* are output ports. The test bench for the full adder contains no ports as shown below because it does not communicate with the external environment.

```
module full_adder_tb;
```

Input ports Input ports allow signals to enter the module from external sources. The width of the input port is declared within the module. The size of the input port can be declared as either a scalar such as *a*, *b*, *cin* or as a vector such as *[3:0] a, b*, where *a* and *b* are the augend and addend inputs, respectively, of a 4-bit adder. The format of the declarations shown below is the same for both behavioral and structural modeling. The input ports are declared as type **reg** for test benches with a specified width, either scalar or vector.

```
input [3:0] a, b;        //declared as 4-bit vectors
input cin;               //declared as a scalar

reg [3:0] a, b;          //inputs are reg for test benches
reg cin;
```

Output ports Output ports are those that allow signals to exit the module to external destinations. The width of the output port is declared within the module. The output ports are declared as type **wire** for test benches with a specified width, either scalar or vector. The format for output ports is shown below.

```
output [3:0] sum;        //declared as 4-bit vectors
output cout;             //declared as a scalar

wire [3:0] sum;          //outputs are wire for test benches
wire cout;
```

Inout ports An **inout** port is bidirectional — it transfers signals to and from the module depending on the value of a direction control signal. Ports of type **inout** are declared internally as type **wire**; externally, they connect to nets of type **wire**. Since port declarations are implicitly declared as type **wire**, it is not necessary to explicitly declare a port as **wire**. However, an output can also be redeclared as a **reg** type variable if it is used within an **always** statement or an **initial** statement.

Port connection rules A port is an entry into a module from an external source. It connects the external unit to the internal logic of the module. When a module is instantiated within another module, certain rules apply. An error message is indicated if the port connection rules are not followed.

Input ports must always be of type **wire** (net) internally except for test benches; externally, input ports can be **reg** or **wire**. Output ports can be of type **reg** or **wire** internally; externally, output ports must always be connected to a **wire**. The input port names can be different, but the net (**wire**) names connecting the input ports must be the same. When making intermodule port connections, it is permissible to connect ports of different widths. Port width matching occurs by right justification or truncation.

1.8.3 Design Examples

Examples will be presented in this section that illustrate the structural modeling technique for combinational logic. These examples include logic equations, a binary-to excess-3 code converter, an adder, and a comparator. Each example will be completely designed in detail and will include appropriate theory where applicable.

Example 1.26 A combinational logic circuit will be designed using structural modeling that will implement the following equation:

$$z_1 = x_1'x_2x_3' + x_1'x_3x_4' + x_1x_2'x_3'x_4 + x_1x_2'x_3x_4' + x_1'x_2'x_3'x_4' \quad (1.18)$$

The equation will first be minimized. This can be achieved either by utilizing Boolean algebra or by a Karnaugh map. A Karnaugh map will be used in this example and is shown in Figure 1.116.

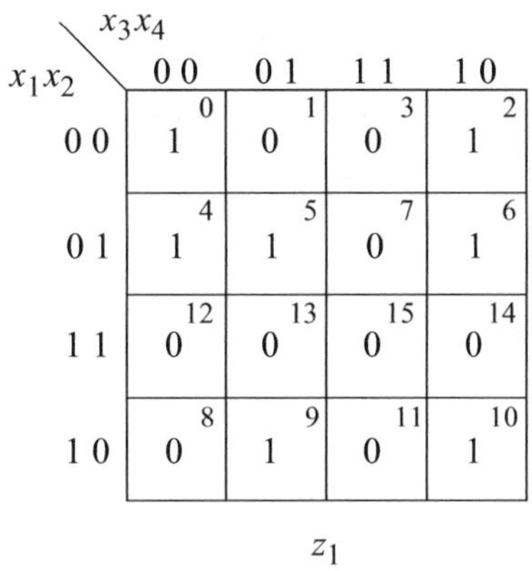

x_1x_2 \ x_3x_4	0 0	0 1	1 1	1 0
0 0	(0) 1	(1) 0	(3) 0	(2) 1
0 1	(4) 1	(5) 1	(7) 0	(6) 1
1 1	(12) 0	(13) 0	(15) 0	(14) 0
1 0	(8) 0	(9) 1	(11) 0	(10) 1

z_1

Figure 1.116 Karnaugh map for Equation 1.18.

Equation 1.18 is minimized as shown below in Equation 1.19.

$$z_1 = x_1'x_4' + x_1'x_2x_3' + x_2'x_3x_4' + x_1x_2'x_3'x_4 \quad (1.19)$$

The logic diagram is shown in Figure 1.117 showing the instantiated logic gates and the net names. The Verilog design module for Equation 1.19 will use the following modules that were designed using dataflow modeling: one two-input AND gate shown in Figure 1.118; two three-input AND gates of the type shown in Figure 1.119; one four-input AND gate shown in Figure 1.120; and one four-input OR gate shown in Figure 1.121

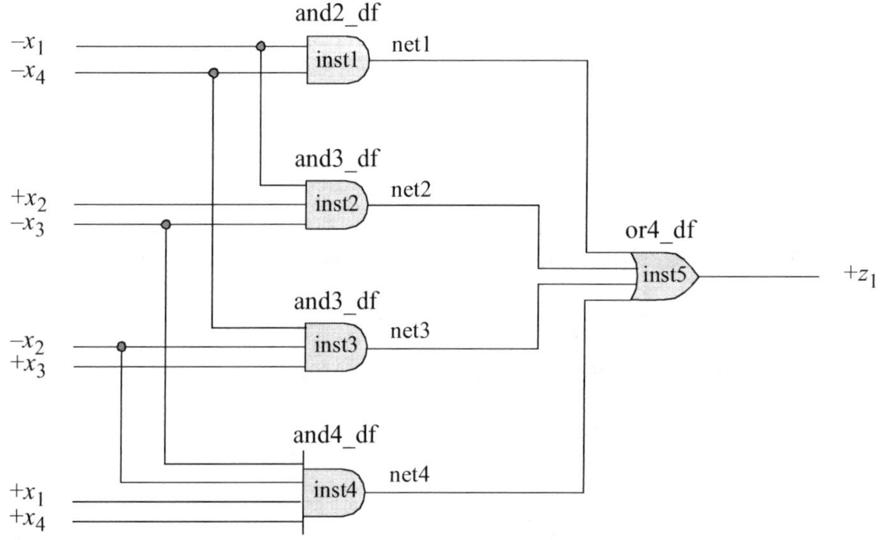

Figure 1.117 Logic diagram for Equation 1.19.

```
//dataflow 2-input and gate

module and2_df (x1, x2, z1);

//list inputs and output
input x1, x2;
output z1;

//define signals as wire for dataflow
wire x1, x2;
wire z1;

//continuous assign for dataflow
assign z1 = x1 & x2;

endmodule
```

Figure 1.118 Two-input AND gate designed using dataflow modeling.

```
//and3 dataflow

module and3_df (x1, x2, x3, z1);

//list inputs and output
input x1, x2, x3;
output z1;

//define signals as wire for dataflow
wire x1, x2, x3;
wire z1;

//continuous assign for dataflow
assign z1 = x1 & x2 & x3;

endmodule
```

Figure 1.119 Three-input AND gate designed using dataflow modeling.

```
//dataflow 4-input and gate

module and4_df (x1, x2, x3, x4, z1);

//list all inputs and outputs
input x1, x2, x3, x4;
output z1;

//define signals as wire (optional
wire x1, x2, x3, x4;
wire z1;

//continuous assign used for dataflow
assign z1 = (x1 & x2 & x3 & x4);

endmodule
```

Figure 1.120 Four-input AND gate designed using dataflow modeling.

```
//dataflow or4

module or4_df (x1, x2, x3, x4, z1);

//list all inputs and outputs
input x1, x2, x3, x4;
output z1;

//define signals as wire (optional
wire x1, x2, x3, x4;
wire z1;

//continuous assign used for dataflow
assign z1 = x1 | x2 | x3 | x4;

endmodule
```

Figure 1.121 Four-input OR gate designed using dataflow modeling.

Equation 1.19 will now be designed using structural modeling by instantiating the logic gates that were designed using dataflow modeling. The design module is shown in Figure 1.122. The test bench module is shown in Figure 1.123 and the outputs are shown in Figure 1.124. The 1s in the outputs are in the same location as the 1s in the minterm locations of the Karnaugh map of Figure 1.16.

```
//structural for the following logic equation
//z1 = x1'x4' + x1'x2x3' + x2'x3x4' + x1x2'x3'x4

module logic_equation (x1, x2, x3, x4, z1);

//list inputs and outputs
input x1, x2, x3, x4;
output z1;

//define internal nets
wire net, net2, net3, net4;

//instantiate the logic gates
and2_df inst1 (
   .x1(~x1),
   .x2(~x4),
   .z1(net1)
   );
                              //continued on next page
```

Figure 1.122 Design module for Equation 1.19.

```
and3_df inst2 (
    .x1(~x1),
    .x2(x2),
    .x3(~x3),
    .z1(net2)
    );

and3_df inst3 (
.x1(~x2),
    .x2(x3),
    .x3(~x4),
    .z1(net3)
    );

and4_df inst4 (
    .x1(x1),
    .x2(~x2),
    .x3(~x3),
    .x4(x4),
    .z1(net4)
    );

or4_df inst5 (
    .x1(net1),
    .x2(net2),
    .x3(net3),
    .x4(net4),
    .z1(z1)
    );

endmodule
```

Figure 1.122 (Continued)

```
//test bench for logic equation

module logic_equation_tb;

reg x1, x2, x3, x4;  //inputs are reg for test bench
wire z1;             //outputs are wire for test bench

                              //continued on next page
```

Figure 1.123 Test bench module for Equation 1.19.

```
//apply input vectors and display variables
initial
begin: apply_stimulus
   reg [4:0] invect;
   for (invect = 0; invect < 16; invect = invect + 1)
      begin
         {x1, x2, x3, x4} = invect [4:0];
         #10 $display ("x1 x2 x3 x4 = %b, z1 = %b",
                        {x1, x2, x3, x4}, z1);
      end
end

//instantiate the module into the test bench
logic_equation inst1 (x1, x2, x3, x4, z1);

endmodule
```

Figure 1.123 (Continued)

```
x1 x2 x3 x4 = 0000, z1 = 1
x1 x2 x3 x4 = 0001, z1 = 0
x1 x2 x3 x4 = 0010, z1 = 1
x1 x2 x3 x4 = 0011, z1 = 0
x1 x2 x3 x4 = 0100, z1 = 1
x1 x2 x3 x4 = 0101, z1 = 1
x1 x2 x3 x4 = 0110, z1 = 1
x1 x2 x3 x4 = 0111, z1 = 0
x1 x2 x3 x4 = 1000, z1 = 0
x1 x2 x3 x4 = 1001, z1 = 1
x1 x2 x3 x4 = 1010, z1 = 1
x1 x2 x3 x4 = 1011, z1 = 0
x1 x2 x3 x4 = 1100, z1 = 0
x1 x2 x3 x4 = 1101, z1 = 0
x1 x2 x3 x4 = 1110, z1 = 0
x1 x2 x3 x4 = 1111, z1 = 0
```

Figure 1.124 Outputs for Equation 1.19.

Example 1.27 An equation will be obtained for a logic circuit that will generate a logic 1 on output z_1 if a 4-bit unsigned binary number $N = x_1x_2x_3x_4$ satisfies the following criteria, where x_4 is the low-order bit:

$$3 < N \leq 7 \text{ or } 12 \leq N < 15$$

Built-in primitive gates will be used for the design module. Then the test bench will be generated and the outputs displayed. The equation will be obtained using the Karnaugh map shown in Figure 1.125, which was created from the number range for the variable N. The equation for z_1 is shown in Equation 1.20.

x_1x_2 \ x_3x_4	0 0	0 1	1 1	1 0
0 0	(0) 0	(1) 0	(3) 0	(2) 0
0 1	(4) 1	(5) 1	(7) 1	(6) 1
1 1	(12) 1	(13) 1	(15) 0	(14) 1
1 0	(8) 0	(9) 0	(11) 0	(10) 0

z_1

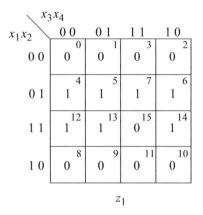

Figure 1.125 Karnaugh map for Example 1.27.

$$\begin{aligned} z_1 &= x_1'x_2 + x_2x_3' + x_2x_4' \\ &= x_2(x_1' + x_3' + x_4') \end{aligned} \tag{1.20}$$

The logic diagram is shown in Figure 1.126 showing the instantiation names and the net names. The design module is shown in Figure 1.127. The test bench module is shown in Figure 1.128 and the outputs are shown in Figure 1.129.

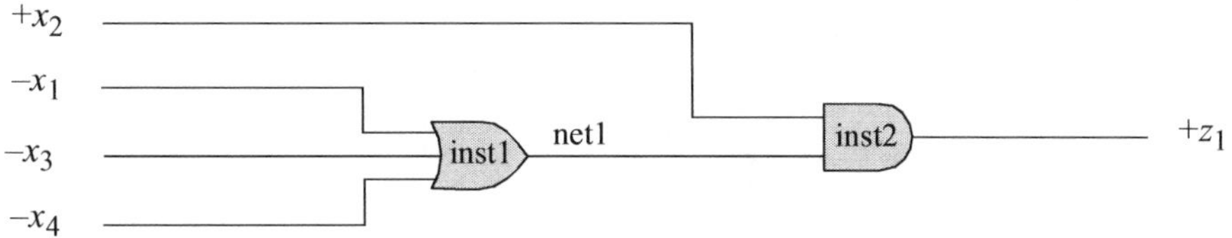

Figure 1.126 Logic diagram for Example 1.27 using Equation 1.20.

```
//structural for a number in the following range
//x2(x1' + x3' + x4')
module number_range (x1, x2, x3, x4, z1);

input x1, x2, x3, x4;    //list inputs and output
output z1;

or    inst1 (net1, ~x1, ~x3, ~x4);
and   inst2 (z1, x2, net1);
endmodule
```

Figure 1.127 Design module for Example 1.27.

```
//test bench for number_range

module number_range_tb;

reg x1, x2, x3, x4;  //inputs are reg for test bench
wire z1;             //outputs are wire for test bench

//apply input vectors and display variables
initial
begin: apply_stimulus
   reg [4:0] invect;
   for (invect = 0; invect < 16; invect = invect + 1)
      begin
         {x1, x2, x3, x4} = invect [4:0];
         #10 $display ("x1 x2 x3 x4 = %b, z1 = %b",
                        {x1, x2, x3, x4}, z1);
      end
end

//instantiate the module into the test bench
number_range inst1 (x1, x2, x3, x4, z1);

endmodule
```

Figure 1.128 Test bench module for Example 1.27.

```
x1 x2 x3 x4 = 0000, z1 = 0
x1 x2 x3 x4 = 0001, z1 = 0
x1 x2 x3 x4 = 0010, z1 = 0
x1 x2 x3 x4 = 0011, z1 = 0
x1 x2 x3 x4 = 0100, z1 = 1
x1 x2 x3 x4 = 0101, z1 = 1
x1 x2 x3 x4 = 0110, z1 = 1
x1 x2 x3 x4 = 0111, z1 = 1
x1 x2 x3 x4 = 1000, z1 = 0
x1 x2 x3 x4 = 1001, z1 = 0
x1 x2 x3 x4 = 1010, z1 = 0
x1 x2 x3 x4 = 1011, z1 = 0
x1 x2 x3 x4 = 1100, z1 = 1
x1 x2 x3 x4 = 1101, z1 = 1
x1 x2 x3 x4 = 1110, z1 = 1
x1 x2 x3 x4 = 1111, z1 = 0
```

Figure 1.129 Outputs for Example 1.27.

Example 1.28 This example converts a 4-bit binary number to a 4-bit excess-3 number. The binary and excess-3 codes are shown in Table 1.16, where the binary bit x_4 and the excess-3 bit z_4 are the low-order bits of their respective codes. When the number three is added to the binary number 1101 (13_{10}), the result exceeds four bits 1 0000 (16_{10}), where the high-order bit represents a carry out. The same is true for the binary numbers 1110 and 1111, which yield 1 0001 (17_{10}) and 1 0010 (18_{10}), respectively.

Table 1.16 Binary-to-Excess-3 Code Conversion

Binary Code				Excess-3 Code			
x_1	x_2	x_3	x_4	z_1	z_2	z_3	z_4
0	0	0	0	0	0	1	1
0	0	0	1	0	1	0	0
0	0	1	0	0	1	0	1
0	0	1	1	0	1	1	0
0	1	0	0	0	1	1	1
0	1	0	1	1	0	0	0
0	1	1	0	1	0	0	1
0	1	1	1	1	0	1	0
1	0	0	0	1	0	1	1
1	0	0	1	1	1	0	0
1	0	1	0	1	1	0	1
1	0	1	1	1	1	1	0
1	1	0	0	1	1	1	1
1	1	0	1	0	0	0	0
1	1	1	0	0	0	0	1
1	1	1	1	0	0	1	0

The Karnaugh maps that indicate the conversion from binary to excess-3 are shown in Figure 1.130, as obtained from Table 1.16. The equations are shown in Equation 1.22.

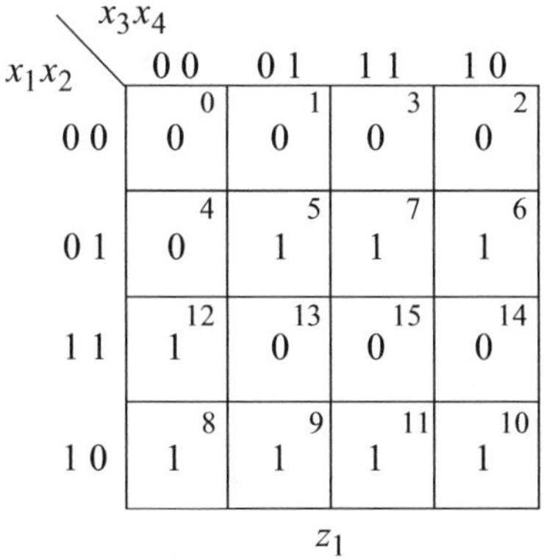

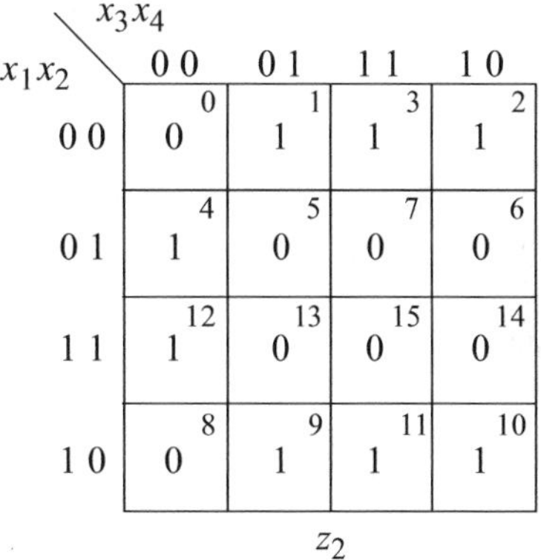

Figure 1.130 Karnaugh maps for Example 1.28. (Continued on next page)

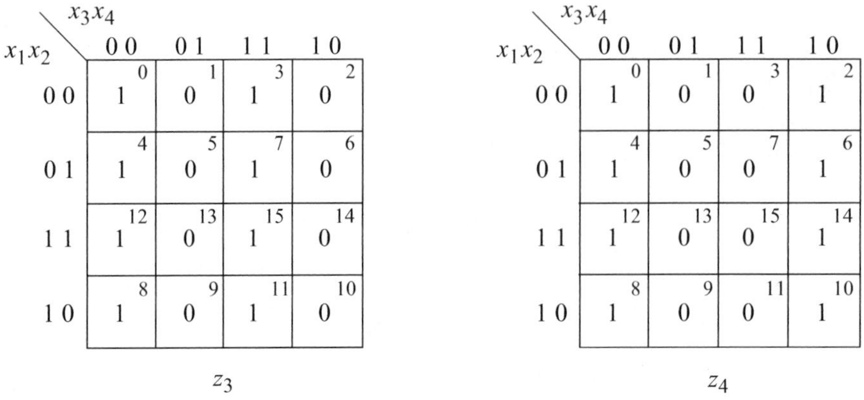

x_1x_2 \ x_3x_4	0 0	0 1	1 1	1 0
0 0	1 (0)	0 (1)	1 (3)	0 (2)
0 1	1 (4)	0 (5)	1 (7)	0 (6)
1 1	1 (12)	0 (13)	1 (15)	0 (14)
1 0	1 (8)	0 (9)	1 (11)	0 (10)

z_3

x_1x_2 \ x_3x_4	0 0	0 1	1 1	1 0
0 0	1 (0)	0 (1)	0 (3)	1 (2)
0 1	1 (4)	0 (5)	0 (7)	1 (6)
1 1	1 (12)	0 (13)	0 (15)	1 (14)
1 0	1 (8)	0 (9)	0 (11)	1 (10)

z_4

Figure 1.130 (Continued)

$$z_1 = \overset{\text{net1}}{x_1'x_2x_3} + \overset{\text{net2}}{x_1'x_2x_4} + \overset{\text{net3}}{x_1x_3'x_4'} + \overset{\text{net4}}{x_1x_2'}$$

$$z_2 = \overset{\text{net6}}{x_2x_3'x_4'} + \overset{\text{net7}}{x_2'x_3} + \overset{\text{net8}}{x_2'x_4}$$

$$z_3 = x_3'x_4' + x_3x_4 = \overset{\text{net10}}{(x_3 \oplus x_4)'}$$

$$z_4 = x_4' \tag{1.22}$$

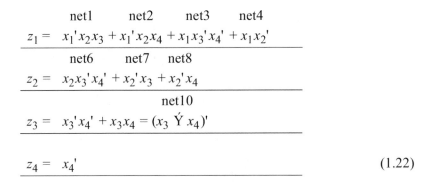

The design module to convert from the binary code to the excess-3 code using dataflow modeling is shown in Figure 1.131. The test bench is shown in Figure 1.132 and the outputs are shown in Figure 1.133.

```
//structural binary-to-excess-3

module binary_to_excess3 (x1, x2, x3, x4, z1, z2, z3, z4);

//list inputs and outputs
input x1, x2, x3, x4;
output z1, z2, z3, z4;

//define internal nets
wire net1, net2, net3, net4, net5, net6, net7, net8, net9,
            net10;                //continued on next page
```

Figure 1.131 Design module for the binary-to-excess-3 code converter.

```
//----------------------------------------------------------
//instantiate the logic for output z1
and3_df inst1 (~x1, x2, x3, net1);

and3_df inst2 (~x1, x2, x4, net2);

and3_df inst3 (x1, ~x3, ~x4, net3);

and2_df inst4 (x1, ~x2, net4);

or4_df inst5 (net1, net2, net3, net4, z1);

//----------------------------------------------------------
//instantiate the logic for output z2
and3_df inst6 (x2, ~x3, ~x4, net6);

and2_df inst7 (~x2, x3, net7);

and2_df inst8 (~x2, x4, net8);

or3_df inst9 (net6, net7, net8, z2);

//----------------------------------------------------------
//instantiate the logic for output z3
xnor2_df inst10 (x3, x4, z3);

//----------------------------------------------------------
//instantiate the logic for output z4
assign z4 = ~x4;

endmodule
```

Figure 1.131 (Continued)

```
//test bench for binary-to-excess3

module binary_to_excess3_tb;

//inputs are reg for test bench
//outputs are wire for test bench
reg x1, x2, x3, x4;
wire z1, z2, z3, z4;

                                //continued on next page
```

Figure 1.132 Test bench module for the binary-to-excess-3 code converter.

```
//apply input vectors and display variables
initial
begin: apply_stimulus
   reg [4:0] invect;
   for (invect = 0; invect < 16; invect = invect + 1)
      begin
         {x1, x2, x3, x4} = invect [4:0];
         #10 $display ("x1 x2 x3 x4 = %b, z1 z2 z3 z4 = %b",
                        {x1, x2, x3, x4}, {z1, z2, z3, z4});
      end
end

//instantiate the module into the test bench
binary_to_excess3 inst1 (x1, x2, x3, x4, z1, z2, z3, z4);

endmodule
```

Figure 1.132 (Continued)

```
x1 x2 x3 x4 = 0000, z1 z2 z3 z4 = 0011
x1 x2 x3 x4 = 0001, z1 z2 z3 z4 = 0100
x1 x2 x3 x4 = 0010, z1 z2 z3 z4 = 0101
x1 x2 x3 x4 = 0011, z1 z2 z3 z4 = 0110
x1 x2 x3 x4 = 0100, z1 z2 z3 z4 = 0111
x1 x2 x3 x4 = 0101, z1 z2 z3 z4 = 1000
x1 x2 x3 x4 = 0110, z1 z2 z3 z4 = 1001
x1 x2 x3 x4 = 0111, z1 z2 z3 z4 = 1010
x1 x2 x3 x4 = 1000, z1 z2 z3 z4 = 1011
x1 x2 x3 x4 = 1001, z1 z2 z3 z4 = 1100
x1 x2 x3 x4 = 1010, z1 z2 z3 z4 = 1101
x1 x2 x3 x4 = 1011, z1 z2 z3 z4 = 1110
x1 x2 x3 x4 = 1100, z1 z2 z3 z4 = 1111
x1 x2 x3 x4 = 1101, z1 z2 z3 z4 = 0000
x1 x2 x3 x4 = 1110, z1 z2 z3 z4 = 0001
x1 x2 x3 x4 = 1111, z1 z2 z3 z4 = 0010
```

Figure 1.133 Outputs for the binary-to-excess-3 code converter.

Example 1.29 This example designs an adder that adds two 4-bit operands using four full adders. A *full adder* is a combinational circuit that adds two operand bits plus a carry-in bit. The carry-in bit represents the carry-out of the previous lower-order stage. A full adder produces two outputs: *sum* and *carry-out*.

The truth table for a full adder is shown in Table 1.17. Operand a_i represents the augend and operand b_i represents the addend. The corresponding equations for the sum and carry-out are listed in Equation 1.23 as obtained directly from the truth table. A block diagram for a full adder for any stage is shown in Figure 1.134, where the inputs are the augend a_i, the addend b_i, and the carry-in from the previous lower-order stage cin_{i-1}. The outputs are the sum, sum_i, and the carry-out, $cout_i$.

Table 1.17 Truth Table for a Full Adder

a_i	b_i	cin_{i-1}	$cout_i$	sum_i
0	0	0	0	0
0	0	1	0	1
0	1	0	0	1
0	1	1	1	0
1	0	0	0	1
1	0	1	1	0
1	1	0	1	0
1	1	1	1	1

$$\begin{aligned} sum_i &= a_i'b_i'cin_{i-1} + a_i'b_icin_{i-1}' + a_ib_i'cin_{i-1}' + a_ib_icin_{i-1} \\ &= a_i \oplus b_i \oplus cin_{i-1} \end{aligned}$$

$$\begin{aligned} cout_i &= a_i'b_icin_{i-1} + a_ib_i'cin_{i-1} + a_ib_icin_{i-1}' + a_ib_icin_{i-1} \\ &= a_ib_i + (a_i \oplus b_i)cin_{i-1} \end{aligned} \quad (1.23)$$

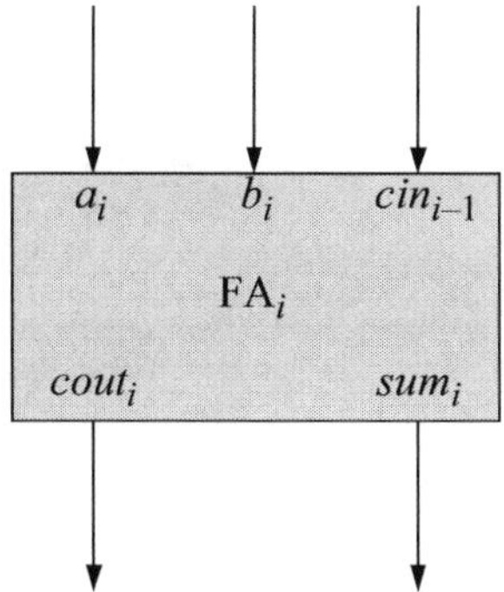

Figure 1.134 Block diagram for a full adder.

The design module for the full adder is shown in Figure 1.135 using dataflow modeling. This module is then instantiated four times into the structural design module of Figure 1.136. The test bench module is shown in Figure 1.137 and the outputs are shown in Figure 1.138.

```
//dataflow full adder
module full_adder (a, b, cin, sum, cout);

//list all inputs and outputs
input a, b, cin;
output sum, cout;

//define wires
wire a, b, cin;
wire sum, cout;

//continuous assign
assign sum = (a ^ b) ^ cin;
assign cout = cin & (a ^ b) | (a & b);

endmodule
```

Figure 1.135 Dataflow design module for a full adder.

```
//structural 4_bit ripple-carry counter
module adder4_struc (a, b, cin, sum, cout);

//define inputs and outputs
input [3:0] a, b;
input cin;
output [3:0] sum;
output cout;

//define internal nets for carries
wire [3:0] c;

assign cout = c[3];

full_adder inst0 (a[0], b[0], cin, sum[0], c[0]);
full_adder inst1 (a[1], b[1], c[0], sum[1], c[1]);
full_adder inst2 (a[2], b[2], c[1], sum[2], c[2]);
full_adder inst3 (a[3], b[3], c[2], sum[3], c[3]);

endmodule
```

Figure 1.136 Structural design module for the 4-bit adder.

```
//test bench for 4-bit ripple-carry adder
module adder4_struc_tb;

//define inputs and outputs
//inputs are reg for test bench
reg [3:0] a, b;
reg cin;

//outputs are wire for test bench
wire [3:0] sum;
wire cout;

initial
$monitor ("a=%b, b=%b, cin=%b, cout=%b, sum=%b",
            a, b, cin, cout, sum);

initial
begin
   #0 a = 4'b0001;   b = 4'b0001;   cin = 1'b0;
   #10a = 4'b0010;   b = 4'b0001;   cin = 1'b1;
   #10a = 4'b0100;   b = 4'b0010;   cin = 1'b0;
   #10a = 4'b0000;   b = 4'b0111;   cin = 1'b1;

   #10a = 4'b1000;   b = 4'b0001;   cin = 1'b1;
   #10a = 4'b0100;   b = 4'b1000;   cin = 1'b0;
   #10a = 4'b0111;   b = 4'b0110;   cin = 1'b1;
   #10a = 4'b1000;   b = 4'b1000;   cin = 1'b0;

   #10a = 4'b0010;   b = 4'b1111;   cin = 1'b1;
   #10a = 4'b1111;   b = 4'b0101;   cin = 1'b0;
   #10a = 4'b1110;   b = 4'b0111;   cin = 1'b1;
   #10a = 4'b1100;   b = 4'b1100;   cin = 1'b0;

   #10a = 4'b1100;   b = 4'b1101;   cin = 1'b1;
   #10a = 4'b1110;   b = 4'b1110;   cin = 1'b0;
   #10a = 4'b1110;   b = 4'b1111;   cin = 1'b1;
   #10a = 4'b1111;   b = 4'b1111;   cin = 1'b1;

   #10   $stop;
end

//instantiate the module into the test bench
adder4_struc inst1 (a, b, cin, sum, cout);

endmodule
```

Figure 1.137 Test bench module for the structural 4-bit adder.

```
a=0001, b=0001, cin=0, cout=0, sum=0010
a=0010, b=0001, cin=1, cout=0, sum=0100
a=0100, b=0010, cin=0, cout=0, sum=0110
a=0000, b=0111, cin=1, cout=0, sum=1000

a=1000, b=0001, cin=1, cout=0, sum=1010
a=0100, b=1000, cin=0, cout=0, sum=1100
a=0111, b=0110, cin=1, cout=0, sum=1110
a=1000, b=1000, cin=0, cout=1, sum=0000

a=0010, b=1111, cin=1, cout=1, sum=0010
a=1111, b=0101, cin=0, cout=1, sum=0100
a=1110, b=0111, cin=1, cout=1, sum=0110
a=1100, b=1100, cin=0, cout=1, sum=1000

a=1100, b=1101, cin=1, cout=1, sum=1010
a=1110, b=1110, cin=0, cout=1, sum=1100
a=1110, b=1111, cin=1, cout=1, sum=1110
a=1111, b=1111, cin=1, cout=1, sum=1111
```

Figure 1.138 Outputs for the structural 4-bit adder.

Example 1.30 This example designs a 3-bit comparator using structural modeling for the following operands:

$$A = a_2 a_1 a_0$$
$$B = b_2 b_1 b_0$$

where a_0 and b_0 are the low-order bits of A and B, respectively. The following outputs will be used:

a_lt_b indicating $A < B$
a_eq_b indicating $A = B$
a_gt_b indicating $A > B$

The equations for the comparator are shown below.

$$(A < B) = a_2{}' b_2 + (a_2 \oplus b_2)' a_1{}' b_1 + (a_2 \oplus b_2)' (a_1 \oplus b_1)' a_0{}' b_0$$
$$(A = B) = (a_2 \oplus b_2)' (a_1 \oplus b_1)' (a_0 \oplus b_0)'$$
$$(A > B) = a_2 b_2{}' + (a_2 \oplus b_2)' a_1 b_1{}' + (a_2 \oplus b_2)' (a_1 \oplus b_1)' a_0 b_0{}'$$

Referring to the equation for ($A < B$), the term *a[2]' b[2]* indicates that if the high-order bits of *a* and *b* are 0 and 1, respectively, then *a* must be less than *b*. If the high-order bits of *a* and *b* are equal, then the relative magnitude of *a* and *b* depends upon the next lower-order bits *a[1]* and *b[1]*. This is indicated by the second term of the equation for ($A < B$).

The structural module instantiates the following dataflow modules: *and2_df*, *xnor2_df*, *and3_df*, *and4_df*, and *or3_df*. The structural design module is shown in Figure 1.139, test bench module is shown in Figure 1.140, and the outputs are shown in Figure 1.141. The test bench applies 12 sets of inputs to demonstrate the relative magnitude of the two operands.

```
//structural 3-bit comparator
module comp3_bit_struc (a, b, a_lt_b, a_eq_b, a_gt_b);

//define inputs and outputs
input [2:0] a, b;
output a_lt_b, a_eq_b, a_gt_b;

//define internal nets
wire net1, net2, net3, net4, net5, net7, net9, net10, net11;

//-----------------------------------------------------
//instantiate the logic for a_lt_b
and2_df inst1 (~a[2], b[2], net1);

xnor2_df inst2 (a[2], b[2], net2);

xnor2_df inst3 (a[1], b[1], net3);

and3_df inst4 (net2, ~a[1], b[1], net4);

and4_df inst5 (net2, net3, ~a[0], b[0], net5);

or3_df inst6 (net1, net4, net5, a_lt_b);

//-----------------------------------------------------
//instantiate the logic for a_eq_b
xnor2_df inst7 (a[0], b[0], net7);

and3_df inst8 (net2, net3, net7, a_eq_b);

//-----------------------------------------------------
//instantiate the logic for a_gt_b
and2_df inst9 (a[2], ~b[2], net9);

and3_df inst10 (net2, a[1], ~b[1], net10);

and4_df inst11 (net2, net3, a[0], ~b[0], net11);

or3_df inst12 (net9, net10, net11, a_gt_b);

endmodule
```

Figure 1.139 Structural design module for the 3-bit comparator.

```
//test bench for structural 3-bit comparator
module comp3_bit_struc_tb;

//inputs are reg for test bench
//outputs are wire for test bench
reg [2:0] a, b;
wire a_lt_b, a_eq_b, a_gt_b;

//display inputs and outputs
initial
$monitor ("a=%b, b=%b, a_lt_b=%b, a_eq_b=%b, a_gt_b=%b",
            a, b, a_lt_b, a_eq_b, a_gt_b);

//apply input vectors
initial
begin

//a_lt_b
   #0    a=3'b001;    b=3'b010;
   #10   a=3'b010;    b=3'b100;
   #10   a=3'b110;    b=3'b111;
   #10   a=3'b100;    b=3'b110;

//a_eq_b
   #10   a=3'b000;    b=3'b000;
   #10   a=3'b010;    b=3'b010;
   #10   a=3'b111;    b=3'b111;
   #10   a=3'b011;    b=3'b011;

//a_gt_b
   #10   a=3'b001;    b=3'b000;
   #10   a=3'b011;    b=3'b010;
   #10   a=3'b101;    b=3'b011;
   #10   a=3'b111;    b=3'b110;

   #10   $stop;
end

//instantiate the module into the test bench
comp3_bit_struc inst1 (a, b, a_lt_b, a_eq_b, a_gt_b);

endmodule
```

Figure 1.140 Test bench module for the 3-bit comparator.

```
a=001, b=010, a_lt_b=1, a_eq_b=0, a_gt_b=0
a=010, b=100, a_lt_b=1, a_eq_b=0, a_gt_b=0
a=110, b=111, a_lt_b=1, a_eq_b=0, a_gt_b=0
a=100, b=110, a_lt_b=1, a_eq_b=0, a_gt_b=0

a=000, b=000, a_lt_b=0, a_eq_b=1, a_gt_b=0
a=010, b=010, a_lt_b=0, a_eq_b=1, a_gt_b=0
a=111, b=111, a_lt_b=0, a_eq_b=1, a_gt_b=0
a=011, b=011, a_lt_b=0, a_eq_b=1, a_gt_b=0

a=001, b=000, a_lt_b=0, a_eq_b=0, a_gt_b=1
a=011, b=010, a_lt_b=0, a_eq_b=0, a_gt_b=1
a=101, b=011, a_lt_b=0, a_eq_b=0, a_gt_b=1
a=111, b=110, a_lt_b=0, a_eq_b=0, a_gt_b=1
```

Figure 1.141 Outputs for the 3-bit comparator.

1.9 Tasks and Functions

Verilog provides tasks and functions that are similar to procedures or subroutines found in other programming languages. These constructs allow a *behavioral* module to be partitioned into smaller segments. Tasks and functions permit modules to execute common code segments that are written once then called when required, thus reducing the amount of code needed. They enhance the readability and maintainability of the Verilog modules.

Tasks and functions are defined within a module and are local to the module. They can be invoked only from a behavioral construct within the module. That is, they are called from an **always** block, an **initial** block, or from other tasks or functions. A function can invoke another function, but not a task. A function must have at least one **input** argument, but does not have **output** or **inout** arguments. The task and function arguments can be considered as the ports of the constructs; however, these ports do connect to the external environment.

A task cannot be invoked from a continuous assignment statement and does not return values to an expression, but places the values on the **output** or **inout** ports. Tasks can contain delays, timing, or event control statements and can execute in nonzero simulation time when event control is applied. A task can invoke other tasks and functions and can have arguments of type **input**, **output**, or **inout**.

1.9.1 Task Declaration

A task is delimited by the keywords **task** and **endtask**. The syntax for a task declaration is as follows:

```
task task_name
    input arguments
    output arguments
    inout  arguments
    task declarations
    local variable declarations
    begin
        statements
    end
endtask
```

Arguments (or parameters) that are of type **input** or **inout** are processed by the task statements; arguments that are of type **output** or **inout**, resulting from the task construct, are passed back to the task invocation statement — the statement that called the task. The keywords **input**, **output**, and **inout** are not ports of the module, they are ports used to pass values between the task invocation statement and the task construct. Additional local variables can be declared within a task, if necessary. Since tasks cannot be synthesized, they are used only in test benches. When a task completes execution, control is passed to the next statement in the module.

1.9.2 Task Invocation

A task can be invoked (or called) from a procedural statement; therefore, it must appear within an **always** or an **initial** block. A task can call itself or be invoked by tasks that it has called. The syntax for a task invocation is as follows, where the expressions are parameters passed to the task:

task_name (expression 1, expression 2, . . . , expression *n*);

Values for arguments of type **output** and **inout** are passed back to the variables in the task invocation statement upon completion of the task. The list of arguments in the task invocation must match the order of **input**, **output**, and **inout** variables in the task declaration. The **output** and **inout** arguments must be of type **reg** because a task invocation is a procedural statement.

Example 1.31 A task module will be generated that performs both arithmetic and logical operations. There are three inputs: *a[7:0]*, *b[7:0]*, and *c[7:0]*, where *a[0]*, *b[0]*, and *c[0]* are the low-order bits of *a*, *b*, and *c*, respectively. There are four outputs: z_1, z_2, z_3, and z_4 that perform the operations shown below.

$$\begin{aligned} z_1 &= (b + c) \mid (a) \\ z_2 &= (a \mathbin{\&} c) + (b) \\ z_3 &= (\sim a + c) \mathbin{\&} (b) \\ z_4 &= (b \mid c) \mathbin{\&} (a) \end{aligned}$$

Figure 1.42 is a block diagram of the module in which the task is embedded. Notice that there are no ports in the module to the external environment. The only ports are in the task which passes variables to the task declaration as **input** ports, as shown below and ports the pass the results back to the task invocation as **output** ports, as shown below.

```
input [7:0] a, b, c;
output [7:0] z1, z2, z3, z4;
```

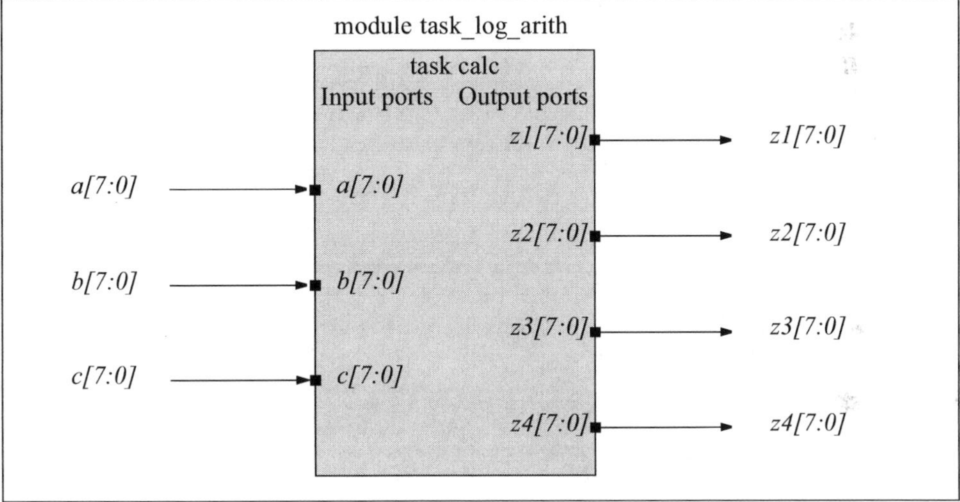

Figure 1.142 Block diagram of the task module of Example 1.31.

The task module is shown in Figure 1.143 in which no ports are listed in the **module** definition. The first set of variables passed to the task declaration called `calc` by the task invocation are shown below as variables *a*, *b*, and *c*. The variables z_1, z_2, z_3, and z_4 are the results that are passed back to the task invocation.

```
a = 8'b1111_1111; b = 8'b0011_1111; c = 8'b0001_1101;

calc (a, b, c, z1, z2, z3, z4);
```

The outputs are shown in Figure 1.144. The module declares 8-bit register vectors *[7:0] a*, *[7:0] b*, and *[7:0] c*. These are redeclared in the task. Output z_1 adds operands *b* and *c* and then performs a bitwise logical OR operation on the sum with operand *a*.

```
//module to illustrate a task
module task_log_arith;

//define input and output ports
reg [7:0] a, b, c;           //input ports
reg [7:0] z1, z2, z3, z4;    //output ports

initial
begin
      a = 8'b1111_1111;    b = 8'b0011_1111;
      c = 8'b0001_1101;
   calc (a, b, c, z1, z2, z3, z4);

      a = 8'b1111_1010;    b = 8'b0011_1100;
      c = 8'b1000_1001;
   calc (a, b, c, z1, z2, z3, z4);

      a = 8'b0011_1110;    b = 8'b0101_1101;
      c = 8'b1110_0001;
   calc (a, b, c, z1, z2, z3, z4);

      a = 8'b0100_1011;    b = 8'b1001_1101;
      c = 8'b1111_0011;
   calc (a, b, c, z1, z2, z3, z4);

end

task calc;
   input [7:0] a, b, c;
   output [7:0] z1, z2, z3, z4;

   begin
      z1 = (b + c) | (a);
      z2 = (a & c) + (b);
      z3 = (~a + c) & (b);
      z4 = (b | c) & (a);

      $display ("a = %b, b = %b, c = %b,
                  z1 = %b, z2 = %b, z3 = %b, z4 = %b",
                     a, b, c, z1, z2, z3, z4);

   end
endtask

endmodule
```

Figure 1.143 Task design module for Example 1.31.

```
a = 11111111, b = 00111111, c = 00011101,
z1 = 11111111, z2 = 01011100, z3 = 00011101, z4 = 00111111

a = 11111010, b = 00111100, c = 10001001,
z1 = 11111111, z2 = 11000100, z3 = 00001100, z4 = 10111000

a = 00111110, b = 01011101, c = 11100001,
z1 = 00111110, z2 = 01111101, z3 = 00000000, z4 = 00111100

a = 01001011, b = 10011101, c = 11110011,
z1 = 11011011, z2 = 11100000, z3 = 10000101, z4 = 01001011
```

Figure 1.144 Outputs for the task module of Figure 1.143.

Example 1.32 A module will be designed that contains a task to count the number of 1s in an 8-bit register *reg_a*. The task returns the number of 1s to a 4-bit register *count*. The task module is shown in Figure 1.145 in which no ports are listed in the **module** definition. The first variable passed to the task declaration called `ctr` by the task invocation is `reg_a = 8'b0000_0000`. The variable `count` is the result that is passed back to the task invocation. The outputs are shown in Figure 1.146.

```
//module to illustrate a task to count the number of 1s
module task_count1s_2;

//define task ports
reg [7:0] reg_a;      //input ports
reg [3:0] count;      //output ports

initial
begin
      reg_a = 8'b0000_0000;    //no 1s
   ctr (reg_a, count);         //invoke the task

      reg_a = 8'b1110_1010;    //five 1s
   ctr (reg_a, count);         //invoke the task

      reg_a = 8'b0111_0001;    //four 1s
   ctr (reg_a, count);         //invoke the task

      reg_a = 8'b1001_1111;    //six 1s
   ctr (reg_a, count);         //invoke the task
                                  //continued on next page
```

Figure 1.145 Task design module to count the number of 1s in a register.

```
      reg_a = 8'b1011_1111;    //seven 1s
   ctr (reg_a, count);         //invoke the task

      reg_a = 8'b1111_1111;    //eight 1s
   ctr (reg_a, count);         //invoke the task
end

task ctr;
   input [7:0] reg_a;
   output [3:0] count;

begin
   count = 0;
   while (reg_a)
      begin
         count = count + reg_a[0];
         reg_a = reg_a >> 1;
      end

   $display ("count = %d", count);
end
endtask

endmodule
```

Figure 1.145 (Continued)

```
count = 0
count = 5
count = 4
count = 6
count = 7
count = 8
```

Figure 1.146 Outputs for the task module that counts the number of 1s.

1.9.3 Function Declaration

Functions are similar to tasks, except that functions return only a single value to the expression from which they are called. Like tasks, functions provide the ability to execute common procedures from within a module. A function can be invoked from a continuous assignment statement or from within a procedural statement and is represented by an operand in an expression.

Functions cannot contain delays, timing, or event control statements and execute in zero simulation time. Although functions can invoke other functions, they are not recursive. Functions cannot invoke a task. Functions must have at least one **input** argument, but cannot have **output** or **inout** arguments.

The syntax for a function declaration is shown below. If the optional *range or type* is omitted, the value returned to the function invocation is a scalar of type **reg**. Functions are delimited by the keywords **function** and **endfunction** and are used to implement combinational logic; therefore, functions cannot contain event controls or timing controls.

```
function [range or type] function name
    input declaration
    other declarations
    begin
        statement
    end
endfunction
```

1.9.4 Function Invocation

A function is invoked from an expression. The function is invoked by specifying the function name together with the input parameters. The syntax is shown below.

function_name (expression 1, expression 2, . . . , expression *n*);

All local registers that are declared within a function are static; that is, they retain their values between invocations of the function. When the function execution is finished, the return value is positioned at the location where the function was invoked. The function module, like tasks, has no ports to communicate with the external environment. The only ports are input ports that receive parameters from the function invocation.

Example 1.33 This example calculates the parity of a 16-bit register and returns one bit indicating whether there is an even number of 1s or an odd number of 1s. If the parity is even, then *parity bit = 1* is printed; if parity is odd, then *parity bit = 0* is printed. That is, a 1 is appended to the register contents if there are an even number of 1s in the register so that all 17 bits contain an odd number of 1s, otherwise a 0 bit is appended. The function module, like tasks, has no ports listed in the **module** definition to communicate with the external environment. The only ports are input ports that receive parameters from the function invocation.

Figure 1.147 shows the block diagram of the module *fctn_parity* with the function *calc_parity* embedded in the module. The design module is shown in Figure 1.148. The variable *contents* is declared as a 16-bit register; the variable *parity* is a scalar register. The statement *parity = calc_parity (16'b1111_0000_1111_0000)* invokes the function *calc_parity* and passes the register contents to the function input port *[15:0] address*. Then the function uses the reduction exclusive-OR operator to determine the parity of the register contents. The parity of the contents is returned to the left-hand side of the *parity* statement, then the parity bit is displayed. The outputs of the module are shown in Figure 1.149.

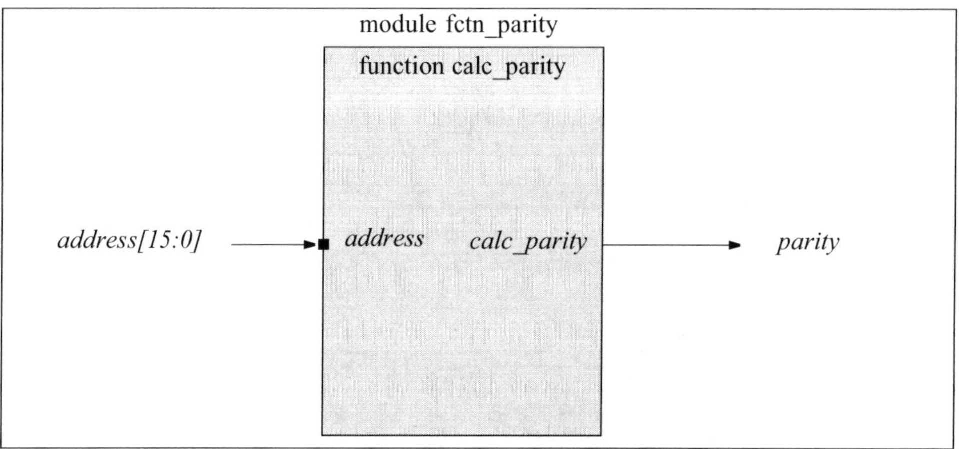

Figure 1.147 Block diagram for the function of Example 1.33.

```
//module to illustrate a function
module fctn_parity;

reg [15:0] contents;
reg parity;

initial
begin
   parity = calc_parity (16'b1111_0000_1111_0000);
      if (parity ==1)
         $display ("parity bit = 0");
      else
         $display ("parity bit = 1");

   parity = calc_parity (16'b1111_0000_1111_0001);
      if (parity ==1)
         $display ("parity bit = 0");
      else
         $display ("parity bit = 1");

                                    //continued on next page
```

Figure 1.148 Function design module to determine the parity of a register.

```
   parity = calc_parity (16'b1111_1111_1111_0000);
      if (parity ==1)
         $display ("parity bit = 0");
      else
         $display ("parity bit = 1");

   parity = calc_parity (16'b1111_1111_1111_1110);
      if (parity ==1)
         $display ("parity bit = 0");
      else
         $display ("parity bit = 1");
end

function calc_parity;
   input [15:0] address;
   begin
      calc_parity = ^address;
   end
endfunction

endmodule
```

Figure 1.148 (Continued)

```
parity bit = 1
parity bit = 0
parity bit = 1
parity bit = 0
```

Figure 1.149 Outputs for the function to determine the parity of a register.

Example 1.34 This example repeats the conversion from the binary code to the excess-3 code of Example 1.28, but uses a function to perform the conversion. The binary and excess-3 codes are reproduced in Table 1.18 for convenience. The excess-3 code is a *nonweighted code* and is obtained by adding three to the 8421 binary code. For example, in Table 1.18 the binary code of 1100 equals 1100 + 0011 = 1111 in the excess-3 code. The binary code of 1101 equals 1101 + 0011 = 1 ← 0000, which yields a 4-bit excess-3 code of 0000.

Table 1.18 Binary-to-Excess-3 Code Conversion

Binary Code				Excess-3 Code			
x_1	x_2	x_3	x_4	z_1	z_2	z_3	z_4
0	0	0	0	0	0	1	1
0	0	0	1	0	1	0	0
0	0	1	0	0	1	0	1
0	0	1	1	0	1	1	0
0	1	0	0	0	1	1	1
0	1	0	1	1	0	0	0
0	1	1	0	1	0	0	1
0	1	1	1	1	0	1	0
1	0	0	0	1	0	1	1
1	0	0	1	1	1	0	0
1	0	1	0	1	1	0	1
1	0	1	1	1	1	1	0
1	1	0	0	1	1	1	1
1	1	0	1	0	0	0	0
1	1	1	0	0	0	0	1
1	1	1	1	0	0	1	0

The excess-3 code is also a *self-complementing code* in which the 1s complement of a code word is identical to the 9s complement of the corresponding 8421 BCD code word in excess-3 notation, as shown below for the decimal number 4. The function design module is shown in Figure 1.150 and the outputs are shown in Figure 1.151.

Excess-3 BCD code	8421 BCD code
4	4
↓	↓
0111	0100
↓ 1s complement	↓ 9s complement
1000	0101
	↓ excess-3
	1000

```
//module to implement a function to convert
//from binary code to excess-3 code
module fctn_excess3a;

reg [7:0] a;
reg [7:0] rslt;

initial
begin
   rslt = excess3 (8'b0000_0011);
      $display ("binary = 0000_0011, excess3 = %b", rslt);

   rslt = excess3 (8'b0011_0000);
      $display ("binary = 0011_0000, excess3 = %b", rslt);

   rslt = excess3 (8'b0000_1111);
      $display ("binary = 0000_1111, excess3 = %b", rslt);

   rslt = excess3 (8'b1111_1111);
      $display ("binary = 1111_1111, excess3 = %b", rslt);

   rslt = excess3 (8'b0000_0001);
      $display ("binary = 0000_0001, excess3 = %b", rslt);

   rslt = excess3 (8'b0000_0010);
      $display ("binary = 0000_0010, excess3 = %b", rslt);

   rslt = excess3 (8'b0000_0011);
      $display ("binary = 0000_0011, excess3 = %b", rslt);

   rslt = excess3 (8'b0000_0100);
      $display ("binary = 0000_0100, excess3 = %b", rslt);
end

function [7:0] excess3;
   input [7:0] a;
   reg [7:0] rslt;
   begin
      rslt = a + 8'b0000_0011;
      excess3 = rslt;
   end
endfunction

endmodule
```

Figure 1.150 Function design module for the conversion from the binary code to the excess-3 code.

```
binary = 0000_0011, excess3 = 00000110
binary = 0011_0000, excess3 = 00110011
binary = 0000_1111, excess3 = 00010010
binary = 1111_1111, excess3 = 00000010

binary = 0000_0001, excess3 = 00000100
binary = 0000_0010, excess3 = 00000101
binary = 0000_0011, excess3 = 00000110
binary = 0000_0100, excess3 = 00000111
```

Figure 1.151 Outputs for the conversion from the binary code to the excess-3 code.

1.10 Problems

1.1 Given the equation shown below, obtain the minimized equation for z_1 in a product-of-sums notation and implement the equation using NAND gate built-in primitives. Obtain the design module, the test bench module, and the outputs. Output z_1 is asserted high.

$$z_1(x_1, x_2, x_3, x_4) = \Sigma_{\mathrm{m}}(1, 4, 7, 9, 11, 13) + \Sigma_{\mathrm{d}}(5, 14, 15)$$

1.2 Obtain the design module using built-in primitives for the equations shown below. Obtain the test bench and outputs.

$$z_1 = (x_1 \oplus x_2)x_3' \qquad z_2 = (x_1 \oplus x_2)' \oplus x_3$$

1.3 Use AND gate and OR gate built-in primitives to implement a circuit in a sum-of-products form that will generate an output z_1 if an input is greater than or equal to 2 and less than 5; and also greater than or equal to 12 and less than 15. Then obtain the design module, test bench module, and outputs.

1.4 Obtain the equation for a logic circuit that will generate a logic 1 on output z_1 if a 4-bit unsigned binary number $N = x_1x_2x_3x_4$ satisfies the following criteria, where x_4 is the low-order bit

$$2 < N \leq 6 \text{ or } 11 \leq N < 14$$

Use NOR user-defined primitives. Obtain the design module, the test bench module, and outputs.

1.5 Obtain the minimal Boolean expression for a logic circuit that generates an output z_1 whenever a 4-bit unsigned binary number N meets the following requirements:

N is an odd number or N is evenly divisible by four.

The format for N is: $N = x_1\ x_2\ x_3\ x_4$, where x_4 is the low-order bit. Then obtain the design module using user-defined primitives, the test bench module, and the outputs.

1.6 Design a modulo-8 counter using the D flip-flop that was designed in the *edge-sensitive user-defined primitives* section of this chapter. Use additional logic gate UDPs as necessary. Obtain the design module, the test bench module, and the outputs.

1.7 Design a comparator using the continuous assignment statement that compares two 2-bit binary operands x_1x_2 and x_3x_4 and generates a high output for z_1 whenever $x_1x_2 \leq x_3x_4$. Design the comparator as a product of sums using NOR logic. Obtain the design module, the test bench module, and outputs.

1.8 Use the continuous assignment statement to execute the six reduction operators. Use all combinations of a 4-bit operand *a[3:0]* for all reduction operators. Obtain the design module, the test bench module, and outputs.

1.9 Implement the following equation using the conditional operator:

$$z_1 = (x_1 < x_2)\ ?\ x_3 : x_4$$

If x_1 is less than x_2, then output z_1 will be assigned the value of x_3, otherwise z_1 will be assigned the value of x_4. Obtain the design module, the test bench module, and outputs.

1.10 Design a 4:1 multiplexer using the conditional operator. The multiplexer inputs are defined as vectors. The select inputs are: *select[1:0],* the data inputs are: *in [3:0]*, and the output is *out*. Obtain the design module, the test bench module, and outputs.

1.11 Use dataflow modeling to illustrate the four relational operators: greater than (>), less than (<), greater than or equal (>=), and less than or equal (<=). Obtain the design module, the test bench module, and outputs.

1.12 Use dataflow modeling to illustrate the three logical operators: the binary logical AND operator (&&), the binary logical OR (| |), and the unary logical negation operator (!). Obtain the design module, the test bench module, and outputs.

1.13 Design a five-function logic unit to show the operation of the five bit-wise operators: AND (&), OR (|), negation (~), exclusive-OR (^), and exclusive-NOR (^ ~ or ~ ^). There will be three 4-bit operands: *a*, *b*, and *c* and one *result* **reg** variable. Generate the test bench and obtain the outputs for the following operations:

AND operation	$= (a \,\&\, c) \,\&\, b$
OR operation	$= (a \mid b) \mid c$
negation operation	$= \sim((a \,\&\, b) \mid c)$
exclusive-OR operation	$= (b \wedge a) \wedge c$
exclusive-NOR operation	$= (a \wedge\sim b) \wedge\sim c$

1.14 Design a binary-to-excess-3 code converter using user-defined primitives of the following types: *udp_and2*, *udp_and3*, *udp_or3*, and *udp_or4*. The binary code is labelled *a*, *b*, *c*, *d*, and the excess-3 code is labelled *w*, *x*, *y*, *z*, where *d* and *z* are the low-order bits of their respective codes. Binary codes above 1100 produce an excess-3 code of 0000. Obtain the design module, the test bench module, and outputs.

1.15 Perform left and right shift operations on the following two unsigned 8-bit operands: *a_reg [7:0]* and *b_reg [7:0]*. Execute a shift left of four bits on *a_reg [7:0]* and a shift right of three bits on *b_reg [7:0]*. Obtain the design module, the test bench module, and outputs.

1.16 Use the conditional statements to design a circuit for the following expression: $z_1 = (x_1 \,\&\, x_4) + (x_2 \,\&\, x_3) + (x_2 \sim\wedge x_4)$. Obtain the design module, the test bench module, and outputs.

1.17 Design a modulo-10 counter using conditional statements. The counter counts in the following sequence: 0000 . . . 1001, 0000. There are two inputs: *clk* and *rst_n*, and one output *count*. The *reset* is active low. Obtain the design module, the test bench module, and outputs.

1.18 Design a module to execute the four logical operations of AND, OR, Exclusive-OR, and Exclusive-NOR using the **case** statement. Obtain the test bench providing four 4-bit vectors for each logical operation and obtain the outputs.

1.19 Design a module to execute the four shift operations of *shift left logical* (SLL), *shift left algebraic* (SLA), *shift right logical* (SRL), and *shift right algebraic* (SRA) using the **case** statement. The operands to be shifted are 8-bit operands. Obtain the test bench providing four shift amounts for each shift operation and obtain the outputs.

1.20 Use the **repeat** loop to increment an integer *count* by three. The process is repeated 20 times. The integer *count* is initialized to zero. Display the output count.

1.21 Design a module using the **case** statement to **rotate** an 8-bit operand left and right. Then obtain the test bench module and the outputs.

1.22 Given the Karnaugh map shown below, design a structural module to implement the sum-of-products expression for output z_1. Instantiate AND gates and OR gates that were designed using dataflow modeling. Obtain the test bench module and the outputs.

x_1x_2 \ x_3x_4	0 0	0 1	1 1	1 0
0 0	0 (0)	0 (1)	0 (3)	0 (2)
0 1	1 (4)	0 (5)	0 (7)	1 (6)
1 1	0 (12)	0 (13)	1 (15)	0 (14)
1 0	1 (8)	1 (9)	0 (11)	0 (10)

z_1

1.23 Use structural modeling to design a logic circuit to generate an output z_1 whenever a 4-bit variable — x_1, x_2, x_3, x_4 — has three or more 1s. Implement the module using AND gates and OR gates that were designed using dataflow modeling. Obtain the test bench and the outputs.

1.24 Design a structural module that will generate an output z_1 if a 4-bit binary number *x[3:0]* has a value that is less than or equal to four or greater than ten. Implement the module using AND gates and OR gates that were designed using dataflow modeling. Obtain the test bench and the outputs.

1.25 Given the Karnaugh map shown below, design two structural modules to obtain the equation for z_1, first as a sum-of-products then as a product-of-sums. Use logic gates that were designed using dataflow modeling. Obtain the test benches and the outputs.

x_1x_2 \ x_3x_4	0 0	0 1	1 1	1 0
0 0	1 (0)	0 (1)	0 (3)	1 (2)
0 1	0 (4)	1 (5)	0 (7)	0 (6)
1 1	0 (12)	1 (13)	1 (15)	1 (14)
1 0	0 (8)	0 (9)	0 (11)	0 (10)

z_1

2

Combinational Logic Design Using Verilog HDL

This chapter provides techniques for designing combinational logic, including logic equations, multiplexers, decoders, encoders, comparators, programmable logic devices, and additional logic devices. Different design methodologies and modeling constructs of the Verilog hardware description language (HDL) will be utilized. Also included in this chapter are the following topics, which should be familiar to the reader and are briefly reviewed: number systems, number representations, Boolean algebra, and minimization techniques.

Combinational logic refers to logic circuits whose present output values depend only upon the present input values. A combinational circuit is a special case of a sequential circuit in which there is no storage capability. The word combinational is used interchangeably with combinatorial.

A combinational logic circuit is comprised of combinational logic elements (or logic primitives), which have one or more inputs and at least one output. The input and output variables are characterized by discrete states such that, at any instant of time, the state of each output is completely determined by the states of the present inputs. A combinational circuit refers to a logic circuit which performs the same fixed mapping of inputs into outputs, regardless of the past input history and may be considered as a one-state sequential machine.

All designs will be completely designed and carried through to completion — nothing is left unfinished or partially designed. The Verilog HDL design examples include the design module, the test bench module, and the outputs obtained from the test bench.

2.1 Number Systems

Numerical data, or operands, are expressed in various positional number systems for each radix. This section will discuss binary, octal, decimal, and hexadecimal positional number systems. A *positional number system* is characterized by a radix (or base) r, which is an integer greater than or equal to 2, and by a set of r digits, which are numbered from 0 to $r - 1$. For example, for radix 8, the digits range from 0 to 7.

In a positional number system, a number is encoded as a vector of n digits in which each digit is weighted according to its position in the vector. An n-bit integer A is represented in a positional number system as follows:

$$A = (a_{n-1} a_{n-2} a_{n-3} \; \ldots \; a_1 a_0) \tag{2.1}$$

where $0 \le a_i \le r - 1$. The high-order and low-order digits are a_{n-1} and a_0, respectively.

2.1.1 Binary Number System

The radix is 2 in the *binary number system*; therefore, only two digits are used: 0 and 1. The low-value digit is 0 and the high-value digit is $(r - 1) = 1$. The binary number system is the most conventional and easily implemented system for internal use in a digital computer; therefore, most digital computers use the binary number system. There is a disadvantage when converting to and from the externally used decimal system; however, this is compensated for by the ease of implementation and the speed of execution in binary of the four basic operations: addition, subtraction, multiplication, and division. The radix point is implied within the internal structure of the computer; that is, there is no specific storage element assigned to contain the radix point.

The weight assigned to each position of a binary number is as follows:

$$2^{n-1} 2^{n-2} \ldots \; 2^3 \; 2^2 \; 2^1 \; 2^0 \bullet 2^{-1} 2^{-2} 2^{-3} \; \ldots \; 2^{-m}$$

where the integer and fraction are separated by the radix point (binary point). The decimal value of the binary number 1101.101_2 is obtained as shown below, where $r = 2$ and $a_i \in \{0,1\}$ for $-m \le i \le n - 1$. Therefore,

$$\begin{array}{ccccccccc} 2^3 & 2^2 & 2^1 & 2^0 & \bullet & 2^{-1} & 2^{-2} & 2^{-3} & \\ 1 & 1 & 0 & 1 & . & 1 & 0 & 1_2 & = (1 \times 2^3) + (1 \times 2^2) + (0 \times 2^1) + (1 \times 2^0) + \\ & & & & & & & & (1 \times 2^{-1}) + (0 \times 2^{-2}) + (1 \times 2^{-3}) \\ & & & & & & & & \\ & & & & & & & & = 13.625_{10} \end{array}$$

2.1.2 Octal Number System

The radix is 8 in the *octal number system*; therefore, eight digits are used, 0 through 7. The low-value digit is 0 and the high-value digit is $(r-1)=7$. The weight assigned to each position of an octal number is as follows:

$$8^{n-1}8^{n-2} \ldots\ 8^3\ 8^2\ 8^1\ 8^0 \bullet 8^{-1}8^{-2}8^{-3} \ldots\ 8^{-m}$$

where the integer and fraction are separated by the radix point (octal point). The decimal value of the octal number 217.6_8 is obtained as shown below, where $r=8$ and $a_i \in \{0,1,2,3,4,5,6,7\}$ for $-m \leq i \leq n-1$. Therefore,

$$\begin{array}{ccccccl} 8^2 & 8^1 & 8^0 & \bullet & 8^{-1} & & \\ 2 & 1 & 7 & . & 6_8 & = & (2 \times 8^2) + (1 \times 8^1) + (7 \times 8^0) + (6 \times 8^{-1}) \\ & & & & & = & 143.75_{10} \end{array}$$

When a count of 1 is added to 7_8, the sum is zero and a carry of 1 is added to the next higher-order column on the left.

2.1.3 Decimal Number System

The radix is 10 in the *decimal number system*; therefore, ten digits are used, 0 through 9. The low-value digit is 0 and the high-value digit is $(r-1)=9$. The weight assigned to each position of a decimal number is as follows:

$$10^{n-1}\,10^{n-2} \ldots\ 10^3\ 10^2\ 10^1\ 10^0 \bullet 10^{-1}\,10^{-2}\,10^{-3} \ldots\ 10^{-m}$$

where the integer and fraction are separated by the radix point (decimal point). The value of 7537_{10} is immediately apparent; however, the value is also obtained as shown below, where $r=10$ and $a_i \in \{0,1,2,3,4,5,6,7,8,9\}$ for $-m \leq i \leq n-1$. That is,

$$\begin{array}{lllll} 10^3 & 10^2 & 10^1 & 10^0 & \\ 6 & 3 & 5 & 7_{10} & = (6 \times 10^3) + (3 \times 10^2) + (5 \times 10^1) + (7 \times 10^0) \end{array}$$

When a count of 1 is added to decimal 9, the sum is zero and a carry of 1 is added to the next higher-order column on the left.

Binary-coded decimal Each decimal digit can be encoded into a corresponding binary number. The highest-valued decimal digit is 9, which requires four bits in the binary representation. Therefore, four binary digits are required to represent each decimal digit. This is shown below, which lists four decimal digits and indicates the corresponding binary-coded decimal (BCD) digits.

Decimal	BCD
6, 9, 12, 124	0110, 1001, 0001 0010, 0001 0010 0100

2.1.4 Hexadecimal Number System

The radix is 16 in the *hexadecimal number system*; therefore, 16 digits are used, 0 through 9 and A through F, where A, B, C, D, E, and F correspond to decimal 10, 11, 12, 13, 14, and 15, respectively. The low-value digit is 0 and the high-value digit is $(r - 1) = 15$ (F). The weight assigned to each position of a hexadecimal number is as follows:

$$16^{n-1}\,16^{n-2}\ \ldots\ 16^3\,16^2\,16^1\,16^0 \bullet 16^{-1}\,16^{-2}\,16^{-3}\ \ldots\ 16^{-m}$$

where the integer and fraction are separated by the radix point (hexadecimal point). The decimal value of the hexadecimal number $6A8C.D416_{16}$ is obtained as shown below, where $r = 16$ and $a_i \in \{0,1,2,3,4,5,6,7,8,9,A,B,C,D,E,F\}$ for $-m \le i \le n - 1$. Therefore,

$$\begin{array}{llllllllll} 16^3 & 16^2 & 16^1 & 16^0 & \bullet & 16^{-1} & 16^{-2} & 16^{-3} & 16^{-4} & \\ 6 & A & 8 & C & . & D & 4 & 1 & 6 & = (6 \times 16^3) + (10 \times 16^2) + (8 \times 16^1) \\ &&&&&&&&& + (12 \times 16^0) + (13 \times 16^{-1}) + (4 \times 16^{-2}) \\ &&&&&&&&& + (1 \times 16^{-3}) + (6 \times 16^{-4}) \\ &&&&&&&&& = 27{,}276.82846_{10} \end{array}$$

When a count of 1 is added to hexadecimal F, the sum is zero and a carry of 1 is added to the next higher-order column on the left.

2.2 Boolean Algebra

In 1854, George Boole introduced a systematic treatment of the logic operations AND, OR, and NOT, which is now called Boolean algebra. This section describes the axioms and theorems that characterize "Boolean algebra". The symbols (or operators) used for the algebra and the corresponding function definitions are listed in Table 2.1. The table also includes the exclusive-OR function, which is characterized by the three operations of AND, OR, and NOT.

Table 2.1 Boolean Operators for Variables x_1 and x_2

Operator	Function	Definition
•	AND	$x_1 \bullet x_2$ (Also $x_1 x_2$)
+	OR	$x_1 + x_2$
'	NOT (negation)	x_1'
⊕	Exclusive-OR	$(x_1 x_2') + (x_1' x_2)$

The AND operator, which corresponds to the Boolean product, is also indicated by the symbol "$\wedge$" ($x_1 \wedge x_2$) or by no symbol if the operation is unambiguous. Thus, $x_1 x_2$, $x_1 \bullet x_2$, and $x_1 \wedge x_2$ are all read as "x_1 AND x_2." The OR operator, which corresponds to the Boolean sum, is also specified by the symbol "$\vee$." Thus, $x_1 + x_2$ and $x_1 \vee x_2$ are both read as "x_1 OR x_2." The symbol for the complement (or negation) operation is usually specified by the prime " ' " symbol immediately following the variable (x_1'), by a bar over the variable ($\overline{x_1}$), or by the symbol "$\neg$" ($\neg x_1$).

Boolean algebra is a deductive mathematical system which can be defined by a set of variables, a set of operators, and a set of axioms (or postulates). An *axiom* is a statement that is universally accepted as true; that is, the statement needs no proof, because its truth is obvious. The axioms of Boolean algebra form the basis from which the theorems and other properties can be derived.

Most axioms and theorems are characterized by two laws. Each law is the dual of the other. The principle of duality specifies that the *dual* of an algebraic expression can be obtained by interchanging the binary operators • and + and by interchanging the identity elements 0 and 1.

Boolean algebra is an algebraic structure consisting of a set of elements B with two binary operators • and + and a unary operator ', such that the following axioms are true, where the notation $x_1 \in X$ is read as "x_1 is an element of the set X":

2.2.1 Axioms

This section presents the seven axioms of Boolean algebra.

Axiom 1: Boolean set definition The set B contains at least two elements x_1 and x_2, where $x_1 \neq x_2$.

Axiom 2: Closure laws For every $x_1, x_2 \in B$,

(a) $x_1 + x_2 \in B$
(b) $x_1 \bullet x_2 \in B$

Axiom 3: Identity laws There exist two unique *identity elements* 0 and 1, where 0 is an identity element with respect to the Boolean sum and 1 is an identity element with respect to the Boolean product. Thus, for every $x_1 \in B$,

(a) $x_1 + 0 = 0 + x_1 = x_1$
(b) $x_1 \bullet 1 = 1 \bullet x_1 = x_1$

Axiom 4: Commutative laws The commutative laws specify that the order in which the variables appear in a Boolean expression is irrelevant — the result is the same. Thus, for every $x_1, x_2 \in B$,

(a) $x_1 + x_2 = x_2 + x_1$
(b) $x_1 \bullet x_2 = x_2 \bullet x_1$

Axiom 5: Associative laws The associative laws state that three or more variables can be combined in an expression using Boolean multiplication or addition and that the order of the variables can be altered without changing the result. Thus, for every $x_1, x_2, x_3 \in B$,

(a) $(x_1 + x_2) + x_3 = x_1 + (x_2 + x_3)$
(b) $(x_1 \bullet x_2) \bullet x_3 = x_1 \bullet (x_2 \bullet x_3)$

Axiom 6: Distributive laws The distributive laws for Boolean algebra are similar, in many respects, to those for college algebra. The interpretation, however, is different and is a function of the Boolean product and the Boolean sum. This is a very useful axiom in minimizing Boolean functions. For every $x_1, x_2, x_3 \in B$,

(a) The operator + is distributive over the operator • such that,
$x_1 + (x_2 \bullet x_3) = (x_1 + x_2) \bullet (x_1 + x_3)$

(b) The operator • is distributive over the operator + such that,
$x_1 \bullet (x_2 + x_3) = (x_1 \bullet x_2) + (x_1 \bullet x_3)$

Axiom 7: Complementation laws For every $x_1 \in B$, there exists an element x_1' (called the complement of x_1), where $x_1' \in B$, such that,

(a) $x_1 + x_1' = 1$
(b) $x_1 \bullet x_1' = 0$

2.2.2 Theorems

This section presents the seven theorems of Boolean algebra, which are derived from the axioms and are listed in pairs, where each relation in the pair is the dual of the other.

Theorem 1: 0 and 1 associated with a variable Every variable in Boolean algebra can be characterized by the identity elements 0 and 1. Thus, for every $x_1 \in B$,

(a) $x_1 + 1 = 1$
(b) $x_1 \bullet 0 = 0$

Theorem 2: 0 and 1 complement The 2-valued Boolean algebra has two distinct identity elements 0 and 1, where $0 \neq 1$. The operations using 0 and 1 are as follows:

$0 + 0 = 0$ $\quad$ $0 + 1 = 1$
$1 \bullet 1 = 1$ $\quad$ $1 \bullet 0 = 0$

A corollary to Theorem 2 specifies that element 1 satisfies the requirements of the complement of element 0, and vice versa. Thus, each identity element is the complement of the other.

(a) $0' = 1$
(b) $1' = 0$

Theorem 3: Idempotent laws Idempotency relates to a nonzero mathematical quantity which, when applied to itself for a binary operation, remains unchanged. Thus, if $x_1 = 0$, then $x_1 + x_1 = 0 + 0 = 0$ and if $x_1 = 1$, then $x_1 + x_1 = 1 + 1 = 1$. Therefore, one of the elements is redundant and can be discarded. The dual is true for the operator •. The idempotent laws eliminate redundant variables in a Boolean expression and can be extended to any number of identical variables. This law is also referred to as the *law of tautology*, which precludes the needless repetition of the variable. For every $x_1 \in B$,

(a) $x_1 + x_1 = x_1$
(b) $x_1 \bullet x_1 = x_1$

Theorem 4: Involution law The involution law states that the complement of a complemented variable is equal to the variable. There is no dual for the involution law. The law is also called the "law of double complementation". Thus, for every $x_1 \in B$,

$$x_1'' = x_1$$

Theorem 5: Absorption law 1 This version of the absorption law states that some 2-variable Boolean expressions can be reduced to a single variable without altering the result. Thus, for every $x_1, x_2 \in B$,

(a) $x_1 + (x_1 \bullet x_2) = x_1$
(b) $x_1 \bullet (x_1 + x_2) = x_1$

Theorem 6: Absorption law 2 This version of the absorption law is used to eliminate redundant variables from certain Boolean expressions. Absorption law 2 eliminates a variable or its complement and is a very useful law for minimizing Boolean expressions.

(a) $x_1 + (x_1' \bullet x_2) = x_1 + x_2$
(b) $x_1 \bullet (x_1' + x_2) = x_1 \bullet x_2$

Theorem 7: DeMorgan's laws DeMorgan's laws are also useful in minimizing Boolean functions. DeMorgan's laws convert the complement of a sum term or a product term into a corresponding product or sum term, respectively. For every $x_1, x_2 \in B$,

(a) $(x_1 + x_2)' = x_1' \bullet x_2'$
(b) $(x_1 \bullet x_2)' = x_1' + x_2'$

Parts (a) and (b) of DeMorgan's laws represent expressions for NOR and NAND gates, respectively. DeMorgan's laws can be generalized for any number of variables, such that,

(a) $(x_1 + x_2 + \ldots + x_n)' = x_1' \bullet x_2' \bullet \ldots \bullet x_n'$
(b) $(x_1 \bullet x_2 \bullet \ldots \bullet x_n)' = x_1' + x_2' + \ldots + x_n'$

When applying DeMorgan's laws to an expression, the operator • takes precedence over the operator +. For example, use DeMorgan's law to complement the Boolean expression $x_1 + x_2x_3$.

$$\begin{aligned}(x_1 + x_2x_3)' &= [x_1 + (x_2x_3)]' \\ &= x_1'(x_2' + x_3')\end{aligned}$$

Note that: $(x_1 + x_2x_3)' \neq x_1' \bullet x_2' + x_3'$.

2.2.3 Other Terms for Boolean Algebra

This section defines the following Boolean terms: minterm, maxterm, product term, sum term, sum of minterms, sum of products, product of maxterms, and product of sums.

Minterm A minterm is the Boolean product of *n* variables and contains all *n* variables of the function exactly once, either true or complemented. For example, for the function $z_1(x_1, x_2, x_3)$, $x_1x_2'x_3$ is a minterm.

Maxterm A maxterm is a Boolean sum of *n* variables and contains all *n* variables of the function exactly once, either true or complemented. For example, for the function $z_1(x_1, x_2, x_3)$, $(x_1 + x_2' + x_3)$ is a maxterm.

Product term A product term is the Boolean product of variables containing a subset of the possible variables or their complements. For example, for the function $z_1(x_1, x_2, x_3)$, $x_1'x_3$ is a product term, because it does not contain all the variables.

Sum term A sum term is the Boolean sum of variables containing a subset of the possible variables or their complements. For example, for the function $z_1(x_1, x_2, x_3)$, $(x_1' + x_3)$ is a sum term, because it does not contain all the variables.

Sum of minterms A sum of minterms is an expression in which each term contains all the variables, either true or complemented. For example,

$$z_1(x_1, x_2, x_3) = x_1'x_2x_3 + x_1x_2'x_3' + x_1x_2x_3$$

is a Boolean expression in a sum-of-minterms form. This particular form is also referred to as a *minterm expansion*, a *standard sum of products*, a *canonical sum of products*, or a *disjunctive normal form*. Since each term is a minterm, the expression for z_1 can be written in a more compact sum-of-minterms form as $z_1(x_1, x_2, x_3) = \Sigma_m(3,4,7)$, where each term is converted to its minterm value. For example, the first term in the expression is $x_1'x_2x_3$, which corresponds to binary 011, representing minterm 3.

Sum of products A sum of products is an expression in which at least one term does not contain all the variables; that is, at least one term is a proper subset of the possible variables or their complements. For example,

$$z_1(x_1, x_2, x_3) = x_1'x_2x_3 + x_2'x_3' + x_1x_2x_3$$

is a sum of products for the function z_1, because the second term does not contain the variable x_1.

Product of maxterms A product of maxterms is an expression in which each term contains all the variables, either true or complemented. For example,

$$z_1(x_1, x_2, x_3) = (x_1 + x_2 + x_3)(x_1 + x_2' + x_3')(x_1' + x_2 + x_3)$$

is a Boolean expression in a product-of-maxterms form. This particular form is also referred to as a *maxterm expansion*, a *standard product of sums*, a *canonical product of sums*, or a *conjunctive normal form*. Since each term is a maxterm, the expression for z_1 can be written in a more compact product-of-maxterms form as $z_1(x_1, x_2, x_3) = \Pi_M(0,3,4)$, where each term is converted to its maxterm value.

Product of sums A product of sums is an expression in which at least one term does not contain all the variables; that is, at least one term is a proper subset of the possible variables or their complements. For example,

$$z_1(x_1, x_2, x_3) = (x_1' + x_2 + x_3)(x_2' + x_3')(x_1 + x_2 + x_3)$$

is a product of sums for the function z_1, because the second term does not contain the variable x_1.

2.3 Logic Equations

This section synthesizes a variety of logic equations using Verilog HDL. The equations can be in a sum-of-products or a product-of-sums and range from simple to complex.

Example 2.1 Equation 2.2 will be minimized as a sum-of-products as shown in Equation 2.3, then implemented using built-in primitives for the design module. The design module is shown in Figure 2.1, the test bench is shown in Figure 2.2, and the outputs are shown in Figure 2.3.

$$z_1 = x_1 x_3' x_4 + [(x_1 + x_2)' + x_3]' + (x_2 \oplus x_4')' \tag{2.2}$$

$$\begin{aligned} z_1 &= x_1 x_3' x_4 + [(x_1 + x_2)' + x_3]' + (x_2 \oplus x_4')' \\ &= x_1 x_3' x_4 + (x_1' x_2' + x_3)' + x_2' x_4' + x_2 x_4 \\ &= x_1 x_3' x_4 + (x_1 + x_2) x_3' + x_2' x_4' + x_2 x_4 \\ &= x_1 x_3' x_4 + x_1 x_3' + x_2 x_3' + x_2' x_4' + x_2 x_4 \end{aligned} \tag{2.3}$$

```
//sop equation using built-in-primitives

module sop_eqtn_bip  (x1, x2, x3, x4, z1);

//define inputs and output
input x1, x2, x3, x4;
output z1;

//design the equation
and   inst1 (net1, x1, ~x3, x4),
      inst2 (net2, x1, ~x3),
      inst3 (net3, x2, ~x3),
      inst4 (net4, ~x2, ~x4),
      inst5 (net5, x2, x4);

or    inst6 (z1, net1, net2, net3, net4, net5);

endmodule
```

Figure 2.1 Design module for the sum-of-products equation for Example 2.1.

```
//test bench for sop equation using built-in-primitives
module sop_eqtn_bip_tb;

//inputs are reg for test bench
//outputs are wire for test bench
reg x1, x2, x3, x4;
wire z1;

//apply input vectors and display variables
initial
begin: apply_stimulus
   reg [4:0] invect;
   for (invect = 0; invect < 16; invect = invect + 1)
      begin
         {x1, x2, x3, x4} = invect [4:0];
         #10 $display ("x1 x2 x3 x4 = %b, z1 = %b",
                              {x1, x2, x3, x4}, z1);
      end
end

//instantiate the module into the test bench
sop_eqtn_bip inst1 (x1, x2, x3, x4, z1);
endmodule
```

Figure 2.2 Test bench module for Figure 2.1.

```
x1 x2 x3 x4 = 0000, z1 = 1
x1 x2 x3 x4 = 0001, z1 = 0
x1 x2 x3 x4 = 0010, z1 = 1
x1 x2 x3 x4 = 0011, z1 = 0

x1 x2 x3 x4 = 0100, z1 = 1
x1 x2 x3 x4 = 0101, z1 = 1
x1 x2 x3 x4 = 0110, z1 = 0
x1 x2 x3 x4 = 0111, z1 = 1

x1 x2 x3 x4 = 1000, z1 = 1
x1 x2 x3 x4 = 1001, z1 = 1
x1 x2 x3 x4 = 1010, z1 = 1
x1 x2 x3 x4 = 1011, z1 = 0

x1 x2 x3 x4 = 1100, z1 = 1
x1 x2 x3 x4 = 1101, z1 = 1
x1 x2 x3 x4 = 1110, z1 = 0
x1 x2 x3 x4 = 1111, z1 = 1
```

Figure 2.3 Outputs for the sum-of-products module of Figure 2.1.

Example 2.2 This example repeats Example 2.1 using built-in-primitives, but converts Equation 2.3 to a product-of-sums expression. Equation 2.3 is plotted on the Karnaugh map of Figure 2.4 and the product-of-sums equation is shown in Equation 2.4. The design module is shown in Figure 2.5, the test bench module is shown in Figure 2.6, and the outputs are shown in Figure 2.7.

x_1x_2 \ x_3x_4	0 0	0 1	1 1	1 0
0 0	1 (0)	0 (1)	0 (3)	1 (2)
0 1	1 (4)	1 (5)	1 (7)	0 (6)
1 1	1 (12)	1 (13)	1 (15)	0 (14)
1 0	1 (8)	1 (9)	0 (11)	1 (10)

z_1

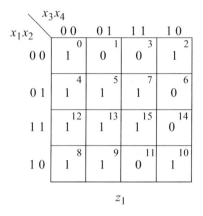

Figure 2.4 Karnaugh map for Example 2.2.

$$z_1 = (x_1 + x_2 + x_4')\,(x_2' + x_3' + x_4)\,(x_2 + x_3' + x_4') \tag{2.4}$$

```
//product-of-sums equation using bip

module pos_eqtn_bip (x1, x2, x3, x4, z1);

//define inputs and output
input x1, x2, x3, x4;
output z1;

//design the equation
or      inst1 (net1, x1, x2, ~x4),
        inst2 (net2, ~x2, ~x3, x4),
        inst3 (net3, x2, ~x3, ~x4);

and     inst4 (z1, net1, net2, net3);

endmodule
```

Figure 2.5 Design module for Example 2.2 using product-of-sums built-in-primitives.

```
//test bench for pos equation using built-in-primitives
module pos_eqtn_bip_tb;

//inputs are reg for test bench
//outputs are wire for test bench
reg x1, x2, x3, x4;
wire z1;

//apply input vectors and display variables
initial
begin: apply_stimulus
   reg [4:0] invect;
   for (invect = 0; invect < 16; invect = invect + 1)
      begin
         {x1, x2, x3, x4} = invect [4:0];
         #10 $display ("x1 x2 x3 x4 = %b, z1 = %b",
                        {x1, x2, x3, x4}, z1);
      end
end

//instantiate the module into the test bench
pos_eqtn_bip inst1 (x1, x2, x3, x4, z1);
endmodule
```

Figure 2.6 Test bench module for Example 2.2.

```
x1 x2 x3 x4 = 0000, z1 = 1
x1 x2 x3 x4 = 0001, z1 = 0
x1 x2 x3 x4 = 0010, z1 = 1
x1 x2 x3 x4 = 0011, z1 = 0

x1 x2 x3 x4 = 0100, z1 = 1
x1 x2 x3 x4 = 0101, z1 = 1
x1 x2 x3 x4 = 0110, z1 = 0
x1 x2 x3 x4 = 0111, z1 = 1

x1 x2 x3 x4 = 1000, z1 = 1
x1 x2 x3 x4 = 1001, z1 = 1
x1 x2 x3 x4 = 1010, z1 = 1
x1 x2 x3 x4 = 1011, z1 = 0

x1 x2 x3 x4 = 1100, z1 = 1
x1 x2 x3 x4 = 1101, z1 = 1
x1 x2 x3 x4 = 1110, z1 = 0
x1 x2 x3 x4 = 1111, z1 = 1
```

Figure 2.7 Outputs for the product-of-sums equation of Example 2.2.

Example 2.3 The minimized expression for z_1 will be obtained in both a sum-of-products and a product-of-sums format using the Karnaugh map shown below. The design modules will use logic gates that were designed using dataflow modeling. The sum-of-product design is shown below. The design module is shown in Figure 2.8, the test bench module is shown in Figure 2.9, and the outputs are shown in Figure 2.10.

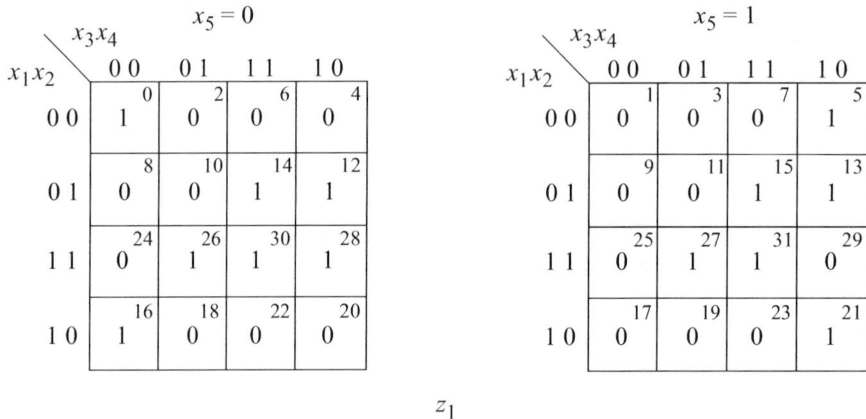

$$z_1 = x_2x_3x_5' + x_1'x_2x_3 + x_1x_2x_4 + x_2'x_3'x_4'x_5' + x_2'x_3x_4'x_5$$

$$z_1 = (x_2 + x_4')(x_3 + x_4 + x_5')(x_1 + x_2' + x_3)(x_2' + x_3 + x_4)(x_2 + x_3' + x_5)(x_1' + x_2' + x_3' + x_4 + x_5')$$

```
//sum-of-products for 5 variables using dataflow modeling
module sop_5var_df (x1, x2, x3, x4, x5, z1);

input x1, x2, x3, x4, x5;     //define inputs and output
output z1;

wire net1, net2, net3, net4, net5;  //define internal nets

//instantiate the logic for output z1
and3_df  inst1 (x2, x3, ~x5, net1);
and3_df  inst2 (~x1, x2, x3, net2);
and3_df  inst3 (x1, x2, x4, net3);
and4_df  inst4 (~x2, ~x3, ~x4, ~x5, net4);
and4_df  inst5 (~x2, x3, ~x4, x5, net5);
or5_df   inst6 (net1, net2, net3, net4, net5, z1);

endmodule
```

Figure 2.8 Design module for Example 2.3 for the sum-of-products form.

```
//test bench for sop_5var_df

module sop_5var_df_tb;

//inputs are reg for test bench
//outputs are wire for test bench
reg x1, x2, x3, x4, x5;
wire z1;

//apply input variables and display variables
initial
begin: apply_stimulus
   reg [5:0] invect;
   for (invect = 0; invect < 32; invect = invect + 1)
      begin
         {x1, x2, x3, x4, x5} = invect [5:0];
         #10 $display ("{x1 x2 x3 x4 x5} = %b, z1 = %b",
                           {x1, x2, x3, x4, x5}, z1);
      end
end

//instantiate the module into the test bench
sop_5var_df inst1 (x1, x2, x3, x4, x5, z1);

endmodule
```

Figure 2.9 Test bench module for Example 2.3 for the sum-of-products.

```
{x1 x2 x3 x4 x5} = 00000, z1 = 1
{x1 x2 x3 x4 x5} = 00001, z1 = 0
{x1 x2 x3 x4 x5} = 00010, z1 = 0
{x1 x2 x3 x4 x5} = 00011, z1 = 0
{x1 x2 x3 x4 x5} = 00100, z1 = 0
{x1 x2 x3 x4 x5} = 00101, z1 = 1
{x1 x2 x3 x4 x5} = 00110, z1 = 0
{x1 x2 x3 x4 x5} = 00111, z1 = 0

{x1 x2 x3 x4 x5} = 01000, z1 = 0
{x1 x2 x3 x4 x5} = 01001, z1 = 0
{x1 x2 x3 x4 x5} = 01010, z1 = 0
{x1 x2 x3 x4 x5} = 01011, z1 = 0
{x1 x2 x3 x4 x5} = 01100, z1 = 1
{x1 x2 x3 x4 x5} = 01101, z1 = 1
{x1 x2 x3 x4 x5} = 01110, z1 = 1
{x1 x2 x3 x4 x5} = 01111, z1 = 1     //continued on next page
```

Figure 2.10 Outputs for Example 2.3 for the sum-of-products.

```
{x1 x2 x3 x4 x5} = 10000, z1 = 1
{x1 x2 x3 x4 x5} = 10001, z1 = 0
{x1 x2 x3 x4 x5} = 10010, z1 = 0
{x1 x2 x3 x4 x5} = 10011, z1 = 0
{x1 x2 x3 x4 x5} = 10100, z1 = 0
{x1 x2 x3 x4 x5} = 10101, z1 = 1
{x1 x2 x3 x4 x5} = 10110, z1 = 0
{x1 x2 x3 x4 x5} = 10111, z1 = 0

{x1 x2 x3 x4 x5} = 11000, z1 = 0
{x1 x2 x3 x4 x5} = 11001, z1 = 0
{x1 x2 x3 x4 x5} = 11010, z1 = 1
{x1 x2 x3 x4 x5} = 11011, z1 = 1
{x1 x2 x3 x4 x5} = 11100, z1 = 1
{x1 x2 x3 x4 x5} = 11101, z1 = 0
{x1 x2 x3 x4 x5} = 11110, z1 = 1
{x1 x2 x3 x4 x5} = 11111, z1 = 1
```

Figure 2.10 (Continued)

The product-of-sums design is shown below using the continuous assignment statement **assign**. The Karnaugh maps and the product-of-sums equation are reproduced below for convenience. The design module is shown Figure 2.11, the test bench module is shown in Figure 2.12, and the outputs are shown in Figure 2.13.

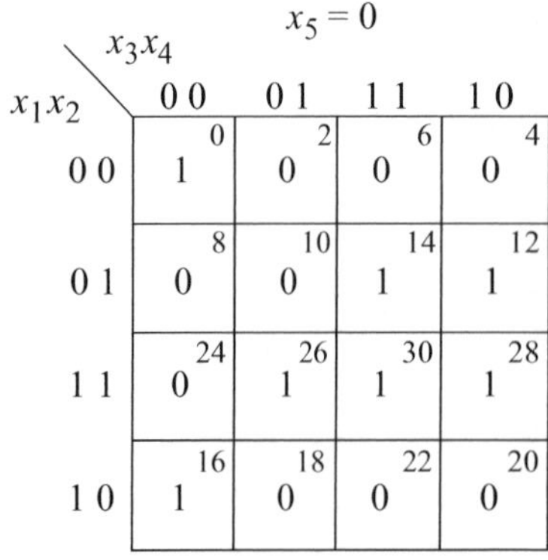

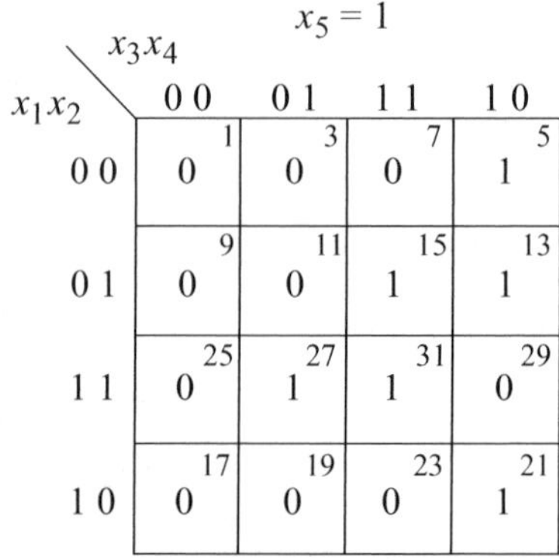

z_1

$$z_1 = (x_2 + x_4')(x_3 + x_4 + x_5')(x_1 + x_2' + x_3)(x_2' + x_3 + x_4)(x_2 + x_3' + x_5)$$
$$(x_1' + x_2' + x_3' + x_4 + x_5')$$

```
//dataflow for product-of-sums equation

module pos_5var_df (x1, x2, x3, x4, x5, z1);

//define inputs and output
input x1, x2, x3, x4, x5;
output z1;

assign z1 = (x2|~x4) & (x3|x4|~x5) & (x1|~x2|x3) & (~x2|x3|x4)
               & (x2|~x3|x5) & (~x1|~x2|~x3|x4|~x5);

endmodule
```

Figure 2.11 Design module for the product-of-sums equation of Example 2.3.

```
//test bench for product-of-sums equation

module pos_5var_df_tb;

//inputs are reg for test bench
//outputs are wire for test bench
reg x1, x2, x3, x4, x5;
wire z1;

//apply input vectors and display variables
initial
begin: apply_stimulus
   reg [5:0] invect;
   for (invect = 0; invect < 32; invect = invect + 1)
      begin
         {x1, x2, x3, x4, x5} = invect [5:0];
         #10 $display ("{x1 x2 x3 x4 x5} = %b, z1 = %b",
                              {x1, x2, x3, x4, x5}, z1);
      end
end

//instantiate the module into the test bench
pos_5var_df inst1 (x1, x2, x3, x4, x5, z1);

endmodule
```

Figure 2.12 Test bench module for the product-of-sums equation of Example 2.3.

```
{x1 x2 x3 x4 x5} = 00000, z1 = 1
{x1 x2 x3 x4 x5} = 00001, z1 = 0
{x1 x2 x3 x4 x5} = 00010, z1 = 0
{x1 x2 x3 x4 x5} = 00011, z1 = 0

{x1 x2 x3 x4 x5} = 00100, z1 = 0
{x1 x2 x3 x4 x5} = 00101, z1 = 1
{x1 x2 x3 x4 x5} = 00110, z1 = 0
{x1 x2 x3 x4 x5} = 00111, z1 = 0

{x1 x2 x3 x4 x5} = 01000, z1 = 0
{x1 x2 x3 x4 x5} = 01001, z1 = 0
{x1 x2 x3 x4 x5} = 01010, z1 = 0
{x1 x2 x3 x4 x5} = 01011, z1 = 0

{x1 x2 x3 x4 x5} = 01100, z1 = 1
{x1 x2 x3 x4 x5} = 01101, z1 = 1
{x1 x2 x3 x4 x5} = 01110, z1 = 1
{x1 x2 x3 x4 x5} = 01111, z1 = 1

{x1 x2 x3 x4 x5} = 10000, z1 = 1
{x1 x2 x3 x4 x5} = 10001, z1 = 0
{x1 x2 x3 x4 x5} = 10010, z1 = 0
{x1 x2 x3 x4 x5} = 10011, z1 = 0

{x1 x2 x3 x4 x5} = 10100, z1 = 0
{x1 x2 x3 x4 x5} = 10101, z1 = 1
{x1 x2 x3 x4 x5} = 10110, z1 = 0
{x1 x2 x3 x4 x5} = 10111, z1 = 0

{x1 x2 x3 x4 x5} = 11000, z1 = 0
{x1 x2 x3 x4 x5} = 11001, z1 = 0
{x1 x2 x3 x4 x5} = 11010, z1 = 1
{x1 x2 x3 x4 x5} = 11011, z1 = 1

{x1 x2 x3 x4 x5} = 11100, z1 = 1
{x1 x2 x3 x4 x5} = 11101, z1 = 0
{x1 x2 x3 x4 x5} = 11110, z1 = 1
{x1 x2 x3 x4 x5} = 11111, z1 = 1
```

Figure 2.13 Outputs for the product-of-sums equation of Example 2.3.

Example 2.4 The function for z_1 shown below will be plotted on a Karnaugh map using x_4 as a map-entered variable. Then the minimized expression for z_1 will be obtained in a sum-of-products notation. The behavioral module, designed using the

always statement, is shown in Figure 2.14. The test bench module is shown in Figure 2.15, and the outputs are shown in Figure 2.16.

$$z_1(x_1, x_2, x_3, x_4) = x_1'x_2'x_3'x_4 + x_1'x_2x_3'x_4 + x_1x_2x_3'x_4 + x_1x_2'x_3x_4 + x_1x_2'x_3x_4'$$

$$z_1(x_1, x_2, x_3, x_4) = x_1'x_2'x_3'(x_4) + x_1'x_2x_3'(x_4) + x_1x_2x_3'(x_4) + x_1x_2'x_3(x_4) + x_1x_2'x_3(x_4')$$

x_1 \ x_2x_3	0 0	0 1	1 1	1 0
0	x_4 (0)	0 (1)	0 (3)	x_4 (2)
1	0 (4)	$x_4 + x_4'$ (5)	0 (7)	x_4 (6)

z_1

$$z_1 = x_1'x_3'x_4 + x_2x_3'x_4 + x_1x_2'x_3$$

```
//behavioral using map-entered variable

module sop_mev_bh (x1, x2, x3, x4, z1);

//define input and output
input x1, x2, x3, x4;
output z1;

//variables are declared as reg in always
reg z1;

//design the logic for output z1
always @ (x1 or x2 or x3 or x4)
   z1 = ~x1 & ~x3 & x4 | x2 & ~x3 & x4 | x1 & ~x2 & x3;

endmodule
```

Figure 2.14 Design module for Example 2.4 using a map-entered variable.

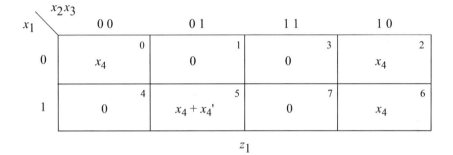

```
//test bench for sop_mev_bh module

module sop_mev_bh_tb;

//inputs are reg for test bench
//outputs are wire for test bench
reg x1, x2, x3, x4;
wire z1;

//apply input vectors and display variables
initial
begin: apply_stimulus
   reg [4:0] invect;
   for (invect = 0; invect < 16; invect = invect + 1)
      begin
         {x1, x2, x3, x4} = invect [4:0];
         #10 $display ("x1 x2 x3 x4 = %b, z1 = %b",
                        {x1, x2, x3, x4}, z1);
      end
end

//instantiate the module into the test bench
sop_mev_bh inst1 (x1, x2, x3, x4, z1);

endmodule
```

Figure 2.15 Test bench module for Example 2.4 using a map-entered variable.

```
x1 x2 x3 x4 = 0000, z1 = 0
x1 x2 x3 x4 = 0001, z1 = 1
x1 x2 x3 x4 = 0010, z1 = 0
x1 x2 x3 x4 = 0011, z1 = 0
x1 x2 x3 x4 = 0100, z1 = 0
x1 x2 x3 x4 = 0101, z1 = 1
x1 x2 x3 x4 = 0110, z1 = 0
x1 x2 x3 x4 = 0111, z1 = 0
x1 x2 x3 x4 = 1000, z1 = 0
x1 x2 x3 x4 = 1001, z1 = 0
x1 x2 x3 x4 = 1010, z1 = 1
x1 x2 x3 x4 = 1011, z1 = 1
x1 x2 x3 x4 = 1100, z1 = 0
x1 x2 x3 x4 = 1101, z1 = 1
x1 x2 x3 x4 = 1110, z1 = 0
x1 x2 x3 x4 = 1111, z1 = 0
```

Figure 2.16 Outputs for Example 2.4 using a map-entered variable.

2.4 Multiplexers

A multiplexer is a logic macro device that allows digital information from two or more data inputs to be directed to a single output. Data input selection is controlled by a set of select inputs that determine which data input is gated to the output. The select inputs are labeled $s_0, s_1, s_2, \ldots, s_i, \ldots, s_{n-1}$, where s_0 is the low-order select input with a binary weight of 2^0 and s_{n-1} is the high-order select input with a binary weight of 2^{n-1}. The data inputs are labeled $d_0, d_1, d_2, \ldots, d_j, \ldots, d_{2^n-1}$. Thus, if a multiplexer has n select inputs, then the number of data inputs will be 2^n and will be labeled d_0 through d_{2^n-1}. For example, if $n = 2$, then the multiplexer has two select inputs s_0 and s_1 and four data inputs d_0, d_1, d_2, and d_3.

Example 2.5 This example designs a 4:1 multiplexer using logic gates that were designed using dataflow modeling. A multiplexer is a logic macro device that allows digital information from two or more data inputs to be directed to a single output. Data input selection is controlled by a set of select inputs that determine which data input is gated to the output. The logic diagram is shown in Figure 2.17. The design module is shown in Figure 2.18, the test bench module is shown in Figure 2.19 and the outputs are shown in Figure 2.20.

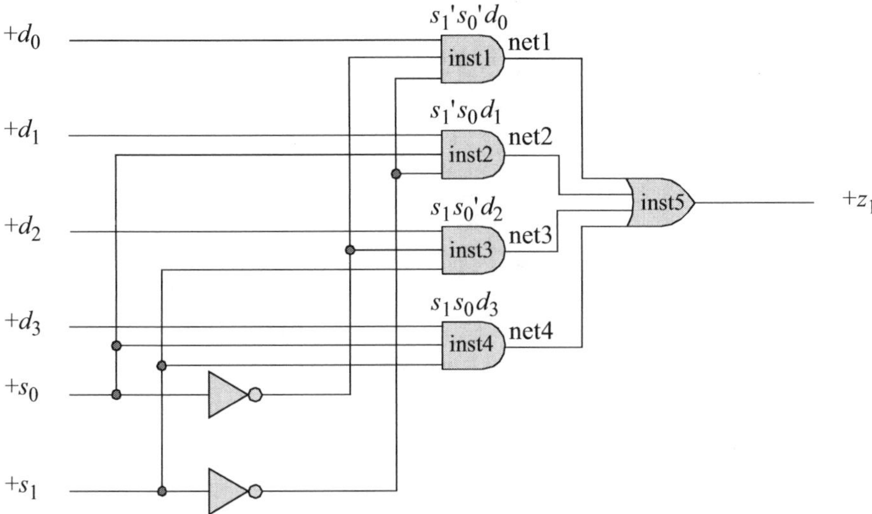

Figure 2.17 Logic diagram for a 4:1 multiplexer.

```
//structural for a 4:1 multiplexer
//using dataflow logic gates

module mux_4to1_struct (s0, s1, d0, d1, d2, d3, z1);

//define inputs and outputs
input s0, s1, d0, d1, d2, d3;
output z1;

//define internal nets
wire net1, net2, net3, net4;

//instantiate the logic gates
and3_df  inst1 (d0, ~s0, ~s1, net1);
and3_df  inst2 (d1, s0, ~s1, net2);
and3_df  inst3 (d2, ~s0, s1, net3);
and3_df  inst4 (d3, s0, s1, net4);

or4_df   inst5 (net1, net2, net3, net4, z1);

endmodule
```

Figure 2.18 Design module for the 4:1 multiplexer.

```
//test bench for 4:1 multiplexer
//using dataflow logic gates

module mux_4to1_struct_tb;


//inputs are reg for test bench
//outputs are wire for test bench
reg s0, s1, d0, d1, d2, d3;;
wire z1;


//display variables
initial
$monitor ("s0 = %b, s1 = %b,
           d0 = %b, d1 = %b, d2 = %b, d3 = %b, z1 = %b",
           s0, s1, d0, d1, d2, d3, z1);

                        //continued on next page
```

Figure 2.19 Test bench module for the 4:1 multiplexer.

```
//apply input vectors
initial
begin
   #0    s0 = 1'b0;  s1 = 1'b0;
         d0 = 1'b1;  d1 = 1'b0;  d2 = 1'b0;  d3 = 1'b0;
   #10   s0 = 1'b0;  s1 = 1'b0;
         d0 = 1'b0;  d1 = 1'b0;  d2 = 1'b0;  d3 = 1'b0;
//------------------------------------------------------------
   #10   s0 = 1'b1;  s1 = 1'b0;
         d0 = 1'b0;  d1 = 1'b1;  d2 = 1'b0;  d3 = 1'b0;
   #10   s0 = 1'b1;  s1 = 1'b0;
         d0 = 1'b0;  d1 = 1'b0;  d2 = 1'b0;  d3 = 1'b0;
//------------------------------------------------------------
   #10   s0 = 1'b0;  s1 = 1'b1;
         d0 = 1'b0;  d1 = 1'b0;  d2 = 1'b1;  d3 = 1'b0;
   #10   s0 = 1'b0;  s1 = 1'b1;
         d0 = 1'b0;  d1 = 1'b0;  d2 = 1'b0;  d3 = 1'b0;
//------------------------------------------------------------
   #10   s0 = 1'b1;  s1 = 1'b1;
         d0 = 1'b0;  d1 = 1'b0;  d2 = 1'b0;  d3 = 1'b1;
   #10   s0 = 1'b1;  s1 = 1'b1;
         d0 = 1'b0;  d1 = 1'b0;  d2 = 1'b0;  d3 = 1'b0;
//------------------------------------------------------------
   #10   s0 = 1'b1;  s1 = 1'b0;
         d0 = 1'b0;  d1 = 1'b1;  d2 = 1'b0;  d3 = 1'b0;
   #10   s0 = 1'b1;  s1 = 1'b0;
         d0 = 1'b0;  d1 = 1'b0;  d2 = 1'b0;  d3 = 1'b0;
//------------------------------------------------------------
   #10   s0 = 1'b1;  s1 = 1'b1;
         d0 = 1'b1;  d1 = 1'b1;  d2 = 1'b1;  d3 = 1'b1;
   #10   s0 = 1'b1;  s1 = 1'b1;
         d0 = 1'b1;  d1 = 1'b1;  d2 = 1'b1;  d3 = 1'b1;
//------------------------------------------------------------
   #10   s0 = 1'b1;  s1 = 1'b1;
         d0 = 1'b1;  d1 = 1'b1;  d2 = 1'b1;  d3 = 1'b0;
   #10   s0 = 1'b1;  s1 = 1'b1;
         d0 = 1'b1;  d1 = 1'b1;  d2 = 1'b1;  d3 = 1'b0;
//------------------------------------------------------------
   #10   $stop;
end

//instantiate the module into the test bench
mux_4to1_struct inst1 (s0, s1, d0, d1, d2, d3, z1);

endmodule
```

Figure 2.19 (Continued)

```
s0 = 0, s1 = 0, | d0 = 1, d1 = 0, d2 = 0, d3 = 0, | z1 = 1
s0 = 0, s1 = 0, | d0 = 0, d1 = 0, d2 = 0, d3 = 0, | z1 = 0

s0 = 1, s1 = 0, | d0 = 0, d1 = 1, d2 = 0, d3 = 0, | z1 = 1
s0 = 1, s1 = 0, | d0 = 0, d1 = 0, d2 = 0, d3 = 0, | z1 = 0

s0 = 0, s1 = 1, | d0 = 0, d1 = 0, d2 = 1, d3 = 0, | z1 = 1
s0 = 0, s1 = 1, | d0 = 0, d1 = 0, d2 = 0, d3 = 0, | z1 = 0

s0 = 1, s1 = 1, | d0 = 0, d1 = 0, d2 = 0, d3 = 1, | z1 = 1
s0 = 1, s1 = 1, | d0 = 0, d1 = 0, d2 = 0, d3 = 0, | z1 = 0

s0 = 1, s1 = 0, | d0 = 0, d1 = 1, d2 = 0, d3 = 0, | z1 = 1
s0 = 1, s1 = 0, | d0 = 0, d1 = 0, d2 = 0, d3 = 0, | z1 = 0

s0 = 1, s1 = 1, | d0 = 1, d1 = 1, d2 = 1, d3 = 1, | z1 = 1
s0 = 1, s1 = 1, | d0 = 1, d1 = 1, d2 = 1, d3 = 0, | z1 = 0
```

Figure 2.20 Outputs for the 4:1 multiplexer.

Example 2.6 This example repeats Example 2.5, but uses the conditional operator to design the 4:1 multiplexer. Recall that the conditional operator has three operands as shown below and can also be nested, also shown below.

conditional_expression **?** true_expression **:** false_expression;

conditional_expression **?** (cond_expr1 **?** true_expr1 **:** false_expr1)
: (cond_expr2 **?** true_expr2 **:** false_expr2);

The design module is shown in Figure 2.21 and the test bench module is shown in Figure 2.22. The outputs are shown in Figure 2.23.

```
//dataflow for 4:1 mux using the conditional operator

module mux_4to1_cond2 (s0, s1, d0, d1, d2, d3, z1);

//define inputs and output
input s0, s1, d0, d1, d2, d3;
output z1;

//use the nested conditional operator
assign z1 = s1 ? (s0 ? d3 : d2) : (s0 ? d1 : d0);

endmodule
```

Figure 2.21 Design module for a 4:1 multiplexer using the conditional operator.

```
//test bench for 4:1 multiplexer
//using the conditional operator

module mux_4to1_cond2_tb;

//inputs are reg for test bench
//outputs are wire for test bench
reg s0, s1, d0, d1, d2, d3;
wire z1;

//display variables
initial
$monitor ("s0 = %b, s1 = %b,
            d0 = %b, d1 = %b, d2 = %b, d3 = %b, z1 = %b",
            s0, s1, d0, d1, d2, d3, z1);

//apply input vectors
initial
begin
   #0     s0 = 1'b0;  s1 = 1'b0;
          d0 = 1'b1;  d1 = 1'b0;  d2 = 1'b0;  d3 = 1'b0;
   #10    s0 = 1'b0;  s1 = 1'b0;
          d0 = 1'b0;  d1 = 1'b0;  d2 = 1'b0;  d3 = 1'b0;
//-----------------------------------------------------------
   #10    s0 = 1'b1;  s1 = 1'b0;
          d0 = 1'b0;  d1 = 1'b1;  d2 = 1'b0;  d3 = 1'b0;
   #10    s0 = 1'b1;  s1 = 1'b0;
          d0 = 1'b0;  d1 = 1'b0;  d2 = 1'b0;  d3 = 1'b0;
//-----------------------------------------------------------
   #10    s0 = 1'b0;  s1 = 1'b1;
          d0 = 1'b0;  d1 = 1'b0;  d2 = 1'b1;  d3 = 1'b0;
   #10    s0 = 1'b0;  s1 = 1'b1;
          d0 = 1'b0;  d1 = 1'b0;  d2 = 1'b0;  d3 = 1'b0;
//-----------------------------------------------------------
   #10    s0 = 1'b1;  s1 = 1'b1;
          d0 = 1'b0;  d1 = 1'b0;  d2 = 1'b0;  d3 = 1'b1;
   #10    s0 = 1'b1;  s1 = 1'b1;
          d0 = 1'b0;  d1 = 1'b0;  d2 = 1'b0;  d3 = 1'b0;
//-----------------------------------------------------------
   #10    s0 = 1'b1;  s1 = 1'b0;
          d0 = 1'b0;  d1 = 1'b1;  d2 = 1'b0;  d3 = 1'b0;
   #10    s0 = 1'b1;  s1 = 1'b0;
          d0 = 1'b0;  d1 = 1'b0;  d2 = 1'b0;  d3 = 1'b0;
//-----------------------------------------------------------
                                    //continued on next page
```

Figure 2.22 Test bench module for the 4:1 multiplexer using the conditional operator.

```
    #10   s0 = 1'b1;  s1 = 1'b1;
          d0 = 1'b1;  d1 = 1'b1;  d2 = 1'b1;  d3 = 1'b1;
    #10   s0 = 1'b1;  s1 = 1'b1;
          d0 = 1'b1;  d1 = 1'b1;  d2 = 1'b1;  d3 = 1'b1;
//------------------------------------------------------------
    #10   s0 = 1'b1;  s1 = 1'b1;
          d0 = 1'b1;  d1 = 1'b1;  d2 = 1'b1;  d3 = 1'b0;
    #10   s0 = 1'b1;  s1 = 1'b1;
          d0 = 1'b1;  d1 = 1'b1;  d2 = 1'b1;  d3 = 1'b0;

    #10   $stop;
end

//instantiate the module into the test bench
 mux_4to1_cond2 inst1 (s0, s1, d0, d1, d2, d3, z1);

endmodule
```

Figure 2.22 (Continued)

s0 = 0, s1 = 0,	d0 = **1**, d1 = 0, d2 = 0, d3 = 0,	z1 = **1**
s0 = 0, s1 = 0,	d0 = 0, d1 = 0, d2 = 0, d3 = 0,	z1 = 0
s0 = 1, s1 = 0,	d0 = 0, d1 = **1**, d2 = 0, d3 = 0,	z1 = **1**
s0 = 1, s1 = 0,	d0 = 0, d1 = 0, d2 = 0, d3 = 0,	z1 = 0
s0 = 0, s1 = 1,	d0 = 0, d1 = 0, d2 = **1**, d3 = 0,	z1 = **1**
s0 = 0, s1 = 1,	d0 = 0, d1 = 0, d2 = 0, d3 = 0,	z1 = 0
s0 = 1, s1 = 1,	d0 = 0, d1 = 0, d2 = 0, d3 = **1**,	z1 = **1**
s0 = 1, s1 = 1,	d0 = 0, d1 = 0, d2 = 0, d3 = 0,	z1 = 0
s0 = 1, s1 = 0,	d0 = 0, d1 = **1**, d2 = 0, d3 = 0,	z1 = **1**
s0 = 1, s1 = 0,	d0 = 0, d1 = 0, d2 = 0, d3 = 0,	z1 = 0
s0 = 1, s1 = 1,	d0 = 1, d1 = 1, d2 = 1, d3 = **1**,	z1 = **1**
s0 = 1, s1 = 1,	d0 = 1, d1 = 1, d2 = 1, d3 = 0,	z1 = 0

Figure 2.23 Outputs for the 4:1 multiplexer using the conditional operator.

Example 2.7 This example uses the continuous assignment statement to design an 8:1 multiplexer. Recall that the continuous assignment statement is used to describe combinational logic where the output of the circuit is evaluated whenever an input changes; that is, the value of the right-hand side expression is *continuously assigned* to the left-hand side net. The syntax for the continuous assignment is shown below and establishes a relationship between a right-hand side expression and a left-hand side net.

assign [delay] lhs_net = rhs_expression

The design module is shown in Figure 2.24. The test bench module is shown in Figure 2.25 and the outputs are shown in Figure 2.26.

```
//dataflow for 8:1 multiplexer using
//the continuous assignment statement

module mux_8to1_assign (sel, data, z1);

//define inputs and outputs
input [2:0] sel;
input [7:0] data;
output z1;

//design the 8:1 multiplexer
assign   z1 =   (~sel[2] & ~sel[1] & ~sel[0] & data[0]) |
                (~sel[2] & ~sel[1] &  sel[0] & data[1]) |
                (~sel[2] &  sel[1] & ~sel[0] & data[2]) |
                (~sel[2] &  sel[1] &  sel[0] & data[3]) |
                ( sel[2] & ~sel[1] & ~sel[0] & data[4]) |
                ( sel[2] & ~sel[1] &  sel[0] & data[5]) |
                ( sel[2] &  sel[1] & ~sel[0] & data[6]) |
                ( sel[2] &  sel[1] &  sel[0] & data[7]);

endmodule
```

Figure 2.24 Design module for an 8:1 multiplexer using the continuous assignment statement.

```
//test bench for 8:1 multiplexer
//using the continuous assignment statement

module mux_8to1_assign_tb;

//inputs are reg for test bench
//outputs wire reg for test bench
reg [2:0] sel;
reg [7:0] data;
wire z1;

//display variables
initial
$monitor ("sel = %b, data = %b, z1 = %b", sel, data, z1);

//apply input vectors
initial
begin
   #0    sel = 3'b000;  data = 8'b0000_0001;
   #10   sel = 3'b001;  data = 8'b0000_0010;
   #10   sel = 3'b010;  data = 8'b0000_0100;
   #10   sel = 3'b011;  data = 8'b0000_1000;
   #10   sel = 3'b100;  data = 8'b0001_0000;
   #10   sel = 3'b101;  data = 8'b0010_0000;
   #10   sel = 3'b110;  data = 8'b0100_0000;
   #10   sel = 3'b111;  data = 8'b1000_0000;

//-----------------------------------------------------
   #10   sel = 3'b000;  data = 8'b1111_1110;
   #10   sel = 3'b001;  data = 8'b1111_1101;
   #10   sel = 3'b010;  data = 8'b1111_1011;
   #10   sel = 3'b011;  data = 8'b1111_0111;
   #10   sel = 3'b100;  data = 8'b1110_1111;
   #10   sel = 3'b101;  data = 8'b1101_1111;
   #10   sel = 3'b110;  data = 8'b1011_1111;
   #10   sel = 3'b111;  data = 8'b0111_1111;
//-----------------------------------------------------

   #10   $stop;

end

//instantiate the module into the test bench
mux_8t01_assign inst1 (sel, data, z1);

endmodule
```

Figure 2.25 Test bench module for the 8:1 multiplexer using the continuous assignment statement.

```
sel = 000, data = 00000001, z1 = 1
sel = 001, data = 00000010, z1 = 1
sel = 010, data = 00000100, z1 = 1
sel = 011, data = 00001000, z1 = 1
sel = 100, data = 00010000, z1 = 1
sel = 101, data = 00100000, z1 = 1
sel = 110, data = 01000000, z1 = 1
sel = 111, data = 10000000, z1 = 1

//-----------------------------------------------------------
sel = 000, data = 11111110, z1 = 0
sel = 001, data = 11111101, z1 = 0
sel = 010, data = 11111011, z1 = 0
sel = 011, data = 11110111, z1 = 0
sel = 100, data = 11101111, z1 = 0
sel = 101, data = 11011111, z1 = 0
sel = 110, data = 10111111, z1 = 0
sel = 111, data = 01111111, z1 = 0
```

Figure 2.26 Outputs for the 8:1 multiplexer using the continuous assignment statement.

Example 2.8 This example repeats Example 2.7 in designing an 8:1 multiplexer, but uses the **case** statement in the design module, which is shown in Figure 2.27. The test bench module is shown in Figure 2.28 and the outputs are shown in Figure 2.29.

```
//8:1 multiplexed using the case statement

module mux_8to1_case4 (sel, data, z1);

//define inputs and outputs
input [2:0] sel;
input [7:0] data;
output z1;

//variables in always are declared as reg
reg z1;

                              //continued on next page
```

Figure 2.27 Design module for an 8:1 multiplexer using the **case** statement.

```
always @ (sel or data)
begin
case (sel)
    (0) : z1 = data [0];
    (1) : z1 = data [1];
    (2) : z1 = data [2];
    (3) : z1 = data [3];
    (4) : z1 = data [4];
    (5) : z1 = data [5];
    (6) : z1 = data [6];
    (7) : z1 = data [7];
    default : z1 = 1'b0;
endcase
end

endmodule
```

Figure 2.27 (Continued)

```
//test bench for 8:1 multiplexer using the case statement
module mux_8to1_case4_tb;

//inputs are reg for test bench
//outputs are wire for test bench
reg [2:0] sel;
reg [7:0] data;
wire z1;

//display variables
initial
$monitor ("sel = %b, data = %b, z1 = %b", sel, data, z1);

//apply input vectors
initial
begin
    #0     sel = 3'b000;  data = 8'b0000_0001;
    #10    sel = 3'b001;  data = 8'b0000_0010;
    #10    sel = 3'b010;  data = 8'b0000_0100;
    #10    sel = 3'b011;  data = 8'b0000_1000;
    #10    sel = 3'b100;  data = 8'b0001_0000;
    #10    sel = 3'b101;  data = 8'b0010_0000;
    #10    sel = 3'b110;  data = 8'b0100_0000;
    #10    sel = 3'b111;  data = 8'b1000_0000;
                                 //continued on next page
```

Figure 2.28 Test bench module for an 8:1 multiplexer using the **case** statement.

```
//------------------------------------------------------
   #10   sel = 3'b000;  data = 8'b1111_1110;
   #10   sel = 3'b001;  data = 8'b1111_1101;
   #10   sel = 3'b010;  data = 8'b1111_1011;
   #10   sel = 3'b011;  data = 8'b1111_0111;
   #10   sel = 3'b100;  data = 8'b1110_1111;
   #10   sel = 3'b101;  data = 8'b1101_1111;
   #10   sel = 3'b110;  data = 8'b1011_1111;
   #10   sel = 3'b111;  data = 8'b0111_1111;
//------------------------------------------------------

   #10   $stop;

end

//instantiate the module into the test bench
mux_8to1_case4 inst1 (sel, data, z1);

endmodule
```

Figure 2.28 (Continued)

```
sel = 000, data = 00000001, z1 = 1
sel = 001, data = 00000010, z1 = 1
sel = 010, data = 00000100, z1 = 1
sel = 011, data = 00001000, z1 = 1
sel = 100, data = 00010000, z1 = 1
sel = 101, data = 00100000, z1 = 1
sel = 110, data = 01000000, z1 = 1
sel = 111, data = 10000000, z1 = 1

//------------------------------------------------------------
sel = 000, data = 11111110, z1 = 0
sel = 001, data = 11111101, z1 = 0
sel = 010, data = 11111011, z1 = 0
sel = 011, data = 11110111, z1 = 0
sel = 100, data = 11101111, z1 = 0
sel = 101, data = 11011111, z1 = 0
sel = 110, data = 10111111, z1 = 0
sel = 111, data = 01111111, z1 = 0
```

Figure 2.29 Outputs for an 8:1 multiplexer using the **case** statement.

2.5 Comparators

A comparator is a logic macro circuit that compares the magnitude of two *n*-bit binary operands. This section designs various comparators to perform a variety of compare operations on different operands.

Example 2.9 This example designs a comparator to determine if a 4-bit vector *a[3:0]* is equal to a 4-bit vector *b[3:0]*. The design module using logic gates that were designed using dataflow modeling is shown in Figure 2.30. The test bench module and the outputs are shown in Figure 2.31 and Figure 2.32, respectively. The statement for equality is shown below from Chapter 1 for 4-bit vectors.

$$(A = B) = (a_3 \oplus b_3)' (a_2 \oplus b_2)' (a_1 \oplus b_1)' (a_0 \oplus b_0)'$$

```
//structural test for equality of 4-bit vectors
module comparator_equal (a, b, equal);

//define inputs and outputs
input [3:0] a, b;
output equal;

//define internal nets
wire net1, net2, net3, net4, net5;

//instantiate the logic to test for equality
xnor2_df  inst1 (a[3], b[3], net1);
xnor2_df  inst2 (a[2], b[2], net2);
xnor2_df  inst3 (a[1], b[1], net3);
xnor2_df  inst4 (a[0], b[0], net4);
and4_df   inst5 (net1, net2, net3, net4, equal);
endmodule
```

Figure 2.30 Design module for the compare for equality example.

```
//test bench for equality test of 4-bit vectors
module comparator_equal_tb;

//inputs are reg for test bench
//outputs are wire for test bench
reg [3:0] a, b;
wire equal;

//display inputs and outputs
initial
$monitor ("a = %b, b = %b, equal = %b", a, b, equal);
                                    //continued on next page
```

Figure 2.31 Test bench module for the compare for equality example.

```
//apply input vectors
initial
begin
   #0    a = 4'b0000;   b = 4'b0000;
   #10   a = 4'b0011;   b = 4'b0010;
   #10   a = 4'b1011;   b = 4'b1011;
   #10   a = 4'b0111;   b = 4'b0011;

   #10   a = 4'b0110;   b = 4'b0110;
   #10   a = 4'b1011;   b = 4'b0010;
   #10   a = 4'b1111;   b = 4'b1111;
   #10   a = 4'b0110;   b = 4'b0011;

   #10   a = 4'b0111;   b = 4'b0111;
   #10   a = 4'b1011;   b = 4'b1010;
   #10   a = 4'b1110;   b = 4'b1110;
   #10   a = 4'b0110;   b = 4'b0111;

   #10   a = 4'b1100;   b = 4'b1100;
   #10   a = 4'b1011;   b = 4'b1110;
   #10   a = 4'b1111;   b = 4'b1111;
   #10   a = 4'b1110;   b = 4'b0111;

   #10   $stop;
end

//instantiate the module into the test bench
comparator_equal inst1 (a, b, equal);

endmodule
```

Figure 2.31 (Continued)

```
a = 0000, b = 0000, equal = 1
a = 0011, b = 0010, equal = 0
a = 1011, b = 1011, equal = 1
a = 0111, b = 0011, equal = 0

a = 0110, b = 0110, equal = 1
a = 1011, b = 0010, equal = 0
a = 1111, b = 1111, equal = 1
a = 0110, b = 0011, equal = 0

                             //continued on next page
```

Figure 2.32 Outputs for the compare for equality example.

```
a = 0111, b = 0111, equal = 1
a = 1011, b = 1010, equal = 0
a = 1110, b = 1110, equal = 1
a = 0110, b = 0111, equal = 0

a = 1100, b = 1100, equal = 1
a = 1011, b = 1110, equal = 0
a = 1111, b = 1111, equal = 1
a = 1110, b = 0111, equal = 0
```

Figure 2.32 Continued)

Example 2.10 This example repeats Example 2.9, but uses built-in primitives in the design. The statement for equality for 4-bit vectors is reproduced below for convenience.

$$(A = B) = (a_3 \oplus b_3)' (a_2 \oplus b_2)' (a_1 \oplus b_1)' (a_0 \oplus b_0)'$$

The design module is shown in Figure 2.33. The test bench module and the outputs are shown in Figure 2.34 and Figure 2.35, respectively.

```
//gate-level module to test for equality of 4-bit vectors
//using built-in primitives

module comparator_equal_bip (a, b, equal);


//define inputs and outputs
input [3:0] a, b;
output equal;


//design the 4-bit comparator
xnor  inst1 (net1, a[3], b[3]);
xnor  inst2 (net2, a[2], b[2]);
xnor  inst3 (net3, a[1], b[1]);
xnor  inst4 (net4, a[0], b[0]);

and   inst5 (equal, net1, net2, net3, net4);


endmodule
```

Figure 2.33 Design module using built-in primitives to compare for the equality of two vectors.

```
//test bench for equality test of 4-bit vectors

module comparator_equal_bip_tb;


//inputs are reg for test bench
//outputs are wire for test bench
reg [3:0] a, b;
wire equal;


//display inputs and outputs
initial
$monitor ("a = %b, b = %b, equal = %b", a, b, equal);


//apply input vectors
initial
begin
   #0    a = 4'b0000;   b = 4'b0000;
   #10   a = 4'b0011;   b = 4'b0010;
   #10   a = 4'b1011;   b = 4'b1011;
   #10   a = 4'b0111;   b = 4'b0011;

   #10   a = 4'b0110;   b = 4'b0110;
   #10   a = 4'b1011;   b = 4'b0010;
   #10   a = 4'b1111;   b = 4'b1111;
   #10   a = 4'b0110;   b = 4'b0011;

   #10   a = 4'b0111;   b = 4'b0111;
   #10   a = 4'b1011;   b = 4'b1010;
   #10   a = 4'b1110;   b = 4'b1110;
   #10   a = 4'b0110;   b = 4'b0111;

   #10   a = 4'b1100;   b = 4'b1100;
   #10   a = 4'b1011;   b = 4'b1110;
   #10   a = 4'b1111;   b = 4'b1111;
   #10   a = 4'b1110;   b = 4'b0111;

   #10   $stop;
end

//instantiate the module into the test bench
comparator_equal_bip inst1 (a, b, equal);

endmodule
```

Figure 2.34 Test bench module to compare for the equality of two vectors.

```
a = 0000, b = 0000, equal = 1
a = 0011, b = 0010, equal = 0
a = 1011, b = 1011, equal = 1
a = 0111, b = 0011, equal = 0

a = 0110, b = 0110, equal = 1
a = 1011, b = 0010, equal = 0
a = 1111, b = 1111, equal = 1
a = 0110, b = 0011, equal = 0

a = 0111, b = 0111, equal = 1
a = 1011, b = 1010, equal = 0
a = 1110, b = 1110, equal = 1
a = 0110, b = 0111, equal = 0

a = 1100, b = 1100, equal = 1
a = 1011, b = 1110, equal = 0
a = 1111, b = 1111, equal = 1
a = 1110, b = 0111, equal = 0
```

Figure 2.35 Outputs to compare for the equality of two vectors.

Example 2.11 This example designs a comparator to detect when a 4-bit vector *a[3:0]* is less than a 4-bit vector *b[3:0]* using the continuous assignment statement. Recall that the *continuous assignment* statement models dataflow behavior and is used to design combinational logic without using gates and interconnecting nets. Continuous assignment statements provide a Boolean correspondence between the right-hand side expression and the left-hand side target. The continuous assignment statement uses the keyword **assign** and has the following syntax with optional drive strength and delay:

assign [drive_strength] [delay] left-hand side target = right-hand side expression

The equation for determining if one vector is less than another vector is shown below for 4-bit vectors.

$$
\begin{aligned}
(A < B) = {} & a_3'\, b_3 + \\
& (a_3 \oplus b_3)'\, a_2'\, b_2 + \\
& (a_3 \oplus b_3)'\, (a_2 \oplus b_2)'\, a_1'\, b_1 + \\
& (a_3 \oplus b_3)'\, (a_2 \oplus b_2)'\, (a_1 \oplus b_1)'\, a_0'\, b_0
\end{aligned}
$$

The design module is shown in Figure 2.36. The test bench module is shown in Figure 2.37, which takes the design through a sequence of input vectors for operands *a[3:0]* and *b[3:0]*. The outputs are shown in Figure 2.38.

```
//dataflow using assign to detect if a < b

module a_lt_b_assign (a, b, a_lt_b);

//list inputs and output
input [3:0] a, b;
output a_lt_b;

//design the 4-bit comparator
assign net1 = (~a[3] & b[3]);

assign net2 = (a[3] ^~ b[3]) & (~a[2] & b[2]);

assign net3 = (a[3] ^~ b[3]) & (a[2] ^~ b[2]) & (~a[1] & b[1]);

assign net4 = (a[3] ^~ b[3]) & (a[2] ^~ b[2]) & (a[1] ^~ b[1])
                & (~a[0] & b[0]);

assign a_lt_b = net1 | net2 | net3 | net4;

endmodule
```

Figure 2.36 Design module to detect if *a[3:0]* is less than *b[3:0]*.

```
//test bench for equality test of 4-bit vectors
module a_lt_b_assign_tb;

//inputs are reg for test bench
//outputs are wire for test bench
reg [3:0] a, b;
wire a_lt_b;

//display inputs and outputs
initial
$monitor ("a = %b, b = %b, a_lt_b = %b", a, b, a_lt_b);

//apply input vectors
initial
begin
   #0    a = 4'b0000;    b = 4'b0001;
   #10   a = 4'b0011;    b = 4'b0010;
   #10   a = 4'b0011;    b = 4'b1011;
   #10   a = 4'b0111;    b = 4'b0011;
                                     //continued on next page
```

Figure 2.37 Test bench module to detect if *a[3:0]* is less than *b[3:0]*.

```
    #10   a = 4'b0110;   b = 4'b0111;
    #10   a = 4'b0011;   b = 4'b0010;
    #10   a = 4'b1110;   b = 4'b1111;
    #10   a = 4'b0110;   b = 4'b0011;

    #10   a = 4'b0110;   b = 4'b0111;
    #10   a = 4'b1011;   b = 4'b1010;
    #10   a = 4'b1110;   b = 4'b1111;
    #10   a = 4'b1110;   b = 4'b0111;

    #10   a = 4'b1000;   b = 4'b1100;
    #10   a = 4'b1011;   b = 4'b1010;
    #10   a = 4'b1110;   b = 4'b1111;
    #10   a = 4'b1110;   b = 4'b0111;

    #10   $stop;
end

//instantiate the module into the test bench
a_lt_b_assign inst1 (a, b, a_lt_b);

endmodule
```

Figure 2.37 (Continued)

```
a = 0000, b = 0001, a_lt_b = 1
a = 0011, b = 0010, a_lt_b = 0
a = 0011, b = 1011, a_lt_b = 1
a = 0111, b = 0011, a_lt_b = 0

a = 0110, b = 0111, a_lt_b = 1
a = 0011, b = 0010, a_lt_b = 0
a = 1110, b = 1111, a_lt_b = 1
a = 0110, b = 0011, a_lt_b = 0

a = 0110, b = 0111, a_lt_b = 1
a = 1011, b = 1010, a_lt_b = 0
a = 1110, b = 1111, a_lt_b = 1
a = 1110, b = 0111, a_lt_b = 0

a = 1000, b = 1100, a_lt_b = 1
a = 1011, b = 1010, a_lt_b = 0
a = 1110, b = 1111, a_lt_b = 1
a = 1110, b = 0111, a_lt_b = 0
```

Figure 2.38 Outputs to detect if *a[3:0]* is less than *b[3:0]*.

Example 2.12 This example designs a comparator to determine if a 4-bit vector *a[3:0]* is greater than a 4-bit vector *b[3:0]* using the behavioral conditional statements **if-else if**. These statements are summarized below as reproduced from Chapter 1.

//nested **if-else if** //choice of multiple statements. One is executed.
if (expression1) statement1; //if expression1 is true, then statement1 is executed.
else if (expression2) statement2; //if expression2 is true, then statement2 is executed.
else if (expression3) statement3; //if expression3 is true, then statement3 is executed.
else default statement;

The equation for determining if one vector is greater than another vector is shown below for 4-bit vectors.

$$
\begin{aligned}
(A > B) = {} & a_3\, b_3' + \\
& (a_3 \oplus b_3)'\, a_2\, b_2' + \\
& (a_3 \oplus b_3)'\, (a_2 \oplus b_2)'\, a_1\, b_1' + \\
& (a_3 \oplus b_3)'\, (a_2 \oplus b_2)'\, (a_1 \oplus b_1)'\, a_0\, b_0'
\end{aligned}
$$

The behavioral design module is shown in Figure 2.39. The test bench module and the outputs are shown in Figures 2.40 and 2.41, respectively.

```
//behavioral if-else if for comparator to detect if A > B
module a_gt_b_cond (a, b, a_gt_b);

input [3:0] a, b;     //define inputs and outputs
output a_gt_b;
reg a_gt_b;

always @ (a or b)     //design the 4-bit comparator
begin
   if (a[3] & ~b[3])
            a_gt_b = 1'b1;

   else if ((a[3] ^~ b[3]) & (a[2] & ~b[2]))
            a_gt_b = 1'b1;

   else if ((a[3] ^~ b[3]) & (a[2] ^~ b[2]) & (a[1] & ~b[1]))
            a_gt_b = 1'b1;

   else if ((a[3] ^~ b[3]) & (a[2] ^~ b[2]) & (a[1] ^~ b[1]) &
            (a[0] & ~b[0]))
            a_gt_b = 1'b1;

   else a_gt_b = 1'b0;
end
endmodule
```

Figure 2.39 Design module to determine if *a[3:0]* is greater than *b[3:0]*.

```
//test bench to detect if A > B

module a_gt_b_cond_tb;

//inputs are reg for test bench
//outputs are wire for test bench
reg [3:0] a, b;
wire a_gt_b;

//display inputs and outputs
initial
$monitor ("a = %b, b = %b, a_gt_b = %b", a, b, a_gt_b);

//apply input vectors
initial
begin
   #0    a = 4'b0000;   b = 4'b0001;
   #10   a = 4'b0011;   b = 4'b0010;
   #10   a = 4'b0011;   b = 4'b1011;
   #10   a = 4'b0111;   b = 4'b0011;

   #10   a = 4'b0110;   b = 4'b0111;
   #10   a = 4'b0011;   b = 4'b0010;
   #10   a = 4'b1110;   b = 4'b1111;
   #10   a = 4'b0110;   b = 4'b0011;

   #10   a = 4'b0110;   b = 4'b0111;
   #10   a = 4'b1011;   b = 4'b1010;
   #10   a = 4'b1110;   b = 4'b1111;
   #10   a = 4'b1110;   b = 4'b0111;

   #10   a = 4'b1000;   b = 4'b1100;
   #10   a = 4'b1011;   b = 4'b1010;
   #10   a = 4'b1110;   b = 4'b1111;
   #10   a = 4'b1110;   b = 4'b0111;

   #10   $stop;
end

//instantiate the module into the test bench
a_gt_b_cond inst1 (a, b, a_gt_b);

endmodule
```

Figure 2.40 Test bench module to determine if *a[3:0]* is greater than *b[3:0]*.

The structural design module using built-in primitives is shown in Figure 2.43 illustrating the design for the input logic, the AND array, and the OR array. The test bench module is shown in Figure 2.44 and the outputs are shown in Figure 2.45.

```
//structural prom to generate four equations
//z1 = x1' x2' + x1 x2'
//z2 = x1' x2' + x1' x2
//z3 = x1' x2 + x1 x2'
module prom3 (x1, x2, z1, z2, z3);

input x1, x2;      //define inputs and outputs
output z1, z2, z3;

//define internal nets
wire net1, net2, net3, net4, net5, net6, net7, net8;

//define the input logic
buf (net1, x1);
not (net2, x1);

buf (net3, x2);
not (net4, x2);

//define the logic for the and array
and    (net5, net2, net4),
       (net6, net2, net3),
       (net7, net1, net4),
       (net8, net1, net3);

//define the logic for the or array
or     (z1, net5, net7),
       (z2, net5, net6),
       (z3, net6, net7);

endmodule
```

Figure 2.43 Structural design module for the PROM of Figure 2.42.

```
//test bench for the structural prom3 module
module prom3_tb;

//inputs are reg for test bench
//outputs are wire for test bench
reg x1, x2;
wire z1, z2, z3;                //continued on next page
```

Figure 2.44 Test bench module for the PROM of Figure 2.42.

```
//display variables
initial
$monitor ("x1 x2 = %b, z1 z2 z3 = %b",
            {x1, x2}, {z1, z2, z3});

//apply input vectors
initial
begin
   #0     x1 = 1'b0;x2 = 1'b0;
   #10    x1 = 1'b0;x2 = 1'b1;
   #10    x1 = 1'b1;x2 = 1'b0;
   #10    x1 = 1'b1;x2 = 1'b1;

   #10    $stop;
end

//instantiate the module into the test bench
prom3 inst1 (x1, x2, z1, z2, z3);
endmodule
```

Figure 2.44 (Continued)

```
x1 x2 = 00, z1 z2 z3 = 110
x1 x2 = 01, z1 z2 z3 = 011
x1 x2 = 10, z1 z2 z3 = 101
x1 x2 = 11, z1 z2 z3 = 000
```

Figure 2.45 Outputs for the PROM of Figure 2.42.

Example 2.14 PROMs can be used to implement sequential logic also. However, this chapter concentrates on combinational logic. The truth table to implement a sum-of-minterms combinational circuit is shown in Table 2.3. The equations that represent the sum of minterms are shown in Equation 2.5. The PROM organization is shown in Figure 2.46.

Table 2.3 Truth Table for the PROM of Example 2.14

Address Inputs $x_1\ x_2$	Outputs $z_1\ z_2\ z_3\ z_4$
0 0	1 1 0 0
0 1	0 1 1 0
1 0	0 0 1 1
1 1	1 0 0 1

$$z_1 = x_1'x_2' + x_1x_2$$
$$z_2 = x_1'x_2' + x_1'x_2$$
$$z_3 = x_1'x_2 + x_1x_2'$$
$$z_4 = x_1x_2' + x_1x_2 \tag{2.5}$$

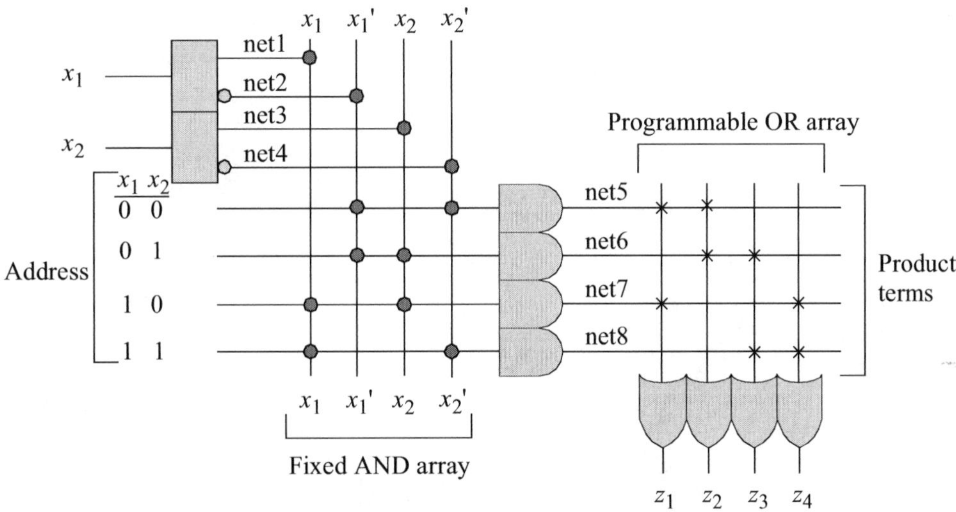

Figure 2.46 PROM organization and programming for Example 2.14.

The structural design module using built-in primitives is shown in Figure 2.47 illustrating the design for the input logic, the AND array, and the OR array. The test bench module is shown in Figure 2.48 and the outputs are shown in Figure 2.49.

```
//structural prom to generate four equations

module prom4 (x1, x2, z1, z2, z3, z4);

//define inputs and outputs
input x1, x2;
output z1, z2, z3, z4;              //continued on next page
```

Figure 2.47 Structural design module for the PROM of Figure 2.46.

```
//define internal nets
wire net1, net2, net3, net4, net5, net6, net7, net8;

//define the input logic
assign   net1 = x1,
         net2 = ~x1,
         net3 = x2,
         net4 = ~x2;

//define the logic for the and array
and (net5, net2, net4);
and (net6, net2, net3);
and (net7, net1, net3);
and (net8, net1, net4);

//define the logic for the or array
or  (z1, net5, net7);
or  (z2, net5, net6);
or  (z3, net6, net8);
or  (z4, net7, net8);

endmodule
```

Figure 2.47 (Continued)

```
//test bench for the structural prom module
module prom4_tb;

//inputs are reg for test bench
//outputs are wire for test bench
reg x1, x2;
wire z1, z2, z3, z4;

initial          //display variables
$monitor ("x1 x2 = %b, z1 z2 z3 z4 = %b", {x1, x2},
             {z1, z2, z3, z4});

initial      //apply input vectors
begin
   #0    x1 = 1'b0;  x2 = 1'b0;
   #10   x1 = 1'b0;  x2 = 1'b1;
   #10   x1 = 1'b1;  x2 = 1'b0;
   #10   x1 = 1'b1;  x2 = 1'b1;
   #10   $stop;
end                                 //continued on next page
```

Figure 2.48 Test bench module for the PROM of Figure 2.46.

```
//instantiate the module into the test bench
prom4 inst1 (x1, x2, z1, z2, z3, z4);

endmodule
```

Figure 2.48 (Continued)

```
x1 x2 = 00, | z1 z2 z3 z4 = 1100
x1 x2 = 01, | z1 z2 z3 z4 = 0110
x1 x2 = 10, | z1 z2 z3 z4 = 0011
x1 x2 = 11, | z1 z2 z3 z4 = 1001
```

Figure 2.49 Outputs for the PROM of Figure 2.46.

2.6.2 Programmable Array Logic

A PAL device confirms to the general structure of a PLD. The number of AND gates and OR gates is variable, depending on the part number of the commercially available PAL. In many cases, the outputs are also fed back through separate buffers (drivers) to the programmable AND array.

Example 2.15 This example designs a structural module using a programmable array logic (PAL) device to implement a 3-bit binary-to-Gray code converter. The conversion table is shown in Table 2.4 and the corresponding Karnaugh maps are shown in Figure 2.50. The PAL device is shown in Figure 2.51.

Table 2.4 Binary-to-Gray Code Conversion

Binary			Gray		
b_1	b_2	b_3	g_1	g_2	g_3
0	0	0	0	0	0
0	0	1	0	0	1
0	1	0	0	1	1
0	1	1	0	1	0
1	0	0	1	1	0
1	0	1	1	1	1
1	1	0	1	0	1
1	1	1	1	0	0

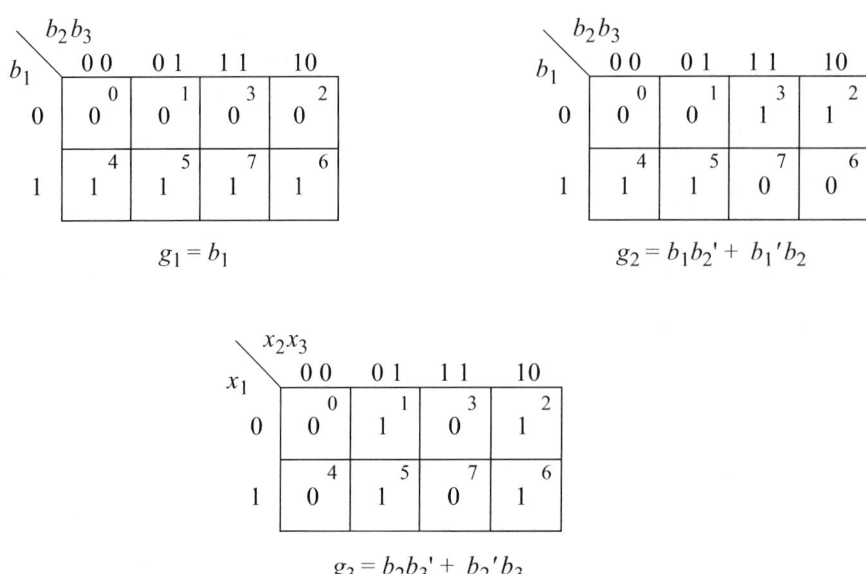

Figure 2.50 Karnaugh maps for the binary-to-Gray code conversion.

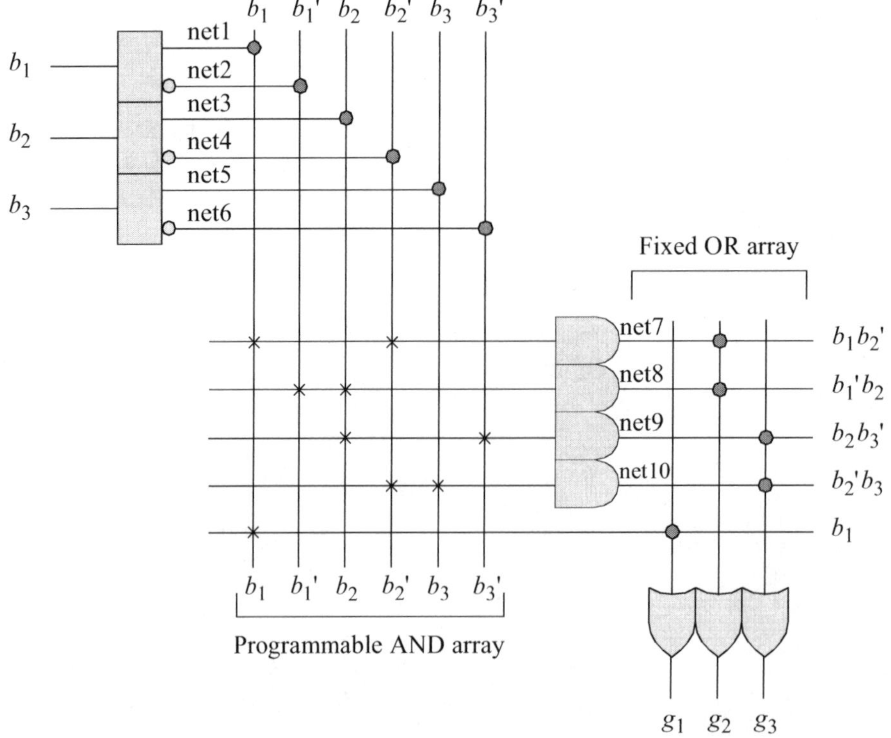

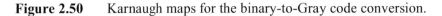

Figure 2.51 PAL device for the binary-to-Gray code conversion.

The design module is shown in Figure 2.52. The test bench module and outputs are shown in Figures 2.53 and 2.54, respectively.

```
//structural pal for binary-to-Gray code converter

module pal7 (b1, b2, b3, g1, g2, g3);

//define inputs and outputs
input b1, b2, b3;
output g1, g2, g3;

//define internal nets
wire net1, net2, net3, net4, net5, net6,
      net7, net8, net9, net10;

//define the input logic
buf(net1, b1);
not(net2, b1);

buf(net3, b2);
not(net4, b2);

buf(net5, b3);
not(net6, b3);

//design the logic for the and array
and    (net7, net1, net4),
       (net8, net2, net3),
       (net9, net3, net6),
       (net10, net4, net5);

//define the logic for the or array
or     (g1, net1),
       (g2, net7, net8),
       (g3, net9, net10);

endmodule
```

Figure 2.52 Structural design module for the binary-to-Gray code converter.

```
//test bench for pal binary-to-Gray code converter

module pal7_tb;

//inputs are reg for test bench
//outputs are wire for test bench
reg b1, b2, b3;
wire g1, g2, g3;

initial     //display variables
$monitor ("b1 b2 b3 = %b, g1 g2 g3 = %b",
            {b1, b2, b3}, {g1, g2, g3});


initial     //apply input vectors
begin
   #0    b1 = 1'b0;  b2 = 1'b0;  b3 = 1'b0;
   #10   b1 = 1'b0;  b2 = 1'b0;  b3 = 1'b1;
   #10   b1 = 1'b0;  b2 = 1'b1;  b3 = 1'b0;
   #10   b1 = 1'b0;  b2 = 1'b1;  b3 = 1'b1;

   #10   b1 = 1'b1;  b2 = 1'b0;  b3 = 1'b0;
   #10   b1 = 1'b1;  b2 = 1'b0;  b3 = 1'b1;
   #10   b1 = 1'b1;  b2 = 1'b1;  b3 = 1'b0;
   #10   b1 = 1'b1;  b2 = 1'b1;  b3 = 1'b1;

   #10   $stop;
end

//instantiate the module into the test bench
pal7 inst1 (b1, b2, b3, g1, g2, g3);

endmodule
```

Figure 2.53 Test bench module for the binary-to-Gray code converter.

```
b1 b2 b3 = 000, g1 g2 g3 = 000
b1 b2 b3 = 001, g1 g2 g3 = 001
b1 b2 b3 = 010, g1 g2 g3 = 011
b1 b2 b3 = 011, g1 g2 g3 = 010

b1 b2 b3 = 100, g1 g2 g3 = 110
b1 b2 b3 = 101, g1 g2 g3 = 111
b1 b2 b3 = 110, g1 g2 g3 = 101
b1 b2 b3 = 111, g1 g2 g3 = 100
```

Figure 2.54 Outputs for the binary-to-Gray code converter.

Example 2.16 This example designs a full adder using a PAL device. A parallel adder that adds two n-bit operands requires n full adders. A *full adder* for stage$_i$ is a combinational circuit that has three inputs: an augend a_i, an addend b_i, and a carry-in cin_i. There are two outputs: a sum labelled sum_i and a carry-out $cout_i$. The truth table for the sum and carry-out functions is shown in Table 2.5 for adding three bits: a, b, and cin and producing two outputs: sum and $cout$.

Table 2.5 Truth Table for Binary Addition

a	*b*	*cin*	*sum*	*cout*
0	0	0	0	0
0	0	1	1	0
0	1	0	1	0
0	1	1	0	1
1	0	0	1	0
1	0	1	0	1
1	1	0	0	1
1	1	1	1	1

Each stage of the addition algorithm must be able to accommodate the carry-in bit c_{i-1} from the immediately preceding lower-order stage. The carry-out of the ith stage is c_i. The sum and carry equations for the full adder are shown in Equation 2.6. The resulting equation for c_i can also be written as $c_i = a_i b_i + (a_i \oplus b_i) c_{i-1}$, although this requires more gate delays.

$$
\begin{aligned}
s_i &= a_i' b_i' c_{i-1} + a_i' b_i c_{i-1}' + a_i b_i' c_{i-1}' + a_i b_i c_{i-1} \\
&= c_{i-1}' (a_i \oplus b_i) + c_{i-1} (a_i \oplus b_i)' \\
&= a_i \oplus b_i \oplus c_{i-1}
\end{aligned}
$$

$$
\begin{aligned}
c_i &= a_i' b_i c_{i-1} + a_i b_i' c_{i-1} + a_i b_i c_{i-1}' + a_i b_i c_{i-1} \\
&= a_i' b_i c_{i-1} + a_i b_i' c_{i-1} + a_i b_i \qquad (2.6)
\end{aligned}
$$

The logic diagram for the full adder using a PAL device is shown in Figure 2.55. The structural design module is shown in Figure 2.56. The test bench module and the outputs are shown in Figures 2.57 and 2.58, respectively.

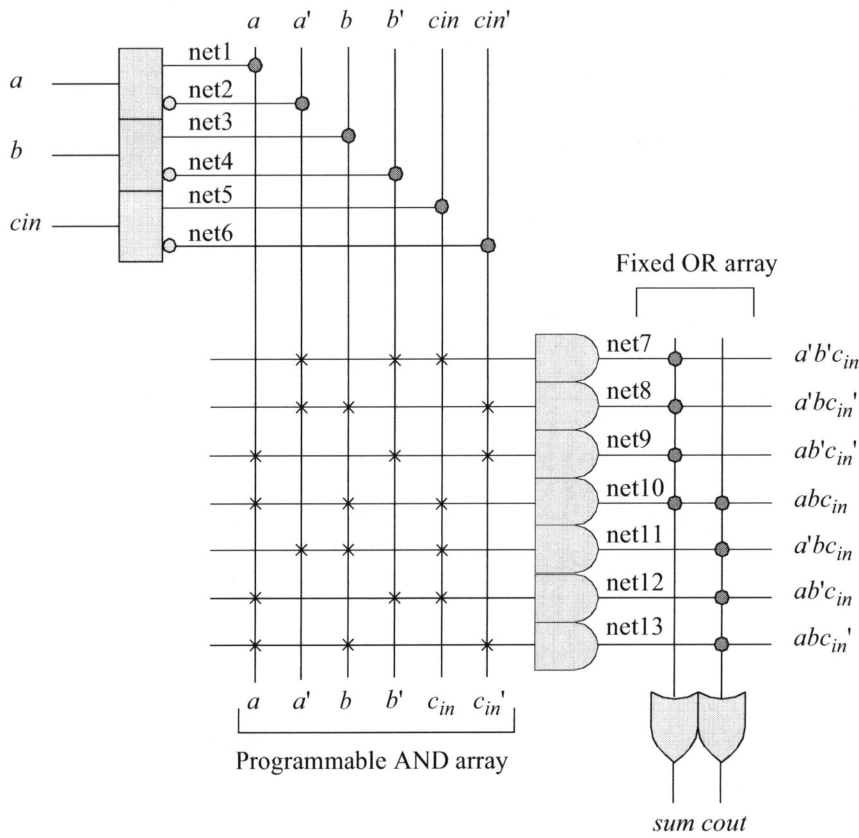

Figure 2.55 Logic diagram for a full adder using a PAL device.

```
//structural pal full adder

module pal_full_adder (a, b, cin, sum, cout);

//define inputs and outputs
input a, b, cin;
output sum, cout;

//define internal nets
wire net1, net2, net3, net4, net5, net6, net7,
      net8, net9, net10, net11, net12, net13;

                                    //continued on next page
```

Figure 2.56 Structural design module using a PAL device for a full adder.

```
//define the input logic
assign   net1 = a,
         net2 = ~a,

         net3 = b,
         net4 = ~b,

         net5 = cin,
         net6 = ~cin;

//define the logic for the and array
and    (net7, net2, net4, net5),
       (net8, net2, net3, net6),
       (net9, net1, net4, net6),
       (net10, net1, net3, net5),
       (net11, net2, net3, net5),
       (net12, net1, net4, net5),
       (net13, net1, net3, net6);

//define the logic for the or array
or     (sum, net7, net8, net9, net10),
       (cout, net10, net11, net12, net13);

endmodule
```

Figure 2.56 (Continued)

```
//test bench for the full adder

module pal_full_adder_tb;

//inputs are reg for test bench
//outputs are wire for test bench
reg a, b, cin;
wire sum, cout;

//display variables
initial
$monitor ("a b cin = %b, sum cout = %b",
       {a, b, cin}, {sum, cout});
                                   //continued on next page
```

Figure 2.57 Test bench module for a full adder using a PAL device.

```
//apply input vectors
initial
begin
    #0     a = 1'b0;    b = 1'b0;    cin = 1'b0;
    #10    a = 1'b0;    b = 1'b0;    cin = 1'b1;
    #10    a = 1'b0;    b = 1'b1;    cin = 1'b0;
    #10    a = 1'b0;    b = 1'b1;    cin = 1'b1;

    #10    a = 1'b1;    b = 1'b0;    cin = 1'b0;
    #10    a = 1'b1;    b = 1'b0;    cin = 1'b1;
    #10    a = 1'b1;    b = 1'b1;    cin = 1'b0;
    #10    a = 1'b1;    b = 1'b1;    cin = 1'b1;

    #10    $stop;
end

//instantiate the module into the test bench
pal_full_adder inst1 (a, b, cin, sum, cout);

endmodule
```

Figure 2.57 (Continued)

```
a b cin = 000, sum cout = 00
a b cin = 001, sum cout = 10
a b cin = 010, sum cout = 10
a b cin = 011, sum cout = 01

a b cin = 100, sum cout = 10
a b cin = 101, sum cout = 01
a b cin = 110, sum cout = 01
a b cin = 111, sum cout = 11
```

Figure 2.58 Outputs for the full adder using a PAL device.

Example 2.17 Outputs z_1, z_2, and z_3 shown in the Karnaugh maps of Figure 2.59, will be implemented using a PAL device. The Boolean equations for the three outputs obtained from the Karnaugh maps are shown in Equation 2.7. Figure 2.60 illustrates a PAL device consisting of three inputs and three outputs that implements the Boolean equations. The design module is shown in Figure 2.61. The test bench module is shown in Figure 2.62 and the outputs are shown in Figure 2.63.

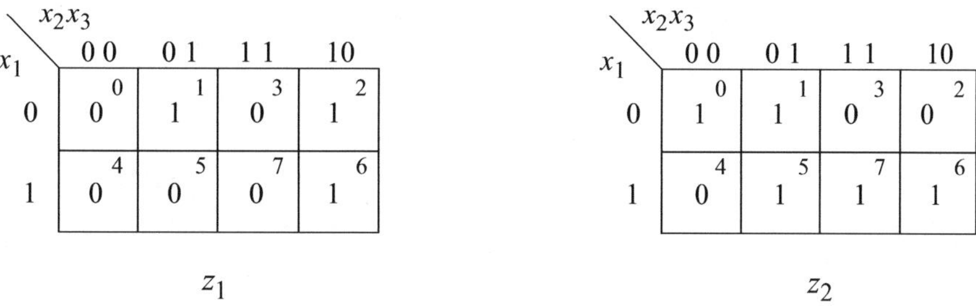

x_1 \ x_2x_3	0 0	0 1	1 1	10
0	0 (0)	1 (1)	0 (3)	1 (2)
1	1 (4)	0 (5)	1 (7)	1 (6)

z_3

Figure 2.59 Karnaugh maps for Example 2.17.

$$z_1 = x_1'x_2'x_3 + x_2x_3'$$

$$z_2 = x_1'x_2' + x_1x_2 + x_2'x_3 \tag{2.7}$$

$$z_3 = x_1'x_2'x_3 + x_2x_3' + x_1x_2 + x_1x_3'$$

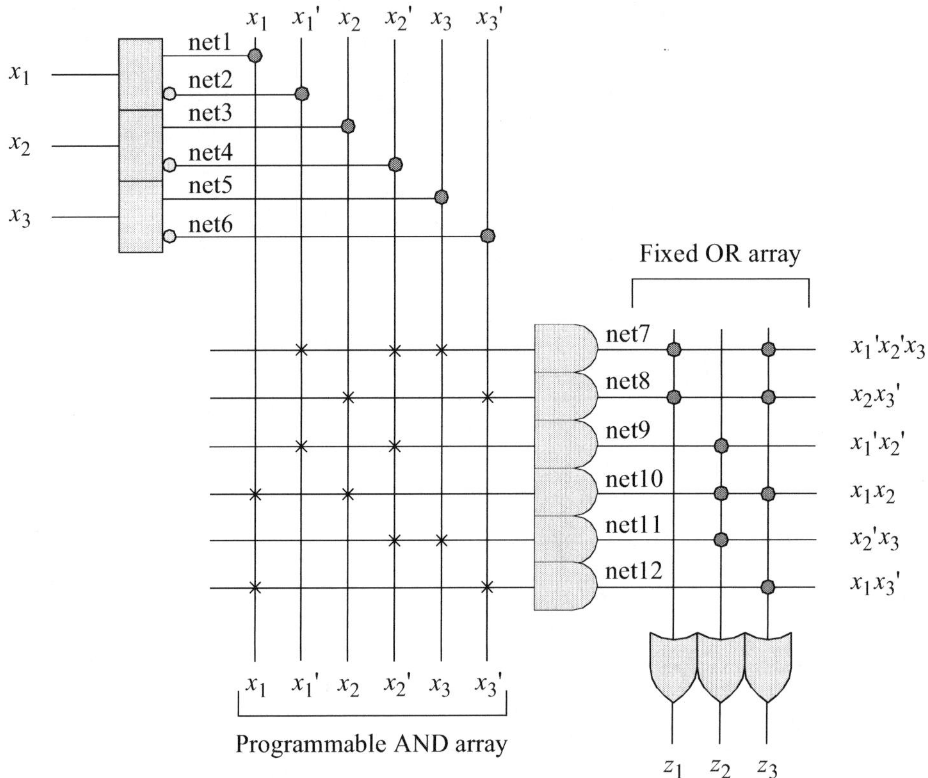

Figure 2.60 PAL device to implement Equation 2.7.

```
//structural pal for sop

module pal_sop (x1, x2, x3, z1, z2, z3);

//define inputs and outputs
input x1, x2, x3;
output z1, z2, z3;

//define internal nets
wire net1, net2, net3, net4, net5, net6, net7,
   net8, net9, net10, net11, net12;

//define the input logic
buf   (net1, x1);
not   (net2, x1);
                              //continued on next page
```

Figure 2.61 Design module for the PAL device to implement Equation 2.7.

```
buf   (net3, x2);
not   (net4, x2);

buf   (net5, x3);
not   (net6, x3);

//design the logic for the and array
and   (net7, net2, net4, net5),
      (net8, net3, net6),
      (net9, net3, net4),
      (net10, net1, net3),
      (net11, net4, net5),
      (net12, net1, net6);

//define the logic for the or array
or    (z1, net7, net8),
      (z2, net9, net10, net11),
      (z3, net7, net8, net10, net12);
endmodule
```

Figure 2.61 (Continued)

```
//test bench for sop pal
module pal_sop_tb;

//inputs are reg for test bench
//outputs are wire for test bench
reg x1, x2, x3;
wire z1, z2, z3;

//display variables
initial
$monitor ("x1 x2 x3 = %b, z1 z2 z3 = %b",
            {x1, x2, x3}, {z1, z2, z3});

//apply input vectors
initial
begin
   #0    x1 = 1'b0;  x2 = 1'b0;  x3 = 1'b0;
   #10   x1 = 1'b0;  x2 = 1'b0;  x3 = 1'b1;
   #10   x1 = 1'b0;  x2 = 1'b1;  x3 = 1'b0;
   #10   x1 = 1'b0;  x2 = 1'b1;  x3 = 1'b1;
                                 //continued on next page
```

Figure 2.62 Test bench module for the PAL device to implement Equation 2.7.

```
    #10   x1 = 1'b1;   x2 = 1'b0;   x3 = 1'b0;
    #10   x1 = 1'b1;   x2 = 1'b0;   x3 = 1'b1;
    #10   x1 = 1'b1;   x2 = 1'b1;   x3 = 1'b0;
    #10   x1 = 1'b1;   x2 = 1'b1;   x3 = 1'b1;

    #10   $stop;
end

//instantiate the module into the test bench
pal_sop inst1 (x1, x2, x3, z1, z2, z3);

endmodule
```

Figure 2.62 (Continued)

```
x1 x2 x3 = 001, z1 z2 z3 = 111
x1 x2 x3 = 010, z1 z2 z3 = 101
x1 x2 x3 = 011, z1 z2 z3 = 000

x1 x2 x3 = 100, z1 z2 z3 = 001
x1 x2 x3 = 101, z1 z2 z3 = 010
x1 x2 x3 = 110, z1 z2 z3 = 111
x1 x2 x3 = 111, z1 z2 z3 = 011
```

Figure 2.63 Outputs for the PAL device to implement Equation 2.7.

2.6.3 Programmable Logic Array

Both the AND array and the OR array are programmable for a PLA. Since both arrays are programmable, the PLA has more programming capability and thus, more flexibility than the PROM or PAL. The output function in a PLA is limited only by the number of AND gates in the AND array, since all AND gates can be programmed to connect to all OR gates. This is in contrast to the output function in a PAL, which is restricted not only by the number of AND gates in the AND array, but also by the fixed connections from the AND array outputs to the OR array.

Example 2.18 This example implements the four outputs z_1, z_2, z_3, and z_4 in Equation 2.8 using a PLA design. There are also three inputs, x_1, x_2, and x_3. The PLA design is shown in Figure 2.64. The structural design module is shown in Figure 2.65. The test bench module and outputs are shown in Figures 2.66 and 2.67, respectively.

$$z_1 = x_1x_2' + x_1'x_2$$

$$z_2 = x_1x_3 + x_1'x_3'$$

$$z_3 = x_1x_2' + x_1'x_2'x_3' + x_1x_3' \qquad (2.8)$$

$$z_4 = x_1x_2x_3 + x_1'x_3$$

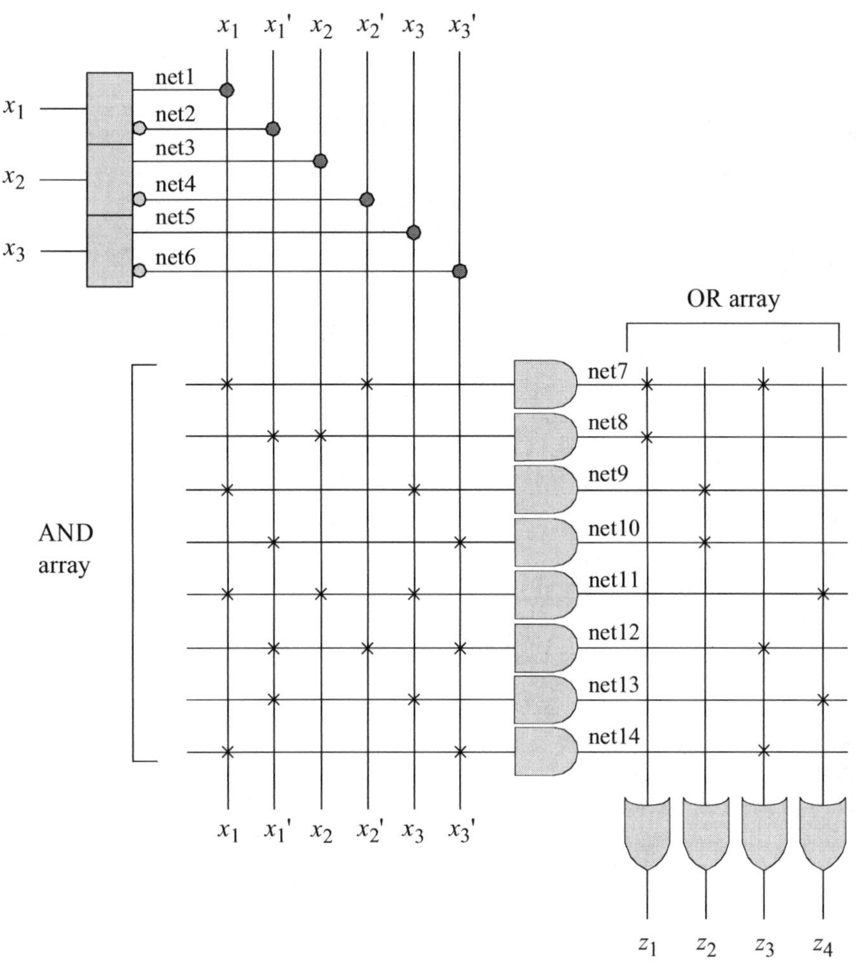

Figure 2.64 Logic diagram for a PLA device to implement Equation 2.8.

```
//structural pla to implement four equations
//z1 = x1x2' + x1'x2
//z2 = x1x3 + x1'x3'
//z3 = x1x2' + x1'x2'x3' + x1x3'
//z4 = x1x2x3 + x1'x3

module pla_4eqtns (x1, x2, x3, z1, z2, z3, z4);

//define inputs and outputs
input x1, x2, x3;
output z1, z2, z3, z4;

//define internal nets
wire net1, net2, net3, net4, net5, net6, net7, net8,
      net9, net10, net11, net12, net13, net14;

//design the input drivers
buf   (net1, x1);
not   (net2, x1);

buf   (net3, x2);
not   (net4, x2);

buf   (net5, x3);
not   (net6, x3);

//design the logic for the and array and the or array for z1
and   (net7, net1, net4),
      (net8, net2, net3);
or    (z1, net7, net8);

//design the logic for the and array and the or array for z2
and   (net9, net1, net5),
      (net10, net2, net6);
or    (z2, net9, net10);

//design the logic for the and array and the or array for z3
and   (net12, net2, net4, net6),
      (net14, net1, net6);
or    (z3, net7, net12, net14);

//design the logic for the and array and the or array for z4
and   (net11, net1, net3, net5),
      (net13, net2, net5);
or    (z4, net11, net13);

endmodule
```

Figure 2.65 Structural design module for Equation 2.8.

```
//test bench to implement four equations

module pla_4eqtns_tb;


//inputs are reg for test bench
//outputs are wire for test bench
reg   x1, x2, x3;
wire  z1, z2, z3, z4;


initial      //display variables
$monitor ("x1 x2 x3 = %b, z1 z2 z3 z4 = %b", {x1, x2, x3},
            {z1, z2, z3, z4});


initial      //apply input vectors
begin
   #0     x1 = 1'b0;  x2 = 1'b0;  x3 = 1'b0;
   #10    x1 = 1'b0;  x2 = 1'b0;  x3 = 1'b1;
   #10    x1 = 1'b0;  x2 = 1'b1;  x3 = 1'b0;
   #10    x1 = 1'b0;  x2 = 1'b1;  x3 = 1'b1;

   #10    x1 = 1'b1;  x2 = 1'b0;  x3 = 1'b0;
   #10    x1 = 1'b1;  x2 = 1'b0;  x3 = 1'b1;
   #10    x1 = 1'b1;  x2 = 1'b1;  x3 = 1'b0;
   #10    x1 = 1'b1;  x2 = 1'b1;  x3 = 1'b1;


   #10    $stop;

end


//instantiate the module into the test bench
pla_4eqtns inst1 (x1, x2, x3, z1, z2, z3, z4);


endmodule
```

Figure 2.66 Test bench module for the PLA device to implement Equation 2.8.

```
x1 x2 x3 = 000, z1 z2 z3 z4 = 0110
x1 x2 x3 = 001, z1 z2 z3 z4 = 0001
x1 x2 x3 = 010, z1 z2 z3 z4 = 1100
x1 x2 x3 = 011, z1 z2 z3 z4 = 1001

x1 x2 x3 = 100, z1 z2 z3 z4 = 1010
x1 x2 x3 = 101, z1 z2 z3 z4 = 1110
x1 x2 x3 = 110, z1 z2 z3 z4 = 0010
x1 x2 x3 = 111, z1 z2 z3 z4 = 0101
```

Figure 2.67 Outputs for the PLA device to implement Equation 2.8.

Example 2.19 A 5-input majority circuit will be designed using a PLA. The output of a majority circuit is a logic 1 if the majority of the inputs is a logic 1; otherwise, the output is a logic 0. Therefore, a majority circuit must have an odd number of inputs in order to have a majority of the inputs be at the same logic level.

A 5-input majority circuit will be designed using the Karnaugh map of Figure 2.68, where a 1 entry indicates that the majority of the inputs is a logic 1. The resulting equation is shown in Equation 2.9 representing the logic for output z_1 in a sum-of-products form. The logic diagram for the PLA device is shown in Figure 2.69. The structural design module using a PLA is shown in Figure 2.70. The test bench module is shown in Figure 2.71 and the outputs are shown in Figure 2.72.

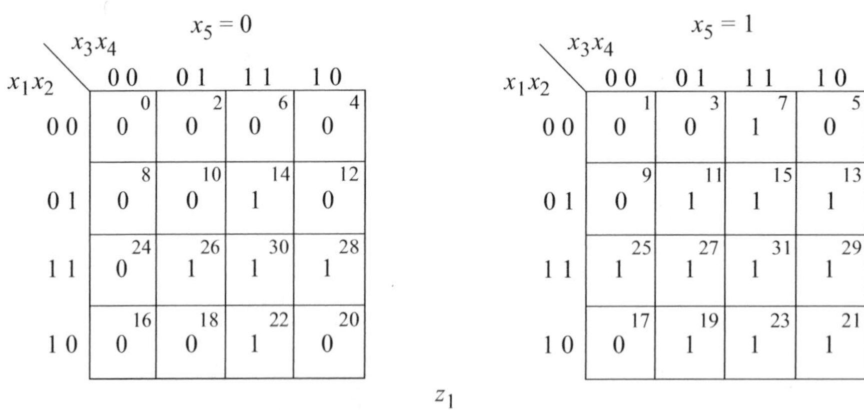

Figure 2.68 Karnaugh map for the majority circuit of Example 2.19.

$$\begin{aligned} z_1 = {} & x_3x_4x_5 + x_2x_3x_5 + x_1x_3x_5 + x_2x_4x_5 + x_1x_4x_5 \\ & + x_1x_2x_5 + x_1x_2x_4 + x_2x_3x_4 + x_1x_3x_4 \end{aligned} \tag{2.9}$$

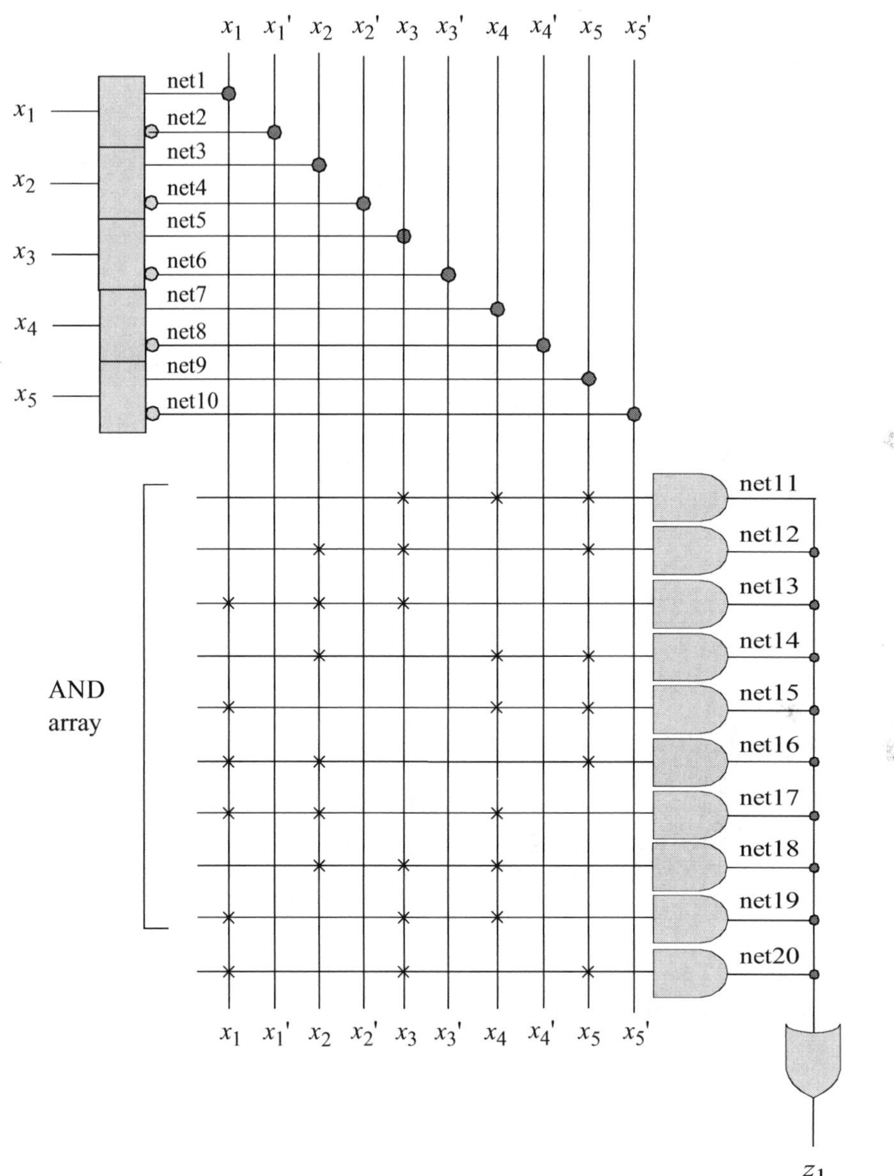

Figure 2.69 Logic diagram for the majority circuit of Example 2.19.

```
//structural for a 5-input majority circuit
module pla_majority (x1, x2, x3, x4, x5, z1);
//define inputs and output
input x1, x2, x3, x4, x5;
output z1;                         //continued on next page
```

Figure 2.70 Structural design module for the majority circuit of Example 2.19.

```
//define internal nets
wire net1, net2, net3, net4, net5, net6, net7, net8,
      net9, net10, net11, net12, net13, net14,
      net15, net16, net17, net18, net19, net20;


//design the input drivers
buf   (net1, x1);
not   (net2, x1);

buf   (net3, x2);
not   (net4, x2);

buf   (net5, x3);
not   (net6, x3);

buf   (net7, x4);
not   (net8, x4);

buf   (net9, x5);
not   (net10, x5);


//define the logic for the and array
and   (net11, net5, net7, net9),
      (net12, net3, net5, net9),
      (net13, net1, net3, net5),
      (net14, net3, net7, net9),
      (net15, net1, net7, net9),
      (net16, net1, net3, net9),
      (net17, net1, net3, net7),
      (net18, net3, net5, net7),
      (net19, net1, net5, net7),
      (net20, net1, net5, net9);


//design the logic for output z1
or    (z1, net11, net12, net13, net14, net15,
         net16, net17, net18, net19, net20);

endmodule
```

Figure 2.70 (Continued)

```
//test bench for 5-input majority circuit
module pla_majority_tb;

//inputs are reg for test bench
//outputs are wire for test bench
reg x1, x2, x3, x4, x5;
wire z1;

//apply input vectors
initial
begin: apply_stimulus
   reg [6:0] invect;
   for (invect = 0; invect < 32; invect = invect + 1)
      begin
         {x1, x2, x3, x4, x5} = invect [6:0];
         #10 $display ("x1 x2 x3 x4 x5 = %b, z1 = %b",
                        {x1, x2, x3, x4, x5}, z1);
      end
end

//instantiate the module into the test bench
pla_majority inst1 (x1, x2, x3, x4, x5, z1);

endmodule
```

Figure 2.71 Test bench module for the majority circuit of Example 2.19.

```
x1 x2 x3 x4 x5 = 00000, z1 = 0
x1 x2 x3 x4 x5 = 00001, z1 = 0
x1 x2 x3 x4 x5 = 00010, z1 = 0
x1 x2 x3 x4 x5 = 00011, z1 = 0
x1 x2 x3 x4 x5 = 00100, z1 = 0
x1 x2 x3 x4 x5 = 00101, z1 = 0
x1 x2 x3 x4 x5 = 00110, z1 = 0
x1 x2 x3 x4 x5 = 00111, z1 = 1

x1 x2 x3 x4 x5 = 01000, z1 = 0
x1 x2 x3 x4 x5 = 01001, z1 = 0
x1 x2 x3 x4 x5 = 01010, z1 = 0
x1 x2 x3 x4 x5 = 01011, z1 = 1
x1 x2 x3 x4 x5 = 01100, z1 = 0
x1 x2 x3 x4 x5 = 01101, z1 = 1
x1 x2 x3 x4 x5 = 01110, z1 = 1
x1 x2 x3 x4 x5 = 01111, z1 = 1        //continued on next page
```

Figure 2.72 Outputs for the majority circuit of Example 2.19.

```
x1 x2 x3 x4 x5 = 10000, z1 = 0
x1 x2 x3 x4 x5 = 10001, z1 = 0
x1 x2 x3 x4 x5 = 10010, z1 = 0
x1 x2 x3 x4 x5 = 10011, z1 = 1
x1 x2 x3 x4 x5 = 10100, z1 = 0
x1 x2 x3 x4 x5 = 10101, z1 = 1
x1 x2 x3 x4 x5 = 10110, z1 = 1
x1 x2 x3 x4 x5 = 10111, z1 = 1

x1 x2 x3 x4 x5 = 11000, z1 = 0
x1 x2 x3 x4 x5 = 11001, z1 = 1
x1 x2 x3 x4 x5 = 11010, z1 = 1
x1 x2 x3 x4 x5 = 11011, z1 = 1
x1 x2 x3 x4 x5 = 11100, z1 = 1
x1 x2 x3 x4 x5 = 11101, z1 = 1
x1 x2 x3 x4 x5 = 11110, z1 = 1
x1 x2 x3 x4 x5 = 11111, z1 = 1
```

Figure 2.72 (Continued)

Example 2.20 This example designs a structural module to convert the Gray code to the corresponding binary code using a PLA device. The Gray and binary codes are shown in Table 2.6. The Karnaugh maps used to obtain the equations for the code converter are shown in Figure 2.73. The equations for the binary vectors are shown in Equation 2.10. The Gray-to-binary logic diagram using a PLA device is shown in Figure 2.74.

The structural design module using a PLA device is shown in Figure 2.75. The test bench module is shown in Figure 2.76 and the outputs are shown in Figure 2.77.

Table 2.6 Gray-to-Binary Code Conversion

Gray			Binary		
g_1	g_2	g_3	b_1	b_2	b_3
0	0	0	0	0	0
0	0	1	0	0	1
0	1	1	0	1	0
0	1	0	0	1	1
1	1	0	1	0	0
1	1	1	1	0	1
1	0	1	1	1	0
1	0	0	1	1	1

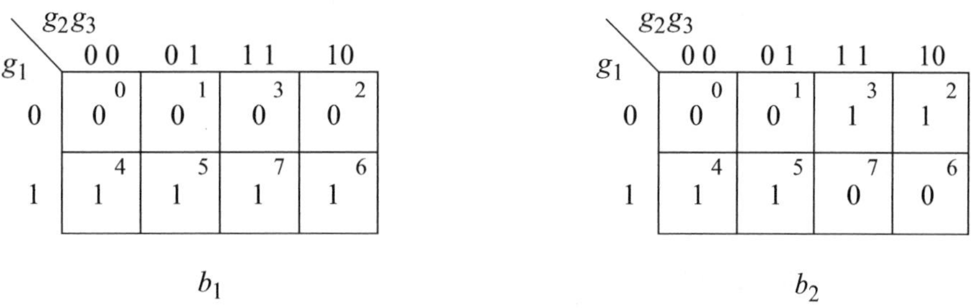

g_1 \ g_2g_3	00	01	11	10
0	0 (0)	1 (1)	0 (3)	1 (2)
1	1 (4)	0 (5)	1 (7)	0 (6)

b_3

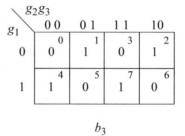

Figure 2.73 Karnaugh maps for the Gray-to-binary code converter.

$$
\begin{aligned}
b_1 &= g_1 \\
b_2 &= g_1' g_2 + g_1 g_2' \\
b_3 &= g_1' g_2' g_3 + g_1' g_2 g_3' + g_1 g_2' g_3' + g_1 g_2 g_3
\end{aligned}
\tag{2.10}
$$

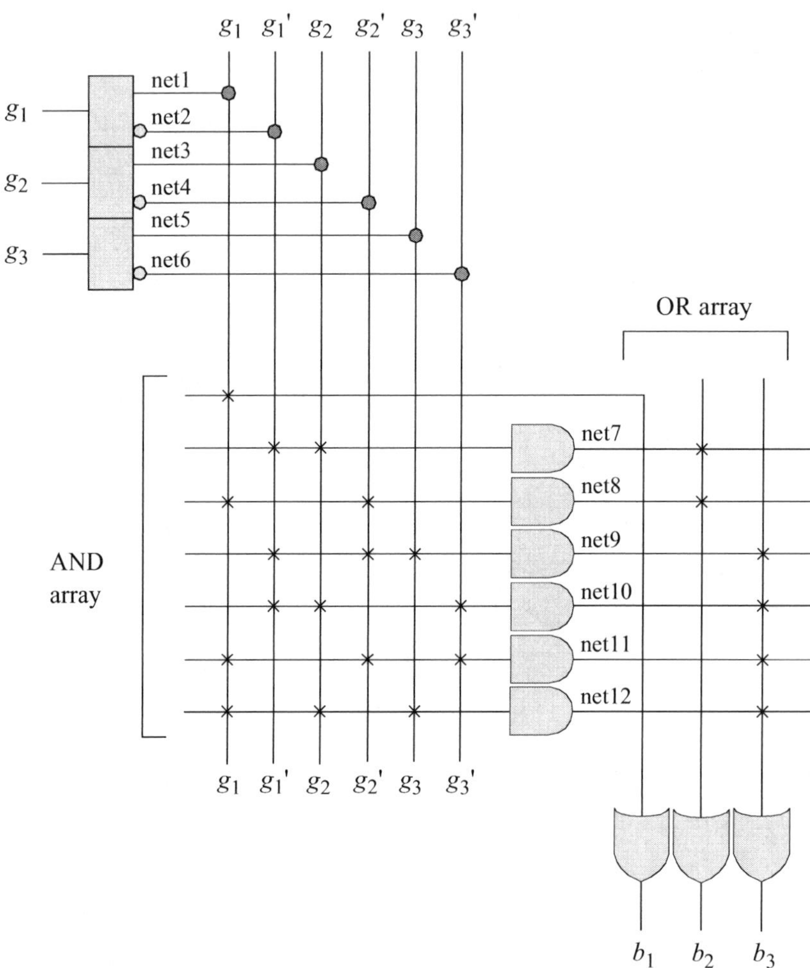

Figure 2.74 PLA device for the Gray-to-binary code converter.

```
//structural for gray-to-binary converter
module pla_gray_to_bin (g1, g2, g3, b1, b2, b3);

//define inputs and outputs
input g1, g2, g3;
output b1, b2, b3;

//define internal nets
wire net1, net2, net3, net4, net5, net6,
      net7, net8, net9, net10, net11, net12;
                                        //continued on next page
```

Figure 2.75 Structural design module for the Gray-to-binary code converter.

```
//define the input drivers
buf    (net1, g1);
not    (net2, g1);

buf    (net3, g2);
not    (net4, g2);

buf    (net5, g3);
not    (net6, g3);

//design the logic for the and array
and    (net7, net2, net3),
       (net8, net1, net4),
       (net9, net2, net4, net5),
       (net10, net2, net3, net6),
       (net11, net1, net4, net6),
       (net12, net1, net3, net5);

//design the logic for the outputs b1, b2, and b3
or     (b1, net1),
       (b2, net7, net8),
       (b3, net9, net10, net11, net12);
endmodule
```

Figure 2.75 (Continued)

```
//test bench for the gray-to-binary converter
module pla_gray_to_bin_tb;

//inputs are reg for test bench
//outputs are wire for test bench
reg g1, g2, g3;
wire b1, b2, b3;

//apply input vectors
initial
begin: apply_stimulus
   reg [4:0] invect;
   for (invect = 0; invect < 8; invect = invect + 1)
      begin
         {g1, g2, g3} = invect [4:0];
         #10 $display ("g1 g2 g3 = %b, b1 b2 b3 = %b",
                           {g1, g2, g3}, {b1, b2, b3});
      end
end                                    //continued on next page
```

Figure 2.76 Test bench module for the Gray-to-binary code converter.

```
//instantiate the module into the test bench
pla_gray_to_bin inst1 (g1, g2, g3, b1, b2, b3);

endmodule
```

Figure 2.76 (Continued)

```
g1 g2 g3 = 000, b1 b2 b3 = 000
g1 g2 g3 = 001, b1 b2 b3 = 001
g1 g2 g3 = 010, b1 b2 b3 = 011
g1 g2 g3 = 011, b1 b2 b3 = 010

g1 g2 g3 = 100, b1 b2 b3 = 111
g1 g2 g3 = 101, b1 b2 b3 = 110
g1 g2 g3 = 110, b1 b2 b3 = 100
g1 g2 g3 = 111, b1 b2 b3 = 101
```

Figure 2.77 Outputs for the Gray-to-binary code converter.

2.7 Additional Design Examples

This section will present several combinational logic design examples utilizing the modeling methods presented in Chapter 1 plus additional techniques. Examples using multiplexers to obtain a minimized design module will also be presented. An *iterative network* will be introduced. An iterative network is an organization of identical cells which are interconnected in an ordered manner with the signals propagating in one direction only. An iterative machine (or network) can consist of combinational logic arranged in a linear array. The Boolean functions obtained in the examples can be minimized and represented in both a sum-of-products form and a product-of-sums form.

Example 2.21 This example uses the Karnaugh map shown in Figure 2.78 to obtain the minimized equation for z_1 in both a sum-of-products form and a product-of-sums form. The sum-of-products expression will be designed using dataflow modeling with the continuous assignment statement utilizing the keyword **assign**. The product-of-sums expression will be designed using built-in-primitives in Example 2.22.

The dataflow design module is shown in Figure 2.79. The test bench module is shown in Figure 2.80 and the outputs are shown in Figure 2.81.

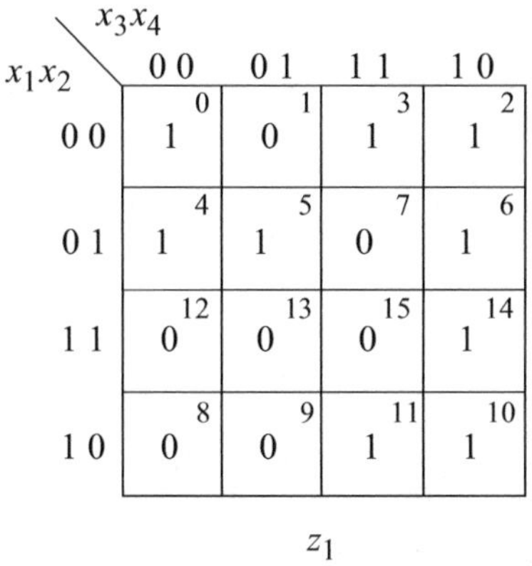

x_1x_2 \ x_3x_4	0 0	0 1	1 1	1 0
0 0	1 (0)	0 (1)	1 (3)	1 (2)
0 1	1 (4)	1 (5)	0 (7)	1 (6)
1 1	0 (12)	0 (13)	0 (15)	1 (14)
1 0	0 (8)	0 (9)	1 (11)	1 (10)

z_1

Figure 2.78 Karnaugh map for Example 2.21.

Sum-of-products form

$$z_1 = x_1'x_4' + x_1'x_2x_3' + x_3x_4' + x_2'x_3$$

Product-of-sums form

$$z_1 = (x_1' + x_3)(x_2' + x_3' + x_4')(x_2 + x_3 + x_4')$$

```
//dataflow for sum-of-products expression
//z1 = x1'x4' + x1'x2x3' + x3x4' + x2'x3

module sop_pos_df_bip (x1, x2, x3, x4, z1);

//define inputs and outputs
input x1, x2, x3, x4;
output z1;

//design logic using the continuous assignment statement
assign z1 = (~x1 & ~x4) | (~x1 & x2 & ~x3) | (x3 & ~x4)
               | (~x2 & x3);

endmodule
```

Figure 2.79 Dataflow design module for Example 2.21.

```
//test bench for sum-of-products equation
module sop_pos_df_bip_tb;

//inputs are reg for test bench
//outputs are wire for test bench
reg x1, x2, x3, x4;
wire z1;

initial     //apply input vectors
begin: apply_stimulus
   reg [4:0] invect;
   for (invect = 0; invect < 16; invect = invect + 1)
      begin
         {x1, x2, x3, x4} = invect [4:0];
         #10 $display ("{x1 x2 x3 x4} = %b, z1 = %b",
                        {x1, x2, x3, x4}, z1);
      end
end

//instantiate the module into the test bench
sop_pos_df_bip inst1 (x1, x2, x3, x4, z1);

endmodule
```

Figure 2.80 Test bench module for Example 2.21.

```
{x1 x2 x3 x4} = 0000, z1 = 1
{x1 x2 x3 x4} = 0001, z1 = 0
{x1 x2 x3 x4} = 0010, z1 = 1
{x1 x2 x3 x4} = 0011, z1 = 1

{x1 x2 x3 x4} = 0100, z1 = 1
{x1 x2 x3 x4} = 0101, z1 = 1
{x1 x2 x3 x4} = 0110, z1 = 1
{x1 x2 x3 x4} = 0111, z1 = 0

{x1 x2 x3 x4} = 1000, z1 = 0
{x1 x2 x3 x4} = 1001, z1 = 0
{x1 x2 x3 x4} = 1010, z1 = 1
{x1 x2 x3 x4} = 1011, z1 = 1

{x1 x2 x3 x4} = 1100, z1 = 0
{x1 x2 x3 x4} = 1101, z1 = 0
{x1 x2 x3 x4} = 1110, z1 = 1
{x1 x2 x3 x4} = 1111, z1 = 0
```

Figure 2.81 Outputs for Example 2.21.

Example 2.22 This example repeats Example 2.21, but uses built-in primitives for the product-of-sums equation of Example 2.21. The Karnaugh map and the product-of-sums equation are reproduced in Figure 2.82 and Equation 2.11 for convenience. The design module is shown in Figure 2.83. The test bench module is shown in Figure 2.84 and the outputs are shown in Figure 2.85.

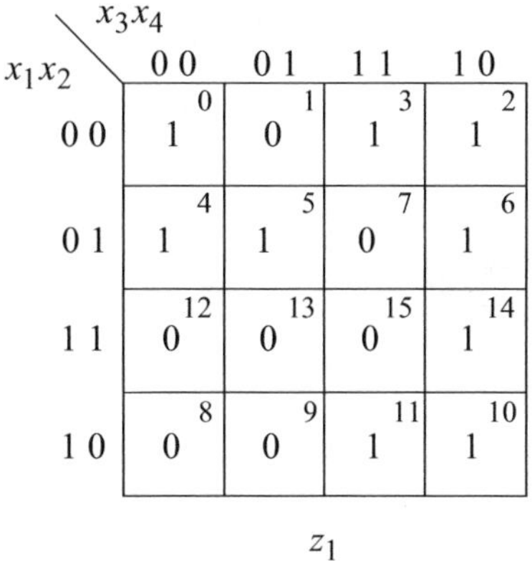

x_1x_2 \ x_3x_4	0 0	0 1	1 1	1 0
0 0	1 (0)	0 (1)	1 (3)	1 (2)
0 1	1 (4)	1 (5)	0 (7)	1 (6)
1 1	0 (12)	0 (13)	0 (15)	1 (14)
1 0	0 (8)	0 (9)	1 (11)	1 (10)

z_1

Figure 2.82 Karnaugh map for Example 2.22.

$$z_1 = (x_1' + x_3)\,(x_2' + x_3' + x_4')\,(x_2 + x_3 + x_4') \tag{2.11}$$

```
//built-in primitives for pos equation
//z1 = (x1' + x3) ( x2' + x3' + x4') (x2 + x3 + x4')
//        net1              net2                 net3

module sop_pos_bip (x1, x2, x3, x4, z1);

//define inputs and output
input x1, x2, x3, x4;
output z1;

//design the logic using built-in primitives
or    inst1 (net1, ~x1, x3),
      inst2 (net2, ~x2, ~x3, ~x4),
      inst3 (net3, x2, x3, ~x4);

and   inst4 (z1, net1, net2, net3);

endmodule
```

Figure 2.83 Design module for Example 2.22.

```
//test bench for pos equation
module sop_pos_bip_tb;

//inputs are reg for test bench
//outputs are wire for test bench
reg x1, x2, x3, x4;
wire z1;

initial          //apply input vectors
begin: apply_stimulus
   reg [4:0] invect;
   for (invect = 0; invect < 16; invect = invect + 1)
      begin
         {x1, x2, x3, x4} = invect [4:0];
         #10 $display ("{x1 x2 x3 x4} = %b, z1 = %b",
                           {x1, x2, x3, x4}, z1);
      end
end

//instantiate the module into the test bench
sop_pos_bip inst1 (x1, x2, x3, x4, z1);

endmodule
```

Figure 2.84 Test bench module for Example 2.22.

```
{x1 x2 x3 x4} = 0000, z1 = 1
{x1 x2 x3 x4} = 0001, z1 = 0
{x1 x2 x3 x4} = 0010, z1 = 1
{x1 x2 x3 x4} = 0011, z1 = 1

{x1 x2 x3 x4} = 0100, z1 = 1
{x1 x2 x3 x4} = 0101, z1 = 1
{x1 x2 x3 x4} = 0110, z1 = 1
{x1 x2 x3 x4} = 0111, z1 = 0

{x1 x2 x3 x4} = 1000, z1 = 0
{x1 x2 x3 x4} = 1001, z1 = 0
{x1 x2 x3 x4} = 1010, z1 = 1
{x1 x2 x3 x4} = 1011, z1 = 1

{x1 x2 x3 x4} = 1100, z1 = 0
{x1 x2 x3 x4} = 1101, z1 = 0
{x1 x2 x3 x4} = 1110, z1 = 1
{x1 x2 x3 x4} = 1111, z1 = 0
```

Figure 2.85 Outputs for Example 2.22.

Example 2.23 This example designs a comparator to compare two 4-bit operands *a[3:0]* and *b[3:0]* to determine if $A < B$ or if $A = B$. The conditional statements **if**, **else if**, and **else** will be used in the design. The equations used for the comparison are shown in Equation 2.12.

$$(A < B) = a_3'\, b_3 + (a_3 \oplus b_3)'\, a_2'\, b_2 + (a_3 \oplus b_3)'\, (a_2 \oplus b_2)'\, a_1'\, b_1 + (a_3 \oplus b_3)'\, (a_2 \oplus b_2)'\, (a_1 \oplus b_1)'\, a_0'\, b_0$$

$$(A = B) = (a_3 \oplus b_3)'\, (a_2 \oplus b_2)'\, (a_1 \oplus b_1)'\, (a_0 \oplus b_0)' \qquad (2.12)$$

The behavioral design module is shown in Figure 2.86. The test bench module is shown in Figure 2.87 and the outputs are shown in Figure 2.88.

```
//behavioral conditional statements to compare
//two operands for less than and equal

module a_lt_eq_b_cond (a, b, a_lt_b, a_eq_b);

//define inputs and outputs
input [3:0] a, b;
output a_lt_b, a_eq_b;

//variables used in always are declared as reg
reg a_lt_b, a_eq_b;

//design the 4-bit comparator for less than
always @ (a or b)
begin
   if (~a[3] & b[3])
      a_lt_b = 1'b1;

   else if ((a[3] ^~ b[3]) & (~a[2] & b[2]))
      a_lt_b = 1'b1;

   else if ((a[3] ^~ b[3]) & (a[2] ^~ b[2]) & (~a[1] & b[1]))
      a_lt_b = 1'b1;

   else if ((a[3] ^~ b[3]) & (a[2] ^~ b[2]) & (a[1] ^~ b[1])
               & (~a[0] & b[0]))
      a_lt_b = 1'b1;

   else a_lt_b = 1'b0;
end                                  //continued on next page
```

Figure 2.86 Behavioral design module for the comparison of two operands.

```
//---------------------------------------------------------
//design the 4-bit comparator for equal
always @ (a or b)
begin
   if ((a[3] ^~ b[3]) & (a[2] ^~ b[2]) & (a[1] ^~ b[1])
         & (a[0] ^~ b[0]))
      a_eq_b = 1'b1;

   else a_eq_b = 1'b0;
end

endmodule
```

Figure 2.86 (Continued)

```
//test bench to detect if A <= B

module a_lt_eq_b_cond_tb;

//inputs are reg for test bench
//outputs are wire for test bench
reg [3:0] a, b;
wire a_lt_b, a_eq_b;

//display inputs and outputs
initial
$monitor ("a = %b, b = %b, a_lt_b = %b, a_eq_b = %b",
            a, b, a_lt_b, a_eq_b);

//apply input vectors
initial
begin
   #0    a = 4'b0000;   b = 4'b0001;
   #10   a = 4'b0011;   b = 4'b0010;
   #10   a = 4'b0011;   b = 4'b1011;
   #10   a = 4'b0111;   b = 4'b0011;

   #10   a = 4'b0101;   b = 4'b0101;

   #10   a = 4'b0110;   b = 4'b0111;
   #10   a = 4'b0011;   b = 4'b0010;
   #10   a = 4'b1110;   b = 4'b1111;      //continued on
   #10   a = 4'b0110;   b = 4'b0011;      //next page
```

Figure 2.87 Test bench module for the comparison of two operands.

```
    #10    a = 4'b1100;    b = 4'b1100;

    #10    a = 4'b0110;    b = 4'b0111;
    #10    a = 4'b1011;    b = 4'b1010;
    #10    a = 4'b1110;    b = 4'b1111;
    #10    a = 4'b1110;    b = 4'b0111;

    #10    a = 4'b0111;    b = 4'b0111;

    #10    a = 4'b1000;    b = 4'b1100;
    #10    a = 4'b1011;    b = 4'b1010;
    #10    a = 4'b1110;    b = 4'b1111;
    #10    a = 4'b1110;    b = 4'b0111;

    #10    a = 4'b0000;    b = 4'b0000;

    #10    $stop;
end

//instantiate the module into the test bench
a_lt_eq_b_cond inst1 (a, b, a_lt_b, a_eq_b);

endmodule
```

Figure 2.87 (Continued)

```
a = 0000, b = 0001, | a_lt_b = 1, a_eq_b = 0
a = 0011, b = 0010, | a_lt_b = 0, a_eq_b = 0
a = 0011, b = 1011, | a_lt_b = 1, a_eq_b = 0
a = 0111, b = 0011, | a_lt_b = 0, a_eq_b = 0
a = 0101, b = 0101, | a_lt_b = 0, a_eq_b = 1


a = 0110, b = 0111, | a_lt_b = 1, a_eq_b = 0
a = 0011, b = 0010, | a_lt_b = 0, a_eq_b = 0
a = 1110, b = 1111, | a_lt_b = 1, a_eq_b = 0
a = 0110, b = 0011, | a_lt_b = 0, a_eq_b = 0
a = 1100, b = 1100, | a_lt_b = 0, a_eq_b = 1

                      //continued on next page
```

Figure 2.88 Outputs for the comparison of two operands.

```
a = 0110, b = 0111, | a_lt_b = 1, a_eq_b = 0
a = 1011, b = 1010, | a_lt_b = 0, a_eq_b = 0
a = 1110, b = 1111, | a_lt_b = 1, a_eq_b = 0
a = 1110, b = 0111, | a_lt_b = 0, a_eq_b = 0
a = 0111, b = 0111, | a_lt_b = 0, a_eq_b = 1


a = 1000, b = 1100, | a_lt_b = 1, a_eq_b = 0
a = 1011, b = 1010, | a_lt_b = 0, a_eq_b = 0
a = 1110, b = 1111, | a_lt_b = 1, a_eq_b = 0
a = 1110, b = 0111, | a_lt_b = 0, a_eq_b = 0
a = 0000, b = 0000, | a_lt_b = 0, a_eq_b = 1
```

Figure 2.88 (Continued)

Example 2.24 A logic circuit will be designed that generates an output z_1 if a 4-bit number *a[3:0]* satisfies the following requirements: *a[3:0]* is greater than 11 or less than 4. The design module will use the continuous assignment statement **assign**. The Karnaugh map is shown in Figure 2.89 and the equation is shown in Equation 2.13. The dataflow design module is shown in Figure 2.90. The test bench module is shown in Figure 2.91 and the outputs are shown in Figure 2.92.

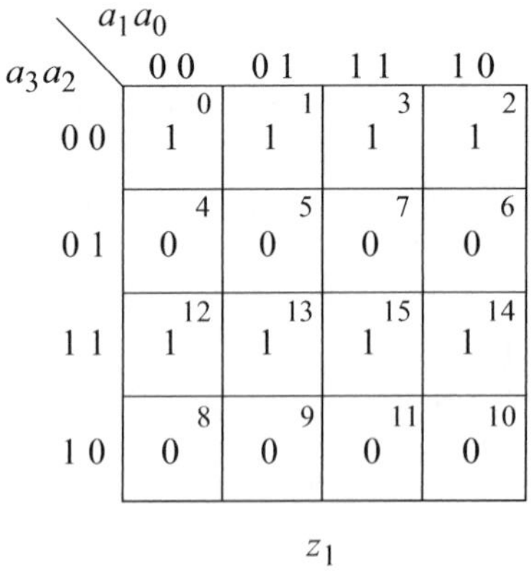

$a_3 a_2$ \ $a_1 a_0$	0 0	0 1	1 1	1 0
0 0	1 (0)	1 (1)	1 (3)	1 (2)
0 1	0 (4)	0 (5)	0 (7)	0 (6)
1 1	1 (12)	1 (13)	1 (15)	1 (14)
1 0	0 (8)	0 (9)	0 (11)	0 (10)

z_1

Figure 2.89 Karnaugh map for Example 2.24

$$z_1 = a_3' a_2' + a_3 a_2 \qquad (2.13)$$

```
//dataflow to detect number in the range >11 and <4

module number_range3 (a, z1);

//define inputs and output
input [3:0] a;
output z1;

//design the dataflow logic using assign
assign z1 = a[3] ^~ a[2];

endmodule
```

Figure 2.90 Dataflow design module to determine if a number is greater than 11 or less than 4.

```
//test bench for number range

module number_range3_tb;


//inputs are reg for test bench
//outputs are wire for test bench
reg [3:0] a;
wire z1;


//apply input vectors
initial
begin: apply_stimulus
   reg [4:0] invect;
   for (invect = 0; invect < 16; invect = invect + 1)
      begin
         a = invect [4:0];
         #10 $display ("a = %b, z1 = %b", a, z1);
      end
end


//instantiate the module into the test bench
number_range3 inst1 (a, z1);

endmodule
```

Figure 2.91 Test bench module to determine if a number is greater than 11 or less than 4.

```
a = 0000, z1 = 1
a = 0001, z1 = 1
a = 0010, z1 = 1
a = 0011, z1 = 1

a = 0100, z1 = 0
a = 0101, z1 = 0
a = 0110, z1 = 0
a = 0111, z1 = 0

a = 1000, z1 = 0
a = 1001, z1 = 0
a = 1010, z1 = 0
a = 1011, z1 = 0

a = 1100, z1 = 1
a = 1101, z1 = 1
a = 1110, z1 = 1
a = 1111, z1 = 1
```

Figure 2.92 Outputs for the dataflow module to determine if a number is greater than 11 or less than 4.

Example 2.25 This example is similar to Example 2.24, but uses the behavioral **if, else if, else** conditional statements. A logic circuit will be designed that generates an output z_1 if a 4-bit number *a[3:0]* satisfies the following requirements:

$$z_1 = a_2{}'\, a_0{}' + a_2\, a_0$$

The Karnaugh map is shown in Figure 2.93. The dataflow design module is shown in Figure 2.94. The test bench module is shown in Figure 2.95 and the outputs are shown in Figure 2.96.

a_3a_2 \ a_1a_0	0 0	0 1	1 1	1 0
0 0	1 (0)	0 (1)	0 (3)	1 (2)
0 1	0 (4)	1 (5)	1 (7)	0 (6)
1 1	0 (12)	1 (13)	1 (15)	0 (14)
1 0	1 (8)	0 (9)	0 (11)	1 (10)

z_1

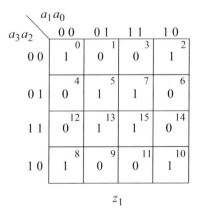

Figure 2.93 Karnaugh map for Example 2.25.

```
//behavioral for the following equation
//z1 = a2' a0' + a2 a0
module number_range4 (a, z1);

input [3:0] a;     //define inputs and output
output z1;

//variables used in always are declared as reg
reg z1;

always @ (a)
begin
   if (~a[2] & ~a[0])
      z1 = 1'b1;

   else if (a[2] & a[0])
      z1 = 1'b1;

   else z1 = 1'b0;
end

endmodule
```

Figure 2.94 Behavioral design module for Example 2.25 using **if**, **else if**, **else**.

```
//test bench for number_range4
module number_range4_tb;

reg [3:0] a;        //inputs are reg for test bench
wire z1;            //outputs are wire for test bench

initial      //apply input vectors
begin: apply_stimulus
   reg [4:0] invect;
   for (invect = 0; invect < 16; invect = invect + 1)
      begin
         a = invect [4:0];
         #10 $display ("a = %b, z1 = %b", a, z1);
      end
end

//instantiate the module into the test bench
number_range4 inst1 (a, z1);

endmodule
```

Figure 2.95 Test bench for Example 2.25 using **if**, **else if**, **else**.

```
a = 0000, z1 = 1
a = 0001, z1 = 0
a = 0010, z1 = 1
a = 0011, z1 = 0

a = 0100, z1 = 0
a = 0101, z1 = 1
a = 0110, z1 = 0
a = 0111, z1 = 1

a = 1000, z1 = 1
a = 1001, z1 = 0
a = 1010, z1 = 1
a = 1011, z1 = 0

a = 1100, z1 = 0
a = 1101, z1 = 1
a = 1110, z1 = 0
a = 1111, z1 = 1
```

Figure 2.96 Outputs for Example 2.25 using **if, else if, else**.

Example 2.26 This example will use the product-of-sums form to design a combinational logic circuit. A product-of-sums is an expression in which at least one term does not contain all the variables; that is, at least one term is a proper subset of the variables or their complements.

The minimal product-of-sums expression can be obtained by combining the 0s in the Karnaugh map to form sum terms in the same manner as the 1s were combined to form product terms. However, since 0s are being combined, each sum term must equal 0. The equation for Example 2.26 is shown in Equation 2.14.

$$z_1(x_1, x_2, x_3, x_4) = \Pi(0, 2, 5, 7, 8. 9, 10, 13, 14) \quad (2.14)$$

$$z_1 = (x_1 + x_2 + x_4)\ (x_1 + x_2' + x_4')\ (x_2' + x_3 + x_4')\ (x_1' + x_3' + x_4)\ (x_1' + x_2 + x_3)$$

The Karnaugh map for the product-of-sums form of Equation 2.14 is shown in Figure 2.97. The logic diagram is shown in Figure 2.98 using AND gates and OR gates. The structural design module is shown in Figure 2.99 using logic gates that were designed using dataflow modeling, such as *or3_df* and *and5_df*. The test bench module and the outputs are shown in Figures 2.100 and 2.101, respectively.

x_1x_2 \ x_3x_4	0 0	0 1	1 1	1 0
0 0	0 (0)	1 (1)	1 (3)	0 (2)
0 1	1 (4)	0 (5)	0 (7)	1 (6)
1 1	1 (12)	0 (13)	1 (15)	0 (14)
1 0	0 (8)	0 (9)	1 (11)	0 (10)

z_1

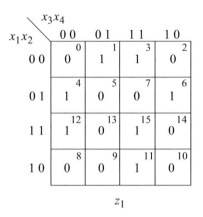

Figure 2.97 Karnaugh map for Example 2.26.

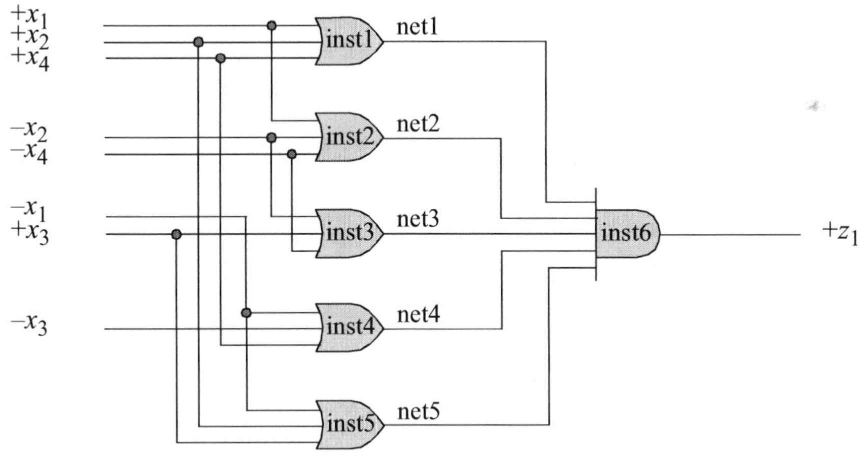

Figure 2.98 Logic diagram for Example 2.26.

```
//structural product-of-sums form for equation
//z1 = (x1 + x2 + x4)(x1 + x2' + x4')(x2' + x3 + x4')
//(x1' + x3' + x4)(x1' + x2 + x3)
module prod_of_sums (x1, x2, x3, x4, z1);

//define the inputs and output
input x1, x2, x3, x4;
output z1;                          //continued on next page
```

Figure 2.99 Structural design module for Example 2.26.

```
//define internal nets
wire net1, net2, net3, net4, net5;

//instantiate the logic gates
or3_df   inst1 (x1, x2, x4, net1);
or3_df   inst2 (x1, ~x2, ~x4, net2);
or3_df   inst3 (~x1, x3, ~x4, net3);
or3_df   inst4 (~x1, ~x3, x4, net4);
or3_df   inst5 (~x1, x2, x3, net5);

and5_df  inst6 (net1, net2, net3, net4, net5, z1);

endmodule
```

Figure 2.99 (Continued)

```
//test bench for product-of-sums

module prod_of_sums_tb;

//inputs are reg for test bench
//outputs are wire for test bench
reg x1, x2, x3, x4;
wire z1;

//apply input vectors
initial
begin: apply_stimulus
reg [4:0] invect;
for (invect = 0; invect < 16; invect = invect + 1)
   begin
      {x1, x2, x3, x4} = invect [4:0];
      #10 $display ("{x1 x2 x3 x4} = %b, z1 = %b",
                     {x1, x2, x3, x4}, z1);
   end
end

//instantiate the module into the test bench
prod_of_sums inst1 (x1, x2, x3, x4, z1);

endmodule
```

Figure 2.100 Test bench module for Example 2.26.

```
{x1 x2 x3 x4} = 0000, z1 = 0
{x1 x2 x3 x4} = 0001, z1 = 1
{x1 x2 x3 x4} = 0010, z1 = 0
{x1 x2 x3 x4} = 0011, z1 = 1

{x1 x2 x3 x4} = 0100, z1 = 1
{x1 x2 x3 x4} = 0101, z1 = 0
{x1 x2 x3 x4} = 0110, z1 = 1
{x1 x2 x3 x4} = 0111, z1 = 0

{x1 x2 x3 x4} = 1000, z1 = 0
{x1 x2 x3 x4} = 1001, z1 = 0
{x1 x2 x3 x4} = 1010, z1 = 0
{x1 x2 x3 x4} = 1011, z1 = 1

{x1 x2 x3 x4} = 1100, z1 = 1
{x1 x2 x3 x4} = 1101, z1 = 0
{x1 x2 x3 x4} = 1110, z1 = 0
{x1 x2 x3 x4} = 1111, z1 = 1
```

Figure 2.101 Outputs for Example 2.26.

Example 2.27 This example will use the *conjunctive normal form* to design the combinational logic circuit of Example 2.26. A conjunctive normal form, also called a "product of maxterms", is an expression in which each term contains all the variables, either true or complemented. The minimal conjunctive normal form can be obtained by combining the individual 0s in the Karnaugh map to form sum terms. The Karnaugh map is reproduced in Figure 2.102 for convenience. The equation is shown in Equation 2.15.

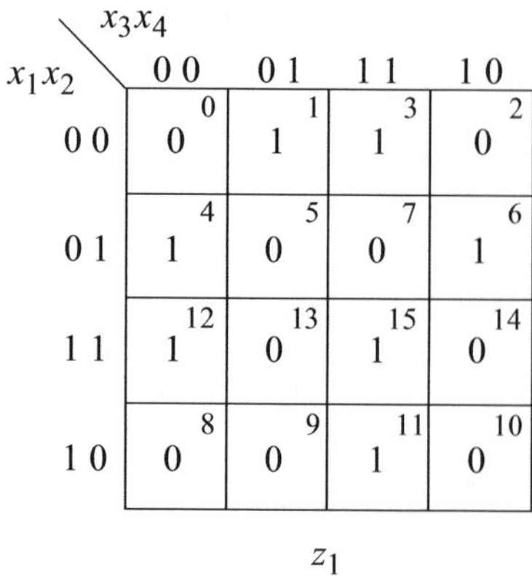

Figure 2.102 Karnaugh map for Example 2.27.

$$z_1(x_1, x_2, x_3, x_4) = \Pi_M(0, 2, 5, 7, 8, 9, 10, 13, 14) \tag{2.15}$$

$$z_1 = (x_1 + x_2 + x_3 + x_4)(x_1 + x_2 + x_3' + x_4)(x_1 + x_2' + x_3 + x_4')$$

$$(x_1 + x_2' + x_3' + x_4')(x_1' + x_2 + x_3 + x_4)(x_1' + x_2 + x_3 + x_4')$$

$$(x_1' + x_2 + x_3' + x_4)(x_1' + x_2' + x_3 + x_4')(x_1' + x_2' + x_3' + x_4)$$

The logic diagram is shown in Figure 2.103 using AND gates and OR gates. The structural design module is shown in Figure 2.104 using logic gates that were designed using dataflow modeling. The test bench module and the outputs are shown in Figures 2.105 and 2.106, respectively.

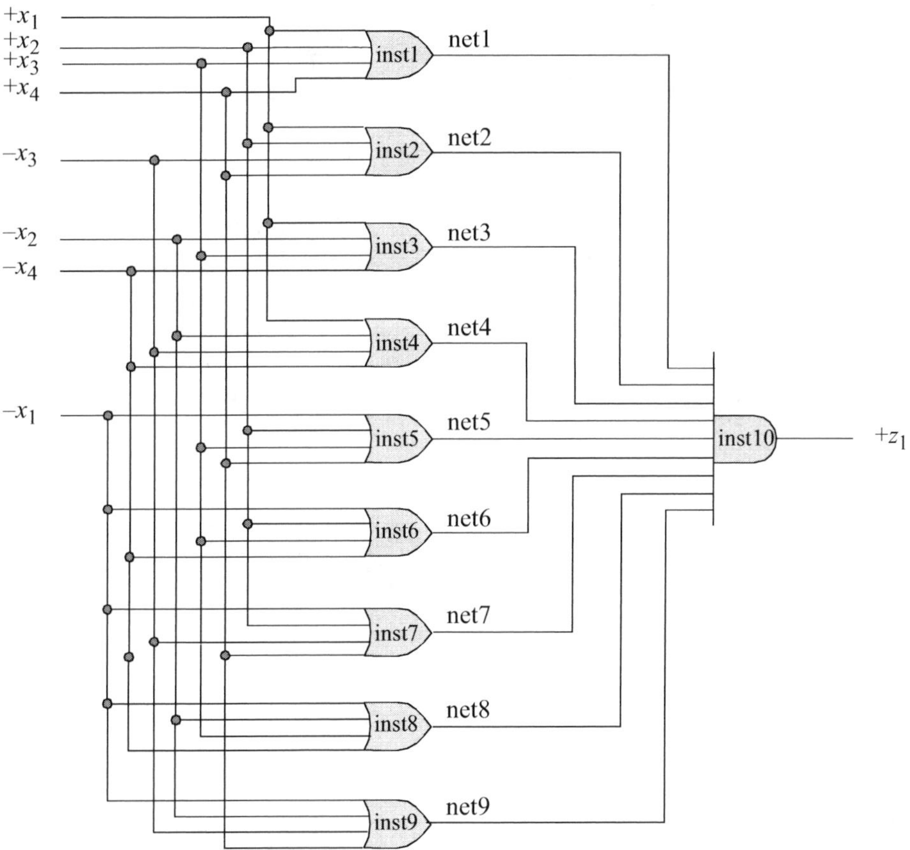

Figure 2.103 Logic diagram for Example 2.27.

```
//built-in primitives for conjunctive normal form
module conjunctive_normal (x1, x2, x3, x4, z1);

input x1, x2, x3, x4;      //define inputs and output
output z1;

//design the logic using built-in primitives
or     inst1 (net1, x1, x2, x3, x4),
       inst2 (net2, x1, x2, ~x3, x4),
       inst3 (net3, x1, ~x2, x3, ~x4),
       inst4 (net4, x1, ~x2, ~x3, ~x4),
       inst5 (net5, ~x1, x2, x3, x4),
       inst6 (net6, ~x1, x2, x3, ~x4),
       inst7 (net7, ~x1, x2, ~x3, x4),
       inst8 (net8, ~x1, ~x2, x3, ~x4),
       inst9 (net9, ~x1, ~x2, ~x3, x4);

and    inst10 (z1, net1, net2, net3, net4, net5,
                   net6, net7, net8, net9);
endmodule
```

Figure 2.104 Design module using built-in primitives for the conjunctive normal form of Example 2.27.

```
//test bench for conjunctive normal equation
module conjunctive_normal_tb;

reg x1, x2, x3, x4;  //inputs are reg, outputs are wire
wire z1;

initial          //apply input vectors
begin: apply_stimulus
   reg [4:0] invect;
   for (invect = 0; invect < 16; invect = invect + 1)
      begin
         {x1, x2, x3, x4} = invect [4:0];
         #10 $display ("{x1 x2 x3 x4} = %b, z1 = %b",
                        {x1, x2, x3, x4}, z1);
      end
end

//instantiate the module into the test bench
conjunctive_normal inst1 (x1, x2, x3, x4, z1);

endmodule
```

Figure 2.105 Test bench for the conjunctive normal form of Example 2.27.

```
{x1 x2 x3 x4} = 0000, z1 = 0
{x1 x2 x3 x4} = 0001, z1 = 1
{x1 x2 x3 x4} = 0010, z1 = 0
{x1 x2 x3 x4} = 0011, z1 = 1

{x1 x2 x3 x4} = 0100, z1 = 1
{x1 x2 x3 x4} = 0101, z1 = 0
{x1 x2 x3 x4} = 0110, z1 = 1
{x1 x2 x3 x4} = 0111, z1 = 0

{x1 x2 x3 x4} = 1000, z1 = 0
{x1 x2 x3 x4} = 1001, z1 = 0
{x1 x2 x3 x4} = 1010, z1 = 0
{x1 x2 x3 x4} = 1011, z1 = 1

{x1 x2 x3 x4} = 1100, z1 = 1
{x1 x2 x3 x4} = 1101, z1 = 0
{x1 x2 x3 x4} = 1110, z1 = 0
{x1 x2 x3 x4} = 1111, z1 = 1
```

Figure 2.106 Outputs for the conjunctive normal form of Example 2.27.

Example 2.28 This example uses a 4:1 multiplexer to indicate how a design module can be minimized by using a multiplexer. First a 4:1 multiplexer will be designed using the continuous assignment **assign** statement, then instantiated into the design module for this example. The design module for the multiplexer is shown in Figure 2.107.

```
//dataflow 4:1 mux

module mux4a_df (s, d, z1);

//define inputs and output
input [1:0] s;
input [3:0] d;
output z1;

assign z1 = (~s[1] & ~s[0] & d[0]) |
            (~s[1] &  s[0] & d[1]) |
            ( s[1] & ~s[0] & d[2]) |
            ( s[1] &  s[0] & d[3]);

endmodule
```

Figure 2.107 A 4:1 multiplexer designed using the **assign** statement.

The equation for this example is shown in Equation 2.16 and plotted on the Karnaugh map of Figure 2.108. The equation for output z_1 is shown in Equation 2.17.

$$z_1(x_1, x_2, x_3, x_4) = \Sigma_m(0, 4, 9, 10, 13, 14) \tag{2.16}$$

$x_1x_2 \backslash x_3x_4$	0 0	0 1	1 1	1 0
0 0	1 (0)	0 (1)	0 (3)	0 (2)
0 1	1 (4)	0 (5)	0 (7)	0 (6)
1 1	0 (12)	1 (13)	0 (15)	1 (14)
1 0	0 (8)	1 (9)	0 (11)	1 (10)

z_1

Figure 2.108 Karnaugh map that is generated from Equation 2.16.

$$z_1(x_1, x_2, x_3, x_4) = x_1'x_3'x_4' + x_1x_3'x_4 + x_1x_3x_4' \tag{2.17}$$

The structural design module is shown in Figure 2.109 using a single 4:1 multiplexer that was designed using dataflow modeling and instantiated into the module. The test bench module is shown in Figure 2.110 and the outputs are shown in Figure 2.111.

```
//structural design module
module func_decomp5 (x1, x2, x3, x4, z1);

//define inputs and output
input x1, x2, x3, x4;
output z1;

//instantiate the 4:1 multiplexer
mux4a_df inst1 ({x1, x3}, ({~x4, x4, 1'b0, ~x4}), z1);

endmodule
```

Figure 2.109 Structural design module for Example 2.28.

```
//test bench for design module
module func_decomp5_tb;

reg x1, x2, x3, x4;  //inputs are reg for test bench
wire z1;             //outputs are wire for test bench

//apply input vectors and display variables
initial
begin: apply_stimulus
   reg [4:0] invect;
   for (invect = 0; invect < 16; invect = invect + 1)
      begin
         {x1, x2, x3, x4} = invect [4:0];
         #10 $display ("x1 x2 x3 x4 = %b, z1 = %b",
                              {x1, x2, x3, x4}, z1);
      end
end

//instantiate the module into the test bench
func_decomp5 inst1 (x1, x2, x3, x4, z1);

endmodule
```

Figure 2.110 Test bench module for Example 2.28.

```
x1 x2 x3 x4 = 0000, z1 = 1
x1 x2 x3 x4 = 0001, z1 = 0
x1 x2 x3 x4 = 0010, z1 = 0
x1 x2 x3 x4 = 0011, z1 = 0

x1 x2 x3 x4 = 0100, z1 = 1
x1 x2 x3 x4 = 0101, z1 = 0
x1 x2 x3 x4 = 0110, z1 = 0
x1 x2 x3 x4 = 0111, z1 = 0

x1 x2 x3 x4 = 1000, z1 = 0
x1 x2 x3 x4 = 1001, z1 = 1
x1 x2 x3 x4 = 1010, z1 = 1
x1 x2 x3 x4 = 1011, z1 = 0

x1 x2 x3 x4 = 1100, z1 = 0
x1 x2 x3 x4 = 1101, z1 = 1
x1 x2 x3 x4 = 1110, z1 = 1
x1 x2 x3 x4 = 1111, z1 = 0
```

Figure 2.111 Outputs for Example 2.28.

Example 2.29 This example illustrates the design of an iterative network to design a single-bit detection circuit. In this example, a typical cell will be designed, then instantiated three times into a higher-level module to detect a single bit in a 3-bit input vector $x[1:3]$. Figure 2.112 shows the internal logic of a typical cell, which will be instantiated three times into the higher-level circuit of Figure 2.113. This network will then be designed by module instantiation using of the typical single-bit cell.

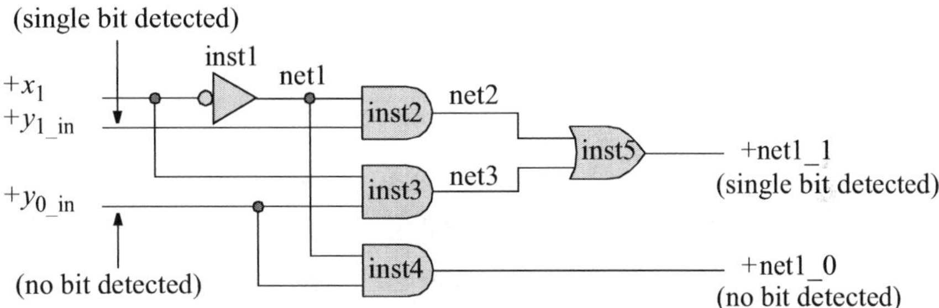

Figure 2.112 Internal logic for a typical cell in the single-bit detection circuit.

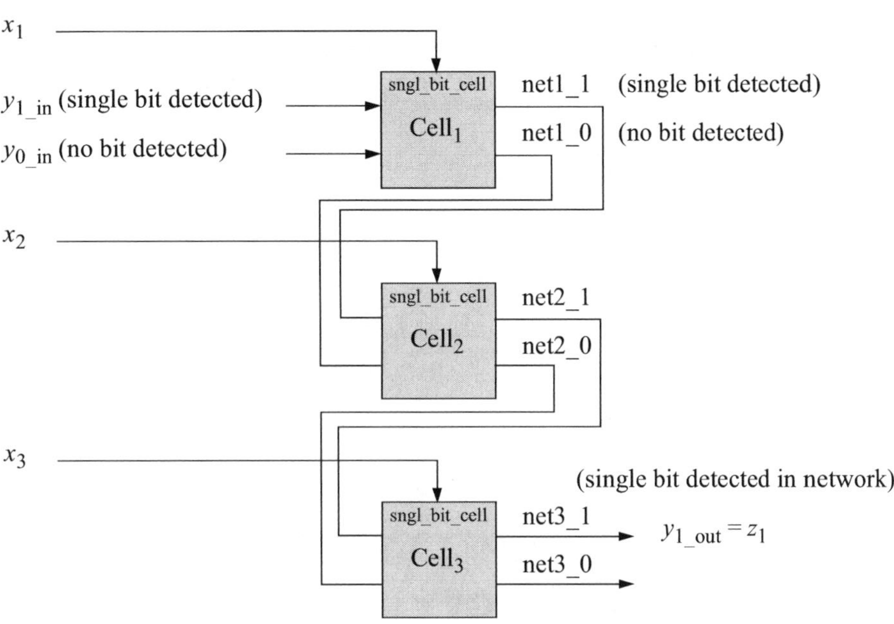

Figure 2.113 Block diagram for the circuit to detect a single bit in a 3-bit input vector $x[1:3]$.

The dataflow design module for a typical cell using built-in primitives is shown in Figure 2.114. The test bench is shown in Figure 2.115 and the outputs are shown in Figure 2.116. The design module for the iterative network is shown in Figure 2.117. The test bench module and the outputs are shown in Figures 2.118 and 2.119, respectively.

```
//typical cell for single-bit detection
module sngl_bit_cell (x1_in, y1_in, y0_in, y1_out, y0_out);

input x1_in, y1_in, y0_in;
output y1_out, y0_out;

not   inst1(net1, x1_in);
and   inst2 (net2, net1, y1_in);
and   inst3 (net3, x1_in, y0_in);
and   inst4 (y0_out, net1, y0_in);
or    inst5 (y1_out, net2, net3);

endmodule
```

Figure 2.114 Dataflow design module for the single-bit detection cell.

```
//test bench for single-bit cell
module sngl_bit_cell_tb;

reg x1_in, y1_in, y0_in;
wire y1_out, y0_out;

initial      //apply input vectors
begin: apply_stimulus
   reg [3:0] invect;
   for  (invect=0; invect<8; invect=invect+1)
      begin
         {x1_in, y1_in, y0_in} = invect [3:0];
         #10  $display ("x1_in y1_in y0_in = %b,
                           y1_out y0_out = %b",
                  {x1_in, y1_in, y0_in}, {y1_out, y0_out});
   end
end

//instantiate the module into the test bench
sngl_bit_cell inst1 (x1_in, y1_in, y0_in, y1_out, y0_out);

endmodule
```

Figure 2.115 Test bench module for the single-bit detection cell.

x1, y1_in, y0_in = 000,	net1_1, net1_0 = 00
x1, y1_in, y0_in = 001,	net1_1, net1_0 = 0**1**
x1, y1_in, y0_in = 010,	net1_1, net1_0 = **1**0
x1, y1_in, y0_in = 011,	net1_1, net1_0 = **11**
x1, y1_in, y0_in = 100,	net1_1, net1_0 = 00
x1, y1_in, y0_in = 101,	net1_1, net1_0 = **1**0
x1, y1_in, y0_in = 110,	net1_1, net1_0 = 00
x1, y1_in, y0_in = 111,	net1_1, net1_0 = **1**0

Figure 2.116 Outputs for the single-bit detection cell.

```
//single-bit detection network designed by instantiation

module sngl_bit_detect3 (x1, x2, x3, z1);

input x1, x2, x3;     //define inputs and output
output z1;

//instantiate the single-bit cell module
//cell 1
sngl_bit_cell  inst1 (x1, 1'b0, 1'b1, net1_1, net1_0);

//cell 2
sngl_bit_cell  inst2 (x2, net1_1, net1_0, net2_1, net2_0);

//cell 3
sngl_bit_cell  inst3 (x3, net2_1, net2_0, z1, 1'b0);

endmodule
```

Figure 2.117 Design module for the single-bit iterative detection network.

```
//test bench for the single-bit detection
//using a typical cell instantiation

module sngl_bit_detect3_tb;

//inputs are reg for test bench
//outputs are wire for test bench
reg x1, x2, x3;
wire z1;
                                     //continued on next page
```

Figure 2.118 Test bench module for the single-bit iterative detection network.

```
//apply input vectors
initial
begin: apply_stimulus
   reg [3:0] invect;
   for (invect = 0; invect < 8; invect = invect + 1)
      begin
         {x1, x2, x3} = invect [3:0];
         #10 $display ("x1 x2 x3 = %b, z1 = %b",
                              {x1, x2, x3}, z1);
      end
end

//instantiate the module into the test bench
sngl_bit_detect3 inst1 (x1, x2, x3, z1);

endmodule
```

Figure 2.118 (Continued)

```
x1 x2 x3 = 000, z1 = 0
x1 x2 x3 = 001, z1 = 1
x1 x2 x3 = 010, z1 = 1

x1 x2 x3 = 011, z1 = 0
x1 x2 x3 = 100, z1 = 1
x1 x2 x3 = 101, z1 = 0
x1 x2 x3 = 110, z1 = 0
x1 x2 x3 = 111, z1 = 0
```

Figure 2.119 Outputs for the single-bit iterative detection network.

2.8 Problems

2.1 Use dataflow modeling to implement the function shown below in a sum-of-products form and also in a product-of-sums form. Obtain the design module, the test bench module, and the outputs. Compare the outputs for both forms.

$$z_1(x_1, x_2, x_3, x_4) = \Sigma_m(0, 1, 6, 7, 11, 12, 13, 15)$$

2.2 A Karnaugh map is shown below using x_5 as a map-entered variable. Obtain the input equations for a nonlinear-select multiplexer using $x_1x_2 = s_1s_0$. A nonlinear-select multiplexer is a smaller multiplexer with fewer data inputs and can be effectively utilized with a corresponding reduction in machine cost.

Use dataflow modeling with the **assign** statement and behavioral modeling with the **case** statement to implement the design module. Provide several combinations of the five variables $x_1x_2x_3x_4x_5$ in the test bench. Obtain the outputs and verify that they conform to the minterm entries of the Karnaugh map.

x_1x_2 \ x_3x_4	0 0	0 1	1 1	1 0
0 0	0 (0)	x_5 (1)	0 (3)	x_5 (2)
0 1	0 (4)	0 (5)	1 (7)	1 (6)
1 1	0 (12)	1 (13)	1 (15)	x_5' (14)
1 0	1 (8)	1 (9)	0 (11)	0 (10)

z_1

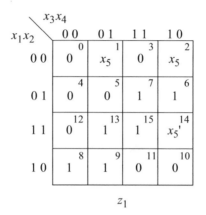

2.3 Design a structural module that will generate a high output z_1 if a 4-bit binary number $x_1x_2x_3x_4$ has a value less than or equal to 4 or greater than 11. Generate a Karnaugh map and obtain the equation for z_1 in a sum-of-products form and for z_2 in a product-of-sums form. Instantiate dataflow modules for the logic gates into the structural module. Obtain the design module, the test bench module for all combinations of the inputs, and the outputs.

2.4 Obtain the Karnaugh map that represents the equation shown below. Then obtain the design module using built-in primitives, the test bench module, and the outputs for the equation shown below. In the same design module obtain the Verilog code for an equivalent equation using only exclusive-OR gates with logic gates that were designed using dataflow modeling.

$$\begin{aligned} z_1 = {} & x_1'x_2'x_3'x_4 + x_1'x_2'x_3x_4' + x_1'x_2x_3'x_4' \\ & + x_1'x_2x_3x_4 + x_1x_2x_3'x_4 + x_1x_2x_3x_4' \\ & + x_1x_2'x_3'x_4' + x_1x_2'x_3x_4 \end{aligned}$$

2.5 Design an octal-to-binary code converter using logic gates that were designed using dataflow modeling. The octal-to-binary conversion table is shown below. Obtain the conversion equations, then design the dataflow module, the test bench module, and obtain the outputs.

Octal Inputs								Binary Outputs		
o[0]	*o[1]*	*o[2]*	*o[3]*	*o[4]*	*o[5]*	*o[6]*	*o[7]*	*b[2]*	*b[1]*	*b[0]*
1	0	0	0	0	0	0	0	0	0	0
0	**1**	0	0	0	0	0	0	0	0	1
0	0	**1**	0	0	0	0	0	0	1	0
0	0	0	**1**	0	0	0	0	0	1	1
0	0	0	0	**1**	0	0	0	1	0	0
0	0	0	0	0	**1**	0	0	1	0	1
0	0	0	0	0	0	**1**	0	1	1	0
0	0	0	0	0	0	0	**1**	1	1	1

2.6 Repeat Problem 2.5 using behavioral modeling with the **case** statement. Obtain the design module, the test bench module, and the outputs.

2.7 Design a 5-input majority circuit using the dataflow continuous **assign** statement. Obtain the Karnaugh map and the equations for the sum-of-products and for the product-of-sums expressions. Obtain the design module for both the sum-of-products expression and the product-of-sums expression. Obtain the test bench module and the outputs.

2.8 Given the Karnaugh map shown below, obtain the equation for output z_1 in a sum-of-products form and for output z_2 in product-of-sums form. Then obtain the design module using logic gates that were designed using dataflow modeling. Obtain the test bench module for all combinations of the five inputs. Obtain the output values for z_1 and z_2.

$x_5 = 0$

x_1x_2 \ x_3x_4	00	01	11	10
00	0 (0)	1 (2)	1 (6)	0 (4)
01	1 (8)	1 (10)	1 (14)	1 (12)
11	0 (24)	1 (26)	0 (30)	0 (28)
10	0 (16)	1 (18)	1 (22)	1 (20)

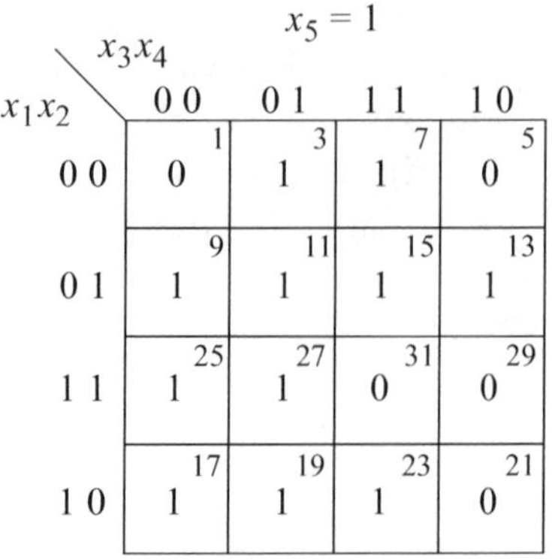

$x_5 = 1$

x_1x_2 \ x_3x_4	00	01	11	10
00	0 (1)	1 (3)	1 (7)	0 (5)
01	1 (9)	1 (11)	1 (15)	1 (13)
11	1 (25)	1 (27)	0 (31)	0 (29)
10	1 (17)	1 (19)	1 (23)	0 (21)

$z_1(z_2)$

2.9 Design a dataflow module for a full adder using logic gates that were designed using dataflow modeling. Recall that a full adder is a combinational circuit that adds two operand bits: the augend *a* and the addend *b* plus a carry-in bit *cin*. The carry-in bit represents the carry-out of the previous lower-order stage. A full adder produces two outputs: a sum bit *sum* and carry-out bit *cout*. The truth table for a full adder is shown below. Obtain the test bench module and the outputs.

a	*b*	*cin*	*cout*	*sum*
0	0	0	0	0
0	0	1	0	1
0	1	0	0	1
0	1	1	1	0
1	0	0	0	1
1	0	1	1	0
1	1	0	1	0
1	1	1	1	1

2.10 Design a 4-bit comparator for two 4-bit unsigned binary operands: *A* [3:0] and *B* [3:0] using behavioral modeling. There are three outputs:

$$A < B,\ A = B,\ A > B$$

Obtain the test bench module and the outputs for 30 input vectors.

2.11 This problem and the next two problems all design a binary-to-Gray code converter using different design techniques: this problem uses dataflow modeling with the continuous **assign** statement; Problem 2.12 uses behavioral modeling with the **always** statement; Problem 2.13 uses behavioral modeling with the **case** statement. The binary-to-Gray code conversion table is shown below. Obtain the test bench module and the outputs.

Binary Code				Gray Code			
b_3	b_2	b_1	b_0	g_3	g_2	g_1	g_0
0	0	0	0	0	0	0	0
0	0	0	1	0	0	0	1
0	0	1	0	0	0	1	1
0	0	1	1	0	0	1	0
0	1	0	0	0	1	1	0
0	1	0	1	0	1	1	1

//continued on next page

Binary Code				Gray Code			
b_3	b_2	b_1	b_0	g_3	g_2	g_1	g_0
0	1	1	0	0	1	0	1
0	1	1	1	0	1	0	0
1	0	0	0	1	1	0	0
1	0	0	1	1	1	0	1
1	0	1	0	1	1	1	1
1	0	1	1	1	1	1	0
1	1	0	0	1	0	1	0
1	1	0	1	1	0	1	1
1	1	1	0	1	0	0	1
1	1	1	1	1	0	0	0

2.12 Repeat Problem 2.11 for the binary-to-Gray code converter using behavioral modeling with the **always** statement. Obtain the test bench module and the outputs.

2.13 Repeat Problem 2.11 for the binary-to-Gray code converter using behavioral modeling with the **case** statement. Obtain the test bench module and the outputs.

2.14 Use structural modeling to design a 4:1 multiplexer using logic gates that were designed using dataflow modeling. Obtain the test bench module and the outputs for 16 combinations of the inputs.

2.15 Design a behavioral module using the **case** statement to design an 8:1 multiplexer. Obtain the test bench module and the outputs for 20 combinations of the inputs.

2.16 Design a behavioral module that adds 5 to a variable *count* to obtain a maximum value of 100 and displays the outputs.

2.17 Plot the following equation on a Karnaugh map, then change the equation to an exclusive-NOR format. Obtain the design module using built-in primitive logic gates. Then obtain the test bench module and the outputs for all combinations of the four variables.

$$z_1 = x_1'x_2'x_3'x_4' + x_1'x_2x_3'x_4 + x_1x_2x_3x_4 + x_1x_2'x_3x_4'$$

2.18 Obtain a minimized equation for z_1 in a sum-of-products representation and for z_2 in a product-of-sums representation for the Karnaugh map shown below, where the outputs are $12 \leq z_1(z_2) < 3$. Then obtain the design module using built-in primitives, the test bench module, and the outputs.

x_1x_2 \ x_3x_4	0 0	0 1	1 1	1 0
0 0	0: 1	1: 1	3: 0	2: 1
0 1	4: 0	5: 0	7: 0	6: 0
1 1	12: 1	13: 1	15: 1	14: 1
1 0	8: 0	9: 0	11: 0	10: 0

3

Sequential Logic Design Using Verilog HDL

3.1 Introduction

This chapter provides techniques for designing sequential logic using Verilog HDL. Sequential logic circuits consist of combinational logic and storage elements, such as *SR* latches, *D* flip-flops, and *JK* flip-flops. They are specified as sequential because the operations of the circuit are executed in sequence. Since these circuits (or sequential machines) contain a finite number of internal states, they are also referred to as *finite state machines*. A *state* is a set of values that is specified at different locations in the state machine.

3.1.1 Definition of a Sequential Machine

A *synchronous sequential machine* is a machine whose present outputs are a function of the present state only or the present state and present inputs. A requirement of a synchronous sequential machine is that state changes occur only when the machine is clocked, either on the positive or negative transition of the clock. Thus, input changes do not affect the present state of the machine until the occurrence of the next active clock transition.

The logic that generates the inputs to the storage elements is called the "δ next-state function", because the next state of the machine is usually determined by the inputs. The outputs generated by the storage elements are called the "λ output function".

In some cases, the output logic may require one or more storage elements, depending on the assertion and deassertion of the output signals.

Some synchronous sequential machines are described by a *state diagram*. A state diagram presents a graphical representation in which the state transitions are more easily followed. The state diagrams are similar to flowchart diagrams in which the transition sequences and thus, the operational characteristics of the machine, are clearly delineated. Two symbols are used: a state symbol and an output symbol.

The *state symbol* is designated by a circle as shown in Figure 3.1. These nodes (or vertices) correspond to the state of the machine; the state name, such as state *a*, is placed inside the circle. The connecting directed lines between states correspond to the allowable state transitions. There are one or more entry paths and one or more exit paths as indicated by the arrows, unless the vertex is a *terminal state*, in which case there is no exit. The symbols $y_1y_2y_3$ represent the names of the storage elements of the machine for that state, which indicate that the storage elements are specified as y_1 is set (1), y_2 is reset (0), and y_3 is set (1). If input x_1 is 0, then the machine proceeds to a particular state. If input x_1 is 1, then the machine proceeds to a different state.

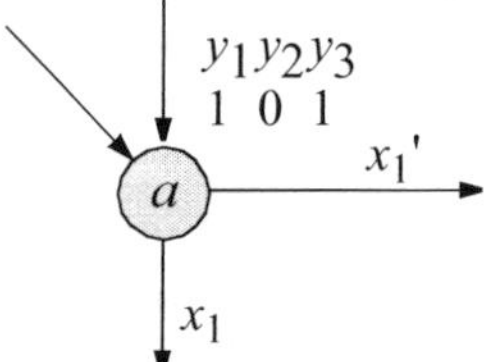

Figure 3.1 The state symbol for a state diagram indicating state *a*.

The *output symbol* is represented by a rectangle and is placed immediately following the state symbol, as shown in Figure 3.2 (a) or placed immediately after an input variable that causes the output to become active as shown in 3.2 (b). Figure 3.2 (a) specifies a Moore machine in which output z_1 is a function of the present state only. Figure 3.2 (b) indicates a Mealy machine in which output z_1 is a function of both the present state and the present input x_1. Moore and Mealy machines will be presented in Sections 3.2.3 and 3.2.4, respectively.

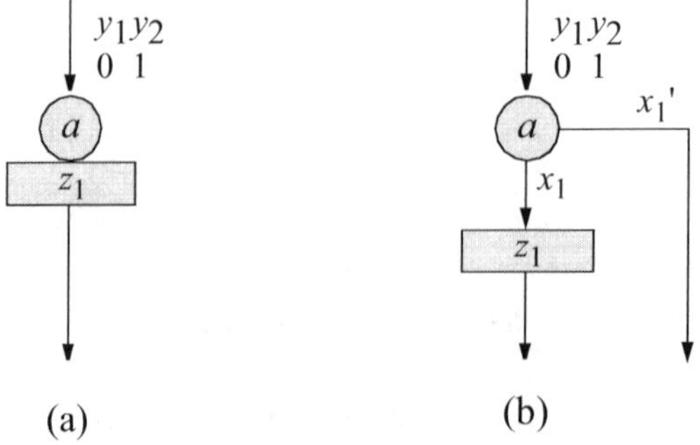

Figure 3.2 State diagram output symbol indicating output z_1.

3.2 Synchronous Sequential Machines

Techniques for synthesizing (designing) synchronous sequential machines are introduced. A detailed procedure is presented to synthesize a synchronous sequential machine from a given set of machine specifications. From this a logic diagram or a list of Boolean functions is obtained from which the sequential machine can be designed.

A synchronous sequential machine requires a state diagram or state table for its precise description. The state diagram depicts the sequence of events that must occur in order for the machine to perform the functions which are defined in the machine specifications.

A proper choice of state code assignments may reduce the number of gates in the δ next-state function logic. Since there are p storage elements, the binary values of these p-tuples can usually be chosen such that the combinational input logic is minimized. A judicious choice of state codes permits more entries in the Karnaugh map to be combined. This results in input equations with fewer terms and fewer variables per term.

3.2.1 Synthesis Procedure

This section develops a detailed method for designing synchronous sequential machines using various types of storage elements. The hierarchical design algorithm is shown below.

1. Develop a state diagram from the problem definition, which may be either a word description and/or a timing diagram.

2. Check for equivalent states and then eliminate redundant states. Equivalent states are presented in Section 3.2.2.

3. Assign state codes for the storage elements in the form of a binary p-tuple, as shown in Section 3.1.1. For example, $y_1y_2y_3 = 101$

4. Determine equivalent states, as described in Section 3.2.2.

5. Select the type of storage element to be used (*SR* latch, *D* flip-flop, *JK* flip-flop) then generate the input maps for the δ next-state function and derive the input equations.

6. Generate the output maps for the λ output function and derive the output equations.

7. Design the logic diagram using the input equations, the storage elements, and the output equations. Then the Verilog design module and test bench can be generated from which the outputs can be obtained.

3.2.2 Equivalent States

At each node in the state diagram, two events occur: the outputs (if applicable) for the present state are generated as a function of the present state only (Moore machine) or the present state and inputs (Mealy machine); the next state is determined as a function of the present state only or the present state and inputs.

Two states Y_i and Y_j of a machine are equivalent if, for every input sequence, the output sequence when started in state Y_i is identical to the output sequence when started in state Y_j or if both states Y_i and Y_j have the same or equivalent next state. When equivalent states have been found, all but one are redundant and should be eliminated before implementing the state diagram with hardware. Two states can be equivalent if they satisfy the following *equivalence relation* properties:

Reflexive For every state Y_i in the machine, $Y_i \equiv Y_i$; that is, Y_i is related to itself.

Symmetric For every pair of states Y_i and Y_j in the machine, if $Y_i \equiv Y_j$, then $Y_j \equiv Y_i$; that is, the order of the relation is not important.

Transitive For any three states Y_i, Y_j, and Y_k in the machine, if $Y_i \equiv Y_j$ and $Y_j \equiv Y_k$, then $Y_i \equiv Y_k$; that is, $Y_i \equiv Y_j \equiv Y_k$.

3.2.3 Moore Machines

Moore machines are synchronous sequential machines in which the output function λ produces an output vector Z_r which is determined by the present state only, and is not a function of the present inputs. The general configuration of a Moore machine is shown in Figure 3.3. The next-state function δ is an $(n + p)$-input, p-output switching function. The output function λ is a p-input, m-output switching function.

If a Moore machine has no data input, then it is referred to as an *autonomous* machine. Autonomous circuits are independent of the inputs. The clock signal is not considered as a data input. An autonomous Moore machine is an important class of synchronous sequential machines, the most common application being a counter. A Moore machine may be synchronous or asynchronous; however, this section pertains to synchronous organizations only.

A Moore machine is a 5-tuple and can be defined as shown in Equation 3.1,

$$M = (X, Y, Z, \delta, \lambda) \tag{3.1}$$

where

X is a nonempty finite set of inputs
Y is a nonempty finite set of states
Z is a nonempty finite set of outputs
$\delta(X, Y) : X \times Y \rightarrow Y$
$\lambda(Y) : Y \rightarrow Z$

The symbols $\delta(X, Y) : X \times Y \rightarrow Y$ specify δ is a function of X and Y, and the symbol $\times$ specifies that X is the Cartesian product of X and Y. The *Cartesian product* of two sets is defined as follows: For any two sets S and T, the Cartesian product of S and T is written as $S \times T$ and is the set of all *ordered pairs* of S and T, where the first member of the ordered pair is an element of S and the second member is an element of T. Thus, the general classification of a synchronous sequential machine M can be defined as the 5-tuple shown in Equation 3.1.

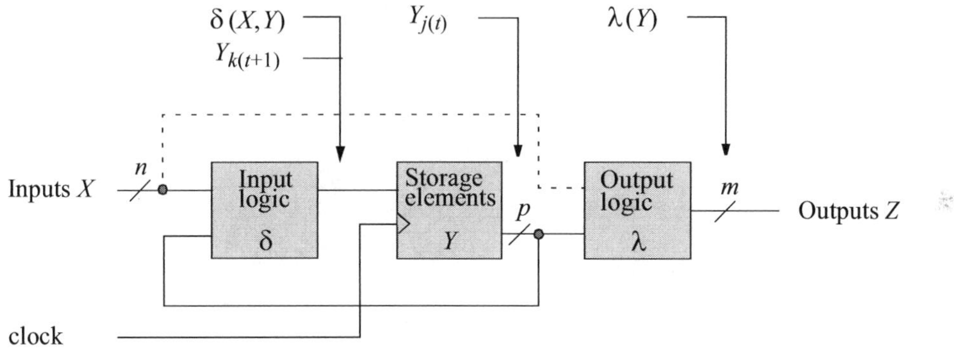

Figure 3.3 Moore synchronous sequential machine in which the outputs are a function of the present state only.

Examples will now be presented to illustrate the design of Moore synchronous sequential machines using Verilog HDL.

Example 3.1 A state diagram for a Moore machine is shown in Figure 3.4, which generates an output z_1 whenever a serial, 3-bit binary word on an input line x_1 is greater than or equal to six. The first bit received in each word is the high-order bit. There is no bit space between words. Figure 3.5 shows the design module using behavioral modeling. The design module uses $A[2:0]$ to replace input line x_1. Output z_1 is asserted whenever the input vector is $A[2:0] = 110$ or 111. The test bench module and the outputs are shown in Figures 3.6 and 3.7, respectively.

When a state has two possible next states, then the two next states should be adjacent (differ by only one variable); that is, if an input causes a state transition from state Y_i to either Y_j or Y_k, then Y_j and Y_k should be assigned adjacent state codes.

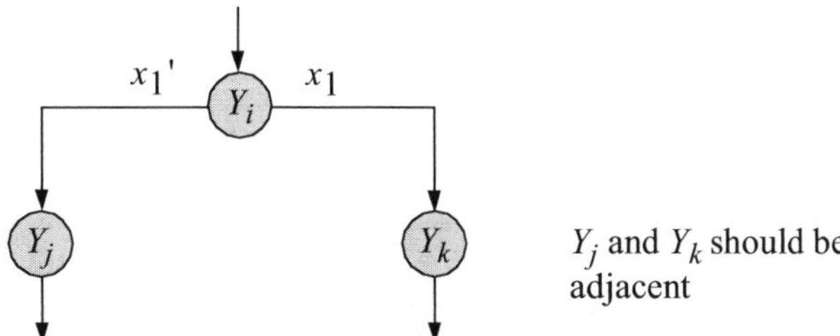

When two states have the same next state, the two states should be adjacent; that is, if Y_i and Y_j both have Y_k as a next state, then Y_i and Y_j should be assigned adjacent state codes.

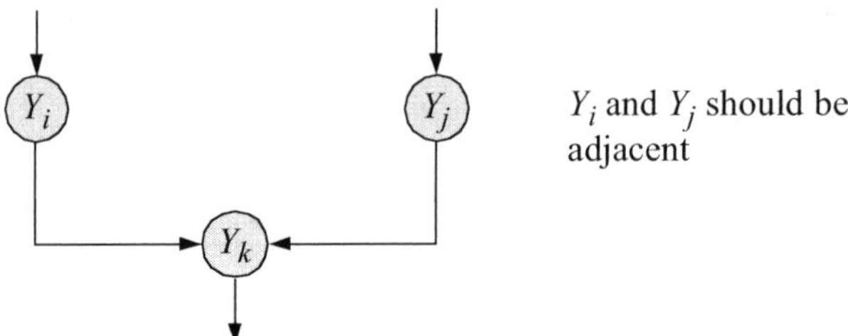

A third rule is useful in minimizing the λ output logic. States which have the same output should have adjacent state code assignments; that is, if states Y_i and Y_j both have z_1 as an output, then Y_i and Y_j should be adjacent. This allows for a larger grouping of 1s in the output map.

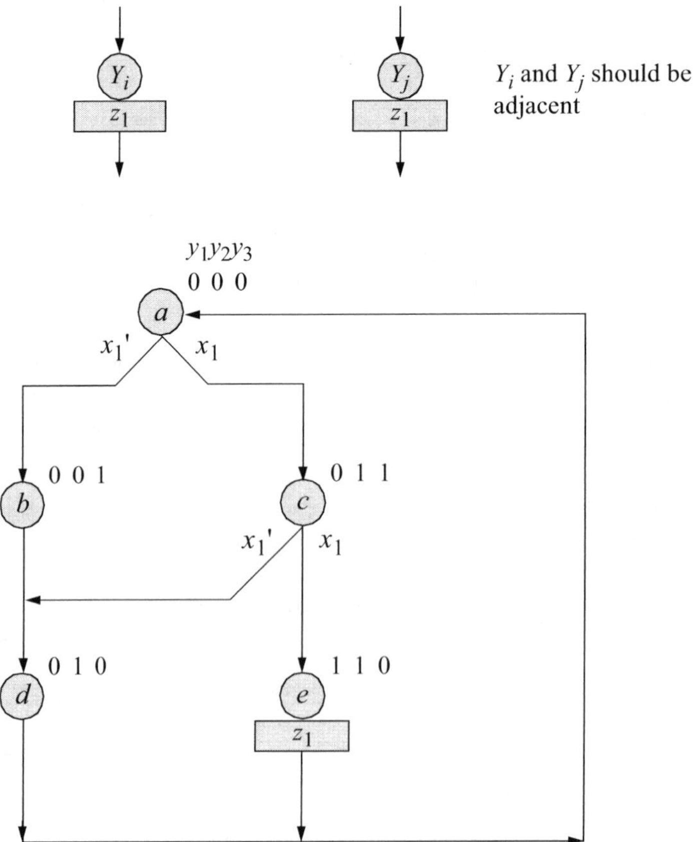

Figure 3.4 State diagram for the Moore synchronous sequential machine of Example 3.1. Unused states are: 100, 101, and 111.

```
//behavioral to determine if a[2:0] >= 6
module a_gt_eq_six (a2, a1, a0, a_gt_eq_six);

input a2, a1, a0;     //define inputs and output
output a_gt_eq_six;

//variables are reg in always
reg a_gt_eq_six;

//determine if a2, a1, a0 >= six
always @ (a2 or a1 or a0)
begin
   if (a2 & a1)
      a_gt_eq_six = 1'b1;

   else
      a_gt_eq_six = 1'b0;
end
endmodule
```

Figure 3.5 Behavioral design module for the Moore synchronous sequential machine of Example 3.1.

```
//test bench to determine if a[2:0] >= 6
module a_gt_eq_six_tb;

reg a2, a1, a0;        //inputs are reg for test bench
wire a_gt_eq_six;      //outputs are wire for test bench

initial      //apply input vectors
begin: apply_stimulus
reg [3:0] invect;
for (invect = 0; invect < 8; invect = invect + 1)
   begin
      {a2, a1, a0} = invect [3:0];
      #10 $display ("a2. a1. a0 = %b, a_gt_eq_six = %b",
                  {a2, a1, a0}, a_gt_eq_six);
   end
end

//instantiate the module into the test bench
a_gt_eq_six inst1 (a2, a1, a0, a_gt_eq_six);
endmodule
```

Figure 3.6 Test bench module for the Moore synchronous sequential machine of Example 3.1.

```
a2. a1. a0 = 000, a_gt_eq_six = 0
a2. a1. a0 = 001, a_gt_eq_six = 0
a2. a1. a0 = 010, a_gt_eq_six = 0
a2. a1. a0 = 011, a_gt_eq_six = 0

a2. a1. a0 = 100, a_gt_eq_six = 0
a2. a1. a0 = 101, a_gt_eq_six = 0
a2. a1. a0 = 110, a_gt_eq_six = 1
a2. a1. a0 = 111, a_gt_eq_six = 1
```

Figure 3.7 Outputs for the Moore synchronous sequential machine of Example 3.1.

Example 3.2 This example uses the state diagram of Example 3.1 for a Moore machine to assert output z_1 in state $y_1y_2y_3 = 110$. This design uses Karnaugh maps, built-in primitives, and *D* flip-flops that were designed using behavioral modeling. The state diagram is reproduced in Figure 3.8 for convenience. The Karnaugh maps are shown in Figure 3.9 and the corresponding equations are shown in Equation 3.3.

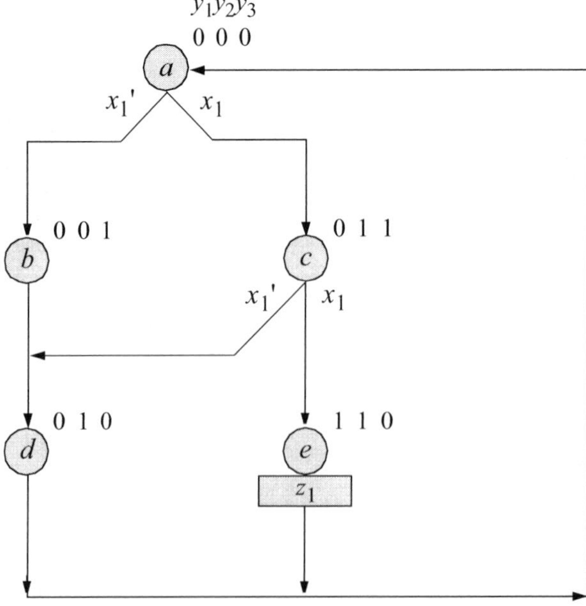

Figure 3.8 State diagram for the Moore synchronous sequential machine of Example 3.2. Unused states are: $y_1y_2y_3 = 100$, 101, and 111.

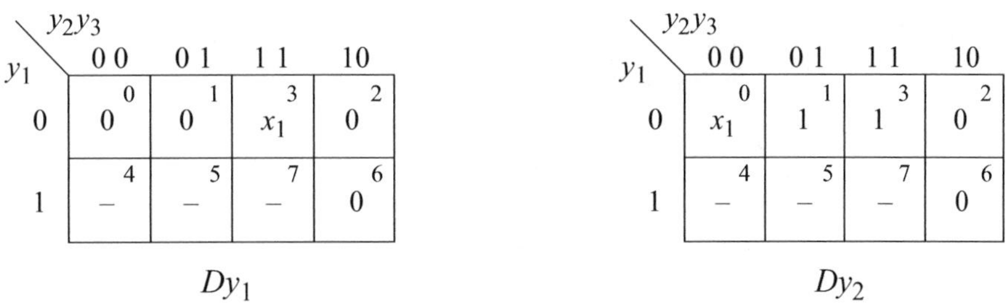

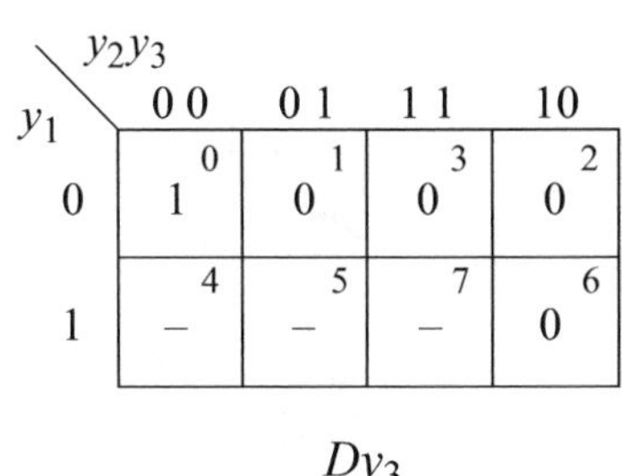

Figure 3.9 Karnaugh maps for Example 3.2.

$$Dy_1 = x_1y_2y_3$$

$$Dy_2 = y_3 + y_2'x_1$$

$$Dy_3 = y_2'y_3' \tag{3.3}$$

The logic diagram is shown in Figure 3.10. The *D* flip-flop that was designed using behavioral modeling is shown in Figure 3.11. The *D* flip-flop will be instantiated three times into the structural design module, each as a single-line instantiation. The sequence for each instantiation will be in the following order: *rst_n*, *clk*, *d*, *q*, which represent the reset input, the clock input, the *D* input to the flip-flop, and the positive output from the flip-flop.

The structural design module for Example 3.2 to assert output z_1 in state $y_1y_2y_3 =$ 110 is shown in Figure 3.12 using built-in primitives and *D* flip-flops. The test bench module is shown in Figure 3.13 using the positive-edge of the clock to cause state changes to occur. The outputs are shown in Figure 3.14.

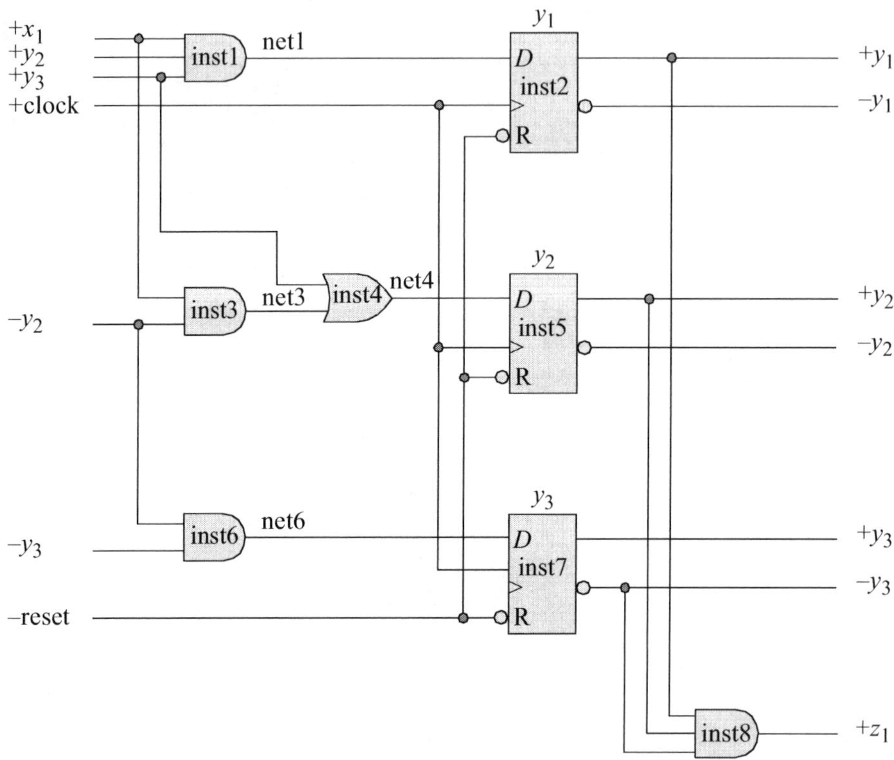

Figure 3.10 Logic diagram for Example 3.2.

```
//behavioral D flip-flop
module d_ff_bh (rst_n, clk, d, q);

input rst_n, clk, d;
output q;

wire rst_n, clk, d;
reg q;

always @ (rst_n or posedge clk)
begin
   if (rst_n == 0)
        q <= 1'b0;
   else q <= d;
end

endmodule
```

Figure 3.11 Behavioral design module for a positive-edge *D* flip-flop.

```
//structural for moore ssm to assert z1 in state y1y2y3 = 110

module assert_z1_state_110 (rst_n, clk, x1, y, z1);

//define inputs and output
input rst_n, clk, x1;
output [1:3] y;
output z1;

//define internal nets
wire net1, net3, net4, net6;

//-----------------------------------------
//instantiate the logic for flip-flop y[1]
and   inst1 (net1, x1, y[2], y[3]);

//instantiate the D flip-flop for y[1]
d_ff_bh inst2 (rst_n, clk, net1, y[1]);

//-----------------------------------------
//instantiate the logic for flip-flop y[2]
and   inst3 (net3, x1, ~y[2]);
or    inst4 (net4, y[3], net3);

//instantiate the D flip-flop for y[2]
d_ff_bh inst5 (rst_n, clk, net4, y[2]);

//-----------------------------------------
//instantiate the logic for flip-flop y[3]
and   inst6 (net6, ~y[2], ~y[3]);

//instantiate the D flip-flop for y[3]
d_ff_bh inst7 (rst_n, clk, net6, y[3]);

//-----------------------------------------
//instantiate the logic for output z1
and   inst8 (z1, y[1], y[2], ~y[3]);

endmodule
```

Figure 3.12 Structural design module for Example 3.2 to assert z_1 in state 110.

```
//test bench for moore ssm to assert z1 in state y1y2y3 = 110
module assert_z1_state_110_tb;

//inputs are reg for test bench
//outputs are wire for test bench
reg rst_n, clk, x1;
wire [1:3] y;
wire z1;

//display variables
initial
$monitor ("x1 = %b, state = %b, z1 = %b", x1, y, z1);

//define clock
initial
begin
   clk = 1'b0;
   forever
      #10 clk = ~clk;
end

//define input sequence
initial
begin
   #0 rst_n = 1'b0;
      x1 = 1'b0;

   #5 rst_n = 1'b1;

   x1 = 1'b1;  @ (posedge clk)   //go to state_b (001)
   x1 = 1'b1;  @ (posedge clk)   //go to state_d (010)
   x1 = 1'b1;  @ (posedge clk)   //go to state_a (000)
   x1 = 1'b1;  @ (posedge clk)   //go to state_c (011)
   x1 = 1'b0;  @ (posedge clk)   //go to state_e (110);
                                 //assert z1

   x1 = 1'b1;  @ (posedge clk)   //go to state_c (011)
   x1 = 1'b0;  @ (posedge clk)   //go to state_d (010)
   x1 = 1'b1;  @ (posedge clk)   //go to state_a (000)

   #10   $stop;
end

//instantiate the module into the test bench
assert_z1_state_110 inst1 (rst_n, clk, x1, y, z1);

endmodule
```

Figure 3.13 Test bench module for Example 3.2.

```
x1 = 0, state = 000, z1 = 0
x1 = 1, state = 001, z1 = 0
x1 = 1, state = 010, z1 = 0
x1 = 1, state = 000, z1 = 0

x1 = 1, state = 011, z1 = 0
x1 = 0, state = 110, z1 = 1
x1 = 1, state = 000, z1 = 0
x1 = 0, state = 011, z1 = 0
x1 = 1, state = 010, z1 = 0
x1 = 1, state = 000, z1 = 0
```

Figure 3.14 Outputs for Example 3.2.

Example 3.3 This example uses the state diagram of Example 3.1 to design a Moore machine to assert output z_1 in state $y_1y_2y_3 = 110$. This design uses behavioral modeling with the **case** statement for comparison. The state diagram is reproduced in Figure 3.15 for convenience. The behavioral model is shown in Figure 3.16. The test bench and outputs are shown in Figures 3.17 and 3.18, respectively.

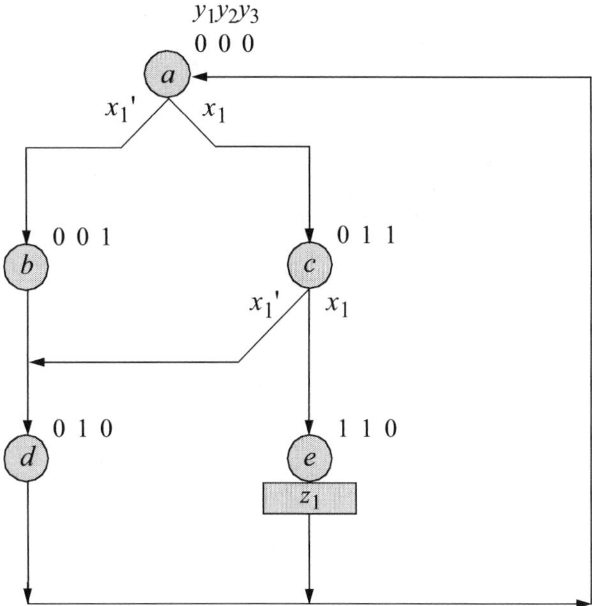

Figure 3.15 State diagram for Example 3.3.

```
//behavioral case to assert z1 in state y1y2y3 = 110
module a_gt_eq_6_case (rst_n, clk, y, x1, z1);

input rst_n, clk, x1;    //define inputs and output
output [2:0] y;
output z1;

reg [2:0] y, next_state;
wire z1;

//assign state codes
parameter   state_a = 3'b000,
            state_b = 3'b001,
            state_c = 3'b011,
            state_d = 3'b010,
            state_e = 3'b110;

assign z1 = (y[2] & y[1]);     //define output

always @ (posedge clk)         //set next state
begin
   if (~rst_n)
      y <= state_a;
   else
      y <= next_state;
end

always @ (y or x1)       //determine next state
begin
   case (y)
      state_a:
         if (x1 == 0) next_state = state_b;
         else next_state = state_c;

      state_b: next_state = state_d;

      state_c:
         if (x1 == 0) next_state = state_d;
         else next_state = state_e;

      state_d: next_state = state_a;

      state_e: next_state = state_a;

      default: next_state = state_a;
   endcase
end

endmodule
```

Figure 3.16 Behavioral design module for Example 3.3.

```
//test bench to assert z1 in state y1y2y3 = 110
module a_gt_eq_6_case_tb;

reg rst_n, clk, x1;       //inputs are reg for test bench
wire [2:0] y;             //outputs are wire for test bench
wire z1;

initial         //display variables
$monitor ("x1 = %b, state = %b, z1 = %b", x1, y, z1);

initial         //define clock
begin
   clk = 1'b0;
   forever
      #10   clk = ~clk;
end

//define input sequence
initial
begin
   #0     rst_n = 1'b0;
          x1 = 1'b0;
   #10    rst_n = 1'b1;

   x1 = 1'b0;@ (posedge clk)
   x1 = 1'b1;@ (posedge clk)
   x1 = 1'b0;@ (posedge clk)

   x1 = 1'b1;@ (posedge clk)
   x1 = 1'b0;@ (posedge clk)
   x1 = 1'b1;@ (posedge clk)

   x1 = 1'b1;@ (posedge clk)
   x1 = 1'b1;@ (posedge clk)
   x1 = 1'b1;@ (posedge clk)

   x1 = 1'b0;@ (posedge clk)

   #10    $stop;

end

//instantiate the module into the test bench
a_gt_eq_6_case inst1 (rst_n, clk, y, x1, z1);

endmodule
```

Figure 3.17 Test bench module for Example 3.3.

```
x1 = 1, state = 000, z1 = 0
x1 = 0, state = 011, z1 = 0
x1 = 1, state = 010, z1 = 0

x1 = 0, state = 000, z1 = 0
x1 = 1, state = 001, z1 = 0
x1 = 1, state = 010, z1 = 0

x1 = 1, state = 000, z1 = 0
x1 = 1, state = 011, z1 = 0
x1 = 0, state = 110, z1 = 1

x1 = 0, state = 000, z1 = 0
```

Figure 3.18 Outputs for Example 3.3.

Example 3.4 The state diagram shown below in Figure 3.19 represents a Moore synchronous sequential machine with three inputs and four outputs. The machine will be designed using behavioral modeling with the **case** statement. The design module is shown in Figure 3.20. The test bench module and the outputs are shown in Figure 3.21 and Figure 3.22, respectively.

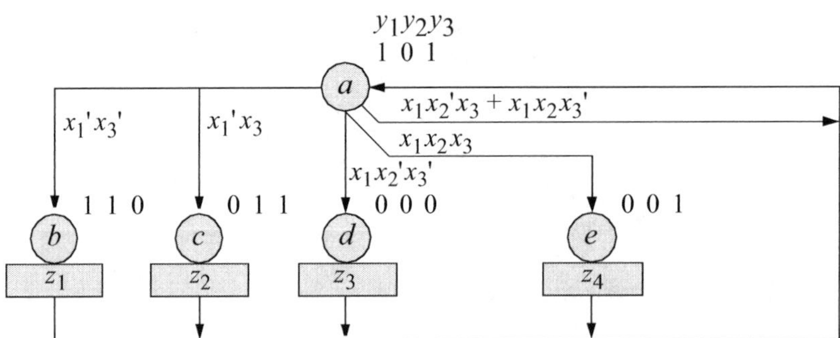

Figure 3.19 State diagram for Example 3.4.

```
//behavioral moore ssm with four outputs
module moore_4_outputs_case (clk, x1, x2, x3, y,
                              z1, z2, z3, z4);

input clk, x1, x2, x3;          //define inputs and outputs
output [2:0] y;
output z1, z2, z3, z4;
                                //continued on next page
```

Figure 3.20 Behavioral design module for Example 3.4.

```
reg [2:0] y, next_state;      //variables are reg in always

//assign state codes
parameter   state_a = 3'b101, //parameter defines a constant
            state_b = 3'b110,
            state_c = 3'b011,
            state_d = 3'b000,
            state_e = 3'b001;

assign   z1 = (y[2] & y[1] & ~y[0]),    //define outputs
         z2 = (~y[2] & y[1] & y[0]),
         z3 = (~y[2] & ~y[1] & ~y[0]),
         z4 = (~y[2] & ~y[1] & y[0]);

//set next state
always @ (posedge clk)
   y <= next_state;

//determine next state
always @ (y or x1 or x2 or x3)
begin
case (y)
   state_a:
      if (x1==0 & x3==0)
         next_state = state_b;
      else if (x1==0 & x3==1)
         next_state = state_c;
      else if (x1==1 & x2==0 & x3==0)
         next_state = state_d;
      else if (x1==1 & x2==1 & x3==1)
         next_state = state_e;
      else if (x1==1 & x2==0 & x3==1)
         next_state = state_a;
      else if (x1==1 & x2==1 & x3==0)
         next_state = state_a;
      else next_state = state_a;

   state_b: next_state = state_a;
   state_c: next_state = state_a;
   state_d: next_state = state_a;
   state_e: next_state = state_a;
   default next_state = state_a;
endcase
end

endmodule
```

Figure 3.20 (Continued)

```
//test bench for moore ssm with four outputs
module moore_4_outputs_case_tb;

//inputs are reg for test bench
//outputs are wire for test bench
reg clk, x1, x2, x3;
wire [2:0] y;
wire z1, z2, z3, z4;

//display variables
initial
$monitor ("x1 x2 x3 = %b, state = %b, z1 z2 z3 z4 = %b",
            {x1, x2, x3}, y, {z1, z2, z3, z4});

//define clock
initial
begin
   clk = 1'b0;
   forever
      #10 clk = ~clk;
end

//apply input vectors
initial
begin
   x1 = 1'b0;x2 = 1'b0;x3 = 1'b0;
   @ (posedge clk)    //go to state_b (110)
   @ (posedge clk)    //go to state_a (101)

   x1 = 1'b0;x2 = 1'b1;x3 = 1'b1;
   @ (posedge clk)    //go to state_c (011)
   @ (posedge clk)    //go to state_a (101)

   x1 = 1'b1;x2 = 1'b0;x3 = 1'b0;
   @ (posedge clk)    //go to state_d (000)
   @ (posedge clk)    //go to state_a (101)

   x1 = 1'b1;x2 = 1'b1;x3 = 1'b1;
   @ (posedge clk)    //go to state_e (001)
   @ (posedge clk)    //go to state_a (101)

   x1 = 1'b1;x2 = 1'b0;x3 = 1'b1;
   @ (posedge clk)    //go to state_a (101)

                                     //continued on next page
```

Figure 3.21 Test bench module for Example 3.4.

```
   x1 = 1'b1;x2 = 1'b1;x3 = 1'b0;
   @ (posedge clk)    //go to state_a (101)

   #10   $stop;
end

//instantiate the module into the test bench
moore_4_outputs_case inst1 (clk, x1, x2, x3, y,
                            z1, z2, z3, z4);

endmodule
```

Figure 3.21 (Continued)

```
x1 x2 x3 = 000, state = 101, z1 z2 z3 z4 = 0000
x1 x2 x3 = 011, state = 110, z1 z2 z3 z4 = 1000

x1 x2 x3 = 011, state = 101, z1 z2 z3 z4 = 0000
x1 x2 x3 = 100, state = 011, z1 z2 z3 z4 = 0100

x1 x2 x3 = 100, state = 101, z1 z2 z3 z4 = 0000
x1 x2 x3 = 111, state = 000, z1 z2 z3 z4 = 0010

x1 x2 x3 = 111, state = 101, z1 z2 z3 z4 = 0000
x1 x2 x3 = 101, state = 001, z1 z2 z3 z4 = 0001

x1 x2 x3 = 110, state = 101, z1 z2 z3 z4 = 0000
```

Figure 3.22 Outputs for Example 3.4.

Example 3.5 A state diagram is shown in Figure 3.23 for a Moore synchronous sequential machine that has three parallel inputs x_1, x_2, and x_3 and three outputs z_1, z_2, and z_3. The inputs represent a 3-bit word. There is one bit space between words. Output z_1 is asserted if the 3-bit word contains a single 1 bit; output z_2 is asserted if the 3-bit word contains two 1 bits; output z_3 is asserted if the 3-bit word contains three 1 bits.

The structural design module will be implemented using D flip-flops and built-in primitives. The Karnaugh maps for the three flip-flops are shown in Figure 3.24 and the corresponding equations are shown in Equation 3.4. The logic diagram is shown in Figure 3.25.

The structural design module is shown in Figure 3.26. The test bench module is shown in Figure 3.27 and the outputs are shown in Figure 3.28.

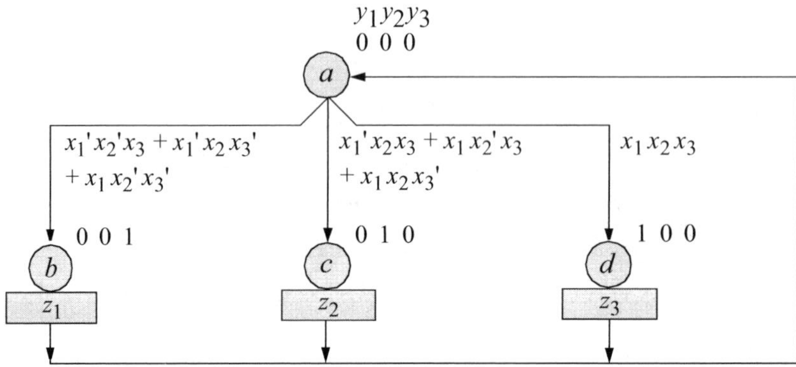

Figure 3.23 State diagram for the Moore machine of Example 3.5. Unused states are: $y_1y_2y_3$ = 011, 101, 110, and 111.

See Equation 3.4 for the meaning of entries (*a*), (*b*), and (*c*).

y_1 \ y_2y_3	00	01	11	10
0	[0] (*a*)	[1] 0	[3] –	[2] 0
1	[4] 0	[5] –	[7] –	[6] –

Dy_1

y_1 \ y_2y_3	00	01	11	10
0	[0] (*b*)	[1] 0	[3] –	[2] 0
1	[4] 0	[5] –	[7] –	[6] –

Dy_2

y_1 \ y_2y_3	00	01	11	10
0	[0] (*c*)	[1] 0	[3] –	[2] 0
1	[4] 0	[5] –	[7] –	[6] –

Dy_3

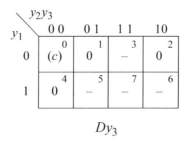

Figure 3.24 Karnaugh maps for the Moore machine of Example 3.5.

(*a*) $Dy_1 = x_1x_2x_3$

(*b*) $Dy_2 = x_1'x_2x_3 + x_1x_2'x_3 + x_1x_2x_3'$

(*c*) $Dy_3 = x_1'x_2'x_3 + x_1'x_2x_3' + x_1x_2'x_3'$ (3.4)

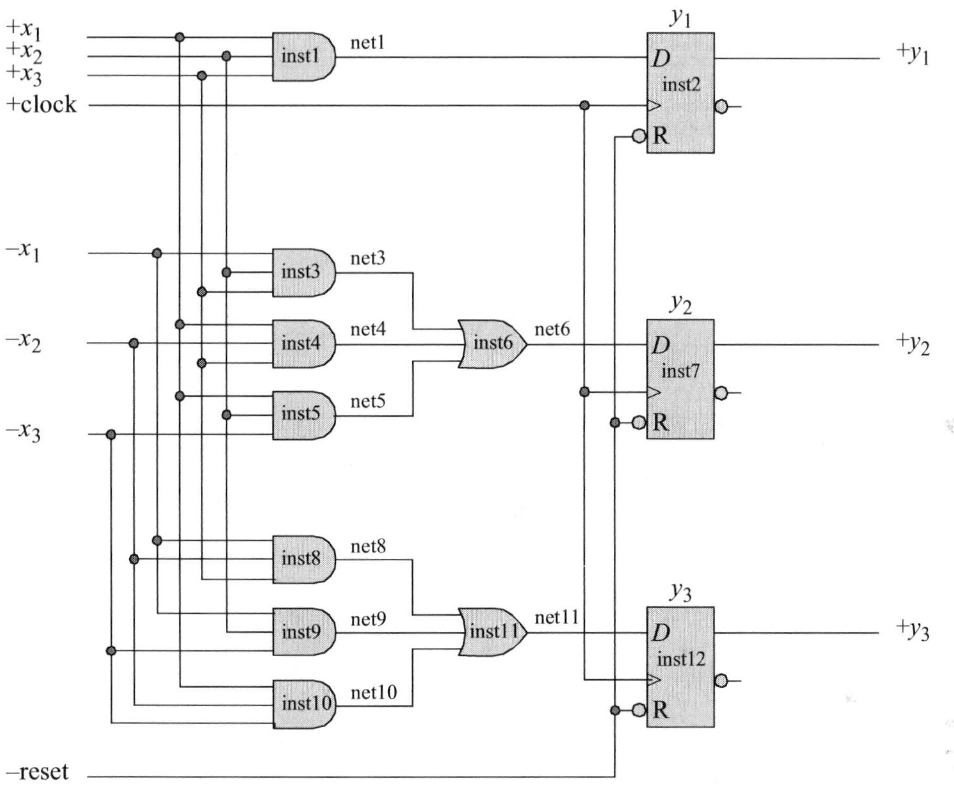

Figure 3.25 Logic diagram for the Moore machine of Example 3.5.

```
//structural for moore with three outputs using D flip-flops

module moore_3_outputs (rst_n, clk, x1, x2, x3, y,
                        z1, z2, z3);

//define inputs and outputs
input rst_n, clk, x1, x2, x3;
output [1:3] y;
output z1, z2, z3;

//define internal nets
wire net1, net3, net4, net5, net6, net8, net9, net10, net11;

                                  //continued on next page
```

Figure 3.26 Structural design module for the Moore machine of Example 3.5.

```
//------------------------------------------------------------
//instantiate the logic for flip-flop y[1]
and   inst1 (net1, x1, x2, x3);

//instantiate the D flip-flop for y[1]
d_ff_bh inst2 (rst_n, clk, net1, y[1]);

//------------------------------------------------------------
//instantiate the logic for flip-flop y[2]
and   inst3 (net3, ~x1, x2, x3),
      inst4 (net4, x1, ~x2, x3),
      inst5 (net5, x1, x2, ~x3);

or    inst6 (net6, net3, net4, net5);

//instantiate the D flip-flop for y[2]
d_ff_bh inst7 (rst_n, clk, net6, y[2]);

//------------------------------------------------------------
//instantiate the logic for flip-flop y[3]
and   inst8 (net8, ~x1, ~x2, x3),
      inst9 (net9, ~x1, x2, ~x3),
      inst10 (net10, x1, ~x2, ~x3);

or    inst11 (net11, net8, net9, net10);

//instantiate the D flip-flop for y[3]
d_ff_bh inst12 (rst_n, clk, net11, y[3]);

//-----------------------------------------------------------
assign   z1 = y[3],
         z2 = y[2],
         z3 = y[1];

endmodule
```

Figure 3.26 (Continued)

```
//test bench for moore with three outputs using D flip-flops
module moore_3_outputs_tb;

reg rst_n, clk, x1, x2, x3;    //inputs are reg
wire  [1:3] y;                 //outputs are wire
wire z1, z2, z3;
```

Figure 3.27 Test bench module for Example 3.5.

```
initial         //display variables
$monitor ("x1 x2 x3 = %b, z1 z2 z3 + %b", {x1, x2, x3},
                {z1, z2, z3});

initial         //define clock
begin
   clk = 1'b0;
   forever
      #10 clk = ~clk;
end

initial         //define input sequence
begin
   #0 rst_n = 1'b0;
      x1 = 1'b0;  x2 = 1'b0;  x3 = 1'b0;
   #5 rst_n = 1'b1;
//-----------------------------------------
   x1 = 1'b0;  x2 = 1'b0;  x3 = 1'b0;
   @ (posedge clk)

   x1 = 1'b0;  x2 = 1'b0;  x3 = 1'b1;
   @ (posedge clk)

   x1 = 1'b0;  x2 = 1'b1;  x3 = 1'b0;
   @ (posedge clk)

   x1 = 1'b0;  x2 = 1'b1;  x3 = 1'b1;
   @ (posedge clk)

   x1 = 1'b1;  x2 = 1'b0;  x3 = 1'b0;
   @ (posedge clk)

   x1 = 1'b1;  x2 = 1'b0;  x3 = 1'b1;
   @ (posedge clk)

   x1 = 1'b1;  x2 = 1'b1;  x3 = 1'b0;
   @ (posedge clk)

   x1 = 1'b1;  x2 = 1'b1;  x3 = 1'b1;
   @ (posedge clk)
   #10   $stop;
end

//instantiate the module into the test bench
moore_3_outputs inst1 (rst_n, clk, x1, x2, x3, y,
            z1, z2, z3);

endmodule
```

Figure 3.27 (Continued)

```
x1 x2 x3 = 000, clock = 0, z1 z2 z3 + 000
x1 x2 x3 = 001, clock = 1, z1 z2 z3 + 000
x1 x2 x3 = 001, clock = 0, z1 z2 z3 + 000
x1 x2 x3 = 010, clock = 1, z1 z2 z3 + 100
x1 x2 x3 = 010, clock = 0, z1 z2 z3 + 100

x1 x2 x3 = 011, clock = 1, z1 z2 z3 + 100
x1 x2 x3 = 011, clock = 0, z1 z2 z3 + 100
x1 x2 x3 = 100, clock = 1, z1 z2 z3 + 010
x1 x2 x3 = 100, clock = 0, z1 z2 z3 + 010
x1 x2 x3 = 101, clock = 1, z1 z2 z3 + 100

x1 x2 x3 = 101, clock = 0, z1 z2 z3 + 100
x1 x2 x3 = 110, clock = 1, z1 z2 z3 + 010
x1 x2 x3 = 110, clock = 0, z1 z2 z3 + 010
x1 x2 x3 = 111, clock = 1, z1 z2 z3 + 010
x1 x2 x3 = 111, clock = 0, z1 z2 z3 + 010

x1 x2 x3 = 111, clock = 1, z1 z2 z3 + 001
```

Figure 3.28   Outputs for Example 3.5.

Example 3.6 This example designs the Moore synchronous sequential machine shown in Figure 3.29 using structural modeling. The design will use D flip-flops that were designed using behavioral modeling and logic gates that were designed using dataflow modeling.

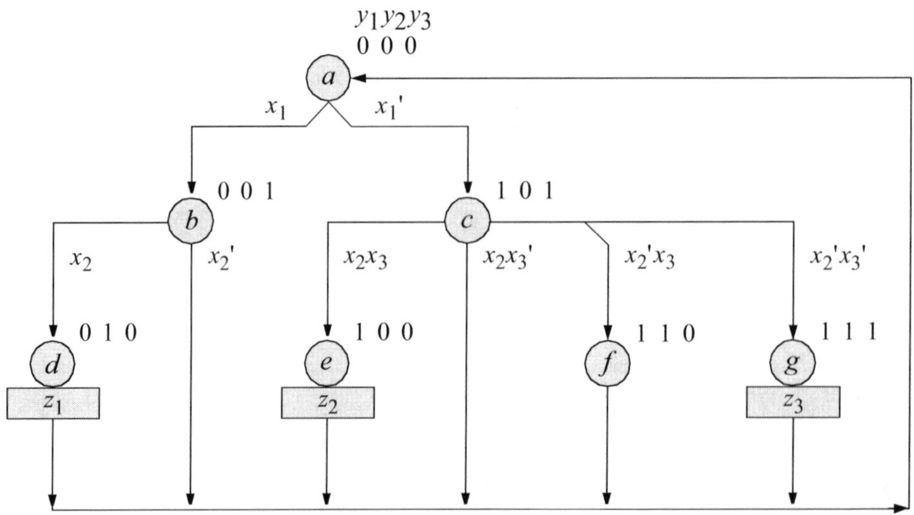

Figure 3.29   State diagram for Example 3.6.

The Karnaugh maps obtained from the state diagram are shown in Figure 3.30. The equations for the *D* flip-flops are shown below the corresponding Karnaugh maps. The equations are reproduced in Figure 3.31 together with the instantiation names and the net names for the structural design module. The structural design module is shown in Figure 3.32. The test bench module and the outputs are shown in Figures 3.33 and 3.34, respectively.

$y_1 \backslash y_2y_3$	0 0	0 1	1 1	1 0
0	(0) x_1'	(1) 0	(3) –	(2) 0
1	(4) 0	(5) $x_2' + x_3$	(7) 0	(6) 0

Dy_1

$$Dy_1 = y_1'y_2'y_3'x_1' + y_1y_2'y_3x_2' + y_1y_2'y_3x_3$$

$y_1 \backslash y_2y_3$	0 0	0 1	1 1	1 0
0	(0) 0	(1) x_2	(3) –	(2) 0
1	(4) 0	(5) x_2'	(7) 0	(6) 0

Dy_2

$$Dy_2 = y_1'y_2'y_3x_2 + y_1y_2'y_3x_2'$$

$y_1 \backslash y_2y_3$	0 0	0 1	1 1	1 0
0	(0) 1	(1) 0	(3) –	(2) 0
1	(4) 0	(5) $x_2'x_3'$	(7) 0	(6) 0

Dy_3

$$Dy_3 = y_1'y_2'y_3' + y_1y_2'y_3x_2'x_3'$$

Figure 3.30 Karnaugh maps for Example 3.6.

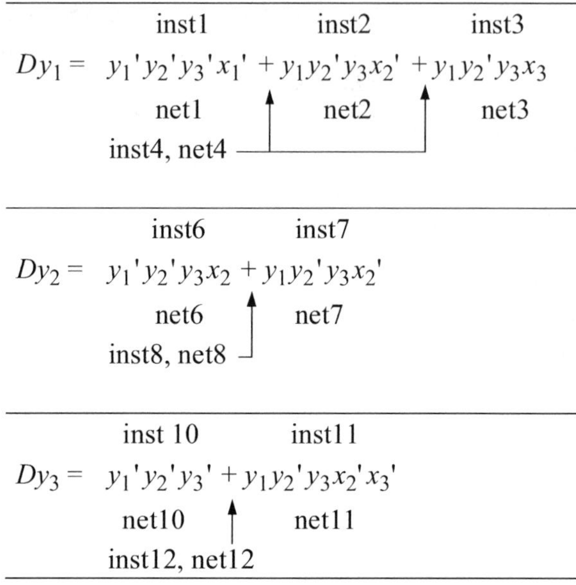

Figure 3.31 Equations with instantiation and net names for Example 3.6.

```
//structural for moore synchronous sequential machine
module moore_ssm31 (rst_n, clk, x1, x2, x3, y, z1, z2, z3);

//define inputs and outputs
input rst_n, clk, x1, x2, x3;
output [1:3] y;
output z1, z2, z3;

//define internal nets
wire net1, net2, net3, net4, net6, net7, net8,
      net10, net11, net12;

//-----------------------------------------------------
//instantiate the logic for flip-flop y[1]
and4_df  inst1 (~y[1], ~y[2], ~y[3], ~x1, net1),
         inst2 (y[1], ~y[2], y[3], ~x2, net2),
         inst3 (y[1], ~y[2], y[3], x3, net3);

or3_df   inst4 (net1, net2, net3, net4);

//instantiate the D flip-flop for y[1]
d_ff_bh  inst5 (rst_n, clk, net4, y[1]);
```

Figure 3.32 Structural design module for Example 3.6.

```
//--------------------------------------------------
//instantiate the logic for flip-flop y[2]
and4_df   inst6 (~y[1], ~y[2], y[3], x2, net6),
          inst7 (y[1], ~y[2], y[3], ~x2, net7);

or2_df    inst8 (net6, net7, net8);

//instantiate the D flip-flop for y[2]
d_ff_bh   inst9 (rst_n, clk, net8, y[2]);

//--------------------------------------------------
//instantiate the logic for flip-flop y[3]
and3_df   inst10 (~y[1], ~y[2], ~y[3], net10);
and5_df   inst11 (y[1], ~y[2], y[3], ~x2, ~x3, net11);

or2_df    inst12 (net10, net11, net12);

//instantiate the D flip-flop for y[3]
d_ff_bh   inst13 (rst_n, clk, net12, y[3]);

//--------------------------------------------------
assign    z1 = ~y[1] & y[2] & ~y[3],
          z2 = y[1] & ~y[2] & ~y[3],
          z3 = y[1] & y[2] & y[3];

endmodule
```

Figure 3.32 (Continued)

```
//test bench for the moore synchronous sequential machine
module moore_ssm31_tb;

//inputs are reg for test bench
//outputs are wire for test bench
reg rst_n, clk, x1, x2, x3;
wire [1:3] y;
wire z1, z2, z3;

//display variables
initial
$monitor ("x1 x2 x3 = %b, state = %b, z1 z2 z3 = %b",
            {x1, x2, x3}, y, {z1, z2, z3});

                                   //continued on next page
```

Figure 3.33 Test bench module for Example 3.6.

```
initial          //define clock
begin
   clk = 1'b0;
   forever
      #10 clk = ~clk;
end

initial          //define input sequence
begin
   #0 rst_n = 1'b0;
   x1 = 1'b1;  x2 = 1'b0;  x3 = 1'b0;

   #5 rst_n = 1'b1;
   x1 = 1'b0;  x2 = 1'b0;  x3 = 1'b0;
//-----------------------------------------------------
   x2 = 1'b1;
   @ (posedge clk)

   x1 = 1'b0;  x2 = 1'b0;  x3 = 1'b0;
   @ (posedge clk)
//-----------------------------------------------------
   x1 = 1'b0;
   @ (posedge clk)

   x2 = 1'b1;x3 = 1'b1;
   @ (posedge clk)
//-----------------------------------------------------
   x1 = 1'b0;
   @ (posedge clk)

   x2 = 1'b0;
   @ (posedge clk)

   x2 = 1'b0;x3 = 1'b0;
   @ (posedge clk)

   x2 = 1'b0;x3 = 1'b0;
   @ (posedge clk)

   #10   $stop;
end

//instantiate the module into the test bench
moore_ssm31 inst1 (rst_n, clk, x1, x2, x3, y, z1, z2, z3);

endmodule
```

Figure 3.33 (Continued)

```
x1 x2 x3 = 100, state = 000, z1 z2 z3 = 000
x1 x2 x3 = 010, state = 001, z1 z2 z3 = 000
x1 x2 x3 = 000, state = 010, z1 z2 z3 = 100
x1 x2 x3 = 000, state = 000, z1 z2 z3 = 000

x1 x2 x3 = 011, state = 101, z1 z2 z3 = 000
x1 x2 x3 = 011, state = 100, z1 z2 z3 = 010
x1 x2 x3 = 001, state = 000, z1 z2 z3 = 000

x1 x2 x3 = 000, state = 101, z1 z2 z3 = 000
x1 x2 x3 = 000, state = 111, z1 z2 z3 = 001
x1 x2 x3 = 000, state = 000, z1 z2 z3 = 000
```

Figure 3.34 Outputs for Example 3.6.

3.2.4 Mealy Machines

Mealy machines are synchronous sequential machines in which the output function λ produces an output vector $Z_{r(t)}$ which is determined by both the present input vector $X_{i(t)}$ and the present state of the machine $Y_{j(t)}$. The general configuration of a Mealy machine is shown in Figure 3.35. A Mealy machine may be synchronous or asynchronous; however, this section pertains to synchronous organizations only.

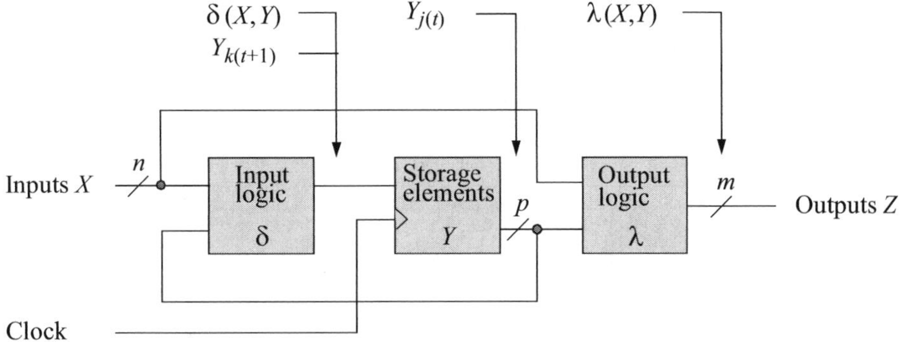

Figure 3.35 Mealy machine in which the outputs are a function of both the present state and the present inputs.

A Mealy machine is a 5-tuple and can be defined as shown in Equation 3.5,

$$M = (X, Y, Z, \delta, \lambda) \tag{3.5}$$

where

X is a nonempty finite set of inputs
Y is a nonempty finite set of states
Z is a nonempty finite set of outputs
$\delta(X, Y) : X \times Y \rightarrow Y$
$\lambda(X, Y) : Y \rightarrow Z$

The Mealy class of synchronous sequential machines is the result of a paper by G. H. Mealy in 1955 on the synthesis of sequential circuits. The definitions for Mealy and Moore machines are the same, except that the outputs of a Mealy machine are a function of both the present inputs and the present state, whereas the outputs of a Moore machine are a function of the present state only. This is the underlying difference between Moore and Mealy machines. A Moore machine, therefore, can be considered as a special case of a Mealy machine.

Example 3.7 The state diagram for a Mealy synchronous sequential machine is shown in Figure 3.36 and will be implemented with *JK* flip-flops in a structural design module. There are three inputs x_1, x_2, and x_3 and one output z_1. There are two state flip-flops y_1 and y_2 that are reset to state a ($y_1y_2 = 11$) and one unused state $y_1y_2 = 00$.

The functional characteristic table for a *JK* flip-flop is shown in Table 3.1 and the excitation table is shown in Table 3.2.

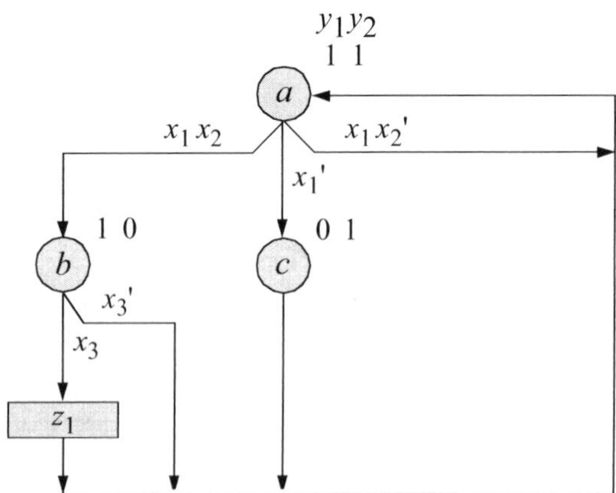

Figure 3.36 State diagram for the Mealy machine of Example 3.7. Unused state is $y_1y_2 = 00$.

The next-state table is shown in Table 3.3 and is obtained directly from the state diagram. For example, consider state a ($y_1y_2 = 11$). Input x_3 does not contribute to a state transition to state b ($y_1y_2 = 10$) or to state c ($y_1y_2 = 01$); therefore, input x_3 is entered as a "don't care" value in the next-state table. If $x_1 = 0$ in state a ($y_1y_2 = 11$), then the machine proceeds to state c ($y_1y_2 = 01$); therefore, the next-state table contains a next state of $y_1y_2 = 01$ whenever $x_1 = 0$.

If $x_1x_2 = 10$ in state a, then the machine remains in state a. If $x_1x_2 = 11$ in state a, then the machine proceeds to state b ($y_1y_2 = 10$), where output z_1 is asserted if $x_3 = 1$, then sequences to state a; otherwise, the machine proceeds to state a without asserting output z_1.

Table 3.1 *JK* Functional Characteristic Table

JK	Function
0 0	No change
0 1	Reset
1 0	Set
1 1	Toggle

Table 3.2 Excitation Table for a *JK* Flip-Flop

Present State $Y_{j(t)}$	Next State $Y_{k(t+1)}$	Data Inputs JK	
0	0	0 –	A dash (–) indicates a "don't care" condition
0	1	1 –	
1	0	– 1	
1	1	– 0	

Table 3.3 Next-State Table for the Mealy Machine of Example 3.7

Present State y_1 y_2	Inputs x_1 x_2 x_3	Next State y_1 y_2	Flip-Flop Inputs Jy_1 Ky_1	Jy_2 Ky_2	Output z_1
0 0	– – –	– –	– –	– –	–
0 1	– – –	1 1	1 –	– 0	0
1 0	– – 0	1 1	– 0	1 –	0
	– – 1	1 1	– 0	1 –	1
1 1	0 0 –	0 1	– 1	– 0	0
	0 1 –	0 1	– 1	– 0	0
	1 0 –	1 1	– 0	– 0	0
	1 1 –	1 0	– 0	– 1	0

The Karnaugh maps for the *JK* flip-flops are shown in Figure 3.37 and the equations are shown in Equation 3.6. The Karnaugh map for output z_1 is shown in Figure 3.38 and the equation for z_1 is shown in Equation 3.7. The logic diagram is shown in Figure 3.39 using AND gates and positive edge-triggered *JK* flip-flops. The design module for a *JK* flip-flop is shown in Figure 3.40. The structural design module is shown in Figure 3.41 using instantiated logic gates that were designed using dataflow modeling and *JK* flip-flops that were designed using behavioral modeling. The test bench module and the outputs are shown in Figures 3.42 and 3.43, respectively.

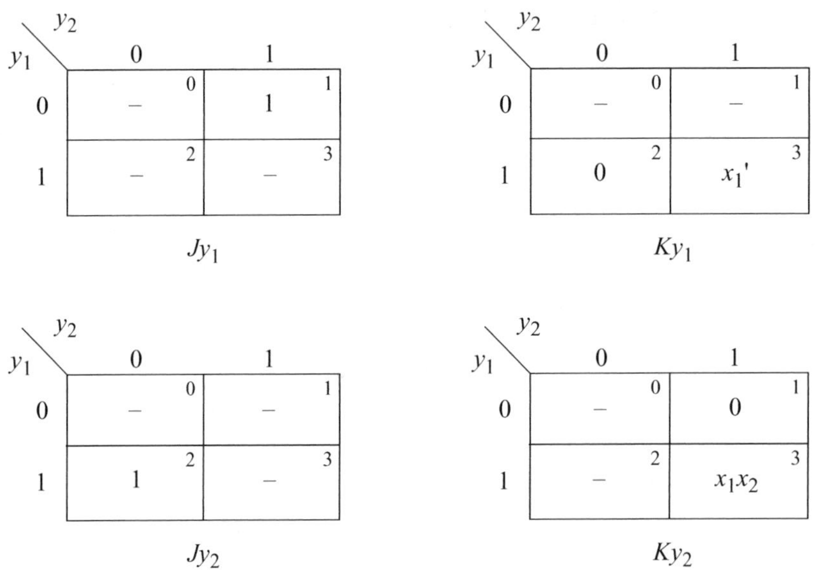

Figure 3.37 Karnaugh maps for the Mealy synchronous sequential machine of Example 3.7.

$$Jy_1 = 1 \qquad Ky_1 = y_2 x_1'$$

$$Jy_2 = 1 \qquad Ky_2 = y_1 x_1 x_2 \tag{3.6}$$

$y_1 \backslash y_2$	0	1
0	– (0)	0 (1)
1	x_3 (2)	0 (3)

z_1

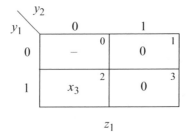

Figure 3.38 Karnaugh map for output z_1.

$$z_1 = y_2' x_3 \tag{3.7}$$

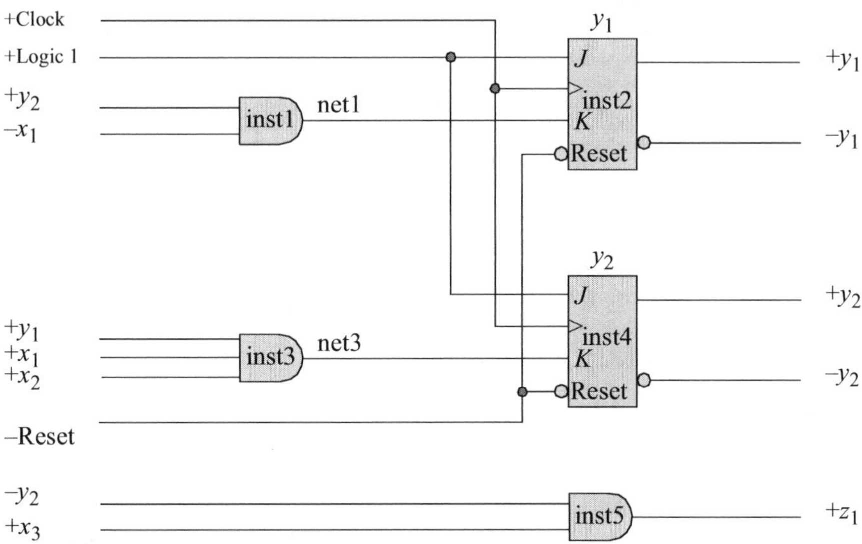

Figure 3.39 Logic diagram for the Mealy machine of Example 3.7.

```
module jk_ff_bh (rst_n, clk, j, k,  q);

input rst_n, clk, j, k;
output q;

wire rst_n, clk, j, k;
reg q;

always @ (posedge clk or negedge rst_n)
begin
   if (~rst_n)     begin     q  <= 1'b0;
end

else if (j==1'b0 && k==1'b0)
   begin     q <= q;       end

else if (j==1'b0 && k==1'b1)
   begin     q  <= 1'b0;     end

else if (j==1'b1 && k==1'b0)
   begin     q  <= 1'b1;     end

else if (j==1'b1 && k==1'b1)
   begin     q  <= ~q;       end
end

endmodule
```

Figure 3.40 Behavioral design module for a *JK* flip-flop.

```
//structural mealy ssm
module mealy_ssm_10a (rst_n, clk, x1, x2, x3, y, z1);

input rst_n, clk, x1, x2, x3;    //define inputs and output
output [1:2] y;
output z1;

wire net1, net3;     //define internal nets
//-------------------------------------------------------
//instantiate the logic for flip-flop y[1]
and2_df  inst1 (y[2], ~x1, net1);

//instantiate the JK flip-flop for y[1]
jk_ff_bh inst2 (rst_n, clk, 1'b1, net1, y[1]);
//-------------------------------------------------------
//instantiate the logic for flip-flop y[2]
and3_df  inst3 (y[1], x1, x2, net3);

jk_ff_bh inst4 (rst_n, clk, 1'b1, net3, y[2]);
//-------------------------------------------------------
//instantiate the logic for the output z1
and2_df  inst5 (~y[2], x3, z1);
endmodule
```

Figure 3.41 Structural design module for the Mealy machine of Example 3.7.

```
//test bench for mealy synchronous sequential machine
module mealy_ssm_10a_tb;

reg rst_n, clk, x1, x2, x3;   //inputs are reg
wire [1:2] y;                 //outputs are wire
wire z1;

initial        //display variables
begin
$monitor ("x1 x2 x3 = %b, state = %b, z1 = %b",
            {x1, x2, x3}, y, z1);
end

initial     //define clock
begin
   clk = 1'b0;
   forever
      #10 clk = ~clk;
end                           //continued on next page
```

Figure 3.42 Test bench module for the Mealy machine of Example 3.7.

```
//define input sequence
initial
begin
   #0    rst_n = 1'b0;      #5    rst_n = 1'b1;
   #10   x1 = 1'b0;  x2 = 1'b0;  x3 = 1'b0;
//-------------------------------------
   x1 = 1'b0;
   @ (posedge clk)
   @ (posedge clk)

   x1 = 1'b1;  x2 = 1'b0;
   @ (posedge clk)

   x1 = 1'b1;  x2 = 1'b1;
   @ (posedge clk)

   x3 = 1'b0;
   @ (posedge clk)

   x1 = 1'b1;  x2 = 1'b1;
   @ (posedge clk)

   x3 = 1'b1;
   @ (posedge clk)

   #10   $stop;
end
//-------------------------------------
//instantiate the module into the test bench
mealy_ssm_10a inst1 (rst_n, clk, x1, x2, x3, y, z1);

endmodule
```

Figure 3.42 (Continued)

```
x1 x2 x3 = xxx, state = 00, z1 = x
x1 x2 x3 = xxx, state = 11, z1 = 0
x1 x2 x3 = 000, state = 11, z1 = 0
x1 x2 x3 = 000, state = 01, z1 = 0

x1 x2 x3 = 100, state = 11, z1 = 0
x1 x2 x3 = 110, state = 11, z1 = 0
x1 x2 x3 = 110, state = 10, z1 = 0
x1 x2 x3 = 110, state = 11, z1 = 0

x1 x2 x3 = 111, state = 10, z1 = 1
x1 x2 x3 = 111, state = 11, z1 = 0
```

Figure 3.43 Outputs for the Mealy machine of Example 3.7.

Example 3.8 A Mealy synchronous sequential machine will be designed using structural modeling with built-in primitives and D flip-flops that were designed using behavioral modeling. There is one output z_1 that is asserted whenever a serial input data line x_1 contains a 3-bit word with an odd number of 1s. There is no space between words.

The state diagram is shown in Figure 3.44 with three unused states: $y_1y_2y_3 = 101$, 110, and 111, which will be assigned values of zero. If two or more flip-flops change value for a state transition sequence, then the machine may momentarily pass through either an unused state or a state in which there is no output — in both cases, there will be no glitch on output z_1.

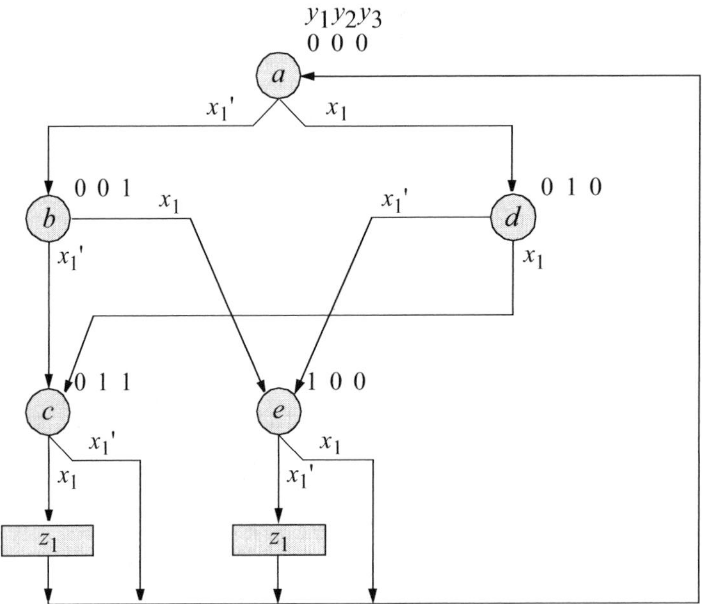

Figure 3.44 State diagram for the Mealy machine of Example 3.8.

The Karnaugh maps for the D flip-flops $y_1y_2y_3$ are shown in Figure 3.45. The equations for the flip-flops are shown in Equation 3.8. The Karnaugh map for output z_1 is shown in Figure 3.46 and the equation for z_1 is shown in Equation 3.9.

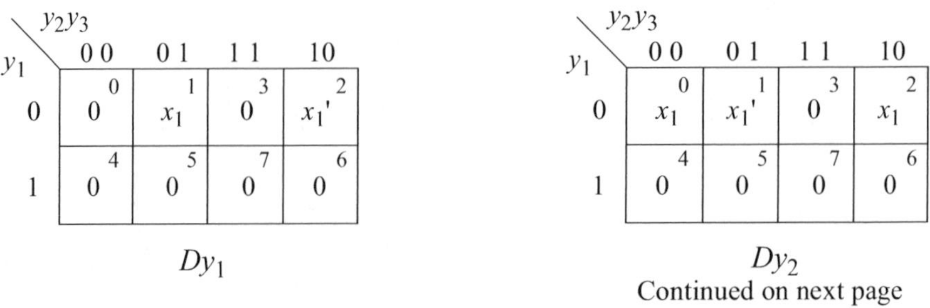

Dy_1

y_1 \ y_2y_3	0 0	0 1	1 1	10
0	(0) 0	(1) x_1	(3) 0	(2) x_1'
1	(4) 0	(5) 0	(7) 0	(6) 0

Dy_2

y_1 \ y_2y_3	0 0	0 1	1 1	10
0	(0) x_1	(1) x_1'	(3) 0	(2) x_1
1	(4) 0	(5) 0	(7) 0	(6) 0

Continued on next page

Figure 3.45 Karnaugh maps for the D flip-flops for Example 3.8.

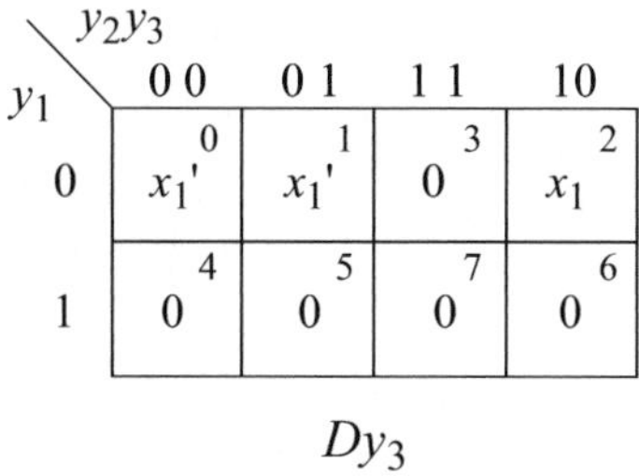

Figure 3.45 (Continued)

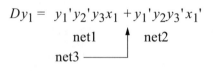

$$Dy_1 = y_1'y_2'y_3x_1 + y_1'y_2y_3'x_1'$$

net1 net2

net3

$$Dy_2 = y_1'y_2'y_3x_1' + y_1'y_2y_3'x_1$$

net4 net5

net6

(3.8)

$$Dy_3 = y_1'y_2'x_1' + y_1'y_2y_3'x_1$$

net7 net8

net9

$y_1 \backslash y_2y_3$	00	01	11	10
0	0 (0)	0 (1)	x_1 (3)	0 (2)
1	x_1' (4)	– (5)	– (7)	– (6)

z_1

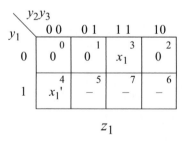

Figure 3.46 Karnaugh map for output z_1.

$$z_1 = y_1x_1' + y_2y_3x_1 \qquad (3.9)$$

net10 net11

z_1

The structural design module is shown in Figure 3.47 using built-in primitives and *D* flip-flops that were designed using behavioral modeling. Refer to page 254 of this chapter for the behavioral design module for the *D* flip-flop. When using built-in primitives, the output signal is listed first, followed by the inputs in any order. The instance name is optional. The test bench module is shown in Figure 3.48 and the outputs are shown in Figure 3.49.

```
//structural for mealy using built-in primitives
module mealy_struc_bip (rst_n, clk, x1, y, z1);

input rst_n, clk, x1;    //define inputs and output
output [1:3] y;
output z1;

//define internal nets
wire net1, net2, net3, net4, net5, net6, net7,
      net8, net9, net10, net11, net12;

//--------------------------------------------------------
//instantiate the logic for flip-flop y[1]
and   (net1, ~y[1], ~y[2], y[3], x1),
      (net2, ~y[1], y[2], ~y[3], ~x1);

or    (net3, net1, net2);

//instantiate the D flip-flop for y[1]
d_ff_bh  inst1 (rst_n, clk, net3, y[1]);

//--------------------------------------------------------
//instantiate the logic for flip-flop y[2]
and   (net4, ~y[1], ~y[2], y[3], ~x1),
      (net5, ~y[1], y[2], ~y[3], x1);

or    (net6, net4, net5);

//instantiate the D flip-flop for y[2]
d_ff_bh  inst2 (rst_n, clk, net6, y[2]);

//--------------------------------------------------------
//instantiate the logic for flip-flop y[3]
and   (net7, ~y[1], ~y[2], ~x1),
      (net8, ~y[1], y[2], ~y[3], x1);

or    (net9, net7, net8);

//instantiate the D flip-flop for y[3]
d_ff_bh  inst3 (rst_n, clk, net9, y[3]);
                                    //continued on next page
```

Figure 3.47 Structural design module for the Mealy machine of Example 3.8.

```
//instantiate the logic for output z1
and    (net10, y[1], ~x1),
       (net11, y[2], y[3], x1);

or     (z1, net10, net11);

endmodule
```

Figure 3.47 (Continued)

```
//test bench for structural mealy using built-in primitives

module mealy_struc_bip_tb;

//inputs are reg for test bench
//outputs are wire for test bench
reg rst_n, clk, x1;
wire [1:3] y;
wire z1;

//display variables
initial
$monitor ("x1 = %b, state = %b, z1 = %b", x1, y, z1);

//define clock
initial
begin
   clk = 1'b0;
   forever
      #10   clk = ~clk;
end

//define input sequence
initial
begin
   #0     rst_n = 1'b0;   x1 = 1'b0;

   #5     rst_n = 1'b1;
//-------------------------------------------
   x1 = 1'b0;   @ (posedge clk)

   x1 = 1'b1;   @ (posedge clk)

   x1 = 1'b0;   @ (posedge clk)
                                          //continued on next page
```

Figure 3.48 Test bench module for the Mealy machine of Example 3.8.

```
    x1 = 1'b1;   @ (posedge clk)

    x1 = 1'b0;   @ (posedge clk)

    x1 = 1'b0;   @ (posedge clk)

    x1 = 1'b1;   @ (posedge clk)

    x1 = 1'b1;   @ (posedge clk)

    x1 = 1'b0;   @ (posedge clk)

    x1 = 1'b0;   @ (posedge clk)

    x1 = 1'b0;   @ (posedge clk)

    #10   $stop;
end

//instantiate the module into the test bench
mealy_struc_bip inst1 (rst_n, clk, x1, y, z1);

endmodule
```

Figure 3.48 (Continued)

```
x1 = 0, state = 000, z1 = 0
x1 = 0, state = 001, z1 = 0
x1 = 1, state = 011, z1 = 1
x1 = 0, state = 000, z1 = 0

x1 = 1, state = 001, z1 = 0
x1 = 0, state = 100, z1 = 1
x1 = 0, state = 000, z1 = 0

x1 = 1, state = 001, z1 = 0
x1 = 1, state = 100, z1 = 0
x1 = 0, state = 000, z1 = 0

x1 = 0, state = 001, z1 = 0
x1 = 0, state = 011, z1 = 0
x1 = 0, state = 000, z1 = 0
```

Figure 3.49 Outputs for the Mealy machine of Example 3.8.

Example 3.9 A Mealy synchronous sequential machine that generates an output z_1 whenever the sequence 1001 is detected on a serial data input line x_1. Overlapping sequences are valid, as shown below, which will assert z_1 three times.

$$. . . 0110\underline{1001}0001\underline{10010010} . . .$$

The state diagram is shown in Figure 3.50. This example will use logic gates that were designed using dataflow modeling and D flip-flops that were designed using behavioral modeling.

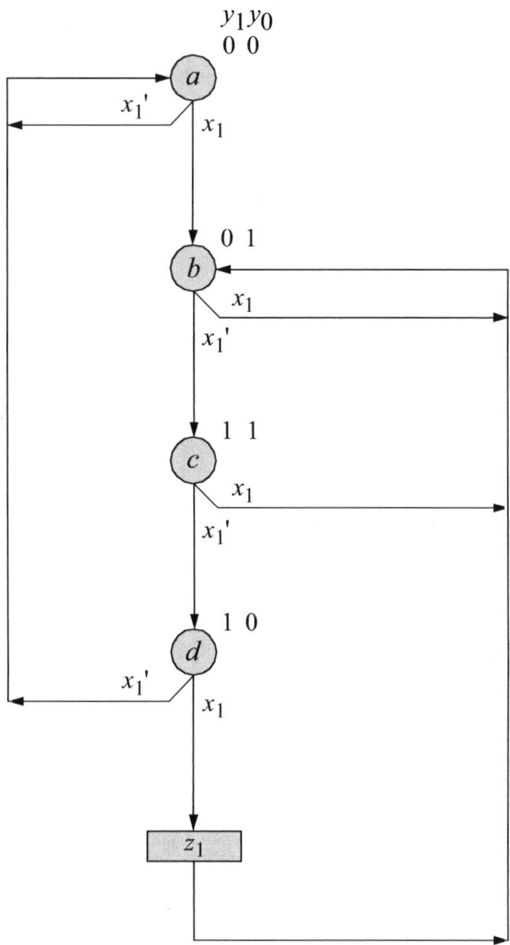

Figure 3.50 State diagram for the Mealy synchronous sequential machine of Example 3.9.

The Karnaugh maps for flip-flops y_1 and y_2 are shown in Figure 3.51. The equations for the flip-flops are shown in Equation 3.10. The equation for output z_1 is shown in Equation 3.11. The structural design module using dataflow logic gates and behavioral D flip-flops is shown in Figure 3.52. The test bench module and the outputs are shown in Figures 3.53 and 3.54, respectively.

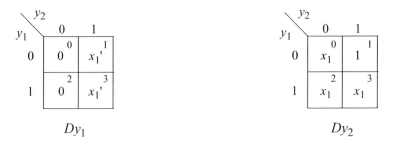

Figure 3.51 Karnaugh maps for Example 3.9.

$$\overset{\text{net1}}{Dy_1 = y_2 x_1'}$$

$$Dy_2 = x_1 + \overset{\text{net3}}{y_1' y_2} \quad \text{(+ is net4)} \qquad (3.10)$$

$$z_1 = y_1 y_2' x_1 \qquad (3.11)$$

```
//structural for mealy to detect the sequence 1001

module mealy_1001_sequence (rst_n, clk, x1, y, z1);

//define inputs and output
input rst_n, clk, x1;
output [1:2] y;
output z1;

//define internal nets
wire net1, net2, net3;

                         //continued on next page
```

Figure 3.52 Structural design module for the Mealy machine of Example 3.9.

```
//-----------------------------------------------
//instantiate the logic for flip-flop y[1]
and2_df  inst1 (y[2], ~x1, net1);

//instantiate the D flip-flop for y[1]
d_ff_bh  inst2 (rst_n, clk, net1, y[1]);

//-----------------------------------------------
//instantiate the logic for flip-flop y[2]
and2_df  inst3 (~y[1], y[2], net3);
or2_df   inst4 (x1, net3, net4);

//instantiate the D flip-flop for y[2]
d_ff_bh  inst5 (rst_n, clk, net4, y[2]);

//-----------------------------------------------
//instantiate the logic for output z1
and3_df  inst6 (y[1], ~y[2], x1, z1);

endmodule
```

Figure 3.52 (Continued)

```
//test bench for mealy_1001_sequence
module mealy_1001_sequence_tb;

reg rst_n, clk, x1;       //inputs are reg for test bench
wire [1:2] y;             //outputs are wire for test bench
wire z1;

//display variables
initial
begin
   $monitor ("x1 = %b, state = %b, z1 = %b", x1, y, z1);
end

//define clock
initial
begin
   clk = 1'b0;
   forever
      #10 clk = ~clk;
end
                                  //continued on next page
```

Figure 3.53 Test bench module for Example 3.9.

```
//define input sequence
initial
begin
   #0 rst_n = 1'b0;x1 = 1'b0;
   #5 rst_n = 1'b1;x1 = 1'b0;

//------------------------------------------------
   x1 = 1'b1;  @ (posedge clk)
   x1 = 1'b0;  @ (posedge clk)
   x1 = 1'b0;  @ (posedge clk)
   x1 = 1'b1;  @ (posedge clk)


   x1 = 1'b0;  @ (posedge clk)
   x1 = 1'b1;  @ (posedge clk)
   x1 = 1'b0;  @ (posedge clk)
   x1 = 1'b0;  @ (posedge clk)

   x1 = 1'b1;  @ (posedge clk)
   x1 = 1'b1;  @ (posedge clk)
   x1 = 1'b1;  @ (posedge clk)
   x1 = 1'b0;  @ (posedge clk)


   x1 = 1'b0;  @ (posedge clk)
   x1 = 1'b1;  @ (posedge clk)
   x1 = 1'b1;  @ (posedge clk)
   x1 = 1'b0;  @ (posedge clk)


   x1 = 1'b0;  @ (posedge clk)
   x1 = 1'b1;  @ (posedge clk)
   x1 = 1'b1;  @ (posedge clk)
   x1 = 1'b1;  @ (posedge clk)


   x1 = 1'b0;  @ (posedge clk)


   #10   $stop;

end


//instantiate the module into the test bench
mealy_1001_sequence inst1 (rst_n, clk, x1, y, z1);

endmodule
```

Figure 3.53 (Continued)

```
x1 = 0, state = 00, z1 = 0
x1 = 1, state = 00, z1 = 0

x1 = 0, state = 01, z1 = 0
x1 = 0, state = 11, z1 = 0

x1 = 1, state = 10, z1 = 1
x1 = 0, state = 01, z1 = 0

x1 = 1, state = 11, z1 = 0
x1 = 0, state = 01, z1 = 0

x1 = 0, state = 11, z1 = 0
x1 = 1, state = 10, z1 = 1

x1 = 1, state = 01, z1 = 0
x1 = 0, state = 01, z1 = 0

x1 = 0, state = 11, z1 = 0
x1 = 1, state = 10, z1 = 1

x1 = 1, state = 01, z1 = 0
x1 = 0, state = 01, z1 = 0

x1 = 0, state = 11, z1 = 0
x1 = 1, state = 10, z1 = 1

x1 = 1, state = 01, z1 = 0
x1 = 0, state = 01, z1 = 0

x1 = 0, state = 11, z1 = 0
```

Figure 3.54 Outputs for Example 3.9.

Example 3.10 A state diagram for a Mealy synchronous sequential machine is shown in Figure 3.55 which has three parallel inputs $x_1x_2x_3$ and two outputs z_1z_2. Output z_1 is asserted if the input sequence is $x_1x_2x_3$ = 000, 111, 000. Output z_2 is asserted if the input sequence is $x_1x_2x_3$ = 111, 000, 111.

There are two unused states: $y_1y_2y_3$ = 110 and 111. The state codes are assigned such that any transition through an unused state will not cause an output to be asserted if the input sequence from state *e* to state *a* is any input value that is not $x_1x_2x_3$ = 000 or 111.

The behavioral design module is shown in Figure 3.56 using the **case** statement with the **if, else if, else** conditional statements. The test bench module is shown in Figure 3.57 and the outputs are shown in Figure 3.58.

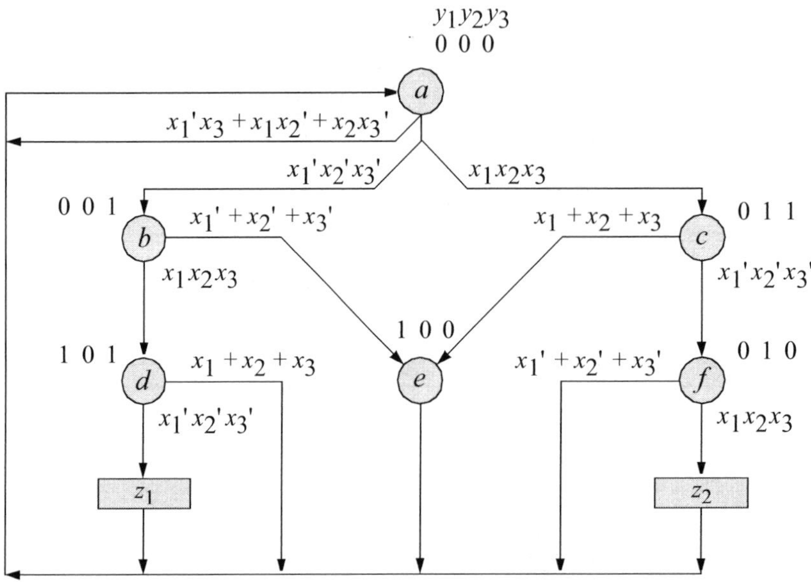

Figure 3.55 State diagram for the Mealy machine of Example 3.10. Unused states are: $y_1y_2y_3$ = 110 and 111.

```
//behavioral mealy three parallel inputs
module mealy_111_000_bh_case (rst_n, clk, x1, x2, x3,
                              y, z1, z2);

//define inputs and outputs
input rst_n, clk, x1, x2, x3;
output [1:3] y;
output z1, z2;

//variables in always are reg
reg [1:3] y, next_state;

//assign state codes
parameter   state_a = 3'b000,
            state_b = 3'b001,
            state_c = 3'b011,
            state_d = 3'b101,
            state_e = 3'b100,
            state_f = 3'b010;
                              //continued on next page
```

Figure 3.56 Behavioral design module for the Mealy synchronous sequential machine of Example 3.10.

```
//set next state
always @ (posedge clk)
begin
   if (~rst_n)
      y = state_a;
   else
      y = next_state;
end

//define outputs
assign   z1 = (y[1] & ~y[2] & y[3] & ~x1 & ~x2 & ~x3),
         z2 = (~y[1] & y[2] & ~y[3] & x1 & x2 & x3);

//determine next state
always @ (y or x1 or x2 or x3)
begin
   case (y)
//--------------------------------------------
      state_a:
         if ((x1==0 & x3==1) | (x1==1 & x2==0)
               | (x2==1 & x3==0))
            next_state = state_a;

         else if (x1==0 & x2==0 & x3==0)
            next_state = state_b;

         else if (x1==1 & x2==1 & x3==1)
            next_state = state_c;

//--------------------------------------------
      state_b:
         if (x1==1 & x2==1 & x3==1)
            next_state = state_d;

      else if (x1==0 | x2==0 | x3==0)
         next_state = state_e;

//--------------------------------------------
      state_c:
         if (x1==1 | x2==1 | x3==1)
            next_state = state_e;

         else if (x1==0 & x2==0 & x3==0)
            next_state = state_f;

//--------------------------------------------
                                    //continued on next page
```

Figure 3.56 (Continued)

```
   state_d:
         if (x1==0 & x2==0 & x3==0)
            next_state = state_a;//assert z1

         else if (x1==1 | x2==1 | x3==1)
            next_state = state_a;
//-------------------------------------------
   state_e: next_state = state_a;
//-------------------------------------------
      state_f:
         if (x1==0 | x2==0 | x3==0)
            next_state = state_a;

         else if (x1==1 & x2==1 & x3==1)
            next_state = state_a;//assert z2
//-------------------------------------------
      default next_state = state_a;
endcase
end

endmodule
```

Figure 3.56 (Continued)

```
//test bench for mealy three parallel inputs
module mealy_111_000_bh_case_tb;

//inputs are reg for test bench
//outputs are wire for test bench
reg rst_n, clk, x1, x2, x3;
wire [1:3] y;
wire z1, z2;

initial      //display variables
$monitor ("x1 x2 x3 = %b, state = %b, z1 = %b, z2 = %b",
            {x1, x2, x3}, y, z1, z2);

initial      //define clock
begin
   clk = 1'b0;
   forever
      #10 clk = ~clk;
end                                   //continued on next page
```

Figure 3.57 Test bench module for the Mealy machine of Example 3.10.

```
//define input sequence
initial
begin
   #0 rst_n = 1'b0;  #5 rst_n = 1'b1;
   #10 x1 = 1'b0; x2 = 1'b0;  x3 = 1'b0;
//------------------------------------------
   x1=1'b0; x3=1'b1;
   @ (posedge clk)   //go to state_a (000)

   x1=1'b1; x2=1'b0;
   @ (posedge clk)   //go to state_a (000)

   x2=1'b1; x3=1'b0;
   @ (posedge clk)   //go to state_a (000)

   x1=1'b0; x2=1'b0; x3=1'b0;
   @ (posedge clk)   //go to state_b (001)
//------------------------------------------
   x1=1'b1; x2=1'b1; x3=1'b1;
   @ (posedge clk)   //go to state_d (101)
//------------------------------------------
   x1=1'b0; x2=1'b0; x3=1'b0;
   @ (posedge clk)   //go to state_a (000), assert z1
//------------------------------------------
   x1=1'b1; x2=1'b1; x3=1'b1;
   @ (posedge clk)   //go to state_c (011)

   x1=1'b0; x2=1'b0; x3=1'b0;
   @ (posedge clk)   //go to state_f (010)

   x1=1'b1; x2=1'b1; x3=1'b1;
   @ (posedge clk)   //go to state_a (000), assert z2
//------------------------------------------
   x1=1'b1; x2=1'b1; x3=1'b1;
   @ (posedge clk)   //go to state_c (011)

   x1=1'b1; x2=1'b0; x3=1'b0;
   @ (posedge clk)   //go to state_e (100)
   @ (posedge clk)
//------------------------------------------
   #20   $stop;
end

//instantiate the module into the test bench
mealy_111_000_bh_case inst1 (rst_n, clk, x1, x2, x3,
                                        y, z1, z2);

endmodule
```

Figure 3.57 (Continued)

```
x1 x2 x3 = xxx, state = xxx, z1 = x, z2 = x
x1 x2 x3 = 001, state = xxx, z1 = 0, z2 = 0
x1 x2 x3 = 101, state = 000, z1 = 0, z2 = 0
x1 x2 x3 = 110, state = 000, z1 = 0, z2 = 0

x1 x2 x3 = 000, state = 000, z1 = 0, z2 = 0
x1 x2 x3 = 111, state = 001, z1 = 0, z2 = 0
x1 x2 x3 = 000, state = 101, z1 = 1, z2 = 0
x1 x2 x3 = 111, state = 000, z1 = 0, z2 = 0

x1 x2 x3 = 000, state = 011, z1 = 0, z2 = 0
x1 x2 x3 = 111, state = 010, z1 = 0, z2 = 1
x1 x2 x3 = 111, state = 000, z1 = 0, z2 = 0

x1 x2 x3 = 100, state = 011, z1 = 0, z2 = 0
x1 x2 x3 = 100, state = 100, z1 = 0, z2 = 0
x1 x2 x3 = 100, state = 000, z1 = 0, z2 = 0
```

Figure 3.58 Outputs for the Mealy machine of Example 3.10.

Example 3.11 This example repeats Example 3.10 for a Mealy machine, but uses a structural design module with built-in primitives and *D* flip-flops. The state diagram for the Mealy machine is reproduced in Figure 3.59 for convenience.

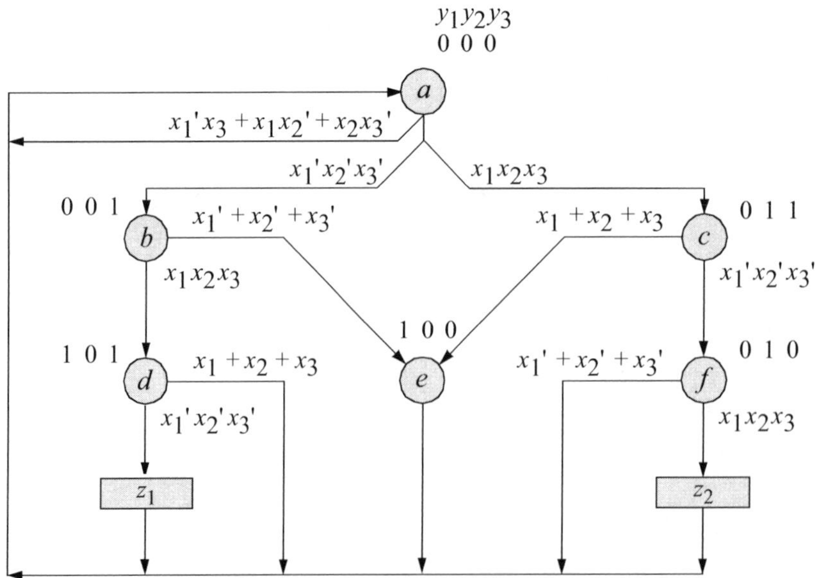

Figure 3.59 State diagram for Mealy machine of Example 3.11.

The equations for flip-flops $y_1y_2y_3$ can be obtained directly from the state diagram. The equations represent the transition paths where the next state for a particular flip-flop is a logic 1. For example, flip-flop y_1 assumes a value of 1 for the following input vectors: from state b ($y_1y_2y_3 = 001$) to state d ($y_1y_2y_3 = 101$), where the inputs are $x_1x_2x_3$; from state b ($y_1y_2y_3 = 001$) to state e ($y_1y_2y_3 = 100$), where the inputs are $x_1' + x_2' + x_3'$; and from state c ($y_1y_2y_3 = 011$) to state e ($y_1y_2y_3 = 100$), where the inputs are $x_1 + x_2 + x_3$. Therefore, the equations for the D input of flip-flop y_1 are shown in Equation 3.12 using built-in primitives.

$$
\begin{aligned}
&\textbf{and} \;\; \text{net1}: y_1'y_2'y_3x_1x_2x_3 \\
&\textbf{or} \;\; \text{net2}: x_1'x_2'x_3' \\
&\textbf{and} \;\; \text{net3}: y_1'y_2'y_3\text{net2} \\
&\textbf{or} \;\; \text{net4}: x_1x_2x_3 \\
&\textbf{and} \;\; \text{net5}: y_1'y_2y_3\text{net4} \\
Dy_1 = \; &\textbf{or} \;\; \text{net6}: \text{net1 net3 net5}
\end{aligned}
\tag{3.12}
$$

The structural design module using built-in primitives and D flip-flops that were designed using behavioral modeling is shown in Figure 3.60. The test bench module and the outputs are shown in Figures 3.61 and 3.62, respectively.

```
//structural for mealy_111_000 with three parallel inputs
module mealy_111_000_struc (rst_n, clk, x1, x2, x3,
                              y, z1, z2);
input rst_n, clk, x1, x2, x3;     //define inputs and outputs
output [1:3] y;
output z1, z2;

//define internal nets
wire net1, net2, net3, net4, net5, net6, net7, net8, net9,
        net10, net11, net12, net13;

//------------------------------------------------------------
//instantiate the logic for flip-flop y[1]
and    inst1 (net1, ~y[1], ~y[2], y[3], x1, x2, x3);
or     inst2 (net2, ~x1, ~x2, ~x3);
and    inst3 (net3, ~y[1], ~y[2], y[3], net2);
or     inst4 (net4, x1, x2, x3);
and    inst5 (net5, ~y[1], y[2], y[3], net4);
or     inst6 (net6, net1, net3, net5);

//instantiate the D flip-flop for y[1]
d_ff_bh   inst7 (rst_n, clk, net6, y[1]);
                                  //continued on next page
```

Figure 3.60 Structural design module for the Mealy machine of Example 3.11.

```
//--------------------------------------------------------------
//instantiate the logic for flip-flop y[2]
and   inst8 (net7, ~y[1], ~y[2], ~y[3], x1, x2, x3),
      inst9 (net8, ~y[1], y[2], y[3], ~x1, ~x2, ~x3);
or    inst10 (net9, net7, net8);

//instantiate the D flip-flop for y[2]
d_ff_bh  inst11 (rst_n, clk, net9, y[2]);

//--------------------------------------------------------------
//instantiate the logic for flip-flop y[3]
and   inst12 (net10, ~y[1], ~y[2], ~y[3], ~x1, ~x2, ~x3),
      inst13 (net11, ~y[1], ~y[2], ~y[3], x1, x2, x3),
      inst14 (net12, ~y[1], ~y[2], y[3], x1, x2, x3);
or    inst15 (net13, net10, net11, net12);

//instantiate the D flip-flop for y[3]
d_ff_bh  inst16 (rst_n, clk, net13, y[3]);

//--------------------------------------------------------------
//define outputs z1 and z2
assign   z1 = (y[1] & ~y[2] & y[3] & ~x1 & ~x2 & ~x3),
         z2 = (~y[1] & y[2] & ~y[3] & x1 & x2 & x3);

endmodule
```

Figure 3.60 (Continued)

```
//test bench for mealy_111_000 with three parallel inputs

module mealy_111_000_struc_tb;

//inputs are reg for test bench
//outputs are wire for test bench
reg rst_n, clk, x1, x2, x3;
wire [1:3] y;
wire z1, z2;

//display variables
initial
$monitor ("x1 x2 x3 = %b, state = %b, z1 = %b, z2 = %b",
            {x1, x2, x3}, y, z1, z2);

                                    //continued on next page
```

Figure 3.61 Test bench module for the Mealy machine of Example 3.11.

```
//define clock
initial
begin
   clk = 1'b0;
   forever
      #10 clk = ~clk;
end

//define input sequence
initial
begin
   #0 rst_n = 1'b0;  x1 = 1'b0;  x2 = 1'b0;  x3 = 1'b0;
   #5 rst_n = 1'b1;
//-------------------------------------------
   x1=1'b0; x3=1'b1;
   @ (posedge clk)   //go to state_a (000)

   x1=1'b1; x2=1'b0;
   @ (posedge clk)   //go to state_a (000)

   x2=1'b1; x3=1'b0;
   @ (posedge clk)   //go to state_a (000)

   x1=1'b0; x2=1'b0; x3=1'b0;
   @ (posedge clk)   //go to state_b (001)

//-------------------------------------------
   x1=1'b1; x2=1'b1; x3=1'b1;
   @ (posedge clk)   //go to state_d (101)

//-------------------------------------------
   x1=1'b0; x2=1'b0; x3=1'b0;
   @ (posedge clk)   //go to state_a (000), assert z1

//-------------------------------------------
   x1=1'b1; x2=1'b1; x3=1'b1;
   @ (posedge clk)   //go to state_c (011)

   x1=1'b0; x2=1'b0; x3=1'b0;
   @ (posedge clk)   //go to state_f (010)

   x1=1'b1; x2=1'b1; x3=1'b1;
   @ (posedge clk)   //go to state_a (000), assert z2

                                  //continued on next page
```

Figure 3.61 (Continued)

```
//------------------------------------------
   x1=1'b1; x2=1'b1; x3=1'b1;
   @ (posedge clk)    //go to state_c (011)

   x1=1'b1; x2=1'b0; x3=1'b0;
   @ (posedge clk)    //go to state_e (100)

   @ (posedge clk)

//------------------------------------------
   #20    $stop;

end

//instantiate the module into the test bench
mealy_111_000_struc inst1 (rst_n, clk, x1, x2, x3, y,
                              z1, z2);

endmodule
```

Figure 3.61 (Continued)

```
x1 x2 x3 = 000, state = 000, z1 = 0, z2 = 0
x1 x2 x3 = 001, state = 001, z1 = 0, z2 = 0
x1 x2 x3 = 101, state = 100, z1 = 0, z2 = 0
x1 x2 x3 = 110, state = 000, z1 = 0, z2 = 0

x1 x2 x3 = 000, state = 000, z1 = 0, z2 = 0
x1 x2 x3 = 111, state = 001, z1 = 0, z2 = 0
x1 x2 x3 = 000, state = 101, z1 = 1, z2 = 0
x1 x2 x3 = 111, state = 000, z1 = 0, z2 = 0

x1 x2 x3 = 000, state = 011, z1 = 0, z2 = 0
x1 x2 x3 = 111, state = 010, z1 = 0, z2 = 1
x1 x2 x3 = 111, state = 000, z1 = 0, z2 = 0
x1 x2 x3 = 100, state = 011, z1 = 0, z2 = 0

x1 x2 x3 = 100, state = 100, z1 = 0, z2 = 0
x1 x2 x3 = 100, state = 000, z1 = 0, z2 = 0
```

Figure 3.62 Outputs for the Mealy machine of Example 3.11.

3.2.5 Synchronous Registers

Synchronous registers are designed using storage elements, such as *D* flip-flops, *JK* flip-flops, and *SR* latches. Each cell of a register stores one bit of binary information. There are many different types of synchronous registers, including parallel-in, parallel-out; parallel-in, serial-out; serial-in, parallel-out; and serial-in, serial-out registers. The next state of a register is usually a direct correspondence to the input vector, whose binary variables connect to the flip-flop data inputs, either directly or through δ next-state logic.

The state of the register is unchanged until the next active clock transition. Some registers may modify the data, such as shifting left or shifting right, where a left shift of one bit corresponds to a multiply-by-two operation and a right shift of one bit corresponds to a divide-by-two operation.

Parallel-in, parallel-out registers The most widely used register is the *parallel-in, parallel-out* (PIPO) register used for temporary storage of binary data. A typical application for a PIPO register is for the temporary storage of data, such as an index to be utilized in determining a memory location, a memory address register to address memory, and as a memory data register to contain information that is sent to or received from memory.

A typical PIPO register is shown in Figure 3.63 using *D* flip-flops. The clock signal loads the register with the $x_1 \ldots x_n$ data inputs on active high transitions. The clock signal is obtained from external logic which allows a single clock pulse to be generated only when the register is to be loaded from a new set of inputs.

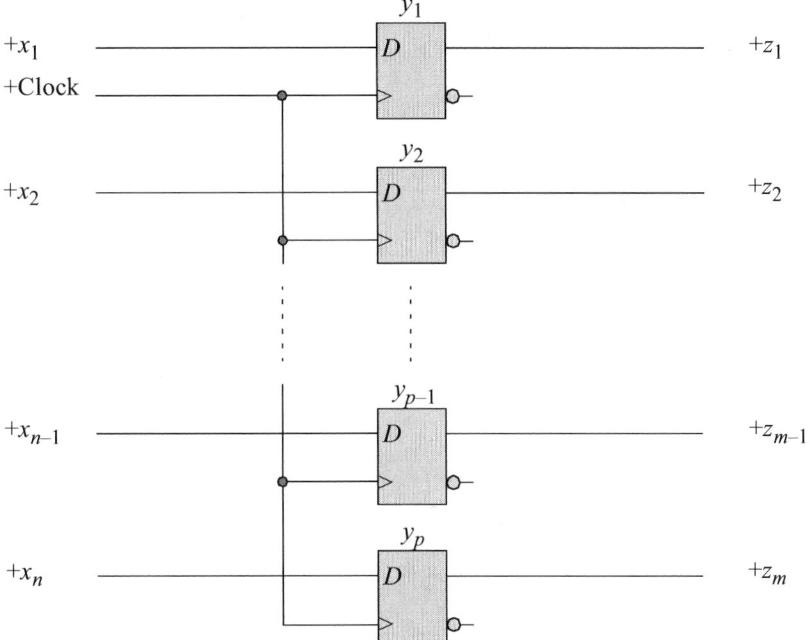

Figure 3.63 Logic diagram for a parallel-in, parallel-out register.

The synthesis procedure is not required for this basic type of register using *D* flip-flops. An alternative approach is to use *JK* flip-flops as the storage elements. In this approach, the register is clocked continuously by the system clock, which is a free-running astable multivibrator. The register is loaded, however, only when a *load* signal is active. When the *load* input is inactive, the data inputs of each flip-flop are $JK = 00$, which causes no change to the state of the machine. Thus, the register remains in its present state until the *load* input changes to an active level. The new input vector X_i then replaces the previous state of the register.

Parallel-in, serial-out registers A *parallel-in, serial-out* (PISO) register accepts binary input data in parallel and generates binary output data in serial form. The binary data can be shifted either left or right under control of a shift direction signal and a clock pulse, which is applied to all flip-flops simultaneously. The register shifts left or right 1 bit position at each active clock transition. Bits shifted out of one end of the register are lost unless the register is cyclic, in which case, the bits are shifted (or rotated) into the other end. The bits that are shifted in are all 0s. The logic diagram for a PISO register is shown in Figure 3.64.

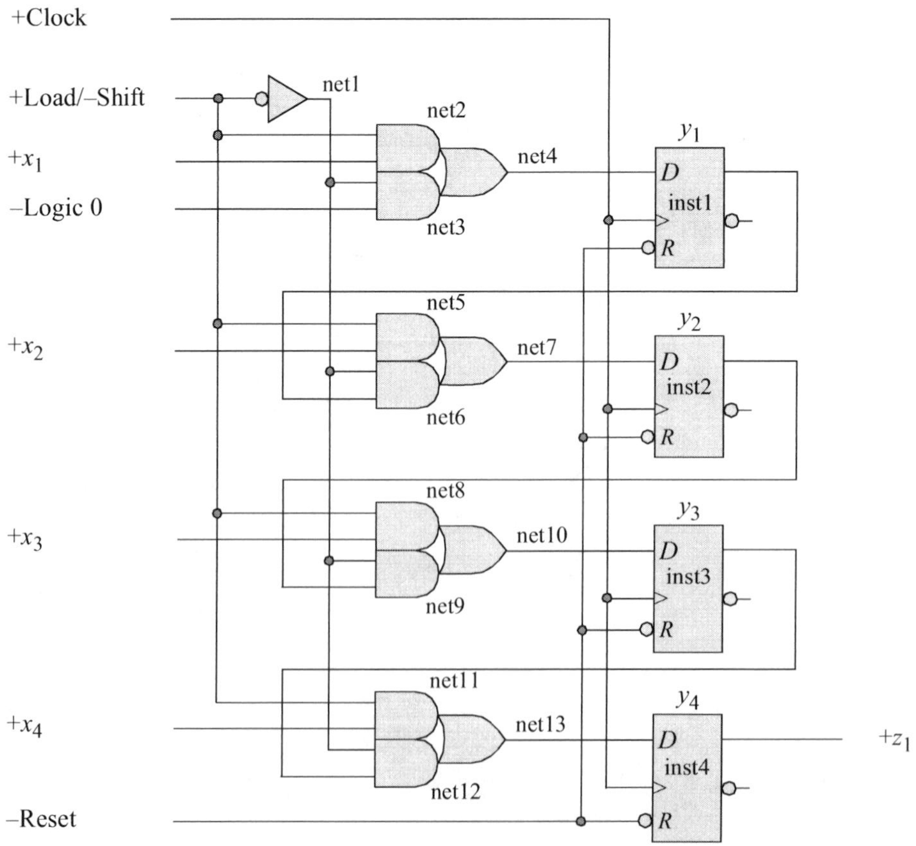

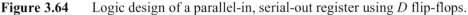

Figure 3.64 Logic design of a parallel-in, serial-out register using *D* flip-flops.

In Figure 3.64, when the *Load* signal is active high the input data is loaded into the shift register. When the *Load* signal is low (*–Shift*), the register is shifted right one bit position at each active clock pulse. The structural design module is shown in Figure 3.65 using built-in primitives and *D* flip-flops that were designed using behavioral modeling. The test bench module and the outputs are shown in Figures 3.66 and 3.67, respectively.

```
//structural for a 4-bit parallel-in, serial-out register

module piso4_struc2 (rst_n, clk, load, x1, x2, x3, x4,
                     y, z1);

//define inputs and output
input rst_n, clk, load, x1, x2, x3, x4;
output [1:4] y;
output z1;


//define internal nets
wire net1, net2, net3, net4, net5, net6, net7, net8,
     net9, net10, net11, net12, net13;


not   (net1, load);

//----------------------------------------------------------
//instantiate the logic for flip-flop y[1]
and   (net2, load, x1),
      (net3, net1, 1'b0);

or    (net4, net2, net3);

//instantiate the D flip-flop for y[1]
d_ff_bh  inst1 (rst_n, clk, net4, y[1]);


//----------------------------------------------------------
//instantiate the logic for flip-flop y[2]
and   (net5, load, x2),
      (net6, net1, y[1]);

or    (net7, net5, net6);

//instantiate the D flip-flop for y[2]
d_ff_bh  inst2 (rst_n, clk, net7, y[2]);

                                  //continued on next page
```

Figure 3.65 Structural design module for the parallel-in, serial-out register.

```
//------------------------------------------------------------
//instantiate the logic for flip-flop y[3]
and    (net8, load, x3),
       (net9, net1, y[2]);

or     (net10, net8, net9);

//instantiate the D flip-flop for y[3]
d_ff_bh  inst3 (rst_n, clk, net10, y[3]);

//------------------------------------------------------------
//instantiate the logic for flip-flop y[4]
and    (net11, load, x4),
       (net12, net1, y[3]);

or     (net13, net11, net12);

//instantiate the D flip-flop for y[4]
d_ff_bh  inst4 (rst_n, clk, net13, y[4]);

//------------------------------------------------------------
//define output z1
assign z1 = y[4];

endmodule
```

Figure 3.65 (Continued)

```
//test bench for 4-bit parallel-in, serial-out register

module piso4_struc2_tb;

//inputs are reg for test bench
//outputs are wire for test bench
reg rst_n, clk, load, x1, x2, x3, x4;
wire [1:4] y;
wire z1;

//display variables
initial
$monitor ("x1 x2 x3 x4 = %b, state = %b, z1 = %b",
            {x1, x2, x3, x4}, y, z1);

                                //continued on next page
```

Figure 3.66 Test bench module for the parallel-in, serial-out register.

```
//define clock
initial
begin
   clk = 1'b0;
   forever
      #10 clk = ~clk;
end

//define input sequence
initial
begin
   #0 rst_n = 1'b0;
      load = 1'b0;
      x1 = 1'b0;  x2 = 1'b0;  x3 = 1'b0;  x4 = 1'b0;

   #5 rst_n = 1'b1;

//----------------------------------------------------------
   #10   x1 = 1'b1;  x2 = 1'b1;  x3 = 1'b1;  x4 = 1'b1;

   #10   load = 1'b1;

   #10   load = 1'b0;

   #100  $stop;
end

//instantiate the module into the test best
piso4_struc2 inst1 (rst_n, clk, load, x1, x2, x3, x4,
                     y, z1);

endmodule
```

Figure 3.66 (Continued)

```
x1 x2 x3 x4 = 0000, state = 0000, z1 = 0
x1 x2 x3 x4 = 1111, state = 0000, z1 = 0
x1 x2 x3 x4 = 1111, state = 1111, z1 = 1
x1 x2 x3 x4 = 1111, state = 0111, z1 = 1
x1 x2 x3 x4 = 1111, state = 0011, z1 = 1
x1 x2 x3 x4 = 1111, state = 0001, z1 = 1
x1 x2 x3 x4 = 1111, state = 0000, z1 = 0
```

Figure 3.67 Outputs for the parallel-in, serial-out register.

Serial-in, parallel-out registers A *serial-in, parallel-out* (SIPO) register is a synchronous iterative network containing p identical cells. Data enters the register from the left and shifts serially to the right through all p stages, one bit position per clock pulse. After p shifts, the register is fully loaded and the bits are transferred in parallel to the destination.

One application of a serial-in, parallel-out register is to deserialize binary data from a single-track peripheral subsystem. The resulting word of parallel bits is placed on the system data bus of the input/output processor and then sent to the central processing unit for processing.

The data input of each flip-flop is connected directly to the output of the preceding flip-flop with the exception of flip-flop y_1, which receives the external serial binary data. Figure 3.68 shows the implementation of a SIPO register using D flip-flops. Each stage of the machine stores the state of the storage element to its immediate left. Data bits at the serial input are changed at the negative clock transition to allow bit x_1 to be stable at the D inputs of flip-flop y_1 before the next active clock transition.

One useful application of a SIPO register is to generate a series of nonoverlapping pulses for system timing. This is accomplished by inserting a NOR gate drawn as the AND function to provide the input to the left storage element. The inputs to the NOR gate are the negative outputs of flip-flops y_1, y_2, and y_3.

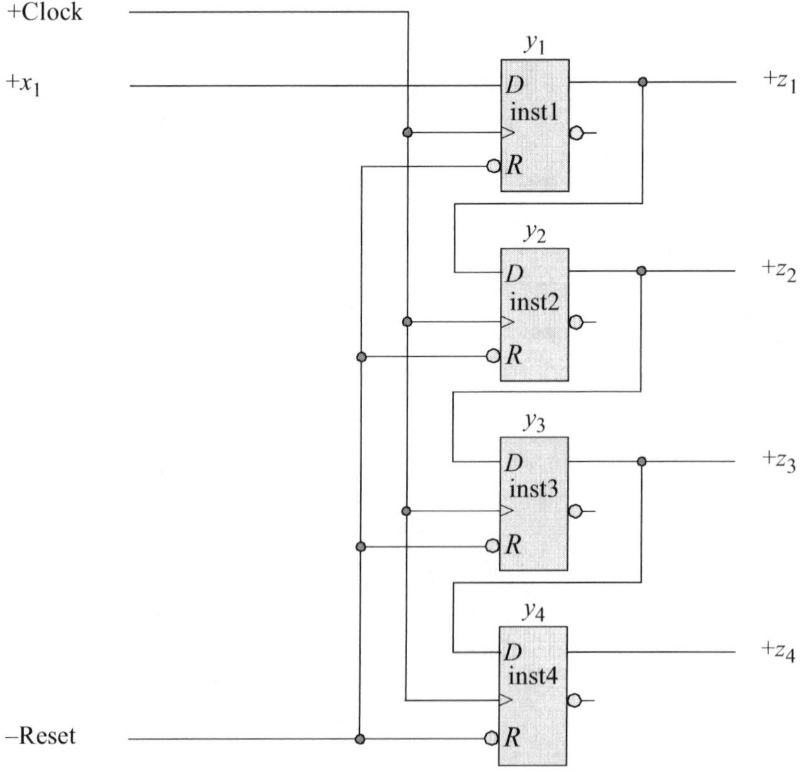

Figure 3.68 Logic diagram for a serial-in, parallel-out register.

The structural design module is shown in Figure 3.69 using *D* flip-flops that were designed using behavioral modeling. The test bench module is shown in Figure 3.70. The symbols #7, #20, etc. represent the time that input x_1 changes value. The sum of all the times indicates the time at that point. For example, the time represented by the fifth time symbol is the sum of the first five time symbols (#0 – #20); that is 50 time units. Input x_1 is assigned a value of 1'b0 at that time. The time units assure that input x_1 will be stabilized before the positive edge of the clock occurs. The test bench takes the machine through an input sequence to generate the output sequence shown below. The outputs are shown in Figure 3.70.

$$z_1z_2z_3z_4 = 0000 - 1111$$

```
//structural for serial-in, parallel-out register

module sipo5_struc (rst_n, clk, x1, y, z1, z2, z3, z4);

//define inputs and outputs
input rst_n, clk, x1;
output [1:4] y;
output z1, z2, z3, z4;

//instantiate flip-flop y[1]
d_ff_bh  inst1 (rst_n, clk, x1, y[1]);

//instantiate flip-flop y[2]
d_ff_bh  inst2 (rst_n, clk, y[1], y[2]);

//instantiate flip-flop y[3]
d_ff_bh  inst3 (rst_n, clk, y[2], y[3]);

//instantiate flip-flop y[4]
d_ff_bh  inst4 (rst_n, clk, y[3], y[4]);

//define outputs z1, z2, z3, and z4
assign   z1 = y[1],
         z2 = y[2],
         z3 = y[3],
         z4 = y[4];

endmodule
```

Figure 3.69 Structural design module for the serial-in, parallel-out register.

```
//test bench for serial-in, parallel-out register
module sipo5_struc_tb;

reg rst_n, clk, x1;  //inputs are reg for test bench
wire [1:4] y;        //outputs are wire for test bench
wire z1, z2, z3, z4;

initial      //display variables
$monitor ("x1 = %b, state = %b, z1 z2 z3 z4 = %b",
            x1, y, {z1, z2, z3, z4});

initial      //define clock
begin
   clk = 1'b0;
   forever
      #10clk = ~clk;
end

initial      //define input sequence
begin
   #0 rst_n = 1'b0;   x1 = 1'b0;
   #3 rst_n = 1'b1;
//------------------------------------------------
   #7    x1 = 1'b1;
   @ (posedge clk)

   #20   x1 = 1'b0;
   @ (posedge clk)

   #20   x1 = 1'b0;
   @ (posedge clk)

   #30   x1 = 1'b0;
   @ (posedge clk)

   #40   x1 = 1'b1;
   @ (posedge clk)

   #10   x1 = 1'b1;
   @ (posedge clk)

   #10   x1 = 1'b1;
   @ (posedge clk)
   #30   $stop;
end

//instantiate the module into the test bench
sipo5_struc inst1 (rst_n, clk, x1, y, z1, z2, z3, z4);
endmodule
```

Figure 3.70 Test bench module for the serial-in, parallel-out register.

```
x1 = 0, state = 0000, z1 z2 z3 z4 = 0000
x1 = 1, state = 1000, z1 z2 z3 z4 = 1000
x1 = 0, state = 0100, z1 z2 z3 z4 = 0100
x1 = 0, state = 0010, z1 z2 z3 z4 = 0010
x1 = 0, state = 0001, z1 z2 z3 z4 = 0001

x1 = 0, state = 0000, z1 z2 z3 z4 = 0000
x1 = 1, state = 1000, z1 z2 z3 z4 = 1000
x1 = 1, state = 1100, z1 z2 z3 z4 = 1100
x1 = 1, state = 1110, z1 z2 z3 z4 = 1110
x1 = 1, state = 1111, z1 z2 z3 z4 = 1111
```

Figure 3.71 Outputs for the serial-in, parallel-out register.

Serial-in, serial-out registers The synthesis of a *serial-in, serial-out* (SISO) register is similar to that of a SIPO register, with the exception that only one output is required. The rightmost flip-flop provides the single output for the register, as shown in Figure 3.72 using D flip-flops.

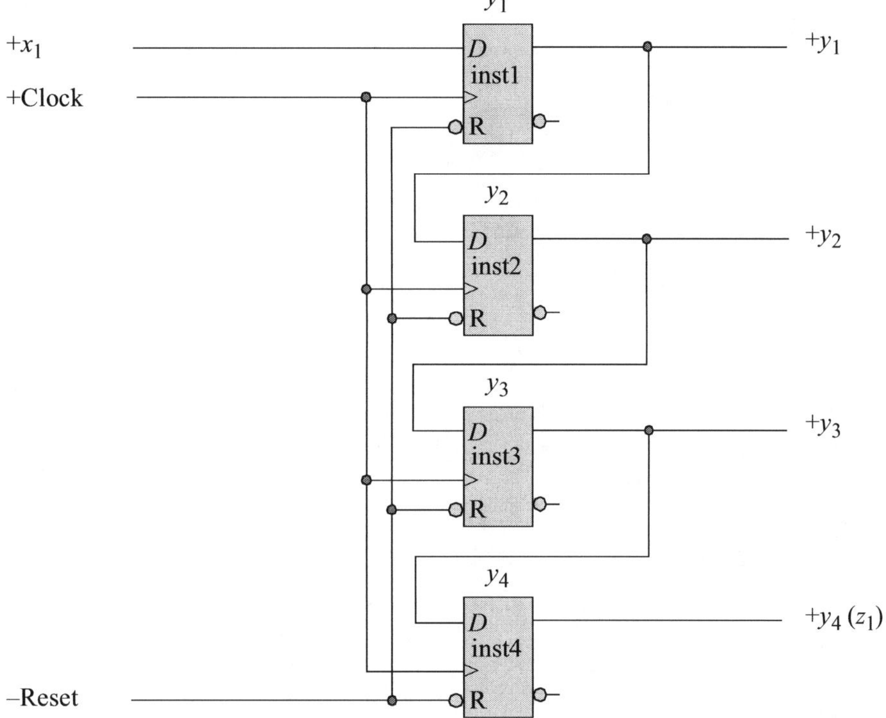

Figure 3.72 Logic diagram for a serial-in, serial-out register.

One application of a SISO register is to deserialize data from a disk drive. A serial bit stream is read from a disk drive and converted into parallel bits by means of a SIPO register. When 8 bits have been shifted into the register, the bytes are shifted in parallel into a matrix of SISO registers, where each bit is shifted into a particular column. The SISO register, in this application, performs the function of a first-in, first-out (FIFO) queue and acts as a buffer between the disk drive and the system input/output (I/O) data bus.

The same implementation of a SISO register matrix can be used as an instruction queue in a CPU instruction pipeline. The CPU prefetches instructions from memory during unused memory cycles and stores the instructions in the FIFO queue. Thus, an instruction stream can be placed in the instruction queue to wait for decoding and execution by the processor. Instruction queueing provides an effective method to increase system throughput.

The structural design module is shown in Figure 3.73 using *D* flip-flops that were designed using behavioral modeling. The test bench module is shown in Figure 3.74. The system function **$time** is used in the test bench to return the current simulation time in nanoseconds. The time is specified whenever a variable changes value. The outputs are shown in Figure 3.75.

```
//structural 4-bit serial-in, serial-out register
module siso4_struc (rst_n, clk, x1, y, z1);

//define inputs and output
input rst_n, clk, x1;
output [1:4] y;
output z1;

//instantiate flip-flop y[1]
d_ff_bh inst1 (rst_n, clk, x1, y[1]);

//instantiate flip-flop y[2]
d_ff_bh inst2 (rst_n, clk, y[1], y[2]);

//instantiate flip-flop y[3]
d_ff_bh inst3 (rst_n, clk, y[2], y[3]);

//instantiate flip-flop y[4]
d_ff_bh inst4 (rst_n, clk, y[3], y[4]);

//define output z1
assign   z1 = y[4];

endmodule
```

Figure 3.73 Structural design module for the serial-in, serial-out register.

```
//test bench for 4-bit serial-in, serial-out register
module siso4_struc_tb;

reg rst_n, clk, x1;       //inputs are reg for test bench
wire [1:4] y;             //outputs are wire for test bench
wire z1;

initial         //display variables
$monitor ($time, "ns, x1 = %b, clk = %b, state = %b,
                z1 = %b", x1, clk, y, z1);

initial         //define clock
begin
   clk = 1'b0;
   forever
      #10 clk = ~clk;
end

initial         //define input sequence
begin
   #0 rst_n = 1'b0;x1 = 1'b0;
   #5 rst_n = 1'b1;
//----------------------------------------------
   #3    x1 = 1'b1;
   #17   x1 = 1'b1;

   #20   x1 = 1'b0;
   #20   x1 = 1'b0;
   #20   x1 = 1'b0;

   #20   x1 = 1'b1;
   #20   x1 = 1'b1;
   #20   x1 = 1'b1;

   #20   x1 = 1'b0;
   #20   x1 = 1'b0;
   #20   x1 = 1'b0;

   #20   x1 = 1'b1;
   #20   x1 = 1'b1;
   #20   x1 = 1'b1;
   #40   $stop;
end

//instantiate the module into the test bench
siso4_struc inst1 (rst_n, clk, x1, y, z1);
endmodule
```

Figure 3.74 Test bench module for the serial-in, serial-out register.

```
0ns,    x1 = 0, clk = 0, state = 0000, z1 = 0
8ns,    x1 = 1, clk = 0, state = 0000, z1 = 0
10ns,   x1 = 1, clk = 1, state = 1000, z1 = 0
20ns,   x1 = 1, clk = 0, state = 1000, z1 = 0
30ns,   x1 = 1, clk = 1, state = 1100, z1 = 0
40ns,   x1 = 1, clk = 0, state = 1100, z1 = 0
45ns,   x1 = 0, clk = 0, state = 1100, z1 = 0
50ns,   x1 = 0, clk = 1, state = 0110, z1 = 0
60ns,   x1 = 0, clk = 0, state = 0110, z1 = 0

70ns,   x1 = 0, clk = 1, state = 0011, z1 = 1
80ns,   x1 = 0, clk = 0, state = 0011, z1 = 1
90ns,   x1 = 0, clk = 1, state = 0001, z1 = 1
100ns,  x1 = 0, clk = 0, state = 0001, z1 = 1
105ns,  x1 = 1, clk = 0, state = 0001, z1 = 1

110ns,  x1 = 1, clk = 1, state = 1000, z1 = 0
120ns,  x1 = 1, clk = 0, state = 1000, z1 = 0
130ns,  x1 = 1, clk = 1, state = 1100, z1 = 0
140ns,  x1 = 1, clk = 0, state = 1100, z1 = 0
150ns,  x1 = 1, clk = 1, state = 1110, z1 = 0
160ns,  x1 = 1, clk = 0, state = 1110, z1 = 0
165ns,  x1 = 0, clk = 0, state = 1110, z1 = 0

170ns,  x1 = 0, clk = 1, state = 0111, z1 = 1
180ns,  x1 = 0, clk = 0, state = 0111, z1 = 1
190ns,  x1 = 0, clk = 1, state = 0011, z1 = 1
200ns,  x1 = 0, clk = 0, state = 0011, z1 = 1
210ns,  x1 = 0, clk = 1, state = 0001, z1 = 1
220ns,  x1 = 0, clk = 0, state = 0001, z1 = 1
225ns,  x1 = 1, clk = 0, state = 0001, z1 = 1
230ns,  x1 = 1, clk = 1, state = 1000, z1 = 0

240ns,  x1 = 1, clk = 0, state = 1000, z1 = 0
250ns,  x1 = 1, clk = 1, state = 1100, z1 = 0
260ns,  x1 = 1, clk = 0, state = 1100, z1 = 0
270ns,  x1 = 1, clk = 1, state = 1110, z1 = 0
280ns,  x1 = 1, clk = 0, state = 1110, z1 = 0

290ns,  x1 = 1, clk = 1, state = 1111, z1 = 1
300ns,  x1 = 1, clk = 0, state = 1111, z1 = 1
```

Figure 3.75 Outputs for the serial-in, serial-out register.

3.2.6 Synchronous Counters

Counters are fundamental hardware devices used in the design of digital systems and have a finite number of states. The λ output logic is usually a function of the present state only; that is, $\lambda(Y_{j(t)})$. The state of the counter is interpreted as an integer with respect to a modulus. The symbol % represents the modulus (remainder/residue) operator. A number A modulo n is defined as the remainder after dividing A by n. Some counters contain a set of binary input variables from which the counter achieves an initial state.

A clock input signal causes the counter flip-flops to change state only at selected discrete intervals of time. Using the clock pulses to initiate state changes, the machine usually counts in either an ascending or descending sequence of states. In most cases counters reset to an initial state of $y_1 y_2 \ldots y_p = 00 \ldots 0$. In general, a p-stage counter counts modulo 2^p.

This section discusses only synchronous counters; asynchronous counters are inherently slow, because of the ripple effect caused by the output of stage y_i functioning as the clock input for stage y_{i+1}.

Modulo-10 counter Modulo-10 counters are extensively used in digital computers when counting is required in radix 10. A modulo-10, or binary-coded decimal (BCD) decade counter, generates ten states in the following sequence: 0000, 0001, 0010, 0011, 0100, 0101, 0110, 0111, 1000, 1001, 0000, ... Thus, each decade requires four flip-flops.

The synthesis of a modulo-10 counter is relatively straightforward. The counter is initially reset to $y[3:0] = 0000$, then increments by one at each active clock transition until a state code of $y[3:0] = 1001$ is reached. At the next active clock transition, the counter sequences to state $y[3:0] = 0000$.

The modulo-10 counter in this section will be designed using behavioral modeling; therefore, there is no need for a state diagram — since the counting sequence is already known — or for a logic diagram. The behavioral design module is shown in Figure 3.76. The test bench module and the outputs are shown in Figures 3.77 and 3.78, respectively.

```
//behavioral modulo-10 counter
module ctr_mod_10_bh (rst_n, clk, y);

//define inputs and outputs
input rst_n, clk;
output [3:0] y;

reg [3:0] y;     //variables are declared as reg in always
                                    //continued on next page
```

Figure 3.76 Behavioral design module for a modulo-10 counter.

```
//define counting sequence
always @ (posedge clk or negedge rst_n)
begin
   if (rst_n == 0)
      y = 4'b0000;
   else
      y = (y + 1) % 10;     //% is the modulus (remainder/
                            //residue) operator
end

endmodule
```

Figure 3.76 (Continued)

```
//test bench for modulo-10 counter
module ctr_mod_10_bh_tb;

reg rst_n, clk;         //inputs are reg for test bench
wire [3:0] y;           //outputs are wire for test bench

initial     //display outputs
$monitor ("count = %b", y);

initial     //define reset
begin
   #0 rst_n = 1'b0;
   #5 rst_n = 1'b1;
end

initial     //define clock
begin
   clk = 1'b0;
   forever
      #10 clk = ~clk;
end

initial     //define length of simulation
begin
   #200 $finish;
end

//instantiate the module into the test bench
ctr_mod_10_bh inst1 (rst_n, clk, y);

endmodule
```

Figure 3.77 Test bench module for the modulo-10 counter.

```
count = 0000
count = 0001
count = 0010
count = 0011
count = 0100

count = 0101
count = 0110
count = 0111
count = 1000
count = 1001

count = 0000
```

Figure 3.78 Outputs for the modulo-10 counter.

Modulo-16 counter A modulo-16 counter will now be designed using D flip-flops and built-in primitives. The counting sequence is: $y_3y_2y_1y_0 = 0000, 0001, 0010, 0011, 0100, 0101, 0110, 0111, 1000, 1001, 1010, 1011, 1100, 1101, 1110, 1111, 0000$. Using the counting sequence shown above, the Karnaugh maps are illustrated in Figure 3.79. The equations for the D flip-flops are shown in Equation 3.13. The logic diagram, obtained from the D flip-flop input equations, is shown in Figure 3.80.

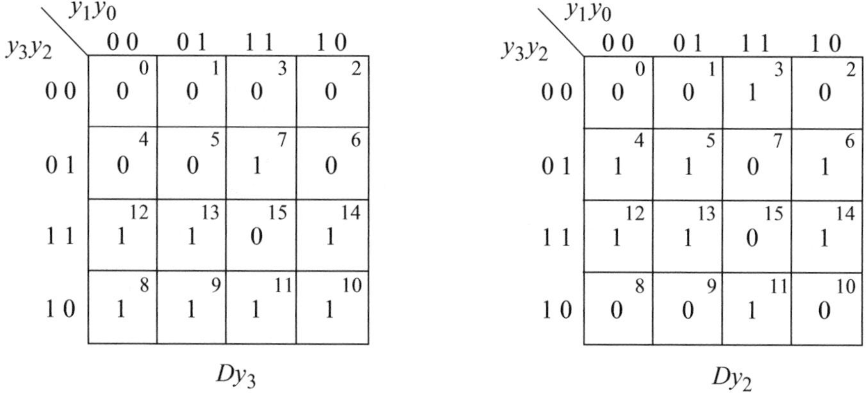

Figure 3.79 Karnaugh maps for the modulo-16 synchronous counter.

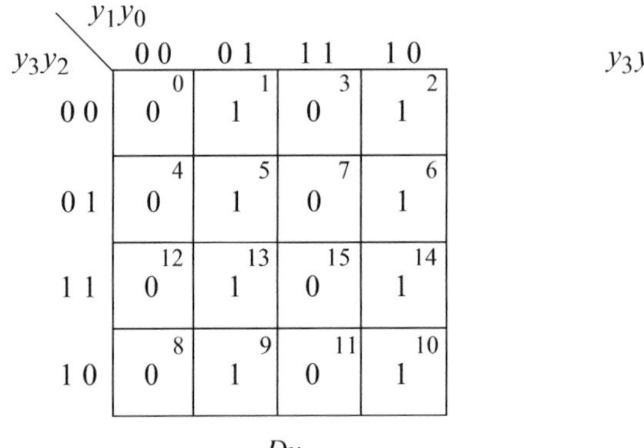

y_3y_2 \ y_1y_0	00	01	11	10
00	1	0	0	1
01	1	0	0	1
11	1	0	0	1
10	1	0	0	1

Dy_0

Figure 3.79 (Continued)

$$Dy_3 = y_3y_2' + y_3y_1' + y_3y_0' + y_3'y_2y_1y_0$$

$$Dy_2 = y_2y_1' + y_2y_0' + y_2'y_1y_0$$

$$Dy_1 = y_1'y_0 + y_1y_0'$$

$$Dy_0 = y_0' \qquad (3.13)$$

Recall that the built-in primitives are multiple-input gates used to describe a net and have one or more scalar inputs, but only one scalar output. The output signal is listed first, followed by the inputs in any order. The outputs are declared as **wire;** the inputs can be declared as either **wire** or **reg**. The gates represent combinational logic functions and can be instantiated into a module, as follows, where the instance name is optional:

gate_type inst1 (output, input_1, input_2, . . . , input_n);

Two or more instances of the same type of gate can be specified in the same construct. Note that only the last instantiation has a semicolon terminating the line. All previous lines are terminated by a comma.

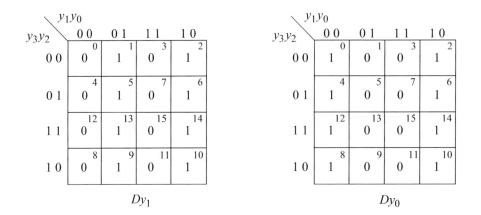

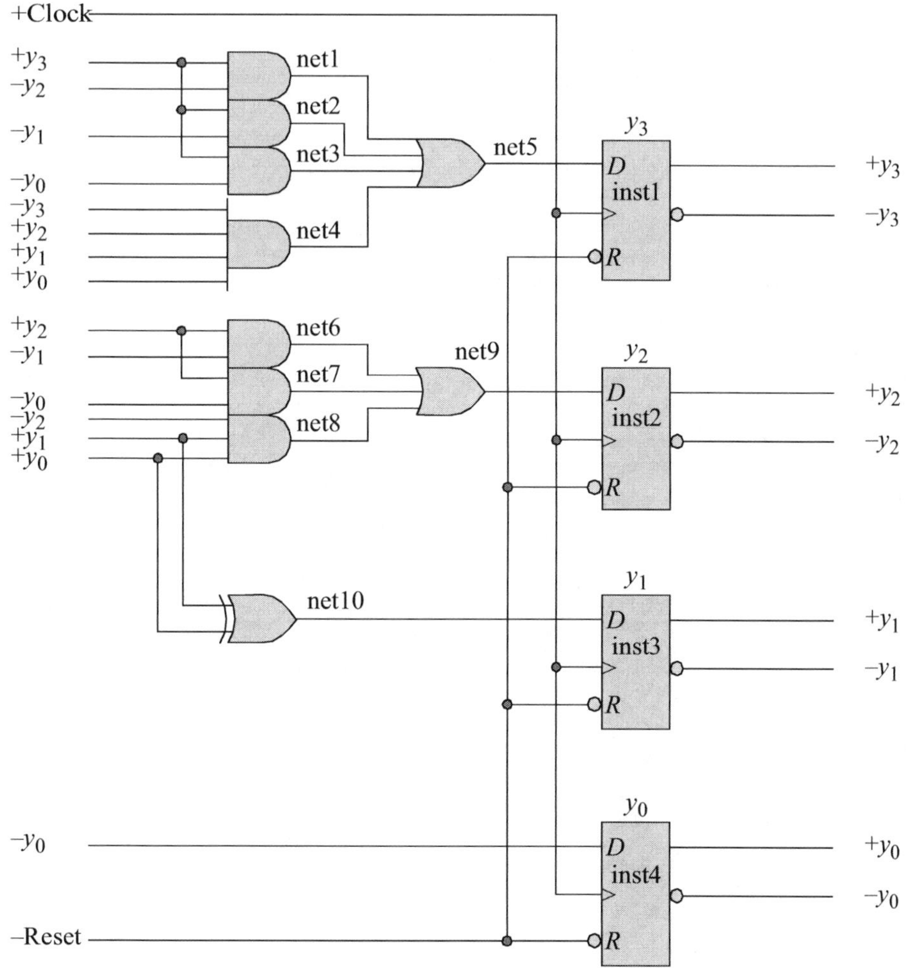

Figure 3.80 Logic diagram for the modulo-16 synchronous counter.

The structural design module is shown in Figure 3.81 using built-in primitives and *D* flip-flops that were designed using behavioral modeling. The test bench module is shown in Figure 3.82 and the outputs are shown in Figure 3.83.

```
//structural for a modulo-16 counter
module ctr_mod16_struc (rst_n, clk, y);

//define inputs and outputs
input rst_n, clk;
output [3:0] y;                  //continued on next page
```

Figure 3.81 Structural design module for a modulo-16 counter.

```
//define internal nets
wire net1, net2, net3, net4, net5, net6,
      net7, net8, net9, net10;


//-------------------------------------------
//instantiate the logic for flip-flop y[3]
and    (net1, y[3], ~y[2]),
       (net2, y[3], ~y[1]),
       (net3, y[3], ~y[0]),
       (net4, ~y[3], y[2], y[1], y[0]);

or     (net5, net1, net2, net3, net4);

//instantiate the D flip-flop for y[3]
d_ff_bh  inst1 (rst_n, clk, net5, y[3]);

//-------------------------------------------
//instantiate the logic for flip-flop y[2]
and    (net6, y[2], ~y[1]),
       (net7, y[2], ~y[0]),
       (net8, ~y[2], y[1], y[0]);

or     (net9, net6, net7, net8);

//instantiate the D flip-flop for y[2]
d_ff_bh  inst2 (rst_n, clk, net9, y[2]);

//-------------------------------------------
//instantiate the logic for flip-flop y[1]
xor    (net10, y[1], y[0]);

//instantiate the D flip-flop for y[1]
d_ff_bh  inst3 (rst_n, clk, net10, y[1]);

//-------------------------------------------
//instantiate the D flip-flop for y[0]
d_ff_bh  inst4 (rst_n, clk, ~y[0], y[0]);

endmodule
```

Figure 3.81 (Continued)

```
//test bench for the modulo-16 counter
module ctr_mod16_struc_tb;

reg rst_n, clk;          //inputs are reg for test bench
wire [3:0] y;            //outputs are wire for test bench

initial          //display outputs
$monitor ("count = %b", y);

//define reset
initial
begin
   #0    rst_n = 1'b0;
   #5    rst_n = 1'b1;
end

//define clock
initial
begin
   clk = 1'b0;
   forever
      #10   clk = ~clk;
end

//define length of simulation
initial
   #300  $stop;

//instantiate the module into the test bench
ctr_mod16_struc inst (rst_n, clk, y);

endmodule
```

Figure 3.82 Test bench module for the modulo-16 counter.

```
count = 0000                  count = 1001
count = 0001                  count = 1010
count = 0010                  count = 1011
count = 0011                  count = 1100
count = 0100                  count = 1101
count = 0101                  count = 1110
count = 0110                  count = 1111
count = 0111                  count = 0000
count = 1000
```

Figure 3.83 Outputs for the modulo-16 counter.

Modulo-8 counter A modulo-8 counter will now be designed using built-in primitives and *D* flip-flops that were designed using behavioral modeling. The counting sequence is: $y_2y_1y_0$ = 000, 001, 010, 011, 100, 101, 110, 111, 000. Using this counting sequence, the Karnaugh maps are illustrated in Figure 3.84. The equations for the *D* flip-flops are shown in Equation 3.14. The logic diagram, obtained from the *D* flip-flop input equations, is shown in Figure 3.85.

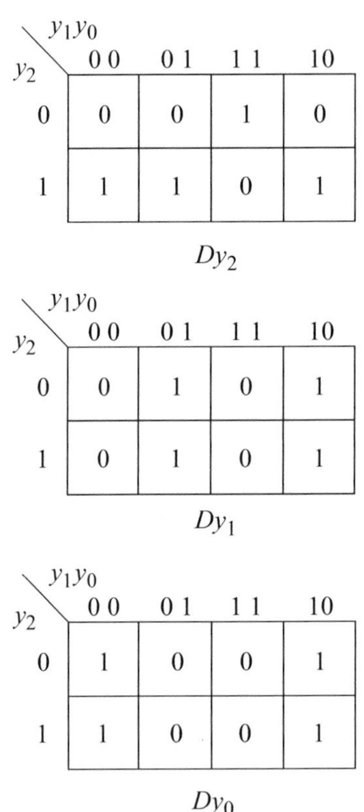

Figure 3.84 Karnaugh maps for the modulo-8 counter.

$$
\begin{aligned}
Dy_2 &= y_2y_1' + y_2y_0' + y_2'y_1y_0 \\
Dy_1 &= y_1'y_0 + y_1y_0' \\
&= y_1 \oplus y_0 \\
Dy_0 &= y_0'
\end{aligned}
\tag{3.14}
$$

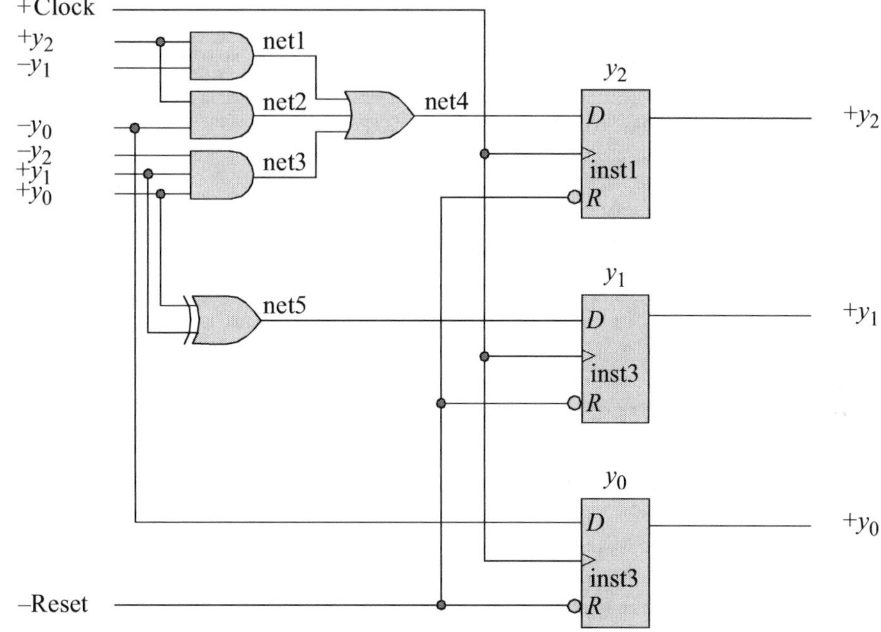

Figure 3.85 Logic diagram for the modulo-8 counter.

The structural design module is shown in Figure 3.86 using built-in primitives and *D* flip-flops that were designed using behavioral modeling. The test bench module is shown in Figure 3.87 and the outputs are shown in Figure 3.88.

```
//structural using bip and D flip-flops
module ctr_mod8 (rst_n, clk, y);

input rst_n, clk;     //define inputs and outputs
output [2:0] y;

wire net1, net2, net3, net4, net5;  //define internal nets

//-----------------------------------------------
//instantiate the logic for flip-flop y[2]
and    (net1, y[2], ~y[1]),
       (net2, y[2], ~y[0]),
       (net3, ~y[2], y[1], y[0]);

or     (net4, net1, net2, net3);
                                    //continued on next page
```

Figure 3.86 Structural design module for the modulo-8 counter.

```
//instantiate the D flip-flop for y[2]
d_ff_bh  inst1 (rst_n, clk, net4, y[2]);

//----------------------------------------------
//instantiate the logic for flip-flop y[1]
xor   (net5, y[0], y[1]);

//instantiate the D flip-flop for y[1]
d_ff_bh  inst2 (rst_n, clk, net5, y[1]);

//----------------------------------------------
//instantiate the logic for flip-flop y[0]
d_ff_bh  inst3 (rst_n, clk, ~y[0], y[0]);

endmodule
```

Figure 3.86 (Continued)

```
//test bench for the modulo-8 counter
module ctr_mod8_tb;

//inputs are reg for test bench
//outputs are wire for test bench
reg rst_n, clk;
wire [2:0] y;

//display outputs
initial
$monitor ("count = %b", y);

//define reset
initial
begin
   #0    rst_n = 1'b0;
   #5    rst_n = 1'b1;
end

//define clock
initial
begin
   clk = 1'b0;
   forever
      #10   clk = ~clk;
end                                //continued on next page
```

Figure 3.87 Test bench module for the modulo-8 counter.

```
//define length of simulation
initial
   #150  $stop;

//instantiate the module into the test bench
ctr_mod8 inst (rst_n, clk, y);

endmodule
```

Figure 3.87 (Continued)

```
count = 000
count = 001
count = 010
count = 011
count = 100
count = 101
count = 110
count = 111
count = 000
```

Figure 3.88 Outputs for the modulo-8 counter.

3.3 Asynchronous Sequential Machines

For an *asynchronous sequential machine* there is no machine clock — state changes occur on the application of input signals only. The synthesis of asynchronous sequential machines is one of the most interesting and certainly the most challenging concepts of sequential machine design. In many situations, a synchronous clock is not available. For example, the interface between an input/output processor (IOP) — or channel — and an input/output (I/O) subsystem control unit is an example of an asynchronous condition.

The control unit requests a word of data during a write operation by asserting an identifying epithet called a "tag-in signal". The channel then places the word on the data bus and asserts an acknowledging tag-out signal. The device control unit accepts the data then de-asserts the in tag, allowing the channel to de-assert the corresponding out tag, completing the data transfer sequence for one word.

An analogous situation occurs for a read operation in which the tag-in signal now indicates that a word is available on the data bus for the channel. The channel accepts the word and responds with the tag-out signal.

The data transfer sequence for the write and read operations was initiated, executed, and completed without utilizing a synchronizing clock signal. This technique permits not only a higher data transfer rate between the channel and an I/O device, but also allows the channel to communicate with I/O devices having a wide range of data transfer rates. The interface control logic in the device control unit is usually implemented as an asynchronous sequential machine.

Asynchronous sequential machines are implemented with Set/Reset (*SR*) latches as the storage elements. Thus, at least one *feedback* path is required in the synthesis of asynchronous machines. Asynchronous machines can be implemented in either a sum-of-products form or in a product-of-sums form.

Techniques will be presented in this chapter to synthesize asynchronous sequential machines irrespective of the varying delays of circuit components. Since there is no system clock, in order to prevent possible race conditions and associated timing problems when two or more inputs change value simultaneously, it will be assumed that only one input variable will change state at a time. This is referred to as a *fundamental-mode model*, further defined with the following characteristics:

1. Only one input will change at a time.
2. No other input will change until the machine has sequenced to a stable state.

A general block diagram for an asynchronous sequential machine is shown in Figure 3.89. The input alphabet X consists of binary input variables $x_1, x_2, \cdots, x_n$ that can change value at any time and are represented as voltage levels rather than pulses. The state alphabet Y is characterized by p storage elements, where $Y_{1e}, Y_{2e}, \cdots, Y_{pe}$ are the *excitation* variables and $y_{1f}, y_{2f}, \cdots, y_{pf}$ are the *feedback* or *secondary* variables. The output alphabet Z is represented by $z_1, z_2, \cdots, z_m$.

Both the δ next-state logic and the λ output logic are composed of combinational logic circuits. The delay element in Figure 3.89 represents the total delay of the machine from the time an input changes until the machine has stabilized in the next state, and is represented as a time delay of Δt. The time correlation between the excitation variables Y_{ie} and the feedback variables y_{if} is specified by Equation 3.15.

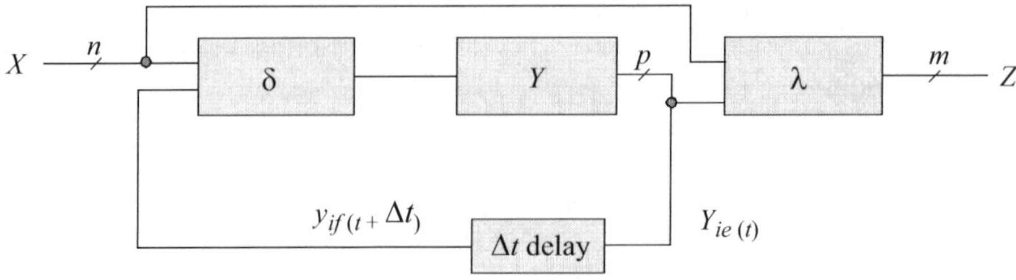

Figure 3.89 General block diagram of an asynchronous sequential machine.

$$y_{if(t+\Delta t)} = Y_{ie(t)} \qquad (3.15)$$

3.3.1 Synthesis Procedure

The machine operation for asynchronous sequential machines is specified by a timing diagram and/or verbal statements. The design procedure is summarized below.

1. **State diagram** The machine specifications are converted into a state diagram. A timing diagram and/or a verbal statement of the machine specifications is converted into a precise delineation which specifies the machine's operation for all applicable input sequences. This step is not a necessary requirement and is usually omitted; however, the state diagram characterizes the machine's operation in a graphical representation and adds completeness to the design procedure.

2. **Primitive flow table** The machine specifications are converted to a state transition table called a "primitive flow table". This is the least methodical step in the synthesis procedure and the most important. The primitive flow table depicts the state transition sequences and output assertions for all valid input vectors. The flow table must correctly represent the machine's operation for all applicable input sequences, even those that are not initially apparent from the machine specifications.

3. **Equivalent states** The primitive flow table may have an inordinate number of rows. The number of rows can be reduced by finding equivalent states and then eliminating redundant states. If the machine's operation is indistinguishable whether commencing in state Y_i or state Y_j, then one of the states is redundant and can be eliminated. The flow table thus obtained is a *reduced primitive flow table.* In order for two stable states to be equivalent, all three of the following conditions must be satisfied:

    1. The same input vector.
    2. The same output value.
    3. The same, or equivalent, next state for all valid input sequences.

4. **Merger diagram** The merger diagram graphically portrays the result of the merging process in which an attempt is made to combine two or more rows of the reduced primitive flow table into a single row. The result of the merging technique is analogous to that of finding equivalent states; that is, the merging process can also reduce the number of rows in the table and, hence, reduce the number of feedback variables that are required. Fewer feedback variables will result in a machine with less logic and, therefore, less cost. Two rows can merge into a single row if the entries in the same column of each row satisfy one of the following three merging rules:

    1. Identical state entries, either stable or unstable.
    2. A state entry and a "don't care."
    3. Two "don't care" entries.

5. **Merged flow table** The merged flow table is constructed from the merger diagram. The table represents the culmination of the merging process in which two or more rows of a primitive flow table are replaced by a single equivalent row which contains one stable state for each merged row.

6. **Excitation maps and equations** An excitation map is generated for each excitation variable. Then the transient states are encoded, where applicable, to avoid critical race conditions. Appropriate assignment of the excitation variables for the transient states can minimize the δ next-state logic for the excitation variables. The operational speed of the machine can also be established at this step by reducing the number of transient states through which the machine must sequence during a cycle. Then the excitation equations are derived from the excitation maps. All static-1 and static-0 hazards are eliminated from the network for a sum-of-products or product-of-sums implementation, respectively. Hazards are defined below.

7. **Output maps and equations** An output map is generated for each machine output. Output values are assigned for all nonstable states so that no transient signals will appear on the outputs. In this step, the speed of circuit operation can also be established. Then the output equations are derived from the output maps, assuring that all outputs will be free of static-1 and static-0 hazards.

8. **Logic diagram** The logic diagram is implemented from the excitation and output equations using an appropriate logic family.

3.3.2 Hazards

A *hazard* can occur when an input variable changes value. Varying propagation delays caused by logic gates, wires, and different path lengths can produce erroneous transient signals on the outputs. These spurious signals are referred to as *hazards*. If the hazard occurs in the feedback path, then an incorrect state transition sequence may result.

When a hazard occurs in the δ next-state logic, the machine may sequence to an invalid next state. If the hazard occurs in the λ output logic, then a glitch may appear on the output signal. An output glitch can cause significant problems. These transitory signals generate a condition which is not specified in the expression for the machine, because Boolean algebra does not take into account the propagation delay of switching circuits. Hazards will be examined and methods presented for detecting and correcting these transient phenomena so that correct operation of an asynchronous sequential machine can be assured.

Figure 3.90 illustrates an example of a combinational circuit with an inherent hazard. The Karnaugh map which represents the circuit is shown in Figure 3.91 and the equation for output z_1 is shown in Equation 3.16. Assume that x_2 changes from 1 to 0. The deassertion of x_2 is immediate. The new value of x_2 propagates to the output along three paths — through AND gate 2 and OR gate 4; also through the inverter and AND gate 3 and OR gate 4. Depending on the circuit delays, a glitch could occur on output z_1.

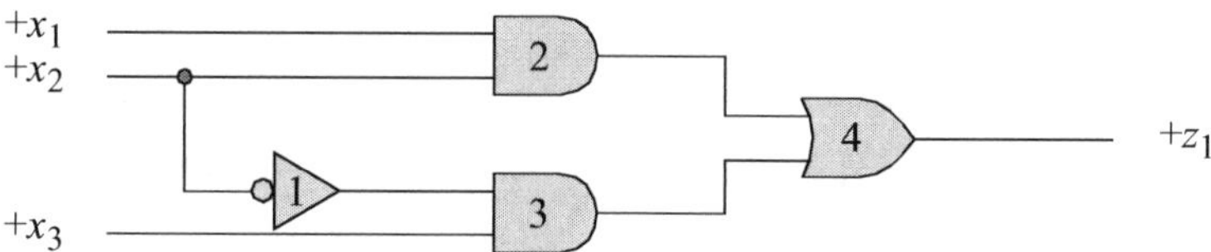

Figure 3.90 Logic circuit which contains a potential static hazard.

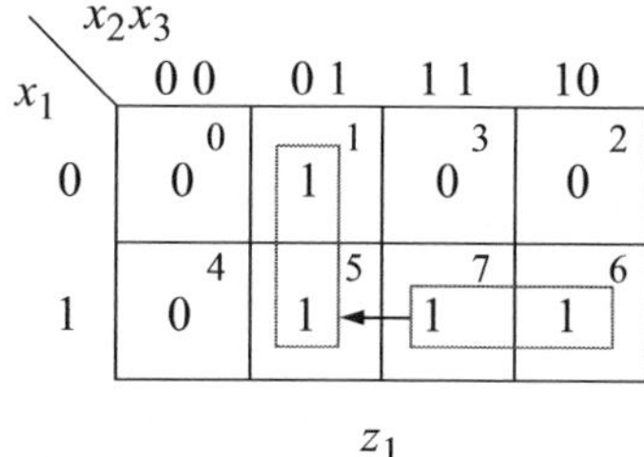

Figure 3.91 Karnaugh map corresponding to the circuit of Figure 3.90.

$$z_1 = x_1x_2 + x_2'x_3 \tag{3.16}$$

The effects of the hazard can be eliminated by adding a third term to the equation for z_1, as shown in Equation 3.17. The output can be made independent of the value of x_2 by including the redundant *prime implicant* x_1x_3, which covers both the initial and terminal state of the transition. A prime implicant is a unique grouping of 1s (an implicant) that does not imply any other grouping of 1s (other implicants). The redundant prime implicant will maintain the output at a constant high level during the transition.

$$z_1 = x_1x_2 + x_2'x_3 + x_1x_3 \tag{3.17}$$

The term x_1x_3 is called a *hazard cover*, since it covers the detrimental effects of the hazard. The effects of a static hazard can be negated by combining adjacent groups of 1s in a Karnaugh map as shown in Figure 3.92. A hazard cover can be applied to a sum-of-products expression or to a product-of-sums expression.

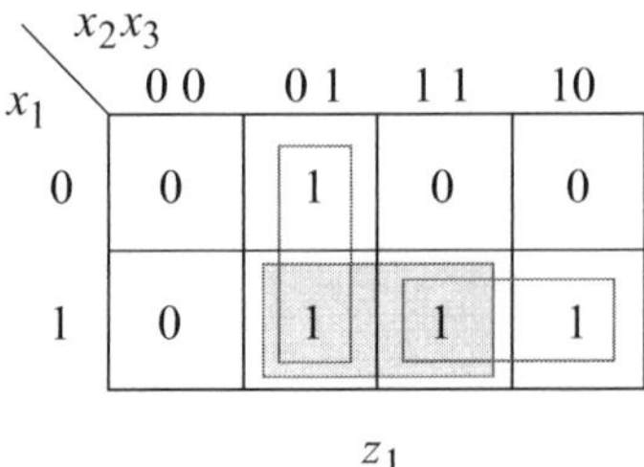

Figure 3.92 Negating the effects of a static hazard by combining adjacent groups of 1s.

3.3.3 Oscillations

An *oscillation* occurs in an asynchronous sequential machine when a single input change results in an input vector in which there is no stable state. Consider the excitation Karnaugh map of Figure 3.93 for Y_{1e}. There are two input variables x_1 and x_2 and one feedback variable y_{1f}. If the machine is in stable state ⓑ and x_1 changes from 0 to 1, then the machine sequences to transient state *c*. Then, after a delay of Δt, the feedback variable becomes equal to the excitation variable and the machine proceeds to transient state *g*. In state *g*, however, the excitation variable $Y_{1e} = 0$, designating state *g* as an unstable (or transient) state, because $y_{1f} \neq Y_{1e}$. After a further delay of Δt, the feedback variable becomes equal to the excitation variable and the machine sequences to state *c*. Since the input vector $x_1x_2 = 11$ provides no stable state, the machine will oscillate between transient states *c* and *g*.

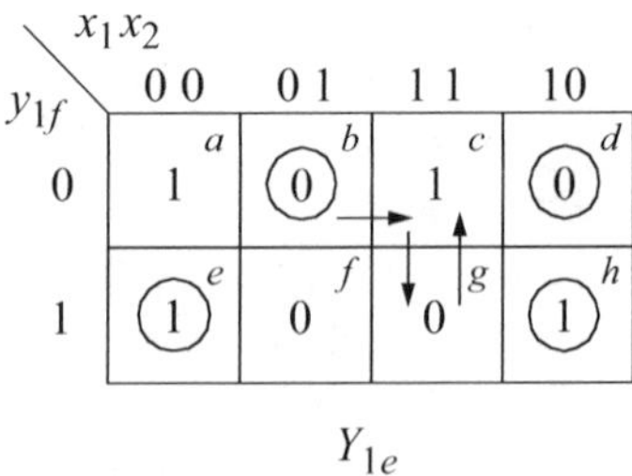

Figure 3.93 Excitation Karnaugh map for an asynchronous sequential machine containing an oscillation.

The excitation Karnaugh map of Figure 3.94 contains *multiple oscillations*, because columns $x_1x_2 = 01$ contain no stable states. There are two input variables x_1 and x_2 and two feedback variables y_{1f} and y_{2f}. The complete set of oscillations is summarized by the expressions shown in Figure 3.95.

$y_{1f}y_{2f}$ \ x_1x_2	0 0	0 1	1 1	1 0
0 0	*a* 10	*b* 01	*c* 10	*d* (00)
0 1	*e* (01)	*f* 00	*g* 11	*h* 11
1 1	*i* 10	*j* 01	*k* (11)	*l* 01
1 0	*m* (10)	*n* 11	*o* 11	*p* 11

$Y_{1e}Y_2$

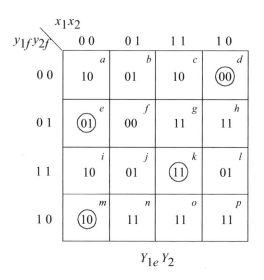

Figure 3.94 Excitation Karnaugh map for an asynchronous sequential machine which produces multiple oscillations.

$(e) \rightarrow f \leftrightarrow b$
$(e) \rightarrow h \leftrightarrow l$
$(k) \rightarrow l \leftrightarrow h$
$(k) \rightarrow j \rightarrow f \leftrightarrow b$
$(m) \rightarrow n \rightarrow j \rightarrow f \leftrightarrow b$
$(m) \rightarrow p \rightarrow l \leftrightarrow h$

Figure 3.95 The complete set of oscillations exhibited by the asynchronous sequential machine represented by the excitation Karnaugh map of Figure 3.94.

An asynchronous sequential machine which has an oscillating characteristic can be used as an astable multivibrator to provide a clock signal to a synchronous sequential machine. An appropriate delay of Δt must be inserted into the network to provide the correct clock frequency. In the synthesis of most asynchronous sequential machines, however, the oscillation phenomenon must be avoided. The machine specifications can be modified slightly such that every input vector will provide at least one stable state. This modification should not drastically alter the general functional operation of the machine.

3.3.4 Races

In both asynchronous sequential machines and synchronous sequential machines, if a change of state occurs between two states with nonadjacent state codes, then the machine may sequence through a transient state before entering the destination stable state. If the sequential machine is a Moore machine, in which the outputs are determined by the present state only, then a transitory erroneous signal may be generated on the output. This glitch results from two or more variables changing state in a single state transition sequence in which the variables change values at different times. There are two types of races: noncritical and critical.

Noncritical races Consider the excitation Karnaugh map of Figure 3.96. There are three paths that exist for noncritical races for the state transition sequence Ⓕ → Ⓞ, depending on the time at which the variables change value. Figure 3.97 illustrates the three possible paths for the sequence Ⓕ → Ⓞ.

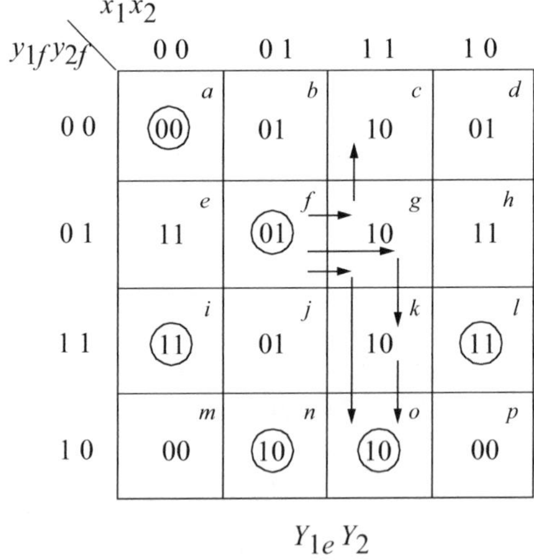

Figure 3.96 Excitation map for an asynchronous sequential machine illustrating noncritical races when input x_1 changes from 0 to 1 in stable state Ⓕ.

Ⓕ → c → o
Ⓕ → k ↔ o
Ⓕ → o

Figure 3.97 The complete set of races exhibited by the asynchronous sequential machine represented by the excitation Karnaugh map of Figure 3.96.

Critical races Consider the excitation Karnaugh map of Figure 3.98. If the machine is presently in state ⓙ and input x_1 changes from 0 to 1, then three possible paths exist depending on the relative propagation delays of the storage elements and associated circuitry. The intended path is from state ⓙ to state ⓒ. Due to differing delay characteristics, however, the machine may terminate the sequence in either state ⓒ or state ⓞ. Figure 3.99 illustrates the three possible state transition sequences.

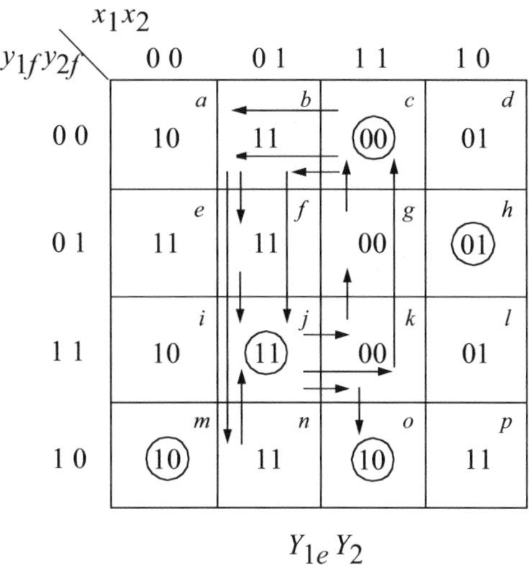

Figure 3.98 Excitation Karnaugh map for an asynchronous sequential machine illustrating a critical race condition when input $\boldsymbol{x_1}$ changes from 0 to 1 in stable state ⓙ.

ⓙ $\rightarrow k \rightarrow o$
ⓙ $\rightarrow g \rightarrow c$
ⓙ $\rightarrow c$

Figure 3.99 The complete set of races exhibited by the asynchronous sequential machine represented by the excitation Karnaugh map of Figure 3.98.

Races can be avoided when it is possible to direct the machine through intermediate unstable states before reaching the destination stable state. This can be achieved by utilizing some of the unspecified entries in the excitation map. Also, it may be possible to add rows to the excitation map without increasing the number of excitation and feedback variables.

3.3.5 Design Examples of Asynchronous Sequential Machines

Various types of asynchronous sequential machines of varying complexity will be designed in this section for Mealy and Moore machines. Different modeling techniques will be incorporated, including built-in primitives, dataflow modeling, behavioral modeling, and structural modeling.

Example 3.12 A Mealy asynchronous sequential machine will be designed that has two inputs x_1 and x_2 and one output z_1. An operational characteristic specifies that input x_1 must envelop all occurrences of the x_2 pulse. Thus, the allowable input vectors are $x_1x_2 = 00$, 10, or 11; the input combination of $x_1x_2 = 01$ will never occur. Output z_1 is to be asserted coincident with the assertion of every second x_2 pulse and is to remain asserted until the deassertion of x_2. A representative timing diagram is shown in Figure 3.100. Although the timing diagram illustrates a valid input sequence to generate an output, other variations are possible and must be considered to adequately represent the operation of the machine for all valid input sequences.

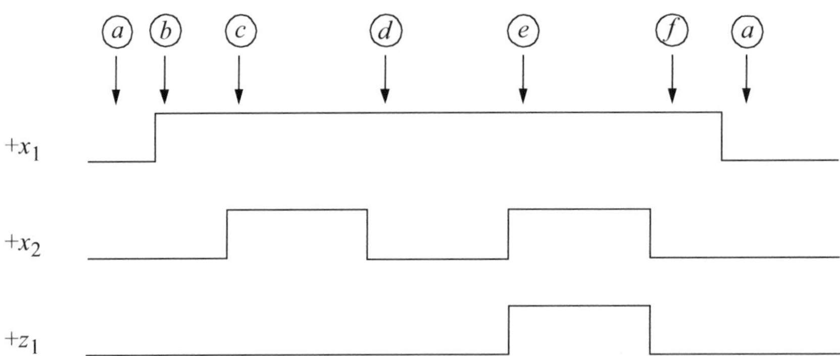

Figure 3.100 Timing diagram for the asynchronous sequential machine of Example 3.12.

A primitive flow table is developed next by beginning at the leftmost section of the timing diagram where $x_1x_2z_1 = 000$ and proceeding left to right assigning a unique stable state name to each different combination of the input vector and the associated output z_1. The primitive flow table is shown in Figure 3.101, which provides a tabular representation of the machine's operation.

The next step is to identify all equivalent stable states and then to eliminate redundant states. In order for two stable states to be equivalent, they must have the same input vector, the same output value, and the same, or equivalent, next state for all valid input sequences. The only possible equivalences exist between stable state pairs {ⓒ, ⓔ}, {ⓑ, ⓓ}, {ⓑ, ⓕ}, and {ⓓ, ⓕ}.

x_1x_2 00	01	11	10	z_1
ⓐ	–	–	b	0
a	–	c	ⓑ	0
–	–	ⓒ	d	0
a	–	e	ⓓ	0
–	–	ⓔ	f	1
a	–	c	ⓕ	0

Figure 3.101 Primitive flow table for the asynchronous sequential machine of Example 3.12.

States ⓒ and ⓔ are not equivalent, because they have different outputs. Stable states ⓑ and ⓓ are not equivalent, because state ⓑ is the precursor of the first x_2 pulse, while stable state ⓓ immediately precedes the second x_2 pulse.

Next, states ⓑ and ⓓ are tested for equivalence. Both have the same input vector ($x_1x_2 = 10$) and both have the same output value ($z_1 = 0$). However, when the input vector changes from $x_1x_2 = 10$ to 11, the next state from state ⓑ is state ⓒ; whereas, the next state from state ⓓ is state ⓔ. Since states ⓒ and ⓔ have already been shown to be nonequivalent, therefore, states ⓑ and ⓓ are not equivalent. The same reasoning applies to stable state pair ⓓ and ⓕ, which are also not equivalent.

Stable state pair ⓑ and ⓕ, however, satisfy all equivalence requirements: Both are entered from the same input vector ($x_1x_2 = 10$); both have identical output values ($z_1 = 0$); and both proceed to the same next stable state ⓒ or ⓐ for an applied input vector of $x_1x_2 = 11$ or 00, respectively. Therefore, stable states ⓑ and ⓕ are equivalent. State ⓕ is redundant and can be eliminated from the primitive flow table. Every occurrence of state f is replaced by equivalent state b. The reduced primitive flow table is shown in Figure 3.102.

The number of rows in a reduced primitive flow table can usually be decreased by merging two or more rows into a single row. Recall the three requirements for merging two rows into a single merged row: Each column in the two rows under consideration must contain identical state names, either stable or unstable, or a state name and an unspecified entry, or two unspecified entries.

In the reduced primitive flow table of Figure 3.102, rows ⓐ and ⓑ can merge, because there is no conflict in any column of the two rows. This merging capability is indicated by a line connecting vertices ⓐ and ⓑ in the merger diagram of Figure 3.103. The only other row with which row ⓐ can merge is row ⓔ — all other rows have a conflict in at least one column. Rows ⓑ, ⓒ, and ⓓ cannot merge with any succeeding row due to conflicting state names in certain columns. The merger diagram of Figure 3.103 yields the following two partitions of maximal compatible sets:

1. {ⓐ, ⓑ}, {ⓒ}, {ⓓ}, {ⓔ}

2. {ⓐ, ⓔ}, {ⓑ}, {ⓒ}, {ⓓ}

x_1x_2 00	01	11	10	z_1
ⓐ	–	–	b	0
a	–	c	ⓑ	0
–	–	ⓒ	d	0
a	–	e	ⓓ	0
–	–	ⓔ	b	1

Figure 3.102 Reduced primitive flow table obtained from the primitive flow table of Figure 3.101.

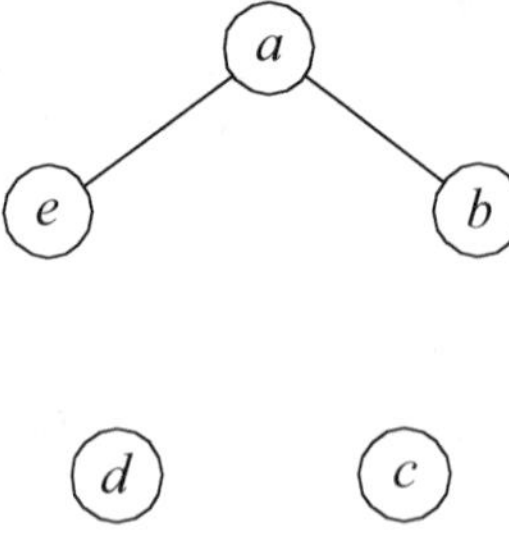

Figure 3.103 Merger diagram for the reduced primitive flow table of Figure 3.102.

All partitions should be analyzed by means of a merged flow table to determine the fewest number of logic gates. The first partition produces the merged flow table shown in Figure 3.104(a). Each row of the merged flow table is generated by transferring the individual rows from the reduced primitive flow table to the merged flow table in accordance with the partition assignments.

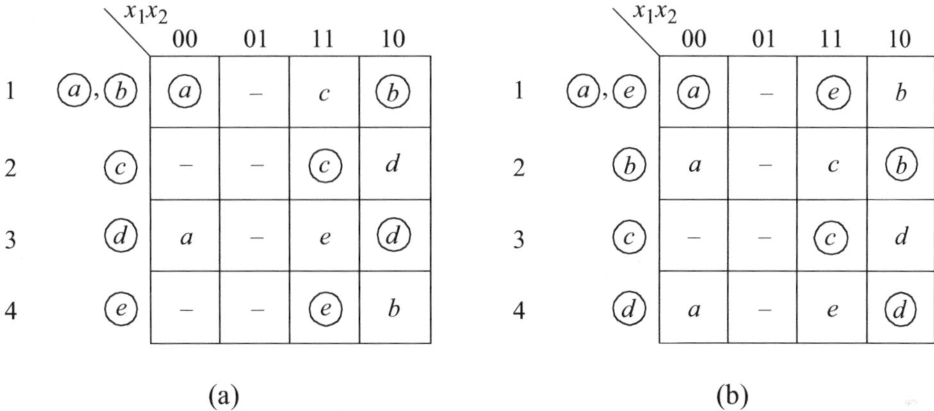

(a)

	x_1x_2	00	01	11	10
1	ⓐ, ⓑ	ⓐ	–	c	ⓑ
2	ⓒ	–	–	ⓒ	d
3	ⓓ	a	–	e	ⓓ
4	ⓔ	–	–	ⓔ	b

(b)

	x_1x_2	00	01	11	10
1	ⓐ, ⓔ	ⓐ	–	ⓔ	b
2	ⓑ	a	–	c	ⓑ
3	ⓒ	–	–	ⓒ	d
4	ⓓ	a	–	e	ⓓ

Figure 3.104 Merged flow tables obtained from the two partitions derived from the merger diagram of Figure 3.103: (a) partition 1: {ⓐ, ⓑ}, {ⓒ}, {ⓓ}, {ⓔ} and (b) partition 2: {ⓐ, ⓔ}, {ⓑ}, {ⓒ}, {ⓓ}.

After enumerating the rows of the merged flow table, all state transition sequences can be identified with reference to individual rows. The state transitions are illustrated in graphical form by means of a transition diagram. The transition diagram for the merged flow table of Figure 3.104(a) is shown in Figure 3.105(a). Row 1 proceeds to row 2 by the sequence ⓑ $\rightarrow c \rightarrow$ ⓒ, as illustrated by the directed line from row 1 to row 2 in Figure 3.105(a).

Notice that row 3 can proceed to two different rows by the following sequences: ⓓ $\rightarrow a \rightarrow$ ⓐ, which represents a transition from row 3 to row 1, and ⓓ $\rightarrow e \rightarrow$ ⓔ, which represents a transition from row 3 to row 4. Thus, in state ⓓ, a change of input vector from $x_1x_2 = 10$ to 00 or 11 results in a transition from row 3 to row 1 or row 4, respectively.

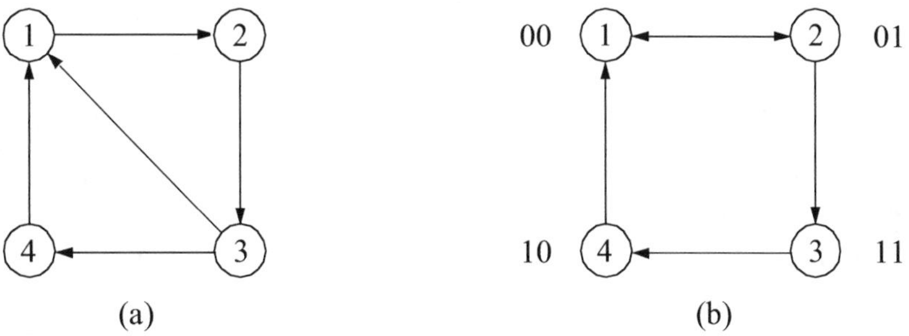

Figure 3.105 Transition diagram for the merged flow tables of Figure 3.104: (a) transition diagram for Figure 3.104(a) and (b) transition diagram for Figure 3.104(b).

The transition diagram of Figure 3.105(a) contains a triangular polygon specified by rows 1, 2, and 3 or by rows 1, 3, and 4. Since three rows cannot all be adjacent, row 3 can proceed to row 1 through row 4, eliminating the need for a line from row 3 to row 1. Providing an additional intermediate state to the cycle from state ⓓ to state ⓐ does not alter the operational characteristics of the machine, but does generate a slightly slower transition.

The transition diagram for the merged flow table of Figure 3.104(b) is depicted in Figure 3.105(b). Since the transition diagram contains no polygons with an odd number of sides, the state transitions do not have to be altered. All transitions proceed through only one transient state. The merged flow table of Figure 3.104(b) and the transition diagram of Figure 3.105(b) will be used to generate the excitation and output equations.

The combined excitation map for excitation variables Y_{1e} and Y_{2e} is shown in Figure 3.106. The stable states are assigned excitation values that are the same as the feedback values of the corresponding rows. It is important to not inadvertently assign excitation values to the "don't care" states that would generate a stable state. The individual excitation maps are shown in Figure 3.107 and the resulting hazard-free excitation equations in Equation 3.18 in a sum-of-products form. The rightmost term in each equation is the hazard cover. The excitation equations are shown in a product-of-sums notation in Equation 3.19.

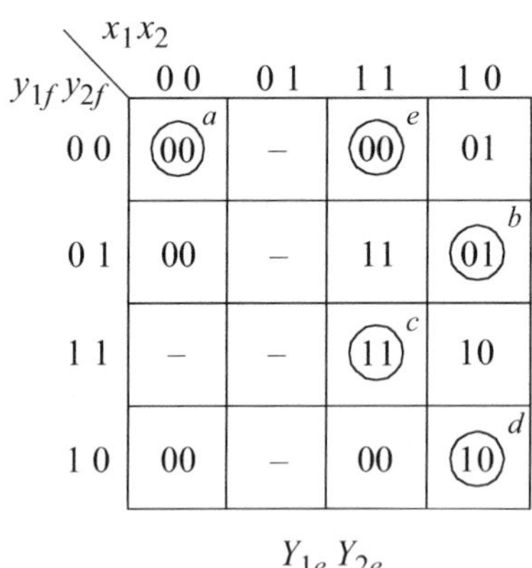

$y_{1f}y_{2f}$ \ x_1x_2	00	01	11	10
00	(00) a	–	(00) e	01
01	00	–	11	(01) b
11	–	–	(11) c	10
10	00	–	00	(10) d

$Y_{1e}\,Y_{2e}$

Figure 3.106 Combined excitation map for the merged flow table of Figure 3.104(b).

The output map is constructed from the merged flow table of Figure 3.104(b) and the reduced primitive flow table of Figure 3.102. The merged flow table indicates the location of the stable states and the reduced primitive flow table specifies the output values of the stable states. The output map is shown in Figure 3.108. The equation for output z_1 is shown in Equation 3.20 as a sum of products and as a product of sums. It is interesting to note that both forms of the equation require not only the same number of logic gates, but also the same number of identical feedback and input variables.

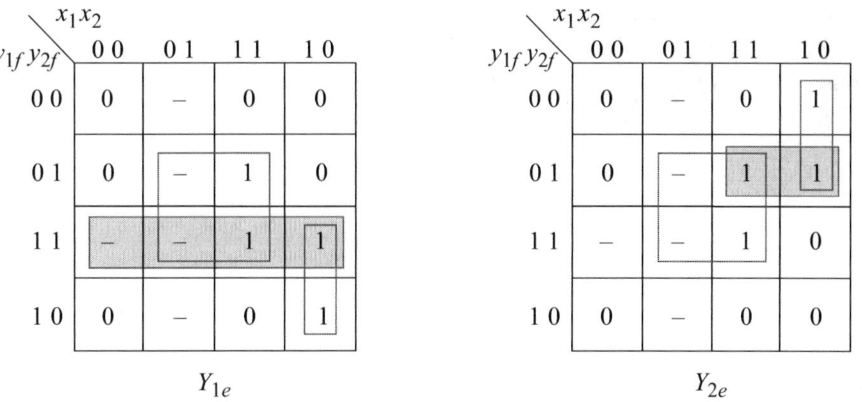

Figure 3.107 Individual excitation maps for Y_{1e} and Y_{2e} obtained from the combined excitation map of Figure 3.106.

$$Y_{1e} = y_{2f}x_2 + y_{1f}x_1x_2' + y_{1f}y_{2f}$$
$$Y_{2e} = y_{2f}x_2 + y_{1f}'x_1x_2' + y_{1f}'y_{2f}x_1 \tag{3.18}$$

$$Y_{1e} = (x_1)(y_{2f} + x_2')(y_{1f} + x_2)$$
$$Y_{2e} = (x_1)(y_{2f} + x_2')(y_{1f}' + x_2) \tag{3.19}$$

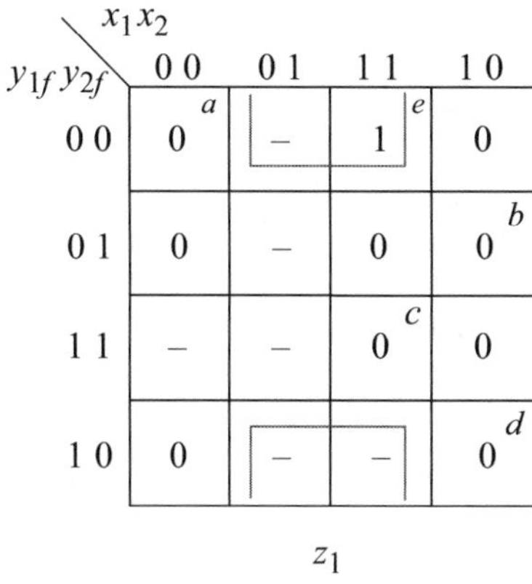

Figure 3.108 Output map for Example 3.12.

$$z_1 = y_{2f}'x_2$$
$$z_1 = (y_{2f}' + x_2) \tag{3.20}$$

The logic diagram is shown in Figure 3.109 in a product-of-sums form. This form requires not only the fewest number of gates, but also the fewest number of inputs per gate. The structural design module using built-in primitives for a product-of-sums form is shown in Figure 3.110. The test bench module and the outputs are shown in Figures 3.111 and 3.112, respectively.

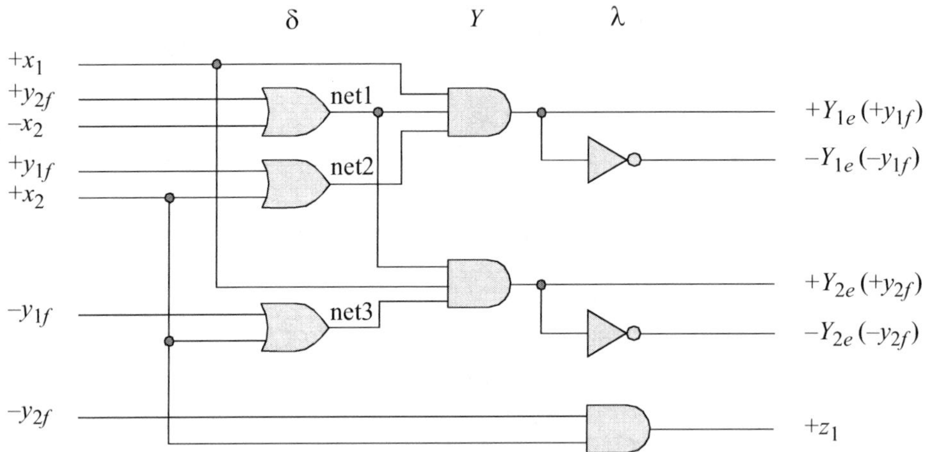

Figure 3.109 Logic diagram for Example 3.12 in a product-of-sums form.

```
//structural using built-in primitives for asm

module asm24_pos_bip (x1, x2, y1e, y2e, z1);

//define inputs and output
input x1, x2;
output y1e, y2e, z1;

//define internal nets
wire net1, net2, net3;

//----------------------------------------
//define the logic for y1e
or      (net1, y2e, ~x2),
        (net2, y1e, x2);

and     (y1e, x1, net1, net2);

                                //continued on next page
```

Figure 3.110 Structural design module for Example 3.12.

```
//-----------------------------------------
//define the logic for y2e
or     (net3, ~y1e, x2);

and    (y2e, net1, x1, net3);

//-----------------------------------------
//define the logic for output z1
and    (z1, ~y2e, x2);

endmodule
```

Figure 3.110 (Continued)

```
//test bench for the asm that uses built-in primitives
module asn24_pos_bip_tb;

//inputs are reg for test bench
//outputs are wire for test bench
reg x1, x2;
wire y1e, y2e, z1;

//display variables
initial
$monitor ("x1 x2 = %b, state = %b, z1 = %b", {x1, x2},
            {y1e, y2e}, z1);

//apply input vectors
initial
begin
   #0    x1 = 1'b0;
         x2 = 1'b0;

   #10   x1=1'b1;     x2=1'b0;     //go to state_b (001)
   #10   x1=1'b0;     x2=1'b0;     //go to state_a (000)
   #10   x1=1'b1;     x2=1'b0;     //go to state_b (001)
   #10   x1=1'b1;     x2=1'b1;     //go to state_c (011)

   #10   x1=1'b1;     x2=1'b0;     //go to state_d (010)
   #10   x1=1'b0;     x2=1'b0;     //go to state_a (000)

   #10   x1=1'b1;     x2=1'b0;     //go to state_b (001)
   #10   x1=1'b1;     x2=1'b1;     //go to state_c (011)
                                   //continued on next page
```

Figure 3.111 Test bench for Example 3.12.

```
   #10   x1=1'b1;    x2=1'b0;    //go to state_d (010)
   #10   x1=1'b0;    x2=1'b0;    //go to state_a (000)

   #10   x1=1'b1;    x2=1'b0;    //go to state_b (001)
   #10   x1=1'b1;    x2=1'b1;    //go to state_c (011)
   #10   x1=1'b1;    x2=1'b0;    //go to state_d (010)
   #10   x1=1'b1;    x2=1'b1;    //go to state_e (110)
                                             //assert z1

   #10   x1=1'b1;    x2=1'b0;    //go to state_f (111)
   #10   x1=1'b1;    x2=1'b1;    //go to state_c (011)
   #10   x1=1'b1;    x2=1'b0;    //go to state_d (010)
   #10   x1=1'b1;    x2=1'b1;    //go to state_e (110)
                                             //assert z1

   #10   x1=1'b1;    x2=1'b0;    //go to state_f (111)
   #10   x1=1'b0;    x2=1'b0;    //go to state_a (000)

   #10   $stop;
end

//instantiate the module into the test bench as a single line
asm24_pos_bip inst1 (x1, x2, y1e, y2e, z1);

endmodule
```

Figure 3.111 (Continued)

```
x1 x2 = 00, state = 00, z1 = 0
x1 x2 = 10, state = 01, z1 = 0
x1 x2 = 00, state = 00, z1 = 0
x1 x2 = 10, state = 01, z1 = 0
x1 x2 = 11, state = 11, z1 = 0
x1 x2 = 10, state = 10, z1 = 0
x1 x2 = 00, state = 00, z1 = 0

x1 x2 = 10, state = 01, z1 = 0
x1 x2 = 11, state = 11, z1 = 0
x1 x2 = 10, state = 10, z1 = 0
x1 x2 = 00, state = 00, z1 = 0

x1 x2 = 10, state = 01, z1 = 0
x1 x2 = 11, state = 11, z1 = 0
x1 x2 = 10, state = 10, z1 = 0
x1 x2 = 11, state = 00, z1 = 1
                                   //continued on next page
```

Figure 3.112 Outputs for Example 3.12.

```
x1 x2 = 10, state = 01, z1 = 0
x1 x2 = 11, state = 11, z1 = 0
x1 x2 = 10, state = 10, z1 = 0
x1 x2 = 11, state = 00, z1 = 1
x1 x2 = 10, state = 01, z1 = 0
x1 x2 = 00, state = 00, z1 = 0
```

Figure 3.112 (Continued)

Example 3.13 Given the reduced primitive flow table shown below in Figure 3.113, a Mealy asynchronous sequential machine will be designed using logic gates that were designed using dataflow modeling. There are no equivalent states because of different outputs. There will be no output glitches. The excitation and output equations will be in a sum-of-products notation.

x_1x_2 00	01	11	10	z_1
(a)	b	–	d	0
a	(b)	c	–	0
–	b	(c)	d	0
a	–	e	(d)	0
–	f	(e)	d	1
g	(f)	c	–	1
(g)	f	–	d	1

Figure 3.113 Reduced primitive flow table for Example 3.13.

The merged flow table is shown in Figure 3.114 and the transition diagram is shown in Figure 3.115. The combined excitation map is shown in Figure 3.116 and the individual excitation maps are shown in Figure 3.117. The excitation equations are shown in Equation 3.21 in a sum-of-products form. The output map for output z_1 is shown in Figure 3.118 and the equation for z_1 is shown in Equation 3.22.

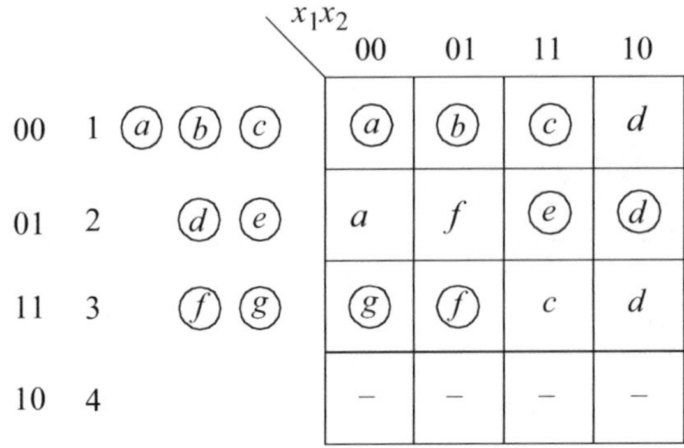

			x_1x_2 00	01	11	10
00	1	(a) (b) (c)	(a)	(b)	(c)	d
01	2	(d) (e)	a	f	(e)	(d)
11	3	(f) (g)	(g)	(f)	c	d
10	4		–	–	–	–

Figure 3.114 Merged flow table for Example 3.13.

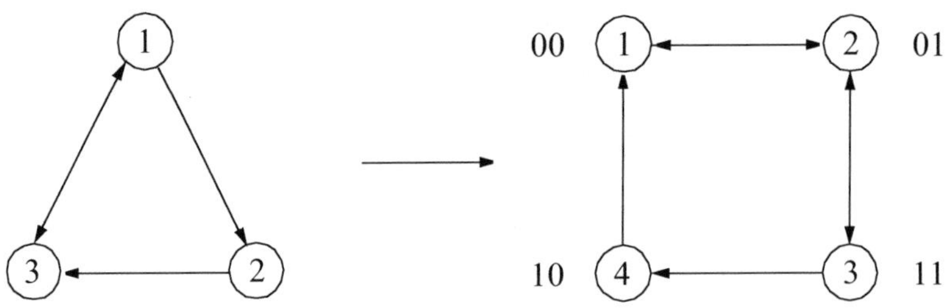

Figure 3.115 Transition diagram for Example 3.13.

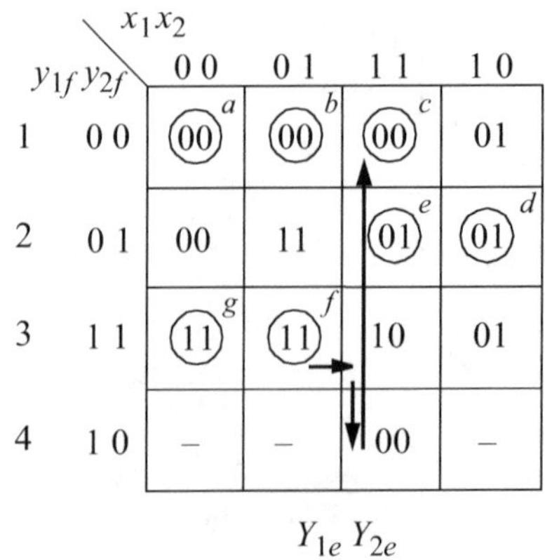

	$y_{1f}y_{2f}$ \ x_1x_2	0 0	0 1	1 1	1 0
1	0 0	(00) a	(00) b	(00) c	01
2	0 1	00	11	(01) e	(01) d
3	1 1	(11) g	(11) f	10	01
4	1 0	–	–	00	–

$Y_{1e}\,Y_{2e}$

Figure 3.116 Combined excitation map for Example 3.13.

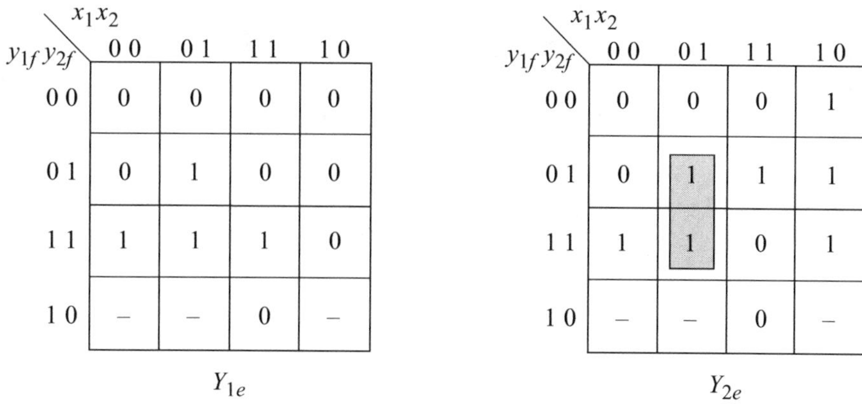

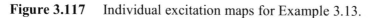

Figure 3.117 Individual excitation maps for Example 3.13.

$$Y_{1e} = y_{1f}x_1' + y_{1f}y_{2f}x_2 + y_{2f}x_1'x_2$$

net1 net2 net3

net4

$$Y_{2e} = x_1x_2' + y_{1f}x_1' + y_{1f}'y_{2f}x_2 + y_{1f}'y_{2f}x_1 + y_{2f}x_1'x_2 \tag{3.21}$$

net5 net6 net7 net8 net9

net10

Hazard cover

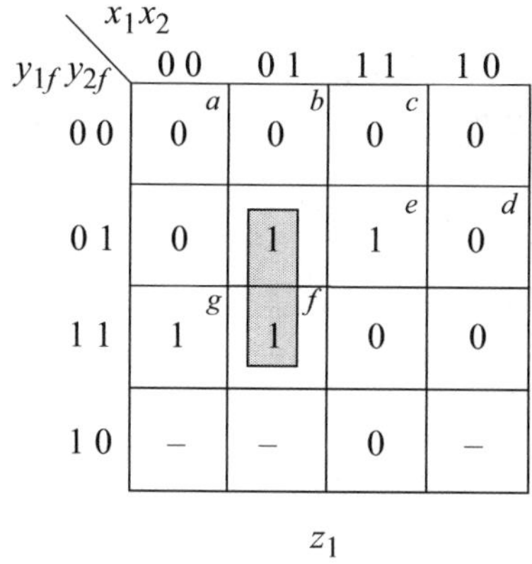

x_1x_2
$y_{1f}y_{2f}$
00 01 11 10
a b c d e f g
00: 0 0 0 0
01: 0 1 1 0
11: 1 1 0 0
10: – – 0 –
z_1

Figure 3.118 Output map for Example 3.13.

$$z_1 = y_{1f}x_1' + y_{1f}'y_{2f}x_2 + y_{2f}x_1'x_2 \text{ (Hazard cover)} \tag{3.22}$$

The structural design module is shown in Figure 3.119 using logic gates that were designed using dataflow modeling. The equations of Equation 3.21 were used to implement the design module. The test bench module is shown in Figure 3.120 and the outputs are shown in Figure 3.121. Use the reduced primitive flow table or the merged flow table to verify the correct operation of the machine.

```
//structural for sum-of-products asm
module asm_struc_df2 (rst_n, x1, x2, y1e, y2e, z1);

//define inputs and output
input rst_n, x1, x2;
output y1e, y2e, z1;

//define internal nets
wire net1, net2, net3, net4, net5, net6,
     net7, net8, net9, net10;

//-------------------------------------------
//instantiate the logic for y1e
and3_df  inst1 (y1e, ~x1, rst_n, net1);

and4_df  inst2 (y1e, y2e, x2, rst_n, net2),
         inst3 (y2e, ~x1, x2, rst_n, net3);

or3_df   inst4 (net1, net2, net3, y1e);

//-------------------------------------------
//instantiate the logic for y2e
and3_df  inst5 (x1, ~x2, rst_n, net5),
         inst6 (y1e, ~x1, rst_n, net6);

and4_df  inst7 (~y1e, y2e, x2, rst_n, net7),
         inst8 (~y1e, y2e, x1, rst_n, net8),
         inst9 (y2e, ~x1, x2, rst_n, net9);

or5_df   inst10(net5, net6, net7, net8, net9, y2e);

//-------------------------------------------
//instantiate the logic for output z1
and2_df  inst11 (y1e, ~x1, net11);
and3_df  inst12 (~y1e, y2e, x2, net12),
         inst13 (y2e, ~x1, x2, net13);

or3_df   inst14 (net11, net12, net13, z1);

endmodule
```

Figure 3.119 Structural design module for the asynchronous sequential machine of Example 3.13.

```
//test bench for sum-of-products asm
module asm_struc_df2_tb;

reg rst_n, x1, x2;    //inputs are reg for test bench
wire y1e, y2e, z1;    //outputs are wire for test bench

initial      //display variables
$monitor ("x1 x2 = %b, state = %b, z1 = %b",
            {x1, x2}, {y1e, y2e}, z1);

//apply input vectors
initial
begin

   #0     x1 = 1'b0;      x2 = 1'b0;      //state_a
          rst_n = 1'b0;
   #5     rst_n = 1'b1;

   #10    x1 = 1'b1;  x2 = 1'b0;  //state_d
   #10    x1 = 1'b1;  x2 = 1'b1;  //state_e, assert z1

   #10    x1 = 1'b0;  x2 = 1'b1;  //state_f, assert z1
   #10    x1 = 1'b0;  x2 = 1'b0;  //state_g, assert z1

   #10    x1 = 1'b1;  x2 = 1'b0;  //state_d
   #10    x1 = 1'b0;  x2 = 1'b0;  //state_a

   #10    x1 = 1'b0;  x2 = 1'b1;  //state_b
   #10    x1 = 1'b1;  x2 = 1'b1;  //state_c

   #10    x1 = 1'b1;  x2 = 1'b0;  //state_d
   #10    x1 = 1'b1;  x2 = 1'b1;  //state_e, assert z1

   #10    x1 = 1'b0;  x2 = 1'b1;  //state_f, assert z1
   #10    x1 = 1'b1;  x2 = 1'b1;  //state_c

   #10    x1 = 1'b1;  x2 = 1'b0;  //state_d
   #10    x1 = 1'b0;  x2 = 1'b0;  //state_a

   #10    $stop;
end

//instantiate the module into the test bench
asm_struc_df2 inst1 (rst_n, x1, x2, y1e, y2e, z1);

endmodule
```

Figure 3.120 Test bench module for the asynchronous sequential machine of Example 3.13.

```
x1 x2 = 00, state = 00, z1 = 0
x1 x2 = 10, state = 01, z1 = 0
x1 x2 = 11, state = 01, z1 = 1
x1 x2 = 01, state = 11, z1 = 1
x1 x2 = 00, state = 11, z1 = 1

x1 x2 = 10, state = 01, z1 = 0
x1 x2 = 00, state = 00, z1 = 0
x1 x2 = 01, state = 00, z1 = 0
x1 x2 = 11, state = 00, z1 = 0
x1 x2 = 10, state = 01, z1 = 0
x1 x2 = 11, state = 01, z1 = 1
x1 x2 = 01, state = 11, z1 = 1

x1 x2 = 11, state = 00, z1 = 0
x1 x2 = 10, state = 01, z1 = 0
x1 x2 = 00, state = 00, z1 = 0
```

Figure 3.121 Outputs for the asynchronous sequential machine of Example 3.13.

Example 3.14 A Moore asynchronous sequential machine will be synthesized, using the continuous assignment statement **assign** that has one input x_1 and one output z_1 that operates according to the timing diagram shown in Figure 3.122. The assertion of input x_1 toggles output z_1. The machine will have no static hazards.

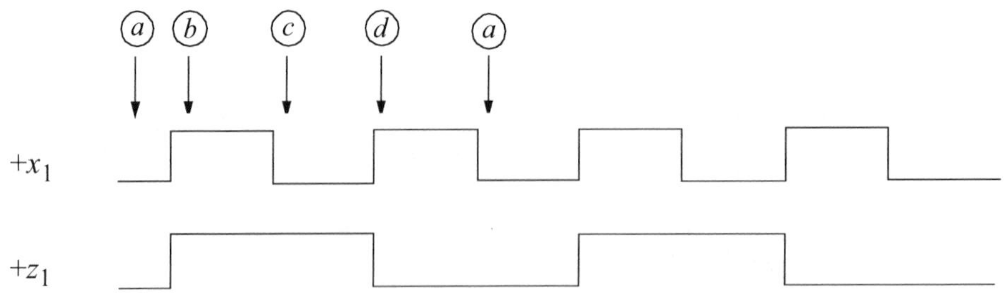

Figure 3.122 Timing diagram for the asynchronous sequential machine of Example 3.14.

The primitive flow table is shown in Figure 3.123. The table has no equivalent states, because no states can merge. The combined excitation map is shown in Figure 3.124 and the individual excitation maps are shown in Figure 3.125. The equations for Y_{1e} and Y_{2e} are shown in Equation 3.23 in a sum-of-products form.

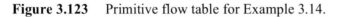

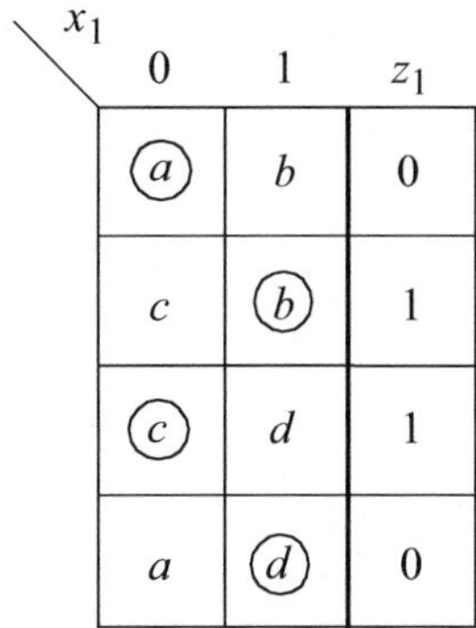

x_1 0	1	z_1
(a)	b	0
c	(b)	1
(c)	d	1
a	(d)	0

Figure 3.123 Primitive flow table for Example 3.14.

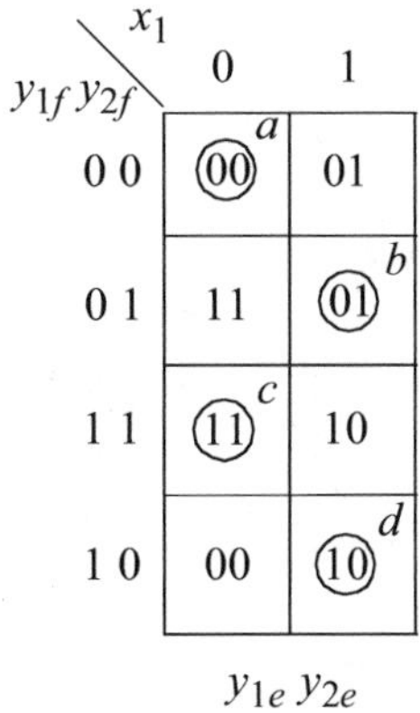

$y_{1f}\,y_{2f}$ \ x_1	0	1
0 0	(00) a	01
0 1	11	(01) b
1 1	(11) c	10
1 0	00	(10) d

$y_{1e}\,y_{2e}$

Figure 3.124 Combined excitation map for Example 3.14.

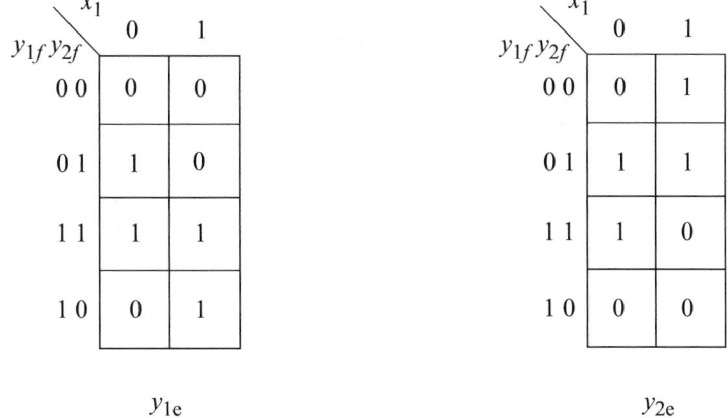

$y_{1f}\,y_{2f}$ \ x_1	0	1
0 0	0	0
0 1	1	0
1 1	1	1
1 0	0	1

y_{1e}

$y_{1f}\,y_{2f}$ \ x_1	0	1
0 0	0	1
0 1	1	1
1 1	1	0
1 0	0	0

y_{2e}

Figure 3.125 Individual excitation maps for Example 3.14.

$$Y_{1e} = \underset{\text{net1}}{y_{1f}x_1} + \underset{\text{net2}}{y_{2f}x_1'} + \underset{\text{net3}}{y_{1f}y_{2f}} \quad \text{(Hazard cover)}$$

$$Y_{2e} = \underset{\text{net4}}{y_{1f}'x_1} + \underset{\text{net5}}{y_{2f}x_1'} + \underset{\text{net6}}{y_{1f}'y_{2f}} \quad \text{(Hazard cover)} \tag{3.23}$$

The output map for z_1 is shown below and the output equation is shown in Equation 3.24. The dataflow design module is shown in Figure 3.126 using the continuous assignment statement **assign**. The test bench module is shown in Figure 3.127 and the outputs are shown in Figure 3.128.

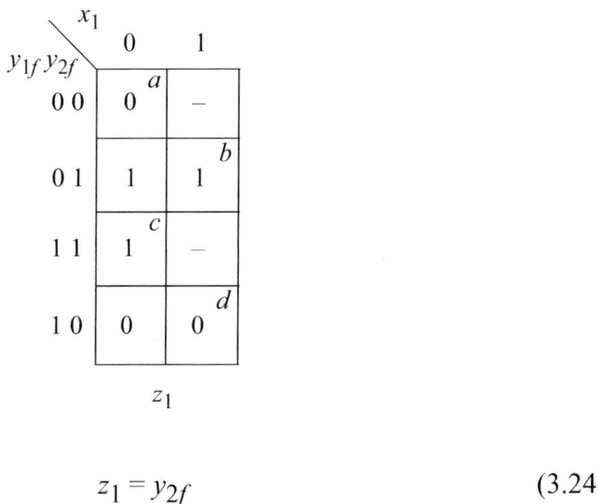

$$z_1 = y_{2f} \tag{3.24}$$

```
//dataflow for asm using the continuous assignment

module asm_toggle2 (rst_n, x1, y1e, y2e, z1);

//define inputs and outputs
input rst_n, x1;
output y1e, y2e, z1;

//define internal nets
wire net1, net2, net3, net4, net5, net6;

                        //continued on next page
```

Figure 3.126 Dataflow design module for Example 3.14.

```
//----------------------------------------
//define the logic for y1e
assign   net1 = y1e & x1 & rst_n,
         net2 = y2e & ~x1 & rst_n,
         net3 = y1e & y2e & rst_n,  //hazard cover
         y1e = net1 | net2 | net3;

//----------------------------------------
//define the logic for y2e
assign   net4 = ~y1e & x1 & rst_n,
         net5 = y2e & ~x1 & rst_n,
         net6 = ~y1e & y2e & rst_n, //hazard cover
         y2e = net4 | net5 | net6;

assign   z1 = net4 | net5 | net6;

endmodule
```

Figure 3.126 (Continued)

```
//test bench for asm using the assign statement
module asm_toggle2_tb;

//inputs are reg for test bench
//outputs are wire for test bench
reg rst_n, x1;
wire y1e, y2e, z1;

//display variables
initial
$monitor ("x1 = %b, state = %b, z1 = %b",
            x1, {y1e, y2e}, z1);

//apply input signals
initial
begin
   #0    rst_n = 1'b0;
         x1 = 1'b0;

   #5    rst_n = 1'b1;

   #10   x1 = 1'b0;      //state_a
   #10   x1 = 1'b1;      //state_b, assert z1
                                  //continued on next page
```

Figure 3.127 Test bench module for Example 3.14.

```
    #10   x1 = 1'b0;      //state_c, assert z1
    #10   x1 = 1'b1;      //state_d, deassert z1

    #10   x1 = 1'b0;      //state_a
    #10   x1 = 1'b1;      //state_b, assert z1

    #10   x1 = 1'b0;      //state_c, assert z1
    #10   x1 = 1'b1;      //state_d, deassert z1

    #10   $stop;

end

//instantiate the module into the test bench
asm_toggle2 inst1 (rst_n, x1, y1e, y2e, z1);

endmodule
```

Figure 3.127 (Continued)

```
x1 = 0, state = 00, z1 = 0
x1 = 1, state = 01, z1 = 1
x1 = 0, state = 11, z1 = 1
x1 = 1, state = 10, z1 = 0

x1 = 0, state = 00, z1 = 0
x1 = 1, state = 01, z1 = 1
x1 = 0, state = 11, z1 = 1
x1 = 1, state = 10, z1 = 0
```

Figure 3.128 Outputs for Example 3.14.

Example 3.15 An asynchronous sequential machine will be designed with Moore and Mealy type outputs using built-in primitives. The machine has two inputs x_1 and x_2 and two outputs z_1 and z_2. The two inputs may overlap, but will not change state simultaneously. Only the following sequences are valid:

$$x_1x_2 = 00 \rightarrow 10 \rightarrow 11 \rightarrow 01 \rightarrow 00$$
$$x_1x_2 = 00 \rightarrow 01 \rightarrow 11 \rightarrow 10 \rightarrow 00$$
$$x_1x_2 = 00 \rightarrow 10 \rightarrow 00$$
$$x_1x_2 = 00 \rightarrow 01 \rightarrow 00$$

Output z_1 is asserted whenever x_1 is active and x_2 is asserted or when x_2 is active and x_1 is asserted. Output z_1 will be de-asserted when either x_1 or x_2 is de-asserted. Output z_2 is asserted coincident with the assertion of z_1 and remains active until the de-assertion of the last active input of an overlapping sequence. A representative timing diagram is shown below in Figure 3.129 and the primitive flow table is shown in Figure 3.130 as obtained from the timing diagram.

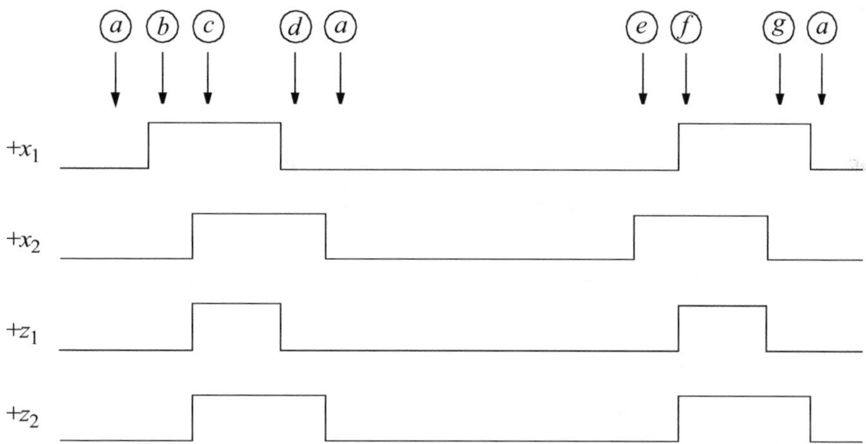

Figure 3.129 Timing diagram for the asynchronous sequential machine of Example 3.15.

x_1x_2 00	01	11	10	z_1	z_2
(a)	e	–	b	0	0
a	–	c	(b)	0	0
–	d	(c)	–	1	1
a	(d)	–	–	0	1
a	(e)	f	–	0	0
–	–	(f)	g	1	1
a	–	–	(g)	0	1

Figure 3.130 Primitive flow table for Example 3.15.

The merger diagram is shown in Figure 3.131. Recall the merging process, which states that two rows can merge into a single row if the entries in the same column of each row satisfy one of the following three merging rules:

1. Identical state entries, either stable or unstable.
2. A state entry and a "don't care."
3. Two "don't care" entries.

Thus the merger diagram of Figure 3.131 yields the following partitions of maximal compatible sets:

{ⓐ}, {ⓑ, ⓒ, ⓓ}, {ⓔ, ⓕ, ⓖ}

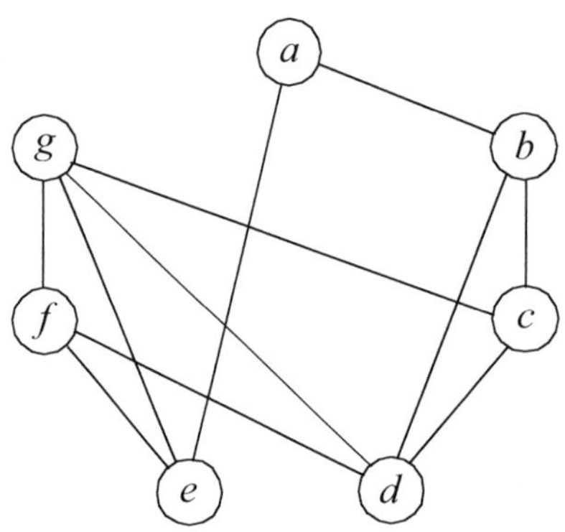

Figure 3.131 Merger diagram for Example 3.15.

The merged flow table illustrating the maximal compatible sets is shown in Figure 3.132. After enumerating the rows of the merged flow table, all state transition sequences can be identified with reference to individual rows. The state transitions are illustrated in graphical form by means of a transition diagram.

	x_1x_2	00	01	11	10
1	ⓐ	ⓐ	e	–	b
2	ⓑ ⓒ ⓓ	a	ⓓ	ⓒ	ⓑ
3	ⓔ ⓕ ⓖ	a	ⓔ	ⓕ	ⓖ
4		–	–	–	–

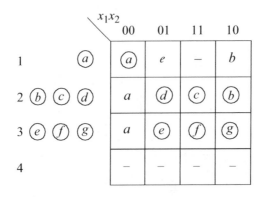

Figure 3.132 Merged flow table for Example 3.15.

The transition diagram for the merged flow table of Figure 3.132 is shown in Figure 3.133. Row 1 proceeds to row 2 by the sequence ⓐ $\to b \to$ ⓑ, as illustrated by the directed line from row 1 to row 2 in Figure 3.133. Similarly, row 1 proceeds to row 3 by the sequence ⓐ $\to e \to$ ⓔ. Also, row 3 proceeds to row 1 by the sequence ⓖ $\to a \to$ ⓐ. The combined excitation map is shown in Figure 3.134 and the individual excitation maps are shown in Figure 3.135.

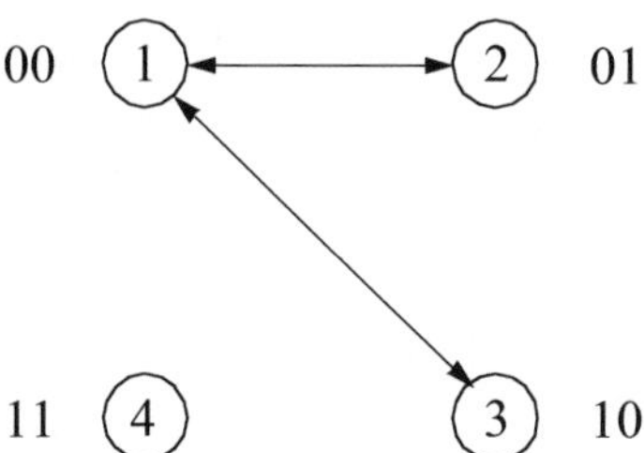

Figure 3.133 Transition diagram for Figure 3.132 of Example 3.15.

	$y_{1f}y_{2f}$ \ x_1x_2	0 0	0 1	1 1	1 0
1	0 0	(00) a	10	–	01
2	0 1	00	(01) d	(01) c	(01) b
4	1 1	–	–	–	–
3	1 0	00	(10) e	(10) f	(10) g

$Y_{1e}\,Y_{2e}$

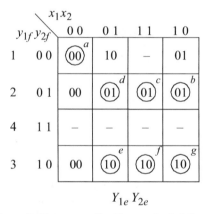

Figure 3.134 Combined excitation map for Example 3.15.

$y_{1f}y_{2f}$ \ x_1x_2	0 0	0 1	1 1	1 0
0 0	0	1	–	0
0 1	0	0	0	0
1 1	–	–	–	–
1 0	0	1	1	1

Y_{1e}

$y_{1f}y_{2f}$ \ x_1x_2	0 0	0 1	1 1	1 0
0 0	0	0	–	1
0 1	0	1	1	1
1 1	–	–	–	–
1 0	0	0	0	0

Y_{2e}

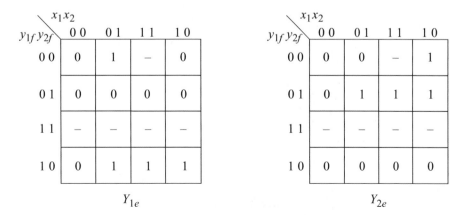

Figure 3.135 Individual excitation maps for Example 3.15.

The excitation equations are shown in Equation 3.25. The output maps for z_1 and z_2 are shown in Figure 3.136 and the output equations are shown in Equation 3.26 in a sum-of-products form.

$$Y_{1e} = y_{1f}x_1 + y_{2f}'x_2 \qquad Y_{2e} = y_{2f}x_2 + y_{1f}'x_1 \tag{3.25}$$

z_1

$y_{1f}y_{2f}$ \ x_1x_2	0 0	0 1	1 1	1 0
0 0	0 *a*	0	–	0
0 1	0	0 *d*	1 *c*	0 *b*
1 1	–	–	–	–
1 0	0	0 *e*	1 *f*	0 *g*

z_2

$y_{1f}y_{2f}$ \ x_1x_2	0 0	0 1	1 1	1 0
0 0	0 *a*	0	–	0
0 1	0	1 *d*	1 *c*	0 *b*
1 1	–	–	–	–
1 0	0	0 *e*	1 *f*	1 *g*

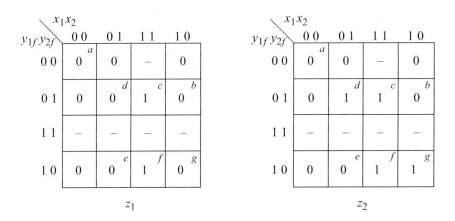

Figure 3.136 Output maps for Example 3.15.

$$z_1 = x_1x_2 \qquad z_2 = y_{2f}x_2 + y_{1f}x_1 \tag{3.26}$$

The dataflow design module for the asynchronous sequential machine is shown in Figure 3.137 using the continuous assignment statement **assign**. The continuous assignment statement models dataflow behavior and provides a Boolean correspondence between the right-hand side expression and the left-hand side target.

The continuous assignment statement assigns a value to a net (**wire**) that has been previously declared. The operands on the right-hand side can be registers, nets, or function calls. The registers and nets can be declared as either scalars or vectors. The test bench module and the outputs are shown in Figures 3.138 and 3.139, respectively.

```
//dataflow for asm using the continuous assignment statement
module asm_assign (rst_n, x1, x2, y1e, y2e, z1, z2);

//define inputs and outputs
input rst_n, x1, x2;
output y1e, y2e, z1, z2;

//define internal nets
wire net1, net2, net3, net4;
                                    //continued on next page
```

Figure 3.137 Dataflow design module for Example 3.15.

```
//----------------------------------------
//design the logic for y1e
assign   net1 = y1e & x1 & rst_n,
         net2 = ~y2e & x2 & rst_n,
         y1e = net1 | net2;

//----------------------------------------
//design the logic for y2e
assign   net3 = y2e & x2 & rst_n,
         net4 = ~y1e & x1 & rst_n,
         y2e = net3 | net4;

//----------------------------------------
//design the logic for outputs z1 and z2
assign   z1 = x1 & x2,
         z2 = (y2e & x2) | (y1e & x1);

endmodule
```

Figure 3.137 (Continued)

```
//test bench for the asm using the assign statement

module asm_assign_tb;

//inputs are reg for test bench
//outputs are wire for test bench
reg rst_n, x1, x2;
wire y1e, y2e, z1, z2;

//display variables
initial
$monitor ("x1x2 = %b, state = %b, z1z2 = %b",
            {x1, x2}, {y1e, y2e}, {z1, z2});

//apply input vectors
initial
begin
   #0    rst_n = 1'b0;
         x1 = 1'b0;
         x2 = 1'b0;

   #5    rst_n = 1'b1;
                                  //continued on next page
```

Figure 3.138 Test bench module for Example 3.15.

```
    #10   x1 = 1'b0;  x2 = 1'b0;
    #10   x1 = 1'b1;  x2 = 1'b0;
    #10   x1 = 1'b1;  x2 = 1'b1;      //z1 = 1, z2 = 1
    #10   x1 = 1'b0;  x2 = 1'b1;      //z1 = 0, z2 = 1
    #10   x1 = 1'b0;  x2 = 1'b0;      //z1 = 0, z2 = 0
    #10   x1 = 1'b0;  x2 = 1'b0;

    #10   x1 = 1'b0;  x2 = 1'b1;      //z1 = 0, z2 = 0
    #10   x1 = 1'b1;  x2 = 1'b1;      //z1 = 1, z2 = 1
    #10   x1 = 1'b1;  x2 = 1'b0;      //z1 = 0, z2 = 1
    #10   x1 = 1'b0;  x2 = 1'b0;      //z1 = 0, z2 = 0
    #10   $stop;
end

//instantiate the module into the test bench
asm_assign inst1 (rst_n, x1, x2, y1e, y2e, z1, z2);
endmodule
```

Figure 3.138 (Continued)

```
x1x2 = 00, state = 00, z1z2 = 00
x1x2 = 10, state = 01, z1z2 = 00
x1x2 = 11, state = 01, z1z2 = 11
x1x2 = 01, state = 01, z1z2 = 01
x1x2 = 00, state = 00, z1z2 = 00

x1x2 = 01, state = 10, z1z2 = 00
x1x2 = 11, state = 10, z1z2 = 11
x1x2 = 10, state = 10, z1z2 = 01
x1x2 = 00, state = 00, z1z2 = 00
```

Figure 3.139 Outputs for Example 3.15.

3.4 Pulse-Mode Asynchronous Sequential Machines

In pulse-mode asynchronous sequential machines, state changes occur on the application of input pulses which trigger the storage elements, rather than on a clock signal. The duration of the pulse is less than the propagation delay of the storage elements and associated logic gates. Thus, an input pulse will initiate a state change, but the completion of the change will not take place until after the corresponding input has been de-asserted. Multiple inputs cannot be active simultaneously.

Unlike a system clock, which has a specified frequency, the input pulses can occur randomly and more than one input pulse can generate an output. If the input pulse is of insufficient duration, then the storage elements may not be triggered and the machine will not sequence to the next state. If the pulse duration is too long, then the pulse will still be active when the machine changes from the present state $Y_{j(t)}$ to the next state $Y_{k(t+1)}$. The storage elements may then be triggered again and sequence the machine to an incorrect next state. If the time between consecutive pulses is too short, then the machine will be triggered while in an unstable condition, resulting in unpredictable behavior.

The pulse width restrictions that are dominant in pulse-mode sequential machines can be eliminated by including *D* flip-flops in the feedback path from the *SR* latches to the δ next-state logic. Providing edge-triggered *D* flip-flops as a constituent part of the implementation negates the requirement of precisely controlled input pulse durations. This is by far the most reliable means of synthesizing pulse-mode machines. The *SR* latches — in conjunction with the *D* flip-flops — form a master-slave configuration.

Figure 3.140 illustrates a block diagram for a pulse-mode asynchronous sequential machine using *SR* latches and *D* flip-flops. The machine is similar in structure to a Moore machine if $\lambda(Y)$ or to a Mealy machine if $\lambda(X,Y)$.

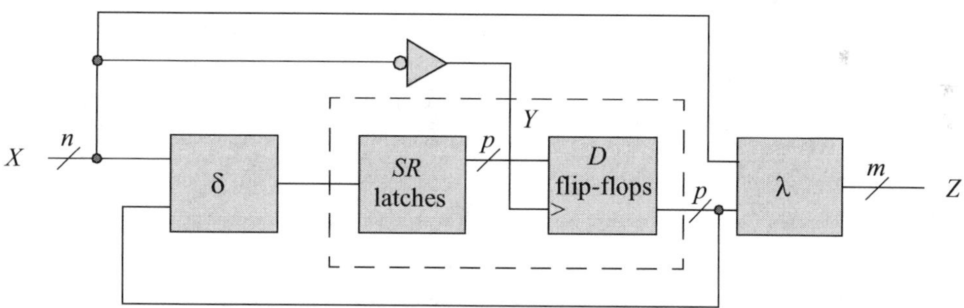

Figure 3.140 General block diagram for pulse-mode sequential machine.

In order for the operation of the machine to be deterministic, some restrictions apply to the input pulses:

1. Input pulses must be of sufficient duration to trigger the storage elements.

2. The time duration of the pulses must be shorter than the minimal propagation delay through the combinational input logic and the storage elements, so that the pulses are de-asserted before the storage elements can again change state.

3. The time duration between successive input pulses must be sufficient to allow the machine to stabilize before application of the next pulse.

4. Only one input pulse can be active at a time.

3.4.1 Synthesis Procedure

Reliability of pulse-mode machines can be increased by inserting delay circuits of an appropriate duration in the output networks of the storage elements. The aggregate delay of the storage elements and the delay circuit must be of sufficient duration so that the input pulse will be de-asserted before the storage element output signals arrive at the δ next-state logic.

Three techniques are commonly used to insert delays in the storage element outputs: An even number of inverters are connected in series with each latch output; a linear delay circuit is connected in series with each latch output; or an edge-triggered *D* flip-flop is connected in series with each latch output.

As stated previously, the flip-flops are set to the state of the latches, but are triggered on the trailing edge of the input pulses. Thus, the flip-flop outputs — and therefore the next state of the machine as represented by the *SR* latch outputs — are received at the δ next-state logic only when the active input pulse has been de-asserted. The *SR* latches and the *D* flip-flops constitute a master-slave relationship and will be the primary means to implement pulse-mode asynchronous sequential machines in this section. *T* flip-flops will also be utilized in the examples to illustrate an alternative method to implement pulse-mode asynchronous sequential machines.

The synthesis procedure will be illustrated in detail in the examples presented in the following sections. The first method will implement pulse-mode machines using *SR* latches with *D* flip-flops in a master-slave configuration. Then *T* flip-flops will be utilized in the synthesis examples. The *T* flip-flops will incorporate a delay circuit to delay the output of the flip-flops from being fed back to the δ input logic before the input signals become de-asserted. Both Moore and Mealy machines will be synthesized in the examples.

3.4.2 *SR* Latches with *D* Flip-Flops as Storage Elements

This section will present the synthesis of pulse-mode asynchronous sequential machines including Moore and Mealy machines. All designs will include the following items where applicable: timing diagrams, state diagrams, Karnaugh maps for the input equations, Karnaugh maps for the output equations, and logic diagrams.

Example 3.16 A Mealy pulse-mode sequential machine will be designed which has two inputs x_1 and x_2 and one output z_1. Output z_1 is asserted coincident with every second x_2 pulse, if and only if the pair of x_2 pulses is immediately preceded by an x_1 pulse.

Structural modeling will be used in the implementation of the Mealy machine using built-in primitives. The storage elements will consist of *SR* latches and *D* flip-flops in a master-slave configuration. The *D* flip-flops were designed using behavioral modeling. A representative timing diagram is shown in Figure 3.141 and the corresponding state diagram is shown in Figure 3.142.

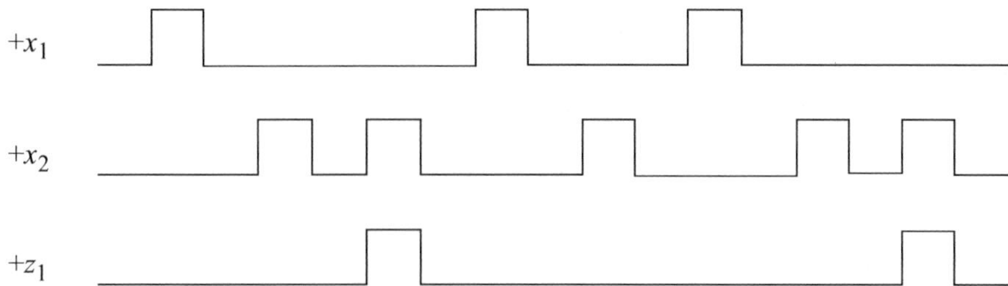

Figure 3.141 Representative timing diagram for the Mealy pulse-mode asynchronous sequential machine of Example 3.16.

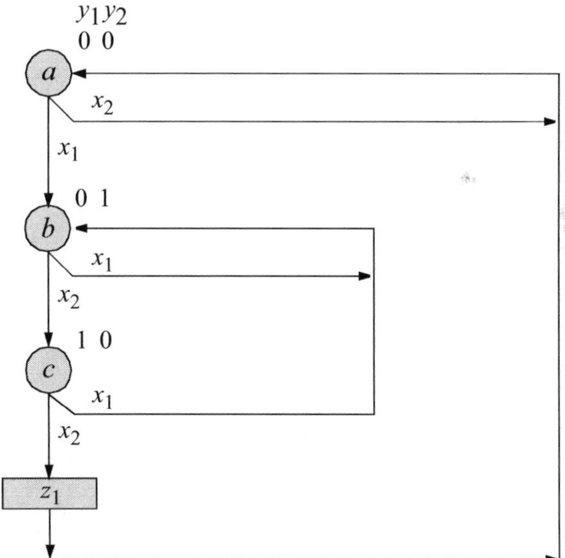

Figure 3.142 State diagram for the Mealy pulse-mode asynchronous sequential machine of Example 3.16.

The Karnaugh maps for the latches of the Mealy machine are shown in Figure 3.143. An entry of *R* specifies a reset condition; an entry of *r* indicates that the machine is to remain in a reset state; an entry of *S* indicates a set condition; and an entry of *s* indicates that the machine is to remain in a set state. The corresponding equations are shown in Equations 3.27 and 3.28.

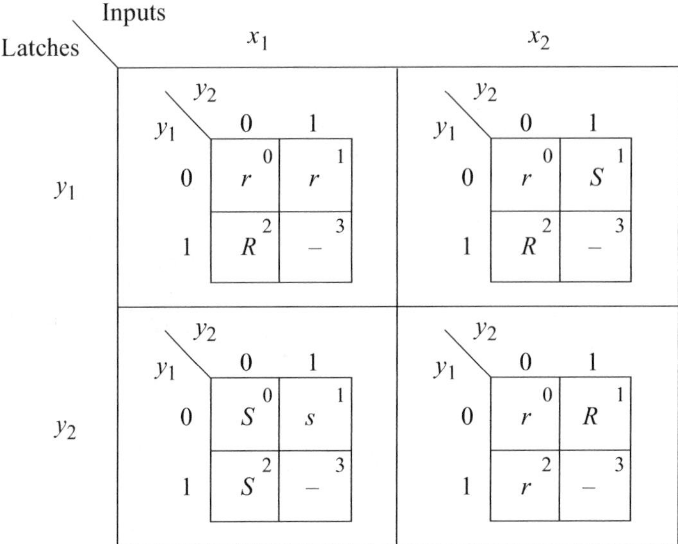

Figure 3.143 Input maps for the Mealy pulse-mode asynchronous sequential machine of Example 3.16.

$$SLy_1 = y_2x_2$$
$$RLy_1 = x_1 + y_1x_2 \quad (3.27)$$

$$SLy_2 = x_1$$
$$RLy_2 = x_2 \quad (3.28)$$

The equation for output z_1 is shown in Equation 3.29. Since only latch y_1 and input x_2 are required to activate output z_1, the equation for z_1 will include only those two variables. The logic diagram is shown in Figure 3.144 which contains D flip-flops that were previously designed using behavioral modeling.

$$z_1 = y_1x_2 \quad (3.29)$$

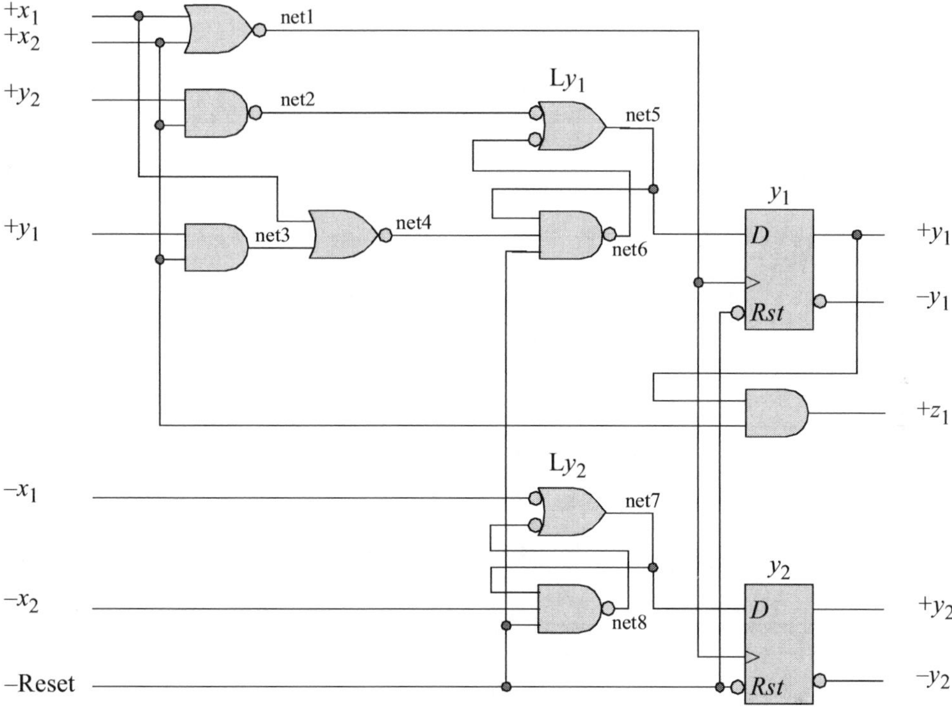

Figure 3.144 Logic diagram for the Mealy pulse-mode asynchronous sequential machine of Example 3.16.

The structural design module using built-in primitives and instantiated *D* flip-flops that were designed using behavioral modeling is shown in Figure 3.145. The test bench module is shown in Figure 3.146 and the outputs are shown in Figure 3.147. The outputs directly correspond to the timing diagram of Figure 3.141.

```
//structural for asm using built-in primitives

module pm_asm_bip (rst_n, x1, x2, y1e, y2e, z1);

//define inputs and outputs
input rst_n, x1, x2;
output y1e, y2e, z1;

//define internal nets
wire net1, net2, net3, net4, net5, net6, net7, net8;
                                    //continued on next page
```

Figure 3.145 Structural design module for the Mealy machine of Example 3.16.

```
//design the clock for the D flip-flops
nor   (net1, x1, x2);

//-------------------------------------
//define the logic for latch Ly1
nand  (net2, y2e, x2);
and   (net3, y1e, x2);
nor   (net4, x1, net3);
nand  (net5, net2, net6),
      (net6, net5, net4, rst_n);

//instantiate the D flip-flop for y1e
d_ff_bh inst1 (rst_n, net1, net5, y1e);  //reset, clk, D, Q)

//-------------------------------------
//define the logic for latch Ly2
nand  (net7, ~x1, net8),
      (net8, net7, ~x2, rst_n);

//instantiate the D flip-flop for y2e
d_ff_bh inst2 (rst_n, net1, net7, y2e);  //reset, clk, D, Q)

//-------------------------------------
//define the logic for output z1
and   (z1, y1e, x2);

endmodule
```

Figure 3.145 (Continued)

```
//test bench for the asm using built-in primitives

module pm_asm_bip_tb;

//inputs are reg for test bench
//outputs are wire for test bench
reg rst_n, x1, x2;
wire y1e, y2e, z1;

//display variables
initial
$monitor ("x1 x2 = %b, state = %b, z1 = %b",
            {x1, x2}, {y1e, y2e}, z1);
                                //continued on next page
```

Figure 3.146 Test bench module for the Mealy machine of Example 3.16.

```
//apply input vectors
initial
begin
   #0     rst_n = 1'b0;
          x1 = 1'b0;  x2 = 1'b0;

   #5     rst_n = 1'b1;

   #10    x1 = 1'b0;  x2 = 1'b1;  //go to state_a (00)
   #10    x1 = 1'b0;  x2 = 1'b0;  //remain in state_a (00)

   #10    x1 = 1'b1;  x2 = 1'b0;  //go to state_b (01)
   #10    x1 = 1'b0;  x2 = 1'b0;  //remain in state_b (01)

   #10    x1 = 1'b1;  x2 = 1'b0;  //remain in state_b (01)
   #10    x1 = 1'b0;  x2 = 1'b0;  //remain in state_b (01)

   #10    x1 = 1'b0;  x2 = 1'b1;  //go to state_c (10)
   #10    x1 = 1'b0;  x2 = 1'b0;  //remain in state_c (10)

   #10    x1 = 1'b0;  x2 = 1'b1;  //assert z1; go to state_a
                                  //(00)
   #10    x1 = 1'b0;  x2 = 1'b0;  //remain in state_a (00)

   #10    x1 = 1'b1;  x2 = 1'b0;  //go to state_b (01)
   #10    x1 = 1'b0;  x2 = 1'b0;  //remain in state_b (01)

   #10    x1 = 1'b0;  x2 = 1'b1;  //go to state_c (10)
   #10    x1 = 1'b0;  x2 = 1'b0;  //remain in state_c (10)

   #10    x1 = 1'b0;  x2 = 1' b1; //assert z1; go to state_a
                                  //(00)
   #10    x1 = 1'b0;  x2 = 1'b0;  //remain in state_a (00)

   #10    x1 = 1'b0;  x2 = 1'b1;  //remain in state_a (00)
   #10    x1 = 1'b0;  x2 = 1'b0;  //remain in state_a (00)

   #10    $stop;
end

//instantiate the module into the test bench
pm_asm_bip inst1 (rst_n, x1, x2, y1e, y2e, z1);

endmodule
```

Figure 3.146 (Continued)

```
x1 x2 = 00, state = 00, z1 = 0
x1 x2 = 01, state = 00, z1 = 0
x1 x2 = 00, state = 00, z1 = 0
x1 x2 = 10, state = 00, z1 = 0
x1 x2 = 00, state = 01, z1 = 0
x1 x2 = 10, state = 01, z1 = 0
x1 x2 = 00, state = 01, z1 = 0
x1 x2 = 01, state = 01, z1 = 0
x1 x2 = 00, state = 10, z1 = 0
x1 x2 = 01, state = 10, z1 = 1

x1 x2 = 00, state = 00, z1 = 0
x1 x2 = 10, state = 00, z1 = 0
x1 x2 = 00, state = 01, z1 = 0
x1 x2 = 01, state = 01, z1 = 0
x1 x2 = 00, state = 10, z1 = 0
x1 x2 = 01, state = 10, z1 = 1

x1 x2 = 00, state = 00, z1 = 0
x1 x2 = 01, state = 00, z1 = 0
x1 x2 = 00, state = 00, z1 = 0
```

Figure 3.147 Outputs for the Mealy machine of Example 3.16.

Example 3.17 The state diagram for a Moore pulse-mode asynchronous sequential machine is shown in Figure 3.148. The input maps are shown in Figure 3.149. The input and output equations are shown in Equation 3.30. The logic diagram is shown in Figure 3.150.

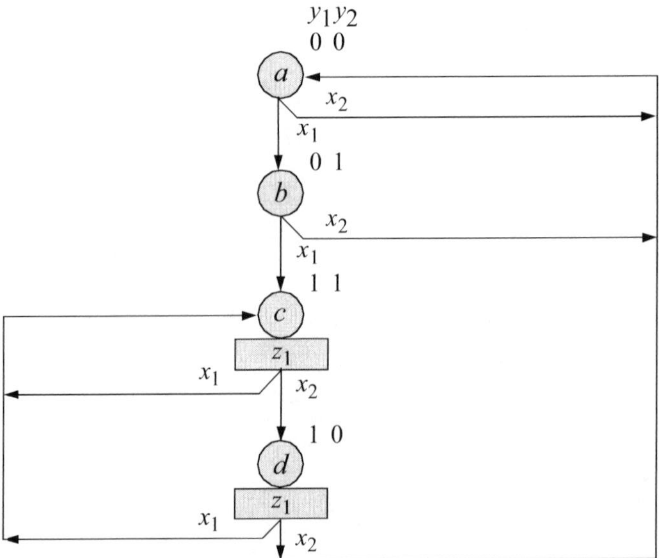

Figure 3.148 State diagram for the Moore machine of Example 3.17.

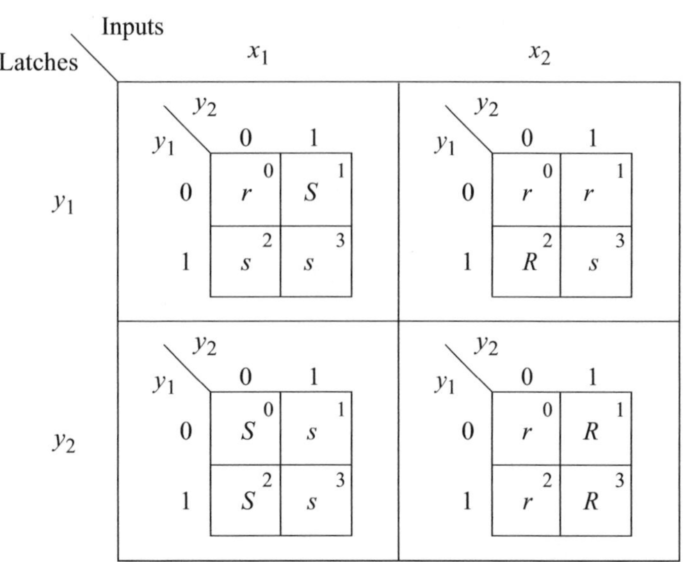

Figure 3.149 Input Karnaugh maps for Example 3.17.

$$\begin{aligned} SLy_1 &= y_2x_1 & RLy_1 &= y_2'x_2 \\ SLy_2 &= x_1 & RLy_2 &= x_2 \\ z_1 &= y_1 \end{aligned} \tag{3.30}$$

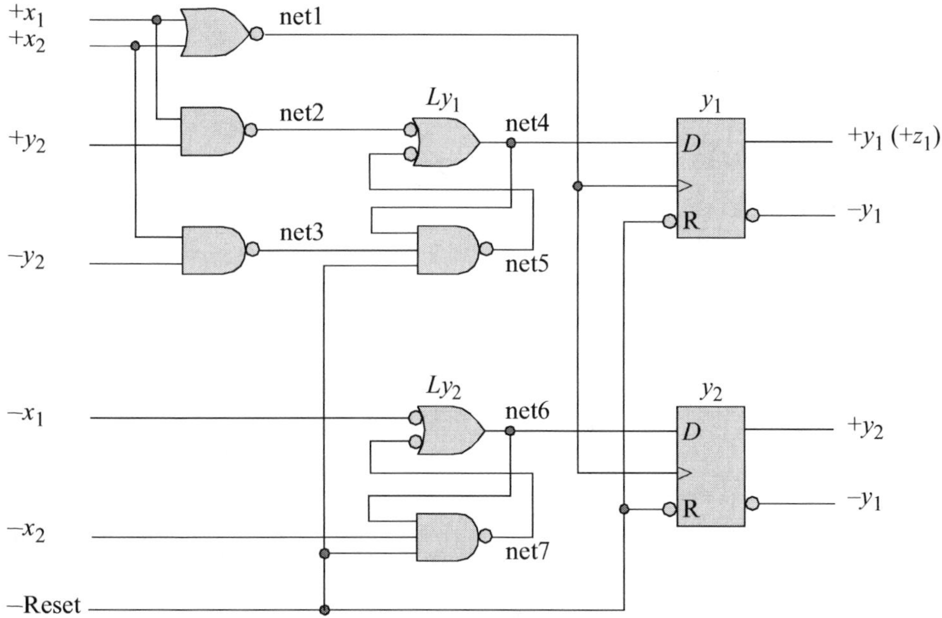

Figure 3.150 Logic diagram for the Moore machine of Example 3.17.

The structural design module using built-in primitives is shown in Figure 3.151. The test bench and outputs are shown in Figures 3.152 and 3.153, respectively.

```
//structural for Moore pulse-mode asm using bip

module pm_asm_moore10 (rst_n, x1, x2, y1e, y2e, z1);

//define inputs and outputs
input rst_n, x1, x2;
output y1e, y2e, z1;

//define internal nets
wire net1, net2, net3, net4, net5, net6, net7;

//define the clock for the D flip-flops
nor   (net1, x1, x2);

//--------------------------------------
//assign the logic for latch Ly1
nand  (net2, x1, y2e),
      (net3, x2, ~y2e),
      (net4, net2, net5),
      (net5, net4, net3, rst_n);

//instantiate the D flip-flop for y1e
d_ff_bh  inst1 (rst_n, net1, net4, y1e);
               //reset, clock, D, Q

//--------------------------------------
//assign the logic for latch Ly2
nand  (net6, ~x1, net7),
      (net7, net6, ~x2, rst_n);

//instantiate the D flip-flop for y2e
d_ff_bh  inst2 (rst_n, net1, net6, y2e);
               //reset, clock, D, Q

assign z1 = y1e;

endmodule
```

Figure 3.151 Structural design module for the Moore machine of Example 3.17.

```
//test bench for pm_asm_moore10
module pm_asm_moore10_tb;

reg rst_n, x1, x2;        //inputs are reg for test bench
wire y1e, y2e, z1;        //outputs are wire for test bench

initial     //display variables
$monitor ("x1 x2 = %b, state = %b, z1 = %b",
            {x1, x2}, {y1e, y2e}, z1);

//apply input vectors
initial
begin
   #0     rst_n = 1'b0;
          x1 = 1'b0;  x2 = 1'b0;

   #5     rst_n = 1'b1;

   #10    x1 = 1'b1;  x2 = 1'b0;      //state_b
   #10    x1 = 1'b0;  x2 = 1'b0;

   #10    x1 = 1'b1;  x2 = 1'b0;      //state_c; assert z1
   #10    x1 = 1'b0;  x2 = 1'b0;

   #10    x1 = 1'b0;  x2 = 1'b1;      //state_d; assert z1
   #10    x1 = 1'b0;  x2 = 1'b0;

   #10    x1 = 1'b1;  x2 = 1'b0;      //state_c; assert z1
   #10    x1 = 1'b0;  x2 = 1'b0;

   #10    x1 = 1'b0;  x2 = 1'b1;      //state_d; assert z1
   #10    x1 = 1'b0;  x2 = 1'b0;

   #10    x1 = 1'b0;  x2 = 1'b1;      //state_a
   #10    x1 = 1'b0;  x2 = 1'b0;

   #10    x1 = 1'b0;  x2 = 1'b1;      //state_a
   #10    x1 = 1'b0;  x2 = 1'b0;

   #10    $stop;
end

//instantiate the module into the test bench
pm_asm_moore10 inst1 (rst_n, x1, x2, y1e, y2e, z1);

endmodule
```

Figure 3.152 Test bench module for the Moore machine of Example 3.17.

```
x1 x2 = 00, state = 00, z1 = 0
x1 x2 = 10, state = 00, z1 = 0
x1 x2 = 00, state = 01, z1 = 0
x1 x2 = 10, state = 01, z1 = 0

x1 x2 = 00, state = 11, z1 = 1
x1 x2 = 01, state = 11, z1 = 1
x1 x2 = 00, state = 10, z1 = 1
x1 x2 = 10, state = 10, z1 = 1

x1 x2 = 00, state = 11, z1 = 1
x1 x2 = 01, state = 11, z1 = 1
x1 x2 = 00, state = 10, z1 = 1
x1 x2 = 01, state = 10, z1 = 1

x1 x2 = 00, state = 00, z1 = 0
x1 x2 = 01, state = 00, z1 = 0
x1 x2 = 00, state = 00, z1 = 0
```

Figure 3.153 Outputs for the Moore machine of Example 3.17.

Example 3.18 The state diagram for a Moore pulse-mode machine is shown in Figure 3.154. The input maps for x_1, x_2, and x_3 are shown in Figure 3.155. The input and output equations are shown in Equation 3.31.

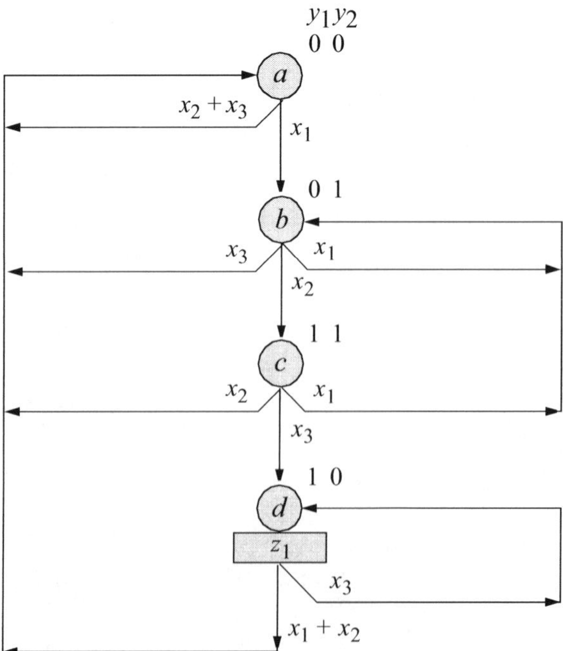

Figure 3.154 State diagram for the Moore machine of Example 3.18.

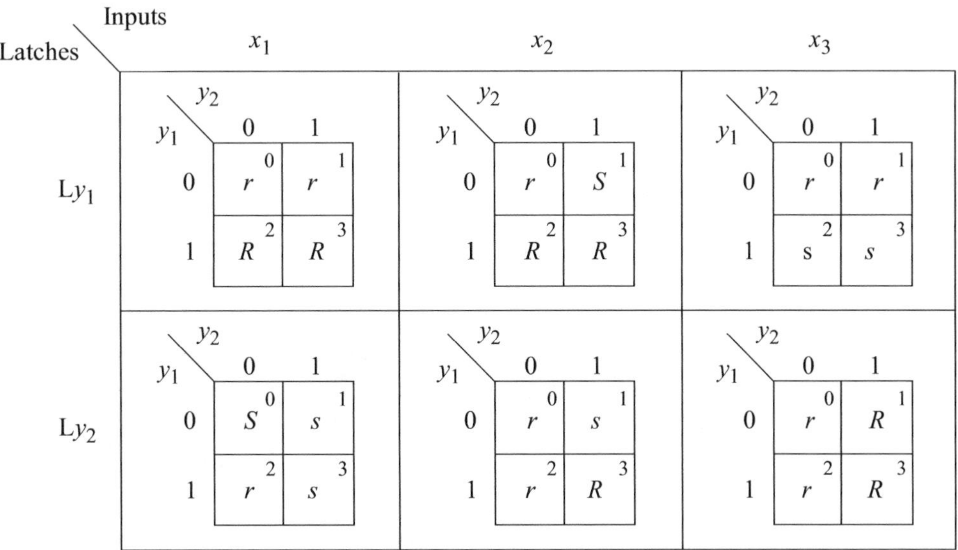

Figure 3.155 Karnaugh maps for the inputs of the Moore machine of Example 3.18.

$$SLy_1 = y_1'y_2x_2 \qquad RLy_1 = x_1 + y_1x_2$$

$$SLy_2 = y_1'x_1 \qquad RLy_2 = y_1x_2 + x_3 \tag{3.31}$$

$$z_1 = y_1y_2'$$

The logic diagram, obtained from the input and output equations, is shown in Figure 3.156 using latches, *D* flip-flops, and logic gates consisting of NOR gates, AND gates, and NAND gates. In the design module, the logic gates will be designed using the continuous assignment statement.

Recall that the continuous assignment statement **assign** is used to describe combinational logic where the output of the circuit is evaluated whenever an input changes; that is, the value of the right-hand side expression is *continuously assigned* to the left-hand side net. Continuous assignments can be used only for nets, not for register variables. A continuous assignment statement establishes a relationship between a right-hand side expression and a left-hand side net. A continuous assignment occurs outside of an **initial** or an **always** statement. The syntax for a continuous assignment statement is

assign <Optional delay> Left-hand side net = Right-hand side expression;

The left-hand side is declared as type **wire** not **reg**. When a variable on the right-hand side changes value, the right-hand side expression is evaluated and the value is assigned to the left-hand side net after the specified delay. The continuous assignment is used to place a value on a net. The *D* flip-flops will be designed using behavioral modeling and instantiated into the design module.

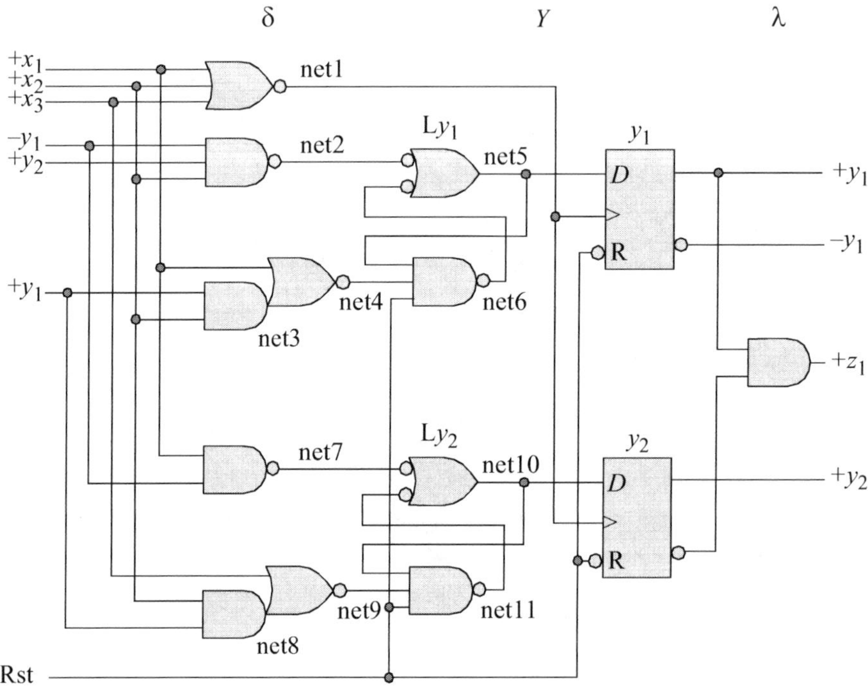

Figure 3.156 Logic diagram for the Moore machine of Example 3.18.

The dataflow design module is shown in Figure 3.157. The test bench module and outputs are shown in Figure 3.158 and Figure 3.159, respectively.

```
//dataflow for pulse-mode asm using assign

module pm_asm_moore11 (rst_n, x1, x2, x3, y1e, y2e, z1);

//define inputs and outputs
input rst_n, x1, x2, x3;
output y1e, y2e, z1;

//define internal nets
wire net1, net2, net3, net4, net5, net6, net7,
      net8, net9, net10, net11;      //continue on next page
```

Figure 3.157 Dataflow design module for Example 3.18.

```
//----------------------------------------------
//define the clock for the D flip-flops
assign   net1 = ~(x1 | x2 | x3);

//----------------------------------------------
//design the logic for latch Ly1
assign   net2 = ~(~y1e & y2e & x2),
         net3 = (y1e & x2),
         net4 = ~(x1 | net3),
         net5 = (~net2 | ~net6),
         net6 = ~(net5 & net4 & rst_n);

//instantiate the D flip-flop for y1e
d_ff_bh inst1 (rst_n, net1, net5, y1e);  //reset, clk, D, Q

//----------------------------------------------
//design the logic for latch Ly2
assign   net7 = ~(x1 & ~y1e),
         net8 = (x2 & y1e & rst_n),
         net9 = ~(x3 | net8),
         net10 = (~net7 | ~net11),
         net11 = ~(net10 & net9 & rst_n);

//instantiate the D flip-flop for y2e
d_ff_bh inst2 (rst_n, net1, net10, y2e);  //reset, clk, D, Q

//design the logic for output z1
assign   z1 = (y1e & ~y2e);

endmodule
```

Figure 3.157 (Continued)

```
//test bench for the moore pulse-mode asm
module pm_asm_moore11_tb;

reg rst_n, x1, x2, x3;      //inputs are reg for test bench
wire y1e, y2e, z1;          //outputs are wire for test bench

//display variables
initial
$monitor ("x1 x2 x3 = %b, state = %b, z1 = %b",
            {x1, x2, x3}, {y1e, y2e}, z1);
                                //continued on next page
```

Figure 3.158 Test bench module for Example 3.18.

```
//apply input vectors
initial
begin
   #0    rst_n = 1'b0;       //reset to state_a
         x1 = 1'b0;  x2 = 1'b0;  x3 = 1'b0;

   #5    rst_n = 1'b1;

   #10   x1 = 1'b1;  x2 = 1'b0;  x3 = 1'b0;  //state_b
   #10   x1 = 1'b0;  x2 = 1'b0;  x3 = 1'b0;

   #10   x1 = 1'b0;  x2 = 1'b1;  x3 = 1'b0;  //state_c
   #10   x1 = 1'b0;  x2 = 1'b0;  x3 = 1'b0;

   #10   x1 = 1'b0;  x2 = 1'b0;  x3 = 1'b1;  //state_d;
   #10   x1 = 1'b0;  x2 = 1'b0;  x3 = 1'b0;  //assert z1

   #10   x1 = 1'b1;  x2 = 1'b0;  x3 = 1'b0;  //state_a
   #10   x1 = 1'b0;  x2 = 1'b0;  x3 = 1'b0;

//----------------------------------------------------------
   #10   x1 = 1'b0;  x2 = 1'b1;  x3 = 1'b0;  //state_a
   #10   x1 = 1'b0;  x2 = 1'b0;  x3 = 1'b0;

   #10   x1 = 1'b1;  x2 = 1'b0;  x3 = 1'b0;  //state_b
   #10   x1 = 1'b0;  x2 = 1'b0;  x3 = 1'b0;

   #10   x1 = 1'b1;  x2 = 1'b0;  x3 = 1'b0;  //state_b
   #10   x1 = 1'b0;  x2 = 1'b0;  x3 = 1'b0;

   #10   x1 = 1'b0;  x2 = 1'b1;  x3 = 1'b0;  //state_c
   #10   x1 = 1'b0;  x2 = 1'b0;  x3 = 1'b0;

   #10   x1 = 1'b0;  x2 = 1'b0;  x3 = 1'b1;  //state_d;
   #10   x1 = 1'b0;  x2 = 1'b0;  x3 = 1'b0;  //assert z1

   #10   x1 = 1'b0;  x2 = 1'b0;  x3 = 1'b1;  //state_d;
   #10   x1 = 1'b0;  x2 = 1'b0;  x3 = 1'b0;  //assert z1


   #10   x1 = 1'b0;  x2 = 1'b1;  x3 = 1'b0;  //state_a
   #10   x1 = 1'b0;  x2 = 1'b0;  x3 = 1'b0;

   #10   x1 = 1'b0;  x2 = 1'b0;  x3 = 1'b1;  //state_a
   #10   x1 = 1'b0;  x2 = 1'b0;  x3 = 1'b0;
                                 //continued on next page
```

Figure 3.158 (Continued)

```
   #10   $stop;
end

//instantiate the module into the test bench
pm_asm_moore11 inst1 (rst_n, x1, x2, x3, y1e, y2e, z1);

endmodule
```

Figure 3.158 (Continued)

```
x1 x2 x3 = 000, state = 00, z1 = 0
x1 x2 x3 = 100, state = 00, z1 = 0
x1 x2 x3 = 000, state = 01, z1 = 0
x1 x2 x3 = 010, state = 01, z1 = 0
x1 x2 x3 = 000, state = 11, z1 = 0
x1 x2 x3 = 001, state = 11, z1 = 0
x1 x2 x3 = 000, state = 10, z1 = 1
x1 x2 x3 = 100, state = 10, z1 = 1
x1 x2 x3 = 000, state = 00, z1 = 0
x1 x2 x3 = 010, state = 00, z1 = 0

x1 x2 x3 = 000, state = 00, z1 = 0
x1 x2 x3 = 100, state = 00, z1 = 0
x1 x2 x3 = 000, state = 01, z1 = 0
x1 x2 x3 = 100, state = 01, z1 = 0
x1 x2 x3 = 000, state = 01, z1 = 0
x1 x2 x3 = 010, state = 01, z1 = 0
x1 x2 x3 = 000, state = 11, z1 = 0
x1 x2 x3 = 001, state = 11, z1 = 0
x1 x2 x3 = 000, state = 10, z1 = 1
x1 x2 x3 = 001, state = 10, z1 = 1
x1 x2 x3 = 000, state = 10, z1 = 1
x1 x2 x3 = 010, state = 10, z1 = 1
x1 x2 x3 = 000, state = 00, z1 = 0

x1 x2 x3 = 001, state = 00, z1 = 0
x1 x2 x3 = 000, state = 00, z1 = 0
```

Figure 3.159 Outputs for the Moore pulse-mode asynchronous sequential machine of Example 3.18.

3.4.3 *T* Flip-Flops as Storage Elements

A toggle (*T*) flip-flop — shown in Figure 3.160 — is a positive-edge-triggered storage device that will now be designed for use in the Verilog design examples in this section. A *T* flip-flop has two inputs: *T* and reset; and two outputs $+y_1$ and $-y_1$. If the flip-flop is reset, then an active pulse on the *T* input will toggle the flip-flop to the set state; if the flip-flop is set, then a pulse on the *T* input will toggle the flip-flop to the reset state.

The *T* flip-flop utilized in these examples incorporates a *D* flip-flop, an exclusive-OR circuit, and a delay circuit as a **buf** built-in primitive, as shown in Figure 3.160. The *T* input connects to the clock input of the *D* flip-flop through a delay circuit, which allows the clock input to be delayed until the signal on the *D* input has stabilized. When *T* has a value of 0, the next state is the same as the present state; when *T* has a value of 1, the next state is the complement of the present state.

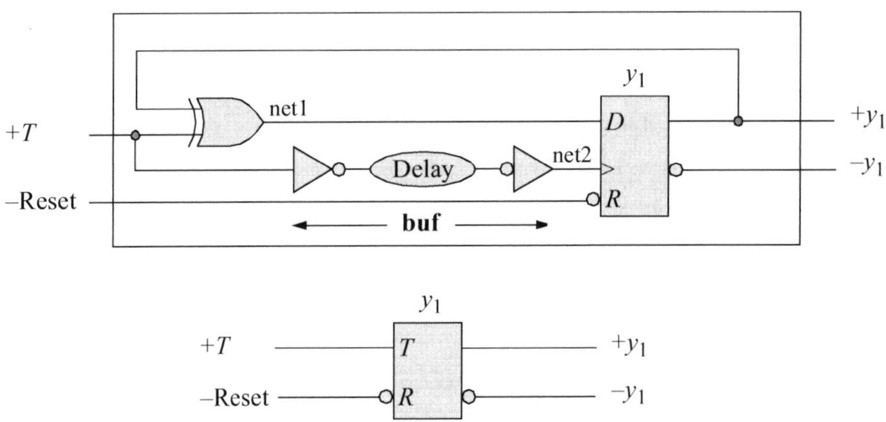

Figure 3.160 A *T* flip-flop.

The design module for the *T* flip-flop is shown in Figure 3.161. The test bench module is shown in Figure 3.162 and the outputs are shown in Figure 3.163.

```
//T flip-flop design using a D flip-flop and an xor
module t_ff_da (rst_n, t, y1);

input rst_n, t;        //define inputs and output
output y1;

wire net1, net2;       //net2 is the T input delayed

//define the logic for the T flip-flop
xor    (net1, t, y1); //flip-flop D input
buf    (net2, t);     //flip-flop clk input delayed
                                    //continued on next page
```

Figure 3.161 Design module for a *T* flip-flop.

```
//instantiate the D flip-flop
d_ff_bh inst1 (rst_n, net2, net1, y1);     //reset, clk, D, Q

endmodule
```

Figure 3.161 (Continued)

```
//test bench for the T-flip-flop
module t_ff_da_tb;

//inputs are reg for test bench
//outputs are wire for test bench
reg rst_n, t;
wire y1;

//display variables
initial
$monitor ($time, "ns, t = %b, y1 = %b", t, y1);

//define input sequence
initial
begin
   #0     rst_n = 1'b0;
          t = 1'b0;
   #5     rst_n = 1'b1;

//------------------------------
   #20    t = 1'b1;
   #10    t = 1'b0;
   #30    t = 1'b1;
   #10    t = 1'b0;
   #20    t = 1'b1;
   #10    t = 1'b0;
   #10    t = 1'b1;

   #10    $stop;

end

//instantiate the module into the test bench as a single line
t_ff_da inst1 (rst_n, t, y1);

endmodule
```

Figure 3.162 Test bench module for the *T* flip-flop.

```
0ns,    t = 0, y1 = 0
25ns,   t = 1, y1 = 1
35ns,   t = 0, y1 = 1
65ns,   t = 1, y1 = 0
75ns,   t = 0, y1 = 0
95ns,   t = 1, y1 = 1
105ns,  t = 0, y1 = 1
115ns,  t = 1, y1 = 0
```

Figure 3.163 Outputs for the *T* flip-flop.

Example 3.19 The state diagram for a Mealy pulse-mode asynchronous sequential machine is shown in Figure 3.164. The machine will be designed using built-in primitives and instantiated *T* flip-flops.

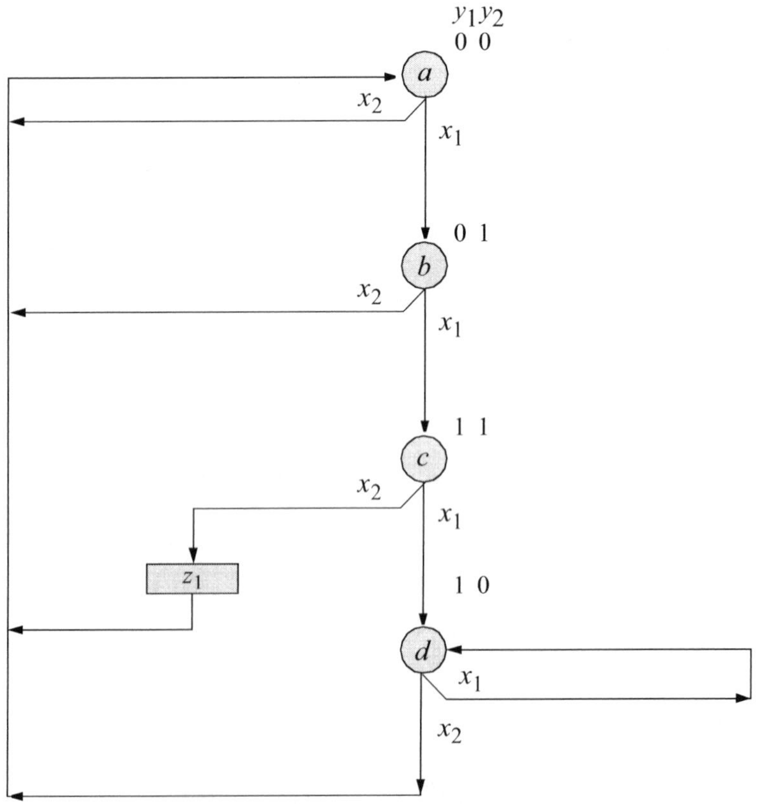

Figure 3.164 State diagram for the Mealy pulse-mode asynchronous sequential machine of Example 3.19.

The Karnaugh maps for the storage elements are shown in Figure 3.165 and the equations for y_1, y_2, and z_1 are shown in Equation 3.32. The design module using built-in primitives and instantiated T flip-flops is shown in Figure 3.166. The test bench module and the outputs are shown in Figures 3.167 and 3.168, respectively.

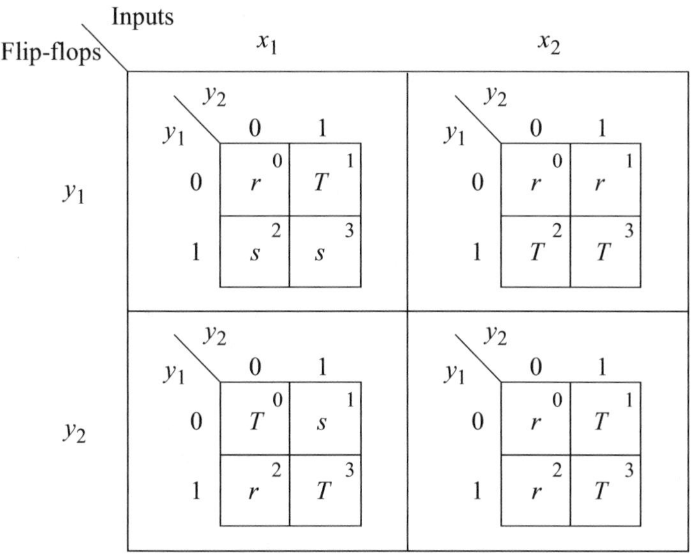

Figure 3.165 Karnaugh maps for T flip-flops y_1 and y_2 for Example 3.19.

net1 net2

$$Ty_1 = y_1'y_2x_1 + y_1x_2$$

net3

net4 net5 net6

$$Ty_2 = y_1'y_2'x_1 + y_1y_2x_1 + y_2x_2 \tag{3.32}$$

net7

$$z_1 = y_1y_2x_2$$

```
//mealy pulse-mode asm using bip and T flip-flops

module pm_asm_mealy_tff (rst_n, x1, x2, y1, y2, z1);

//define inputs and outputs
input rst_n, x1, x2;
output y1, y2, z1;

//define internal nets
wire net1, net2, net3, net4, net5, net6, net7;
wire nety1, nety2;

//------------------------------------------------------------
//design the logic for flip-flop y1
and    (net1, ~y1, y2, x1),
       (net2, y1, x2);

or     (net3, net1, net2);

//instantiate the T flip-flop
t_ff_da inst1 (rst_n, net3, nety1);    //reset, T, Q

buf    #12 (y1, nety1);  //nety1 is the output of the
                         //T flip-flop.  y1 is the output
                         //delayed by 12 time units

//------------------------------------------------------------
//design the logic for flip-flop y2
and    (net4, ~y1, ~y2, x1),
       (net5, y1, y2, x1),
       (net6, y2, x2);

or     (net7, net4, net5, net6);

//instantiate the T flip-flop
t_ff_da inst2 (rst_n, net7, nety2);    //reset, T, Q

buf    #12 (y2, nety2);  //nety2 is the output of the
                         //T flip-flop.  y2 is the output
                         //delayed by 12 time units

//------------------------------------------------------------
//design the logic for output z1
assign   z1 = y1 & y2 & x2;

endmodule
```

Figure 3.166 Design module for Example 3.19 using built-in primitives and instantiated *T* flip-flops.

```
//test bench for the mealy pulse-mode asm

module pm_asm_mealy_tff_tb;

//inputs are reg for test bench
//outputs are wire for test bench
reg rst_n, x1, x2;
wire y1, y2, z1;

//display variables
initial
$monitor ("x1 x2 = %b, state = %b, z1 = %b",
            {x1, x2}, {y1, y2}, z1);

//define input sequence
initial
begin
   #0    rst_n = 1'b0;       //reset to state_a
         x1 = 1'b0;  x2 = 1'b0;

   #5    rst_n = 1'b1;

//-----------------------------------------------------
   #10   x1 = 1'b1;  x2 =1'b0;   //state_b
   #10   x1 = 1'b0;  x2 =1'b0;

   #10   x1 = 1'b1;  x2 =1'b0;   //state_c
   #10   x1 = 1'b0;  x2 =1'b0;

   #10   x1 = 1'b0;  x2 =1'b1;   //state_a
   #10   x1 = 1'b0;  x2 =1'b0;   //assert z1

   #10   x1 = 1'b0;  x2 =1'b0;   //state_a
   #10   x1 = 1'b0;  x2 =1'b0;

//-----------------------------------------------------
   #10   x1 = 1'b1;  x2 =1'b0;   //state_b
   #10   x1 = 1'b0;  x2 =1'b0;

   #10   x1 = 1'b1;  x2 =1'b0;   //state_c
   #10   x1 = 1'b0;  x2 =1'b0;

   #10   x1 = 1'b1;  x2 =1'b0;   //state_d
   #10   x1 = 1'b0;  x2 =1'b0;

   #10   x1 = 1'b1;  x2 =1'b0;   //state_d
   #10   x1 = 1'b0;  x2 =1'b0;      //continued on next page
```

Figure 3.167 Test bench module for Example 3.19.

```
   #10   x1 = 1'b0;  x2 =1'b1;    //state_a
   #10   x1 = 1'b0;  x2 =1'b0;

   #12   $stop;

end

//instantiate the module into the test bench
pm_asm_mealy_tff inst1 (rst_n, x1, x2, y1, y2, z1);

endmodule
```

Figure 3.167 (Continued)

```
x1 x2 = 00, state = 00, z1 = 0
x1 x2 = 10, state = 00, z1 = 0
x1 x2 = 00, state = 00, z1 = 0
x1 x2 = 00, state = 01, z1 = 0
x1 x2 = 10, state = 01, z1 = 0
x1 x2 = 00, state = 01, z1 = 0
x1 x2 = 00, state = 11, z1 = 0
x1 x2 = 01, state = 11, z1 = 1
x1 x2 = 00, state = 11, z1 = 0
x1 x2 = 00, state = 00, z1 = 0
x1 x2 = 10, state = 00, z1 = 0
x1 x2 = 00, state = 00, z1 = 0


x1 x2 = 00, state = 01, z1 = 0
x1 x2 = 10, state = 01, z1 = 0
x1 x2 = 00, state = 01, z1 = 0
x1 x2 = 00, state = 11, z1 = 0
x1 x2 = 10, state = 11, z1 = 0
x1 x2 = 00, state = 11, z1 = 0
x1 x2 = 00, state = 10, z1 = 0
x1 x2 = 10, state = 10, z1 = 0
x1 x2 = 00, state = 10, z1 = 0
x1 x2 = 01, state = 10, z1 = 0
x1 x2 = 00, state = 10, z1 = 0
x1 x2 = 00, state = 00, z1 = 0
```

Figure 3.168 Outputs for Example 3.19.

Example 3.20 The state diagram for a Mealy pulse-mode asynchronous sequential machine is shown in Figure 3.169. The machine will be designed using built-in primitive logic gates and instantiated *T* flip-flops. The Karnaugh maps for the *T* flip-flops are shown in Figure 3.170.

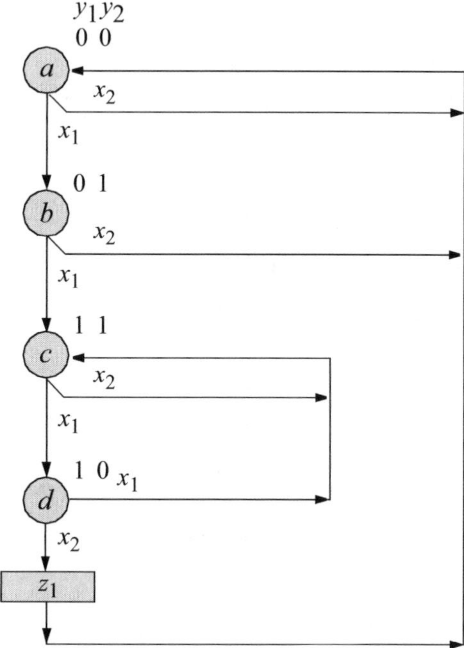

Figure 3.169 State diagram for the Mealy machine of Example 3.20.

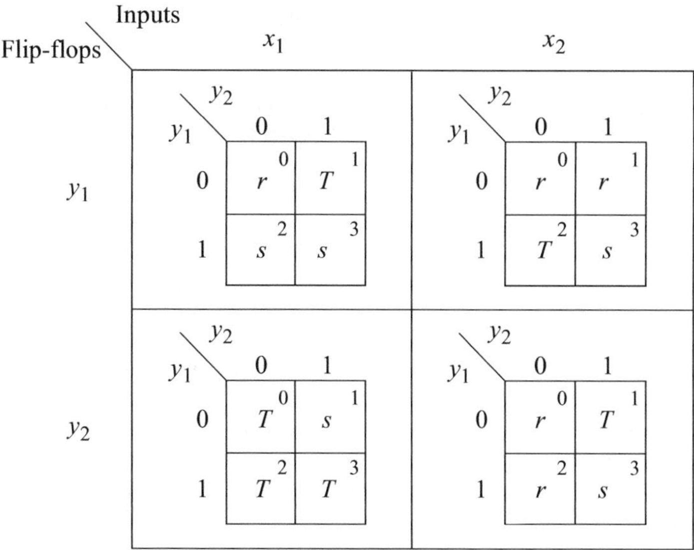

Figure 3.170 Karnaugh maps for the *T* flip-flops of the Mealy machine.

The equations for the T flip-flops and output z_1 are shown in Equation 3.33. The structural design module using built-in primitives and instantiated T flip-flops is shown in Figure 3.171. The test bench module and the outputs are shown in Figures 3.172 and 3.173, respectively.

$$Ty_1 = \underbrace{y_1'y_2x_1}_{\text{net1}} + \underbrace{y_1y_2'x_2}_{\text{net2}} \quad (\text{net3})$$

$$Ty_2 = \underbrace{y_2'x_1}_{\text{net4}} + \underbrace{y_1x_1}_{\text{net5}} + \underbrace{y_1'y_2x_2}_{\text{net6}} \quad (\text{net7}) \tag{3.33}$$

$$z_1 = y_1y_2'x_2$$

```
//mealy pulse-mode asm using bip and T flip-flops

module pm_asm_mealy3_tff (rst_n, x1, x2, y1, y2, z1);

//define inputs and outputs
input rst_n, x1, x2;
output y1, y2, z1;

//define internal nets
wire net1, net2, net3, net4, net5, net6, net7;
wire nety1, nety2;

//-----------------------------------------------------------
//design the logic for flip-flop y1
and    (net1, ~y1, y2, x1),
       (net2, y1, ~y2, x2);

or     (net3, net1, net2);

//instantiate the T flip-flop for y1
t_ff_da inst1 (rst_n, net3, nety1);     //reset, T, Q

buf    #12 (y1, nety1);       //nety1 is the output of the
                              //T flip-flop.  y1 is the output
                              //delayed by 12 time units
                                        //continued on next page
```

Figure 3.171 Structural design module for the Mealy machine of Example 3.20.

```
//design the logic for flip-flop y2
and    (net4, ~y2, x1),
       (net5, y1, x1),
       (net6, ~y1, y2, x2);

or     (net7, net4, net5, net6);

//instantiate the T flip-flop for y2
t_ff_da inst2 (rst_n, net7, nety2);    //reset, T, Q

buf    #12 (y2, nety2);      //nety2 is the output of the
                             //T flip-flop.  y2 is the output
                             //delayed by 12 time units

//------------------------------------------------------------
//design the logic for output z1
assign   z1 = y1 & ~y2 & x2;

endmodule
```

Figure 3.171 (Continued)

```
//test bench for the mealy pulse-mode asm
module pm_asm_mealy3_tff_tb;

//inputs are reg for test bench
//outputs are wire for test bench
reg rst_n, x1, x2;
wire y1, y2, z1;

//display variables
initial
$monitor ("x1 x2 = %b, state = %b, z1 = %b",
            {x1, x2}, {y1, y2}, z1);

//define input sequence
initial
begin
   #0    rst_n = 1'b0;      //reset to state_a
         x1 = 1'b0;  x2 = 'b0;

   #5    rst_n = 1'b1;
                              //continued on next page
```

Figure 3.172 Test bench module for the Mealy machine of Example 3.20.

```
//--------------------------------------------------------
   #10   x1 = 1'b1;  x2 = 1'b0;  //state_b
   #10   x1 = 1'b0;  x2 = 1'b0;

   #10   x1 = 1'b1;  x2 = 1'b0;  //state_c
   #10   x1 = 1'b0;  x2 = 1'b0;

   #10   x1 = 1'b1;  x2 = 1'b0;  //state_d
   #10   x1 = 1'b0;  x2 = 1'b0;

   #10   x1 = 1'b0;  x2 = 1'b1;  //assert z1, state_a
   #10   x1 = 1'b0;  x2 = 1'b0;

//--------------------------------------------------------
   #10   x1 = 1'b0;  x2 = 1'b1;  //state_a
   #10   x1 = 1'b0;  x2 = 1'b0;

   #10   x1 = 1'b1;  x2 = 1'b0;  //state_b
   #10   x1 = 1'b0;  x2 = 1'b0;

   #10   x1 = 1'b0;  x2 = 1'b1;  //state_a
   #10   x1 = 1'b0;  x2 = 1'b0;

   #10   x1 = 1'b1;  x2 = 1'b0;  //state_b
   #10   x1 = 1'b0;  x2 = 1'b0;

   #10   x1 = 1'b1;  x2 = 1'b0;  //state_c
   #10   x1 = 1'b0;  x2 = 1'b0;

   #10   x1 = 1'b0;  x2 = 1'b1;  //state_c
   #10   x1 = 1'b0;  x2 = 1'b0;

   #10   x1 = 1'b1;  x2 = 1'b0;  //state_d
   #10   x1 = 1'b0;  x2 = 1'b0;

   #10   x1 = 1'b0;  x2 = 1'b1;  //assert z1, state_a
   #10   x1 = 1'b0;  x2 = 1'b0;

   #10   $stop;

end

//instantiate the module into the test bench
pm_asm_mealy3_tff inst1 (rst_n, x1, x2, y1, y2, z1);

endmodule
```

Figure 3.172 (Continued)

```
x1 x2 = 00, state = 00, z1 = 0
x1 x2 = 10, state = 00, z1 = 0
x1 x2 = 00, state = 00, z1 = 0
x1 x2 = 00, state = 01, z1 = 0
x1 x2 = 10, state = 01, z1 = 0
x1 x2 = 00, state = 01, z1 = 0
x1 x2 = 00, state = 11, z1 = 0
x1 x2 = 10, state = 11, z1 = 0
x1 x2 = 00, state = 11, z1 = 0
x1 x2 = 00, state = 10, z1 = 0
x1 x2 = 01, state = 10, z1 = 1
x1 x2 = 00, state = 10, z1 = 0
x1 x2 = 00, state = 00, z1 = 0
x1 x2 = 01, state = 00, z1 = 0
x1 x2 = 00, state = 00, z1 = 0

x1 x2 = 10, state = 00, z1 = 0
x1 x2 = 00, state = 00, z1 = 0
x1 x2 = 00, state = 01, z1 = 0
x1 x2 = 01, state = 01, z1 = 0
x1 x2 = 00, state = 01, z1 = 0
x1 x2 = 00, state = 00, z1 = 0
x1 x2 = 10, state = 00, z1 = 0
x1 x2 = 00, state = 00, z1 = 0
x1 x2 = 00, state = 01, z1 = 0
x1 x2 = 10, state = 01, z1 = 0
x1 x2 = 00, state = 01, z1 = 0
x1 x2 = 00, state = 11, z1 = 0
x1 x2 = 01, state = 11, z1 = 0
x1 x2 = 00, state = 11, z1 = 0
x1 x2 = 10, state = 11, z1 = 0
x1 x2 = 00, state = 11, z1 = 0
x1 x2 = 00, state = 10, z1 = 0
x1 x2 = 01, state = 10, z1 = 1
x1 x2 = 00, state = 10, z1 = 0
x1 x2 = 00, state = 00, z1 = 0
```

Figure 3.173 Outputs for the Mealy machine of Example 3.20.

Example 3.21 The state diagram for a Moore pulse-mode asynchronous sequential machine is shown in Figure 3.174. The Karnaugh maps for *T* flip-flops y_1 and y_2 are shown in Figure 3.175. The equations for the *T* flip-flops and output z_1 are shown in Equation 3.34. The logic diagram for the Moore machine is shown in Figure 3.176.

The structural design module using instantiated logic gates that were designed using dataflow modeling and instantiated *T* flip-flops is shown in Figure 3.177. The test bench module and the outputs are shown in Figures 3.178 and 3.179, respectively.

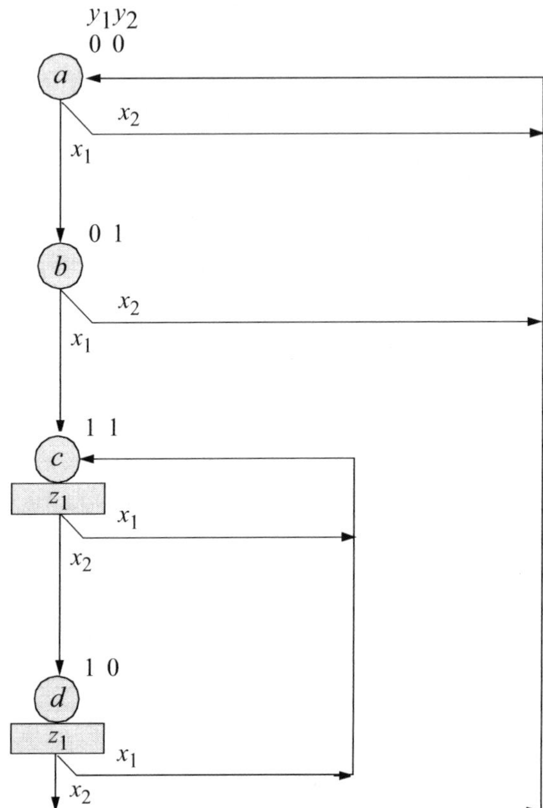

Figure 3.174 State diagram for the Moore machine of Example 3.21.

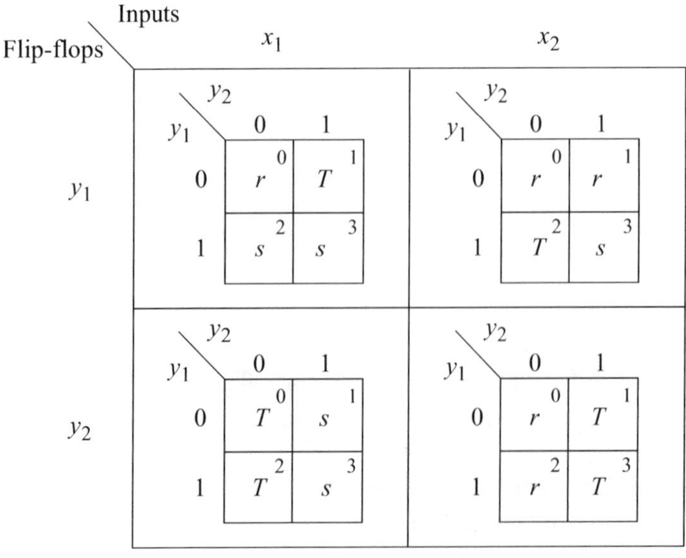

Figure 3.175 Karnaugh maps for the T flip-flops of Example 3.21.

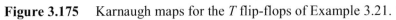

$$Ty_1 = \overbrace{y_1'y_2x_1}^{\text{net1}} + \overbrace{y_1y_2'x_2}^{\text{net2}} \quad (\text{net3})$$

$$Ty_2 = \overbrace{y_2'x_1}^{\text{net4}} + \overbrace{y_2x_2}^{\text{net5}} \quad (\text{net7}) \tag{3.34}$$

$$z_1 = y_1$$

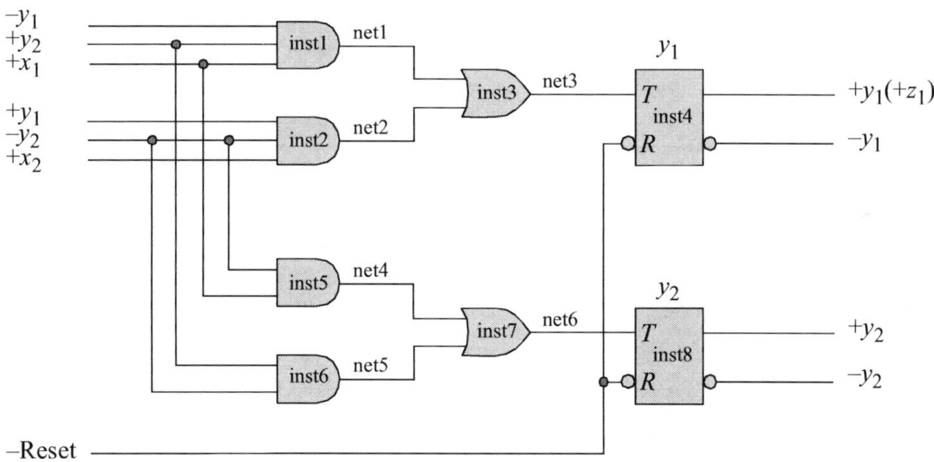

Figure 3.176 Logic diagram for the Moore machine of Example 3.21.

```
//structural for moore pulse-mode asm
module pm_asm_moore_tff (rst_n, x1, x2, y1, y2, z1);

//define inputs and outputs
input rst_n, x1, x2;
output y1, y2, z1;

//define internal nets
wire net1, net2, net3, net4, net5, net6;
wire nety1, nety2;
                              //continued on next page
```

Figure 3.177 Structural design module for the Moore machine of Example 3.21.

```
//-------------------------------------------------------------
//design the logic for T flip-flop y1
and3_df  inst1 (~y1, y2, x1, net1),
         inst2 (y1, ~y2, x2, net2);

or2_df   inst3 (net1, net2, net3);

//instantiate the T flip-flop for y1
t_ff_da inst4 (rst_n, net3, nety1);    //reset, T, Q

buf   #12 (y1, nety1);     //nety1 is the output of the
                           //T flip-flop.  y1 is the output
                           //delayed by 12 time units

//-------------------------------------------------------------
//design the logic for T flip-flop y2
and2_df  inst5 (~y2, x1, net4),
         inst6 (y2, x2, net5);

or2_df   inst7 (net4, net5, net6);

//instantiate the T flip-flop for y2
t_ff_da inst8 (rst_n, net6, nety2);    //reset, T, Q

buf   #12 (y2, nety2);     //nety2 is the output of the
                           //T flip-flop.  y2 is the output
                           //delayed by 12 time units

//-------------------------------------------------------------
//design the logic for output z1
assign   z1 = y1;

endmodule
```

Figure 3.177 (Continued)

```
//test bench for the moore pulse-mode asm
module pm_asm_moore_tff_tb;

//inputs are reg for test bench
//outputs are wire for test bench
reg   rst_n, x1, x2;
wire  y1, y2, z1;
                                   //continued on next page
```

Figure 3.178 Test bench module for the Moore machine of Example 3.21.

```
//display variables
initial
$monitor ("x1 x2 = %b, state = %b, z1 = %b",
            {x1, x2}, {y1, y2}, z1);

//define input sequence
initial
begin
   #0    rst_n = 1'b0;       //reset to state_a
         x1 = 1'b0;  x2 = 1'b0;

   #5    rst_n = 1'b1;

//------------------------------------------------------------
   #10   x1 = 1'b1;  x2= 'b0;     //state_b
   #10   x1 = 1'b0;  x2= 'b0;

   #10   x1 = 1'b1;  x2= 'b0;     //state_c, assert z1
   #10   x1 = 1'b0;  x2= 'b0;

   #10   x1 = 1'b0;  x2= 'b1;     //state_d, assert z1
   #10   x1 = 1'b0;  x2= 'b0;

   #10   x1 = 1'b0;  x2= 'b1;     //state_a
   #10   x1 = 1'b0;  x2= 'b0;

//------------------------------------------------------------
   #10   x1 = 1'b0;  x2= 'b1;     //state_a
   #10   x1 = 1'b0;  x2= 'b0;

   #10   x1 = 1'b1;  x2= 'b0;     //state_b
   #10   x1 = 1'b0;  x2= 'b0;

   #10   x1 = 1'b0;  x2= 'b1;     //state_a
   #10   x1 = 1'b0;  x2= 'b0;

//------------------------------------------------------------
   #10   x1 = 1'b1;  x2= 'b0;     //state_b
   #10   x1 = 1'b0;  x2= 'b0;

   #10   x1 = 1'b1;  x2= 'b0;     //state_c, assert z1
   #10   x1 = 1'b0;  x2= 'b0;

   #10   x1 = 1'b1;  x2= 'b0;     //state_c, assert z1
   #10   x1 = 1'b0;  x2= 'b0;
                                  //continued on next page
```

Figure 3.178 (Continued)

```
    #10   x1 = 1'b0;  x2= 'b1;    //state_d, assert z1
    #10   x1 = 1'b0;  x2= 'b0;

    #10   x1 = 1'b1;  x2= 'b0;    //state_c, assert z1
    #10   x1 = 1'b0;  x2= 'b0;

    #10   x1 = 1'b0;  x2= 'b1;    //state_d, assert z1
    #10   x1 = 1'b0;  x2= 'b0;

    #10   x1 = 1'b0;  x2= 'b1;    //state_a
    #10   x1 = 1'b0;  x2= 'b0;

    #10   $stop;
end

//instantiate the module into the test bench
pm_asm_moore_tff inst1 (rst_n, x1, x2, y1, y2, z1);

endmodule
```

Figure 3.178 (Continued)

```
x1 x2 = 00, state = 00, z1 = 0
x1 x2 = 10, state = 00, z1 = 0
x1 x2 = 00, state = 00, z1 = 0
x1 x2 = 00, state = 01, z1 = 0
x1 x2 = 10, state = 01, z1 = 0
x1 x2 = 00, state = 01, z1 = 0
x1 x2 = 00, state = 11, z1 = 1
x1 x2 = 01, state = 11, z1 = 1
x1 x2 = 00, state = 11, z1 = 1
x1 x2 = 00, state = 10, z1 = 1
x1 x2 = 01, state = 10, z1 = 1
x1 x2 = 00, state = 10, z1 = 1

x1 x2 = 00, state = 00, z1 = 0
x1 x2 = 01, state = 00, z1 = 0
x1 x2 = 00, state = 00, z1 = 0
x1 x2 = 10, state = 00, z1 = 0
x1 x2 = 00, state = 00, z1 = 0
x1 x2 = 00, state = 01, z1 = 0
x1 x2 = 01, state = 01, z1 = 0
x1 x2 = 00, state = 01, z1 = 0
x1 x2 = 00, state = 00, z1 = 0
x1 x2 = 10, state = 00, z1 = 0    //continued on next page
```

Figure 3.179 Outputs for the Moore machine of Example 3.21.

```
x1 x2 = 00, state = 00, z1 = 0
x1 x2 = 00, state = 01, z1 = 0
x1 x2 = 10, state = 01, z1 = 0
x1 x2 = 00, state = 01, z1 = 0
x1 x2 = 00, state = 11, z1 = 1
x1 x2 = 10, state = 11, z1 = 1
x1 x2 = 00, state = 11, z1 = 1
x1 x2 = 01, state = 11, z1 = 1
x1 x2 = 00, state = 11, z1 = 1
x1 x2 = 00, state = 10, z1 = 1
x1 x2 = 10, state = 10, z1 = 1
x1 x2 = 00, state = 10, z1 = 1
x1 x2 = 00, state = 11, z1 = 1
x1 x2 = 01, state = 11, z1 = 1
x1 x2 = 00, state = 11, z1 = 1
x1 x2 = 00, state = 10, z1 = 1
x1 x2 = 01, state = 10, z1 = 1
x1 x2 = 00, state = 10, z1 = 1
x1 x2 = 00, state = 00, z1 = 0
```

Figure 3.179 (Continued)

Example 3.22 The state diagram for a Moore pulse-mode asynchronous sequential machine is shown in Figure 3.180. The machine will be designed using built-in primitives and instantiated *T* flip-flops.

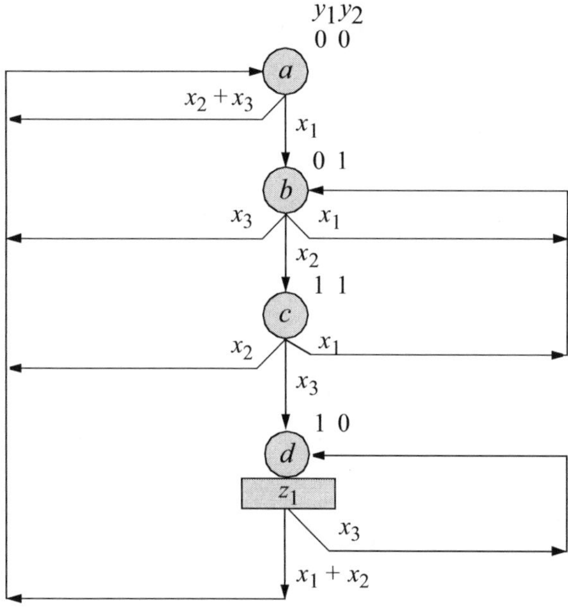

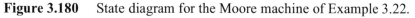

Figure 3.180 State diagram for the Moore machine of Example 3.22.

The Karnaugh maps for the T flip-flops are shown in Figure 3.181. The equations for the T flip-flops and output z_1 are shown in Equation 3.35. The logic diagram for the Moore machine is shown in Figure 3.182 with the net names. The design module for the Moore machine using built-in primitives and T flip-flops is shown in Figure 3.183. The test bench module and the outputs are shown in Figures 3.184 and 3.185, respectively.

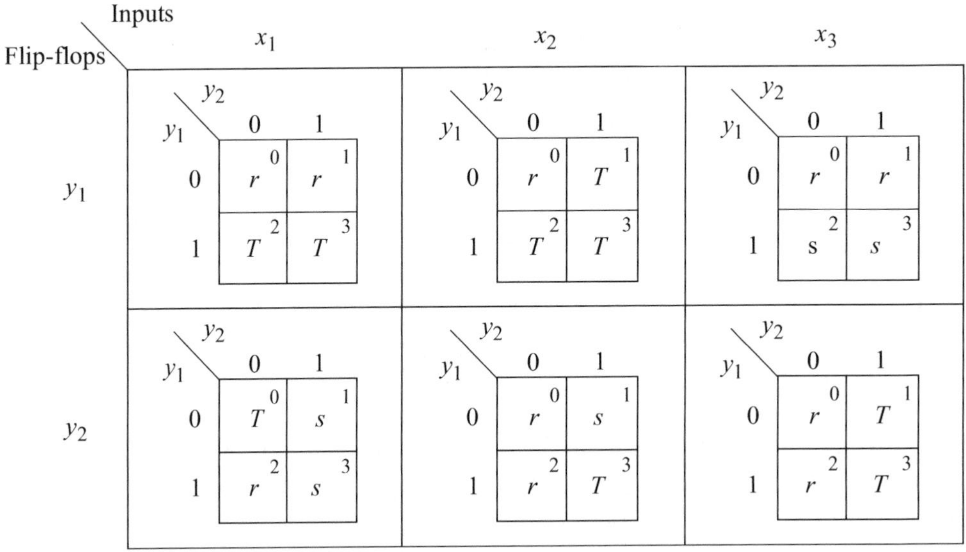

Figure 3.181 Karnaugh maps for the T flip-flops for Example 3.22.

$$Ty_1 = \underbrace{\overbrace{y_1x_1}^{net1} + \overbrace{y_1x_2}^{net2} + \overbrace{y_2x_2}^{net3}}_{net4}$$

$$Ty_2 = \underbrace{\overbrace{y_1'y_2'x_1}^{net5} + \overbrace{y_1y_2x_2}^{net6} + \overbrace{y_2x_3}^{net7}}_{net8} \qquad (3.35)$$

$$z_1 = y_1y_2'$$

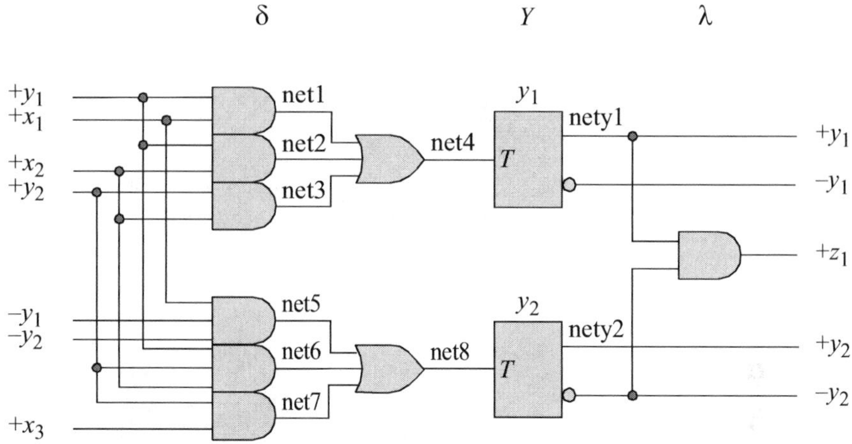

Figure 3.182 Logic diagram for the Moore machine of Example 3.22.

```
//moore pulse-mode asm using bip and T flip-flops

module pm_asm_moore2_tff (rst_n, x1, x2, x3, y1, y2, z1);

//define inputs and outputs
input rst_n, x1, x2, x3;
output y1, y2, z1;

//define internal nets
wire net1, net2, net3, net4, net5, net6, net7;
wire nety1, nety2;

//-----------------------------------------------------------
//design the logic for flip-flop y1
and    (net1, y1, x1),
       (net2, y1, x2),
       (net3, y2, x2);

or     (net4, net1, net2, net3);

//instantiate the T flip-flop
t_ff_da inst1 (rst_n, net4, nety1);     //reset, T, Q

buf    #12 (y1, nety1);   //nety1 is the output of the
                          //T flip-flop.  y1 is the output
                          //delayed by 12 time units
                                  //continued on next page
```

Figure 3.183 Design module using built-in primitives and *T* flip-flops.

```
//-------------------------------------------------------------
//design the logic for flip-flop y2
and   (net5, ~y1, ~y2, x1),
      (net6, y1, y2, x2),
      (net7, y2, x3);

or    (net8, net5, net6, net7);

//instantiate the T flip-flop
t_ff_da inst2 (rst_n, net8, nety2);     //reset, T, Q

buf   #12 (y2, nety2);  //nety2 is the output of the
                        //T flip-flop.  y2 is the output
                        //delayed by 12 time units

//-------------------------------------------------------------
//design the logic for output z1
assign z1 = y1 & ~y2;

endmodule
```

Figure 3.183 (Continued)

```
//test bench for moore pulse-mode asm

module pm_asm_moore2_tff_tb;

//inputs are reg for test bench
//outputs are wire for test bench
reg rst_n, x1, x2, x3;
wire y1, y2, z1;

//display variables
initial
$monitor ("x1 x2 x3 = %b, state = %b, z1 = %b",
            {x1, x2, x3}, {y1, y2}, z1);

//define input sequence
initial
begin
   #0    rst_n = 1'b0;      //reset to state_a
         x1 = 1'b0;  x2 = 1'b0;  x3 = 1'b0;

   #5    rst_n = 1'b1;
                                  //continued on next page
```

Figure 3.184 Test bench module for the Moore machine of Example 3.22.

```
    #10   x1 = 1'b1;  x2 = 1'b0;  x3 = 1'b0;  //state_b
    #10   x1 = 1'b0;  x2 = 1'b0;  x3 = 1'b0;

    #10   x1 = 1'b0;  x2 = 1'b1;  x3 = 1'b0;  //state_c
    #10   x1 = 1'b0;  x2 = 1'b0;  x3 = 1'b0;

    #10   x1 = 1'b0;  x2 = 1'b0;  x3 = 1'b1;  //state_d,
                                               //assert z1
    #10   x1 = 1'b0;  x2 = 1'b0;  x3 = 1'b0;

    #10   x1 = 1'b1;  x2 = 1'b0;  x3 = 1'b0;  //state_a
    #10   x1 = 1'b0;  x2 = 1'b0;  x3 = 1'b0;

//---------------------------------------------------------
    #10   x1 = 1'b0;  x2 = 1'b1;  x3 = 1'b0;  //state_a
    #10   x1 = 1'b0;  x2 = 1'b0;  x3 = 1'b0;

    #10   x1 = 1'b0;  x2 = 1'b0;  x3 = 1'b1;  //state_a
    #10   x1 = 1'b0;  x2 = 1'b0;  x3 = 1'b0;

    #10   x1 = 1'b1;  x2 = 1'b0;  x3 = 1'b0;  //state_b
    #10   x1 = 1'b0;  x2 = 1'b0;  x3 = 1'b0;

    #10   x1 = 1'b1;  x2 = 1'b0;  x3 = 1'b0;  //state_b
    #10   x1 = 1'b0;  x2 = 1'b0;  x3 = 1'b0;

    #10   x1 = 1'b0;  x2 = 1'b1;  x3 = 1'b0;  //state_c
    #10   x1 = 1'b0;  x2 = 1'b0;  x3 = 1'b0;

    #10   x1 = 1'b0;  x2 = 1'b0;  x3 = 1'b1;  //state_d,
                                               //assert z1
    #10   x1 = 1'b0;  x2 = 1'b0;  x3 = 1'b0;

    #10   x1 = 1'b0;  x2 = 1'b0;  x3 = 1'b1;  //state_d,
                                               //assert z1
    #10   x1 = 1'b0;  x2 = 1'b0;  x3 = 1'b0;

    #10   x1 = 1'b1;  x2 = 1'b0;  x3 = 1'b0;  //state_a
    #10   x1 = 1'b0;  x2 = 1'b0;  x3 = 1'b0;

    #12   $stop;
end

//instantiate the module into the test bench
pm_asm_moore2_tff inst1 (rst_n, x1, x2, x3, y1, y2, z1);

endmodule
```

Figure 3.184 (Continued)

```
x1 x2 x3 = 000, state = 00, z1 = 0
x1 x2 x3 = 100, state = 00, z1 = 0
x1 x2 x3 = 000, state = 00, z1 = 0
x1 x2 x3 = 000, state = 01, z1 = 0
x1 x2 x3 = 010, state = 01, z1 = 0
x1 x2 x3 = 000, state = 01, z1 = 0
x1 x2 x3 = 000, state = 11, z1 = 0
x1 x2 x3 = 001, state = 11, z1 = 0
x1 x2 x3 = 000, state = 11, z1 = 0
x1 x2 x3 = 000, state = 10, z1 = 1
x1 x2 x3 = 100, state = 10, z1 = 1
x1 x2 x3 = 000, state = 10, z1 = 1

x1 x2 x3 = 000, state = 00, z1 = 0
x1 x2 x3 = 010, state = 00, z1 = 0
x1 x2 x3 = 000, state = 00, z1 = 0
x1 x2 x3 = 001, state = 00, z1 = 0
x1 x2 x3 = 000, state = 00, z1 = 0
x1 x2 x3 = 100, state = 00, z1 = 0
x1 x2 x3 = 000, state = 00, z1 = 0
x1 x2 x3 = 000, state = 01, z1 = 0
x1 x2 x3 = 100, state = 01, z1 = 0
x1 x2 x3 = 000, state = 01, z1 = 0
x1 x2 x3 = 010, state = 01, z1 = 0
x1 x2 x3 = 000, state = 01, z1 = 0
x1 x2 x3 = 000, state = 11, z1 = 0
x1 x2 x3 = 001, state = 11, z1 = 0
x1 x2 x3 = 000, state = 11, z1 = 0

x1 x2 x3 = 000, state = 10, z1 = 1
x1 x2 x3 = 001, state = 10, z1 = 1
x1 x2 x3 = 000, state = 10, z1 = 1
x1 x2 x3 = 100, state = 10, z1 = 1
x1 x2 x3 = 000, state = 10, z1 = 1

x1 x2 x3 = 000, state = 00, z1 = 0
```

Figure 3.185 Outputs for the Moore machine of Example 3.22.

3.5 Problems

3.1 Design an 8-bit Johnson counter using the **case** statement. A Johnson counter generates a counting sequence in which any two contiguous numbers differ by only one variable, as shown below for a 3-bit Johnson counter. Obtain the behavioral design module, the test bench module, and the outputs.

000, 001, 011, 111, 011, 001, 000

3.2 Design a counter that counts in the sequence shown below using instantiated logic gates designed using dataflow modeling and instantiated D flip-flops designed using behavioral modeling. Obtain the structural design module, the test bench module, and the outputs.

000, 111, 001, 110, 010, 101, 011, 100, 000

3.3 Design a Mealy synchronous sequential machine that will generate an output z_1 whenever the sequence 1001 is detected on a serial input line x_1. Overlapping sequences are valid. For example, the following sequence will assert output z_1 three times: . . . 01101001000110010010 . . . . Use built-in primitives for the logic gates and instantiated D flip-flops for the storage devices. Obtain the structural design module, the test bench module, and the outputs.

3.4 The state diagram for a Moore synchronous sequential machine is shown below with three inputs, x_1, x_2, and x_3. There are two outputs, z_1 and z_2. Obtain the structural design module using built-in primitives and instantiated D flip-flops that were designed using behavioral modeling. Obtain the test bench module and the outputs. Use the **$random** system task for the test bench module to generate a random value for certain inputs when their value can be considered a "don't care" — either 0 or 1. Use *clk'* to gate the outputs to avoid possible glitches.

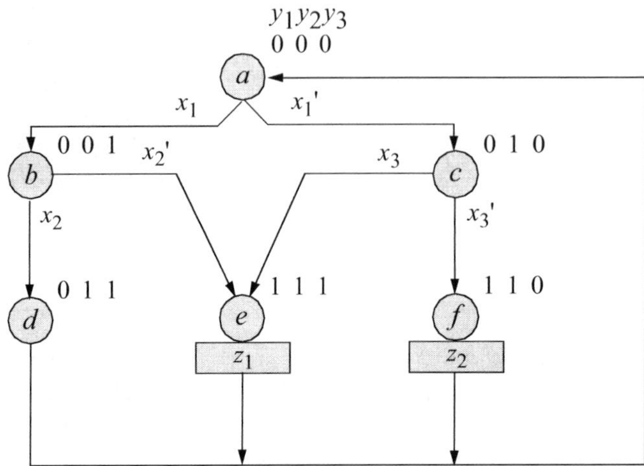

3.5 Given the state diagram shown below for a Moore synchronous sequential machine, design the machine using behavioral modeling with the **case** statement. Obtain the design module, the test bench module, and the outputs. In the design module, the values of the input variables x_1, x_2, and x_3 can be declared as either (x_1==1) or as (x_1) for example, where (x_1) implies a value of 1. The input variables can also be declared as (x_1==0) or as (~x_1), where (~x_1) implies a value of 0.

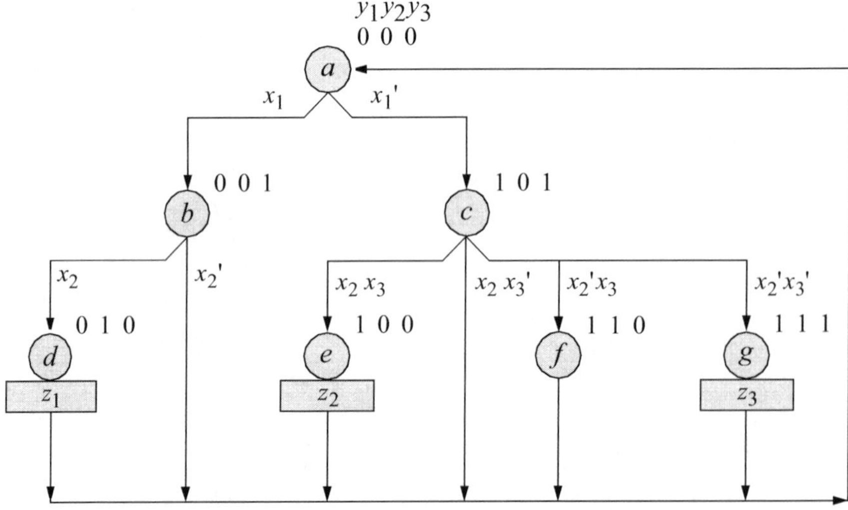

3.6 Given the state diagram shown below for a synchronous sequential machine containing Moore and Mealy outputs, synthesize the machine using linear-select multiplexers and D flip-flops that were designed using behavioral modeling. Obtain the structural design module, the test bench module, and the outputs.

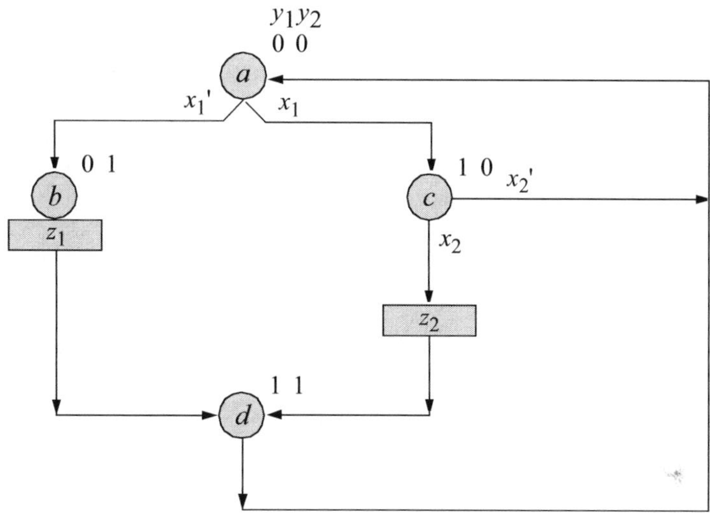

A linear-select multiplexer is one where the flip-flop outputs connect to the multiplexer select inputs in a one-to-one mapping as shown below. The combinational logic which connects to the input of the multiplexer array is either very elementary or nonexistent.

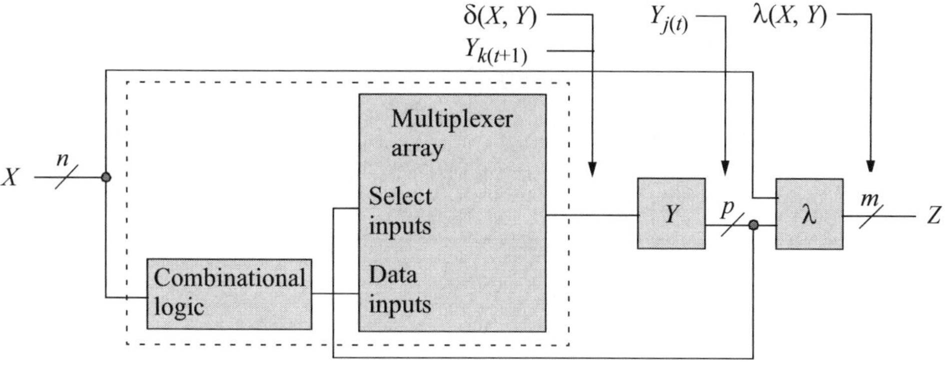

3.7 Given the state diagram for a Mealy synchronous sequential machine shown below, design the machine using the dataflow continuous assign statement **assign**. Obtain the dataflow design module, the test bench module, and the outputs.

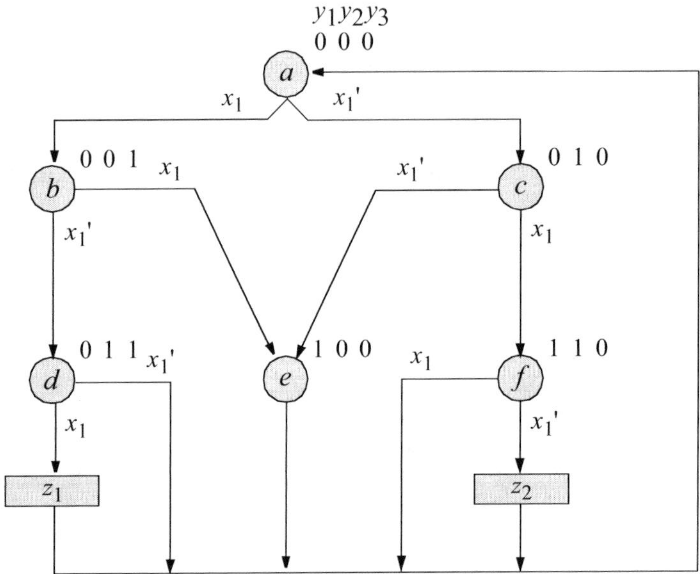

3.8 The timing diagram for an asynchronous sequential machine is shown below. Obtain the primitive flow table, the merger diagram, the merged flow table, the excitation map, and the output map. Design the structural design module using built-in primitives and the test bench module. Obtain the outputs.

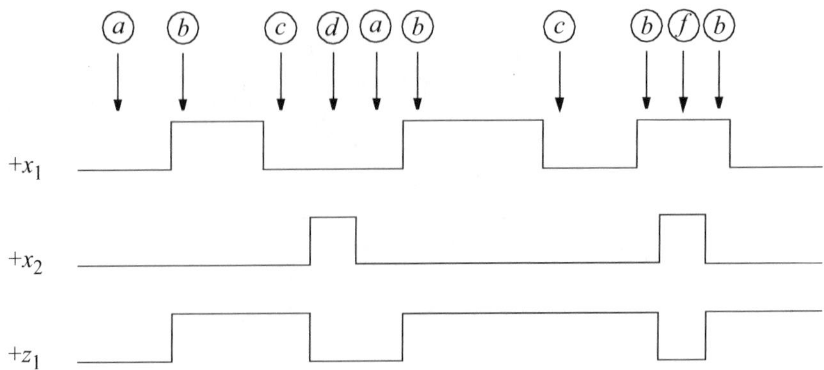

3.9 Synthesize an asynchronous sequential machine using built-in primitives which has two inputs x_1 and x_2 and one output z_1. Output z_1 will be asserted coincident with the assertion of the first x_2 pulse and will remain active for the duration of the first x_2 pulse. The output will be asserted only if the assertion of x_1 precedes the assertion of x_2. Input x_1 will not become de-asserted while x_2 is asserted. The timing diagram is shown below. Obtain the structural design module, the test bench module, and the outputs.

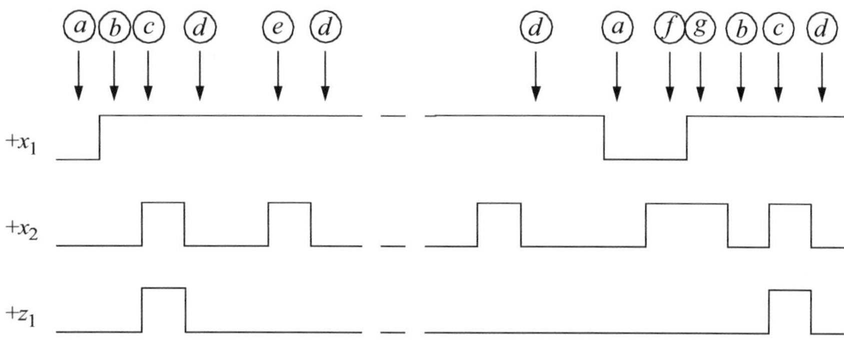

3.10 The timing diagram for a Mealy asynchronous sequential machine is shown below. Design the machine using instantiated logic gates that were designed using dataflow modeling. Obtain the structural design module, the test bench module, and the outputs.

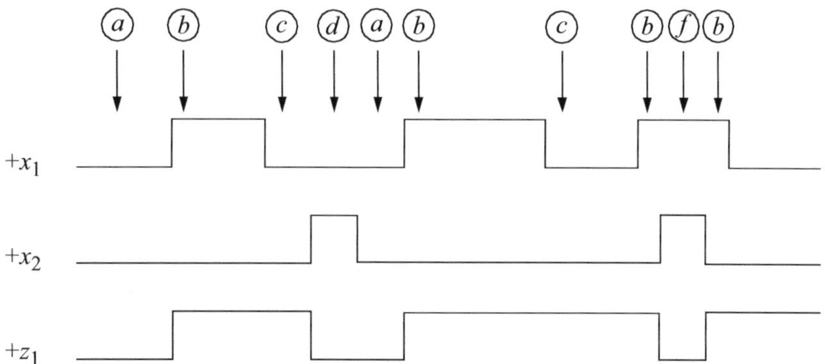

3.11 The timing diagram for a Mealy asynchronous sequential machine is shown below with one input x_1 and two outputs z_1 and z_2. Output z_1 toggles on the rising edge of input x_1. Output z_2 toggles on the falling edge of x_1. Use structural modeling with built-in primitives to obtain the design module. Obtain the test bench module and the outputs.

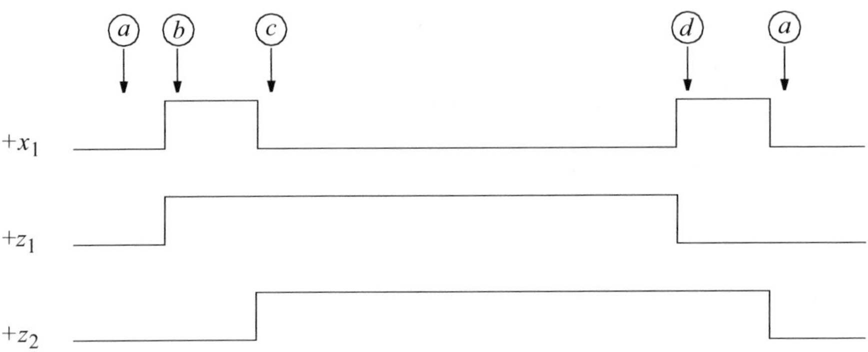

3.12 The waveforms for a Mealy asynchronous sequential machine are shown below with two inputs x_1 and x_2 and one output z_1. Design the machine using dataflow modeling with the **assign** statement. Obtain the design module, the test bench module, and the outputs.

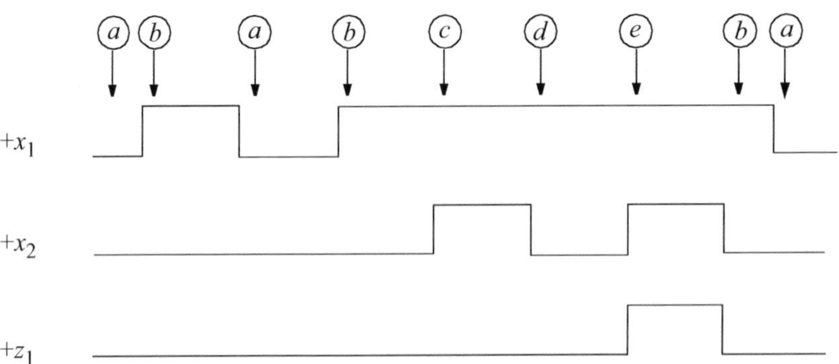

3.13 Given the state diagram shown below for a Moore–Mealy asynchronous sequential machine, design the machine using behavioral modeling with the **case** statement. Then, obtain the test bench and the outputs.

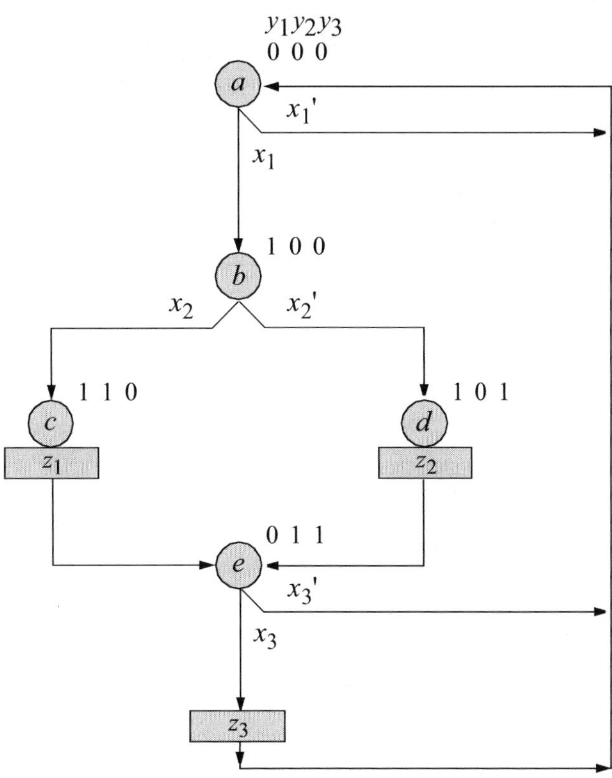

3.14 The state diagram for a Mealy pulse-mode asynchronous sequential machine is shown below. Design the machine using built-in primitives and instantiated D flip-flops. Obtain the structural design module, the test bench module, and the outputs.

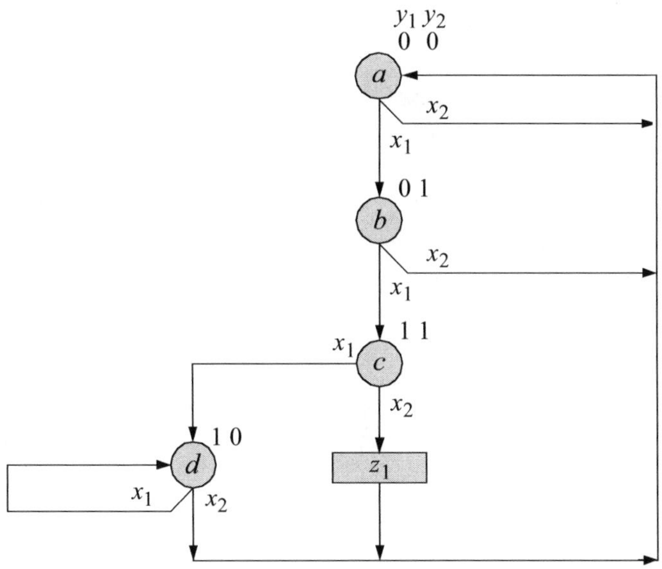

3.15 The state diagram for a Mealy pulse-mode asynchronous sequential machine is shown below. Synthesize the machine using logic gates that were designed using dataflow modeling and D flip-flops that were designed using behavioral modeling. Obtain the structural design module, the test bench module, and the outputs.

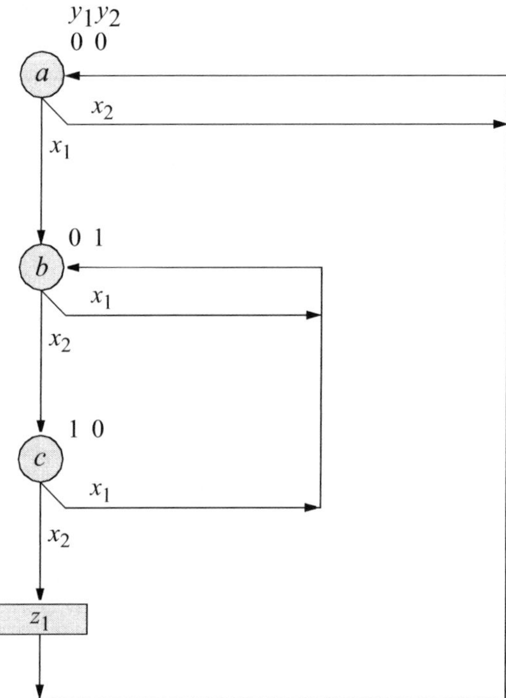

3.16 The state diagram for a Moore pulse-mode asynchronous sequential machine is shown below. Design the machine using built-in primitives and instantiated D flip-flops. Obtain the structural design module, the test bench module, and the outputs.

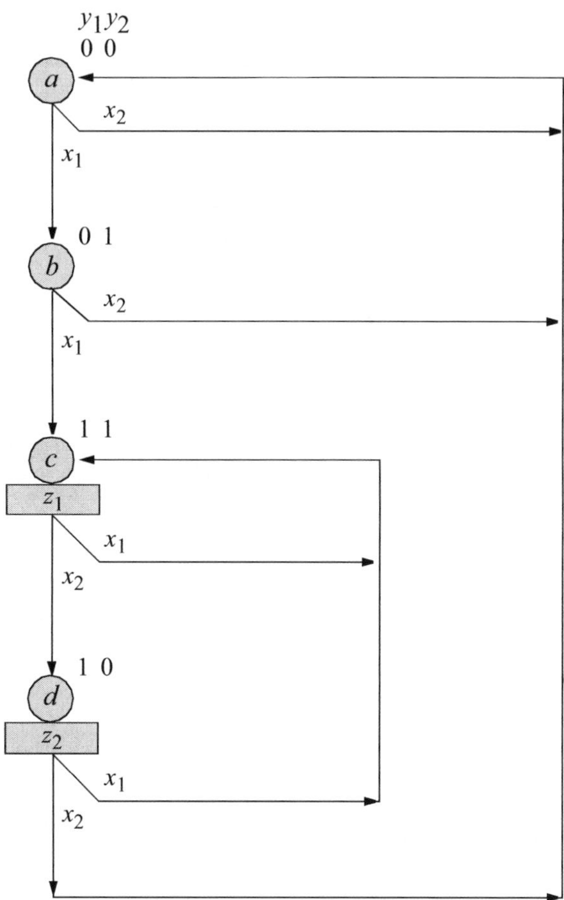

3.17 Repeat Problem 3.16 for a Moore pulse-mode asynchronous sequential machine using the continuous assignment statement **assign** and instantiated D flip-flops. Obtain the dataflow design module, the test bench module, and the outputs.

3.18 A toggle (T) flip-flop will be used in this problem and the next three problems. A T flip-flop has two inputs: T and reset; and two outputs $+y_1$ and $-y_1$. If the flip-flop is reset, then an active pulse on the T input will toggle the flip-flop to the set state; if the flip-flop is set, then a pulse on the T input will toggle the flip-flop to the reset state. Refer to page 372 of Chapter 3 for a description of a T flip-flop.

Design a Moore pulse-mode asynchronous sequential machine according to the state diagram shown below. Obtain the dataflow design module using the continuous **assign** statement, the test bench module, and the outputs.

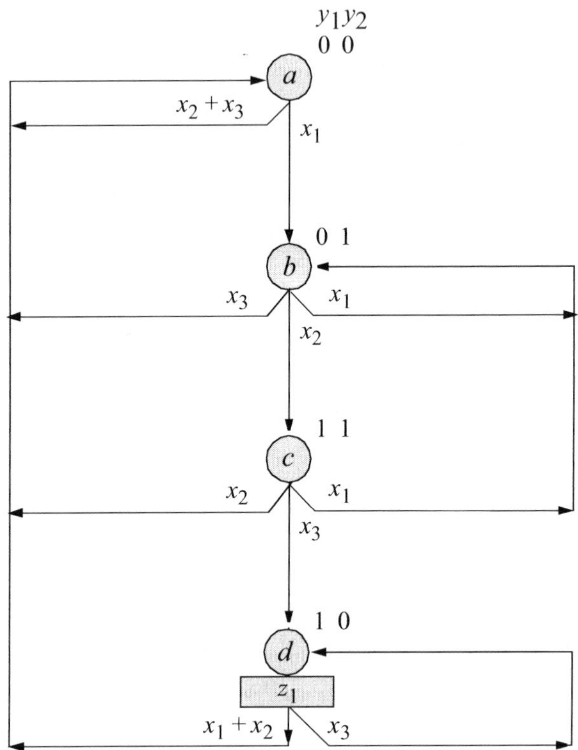

3.19 Given the state diagram shown below for a Moore pulse-mode asynchronous sequential machine, design the machine using structural modeling with built-in primitives and instantiated *T* flip-flops.

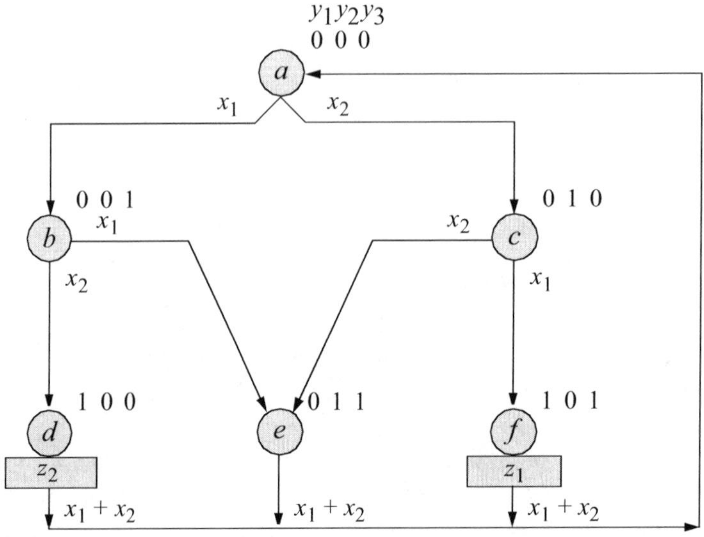

3.20 Given the state diagram for a Mealy pulse-mode asynchronous sequential machine shown below, design a dataflow module using the continuous assignment statement **assign**. Obtain the test bench module and the outputs.

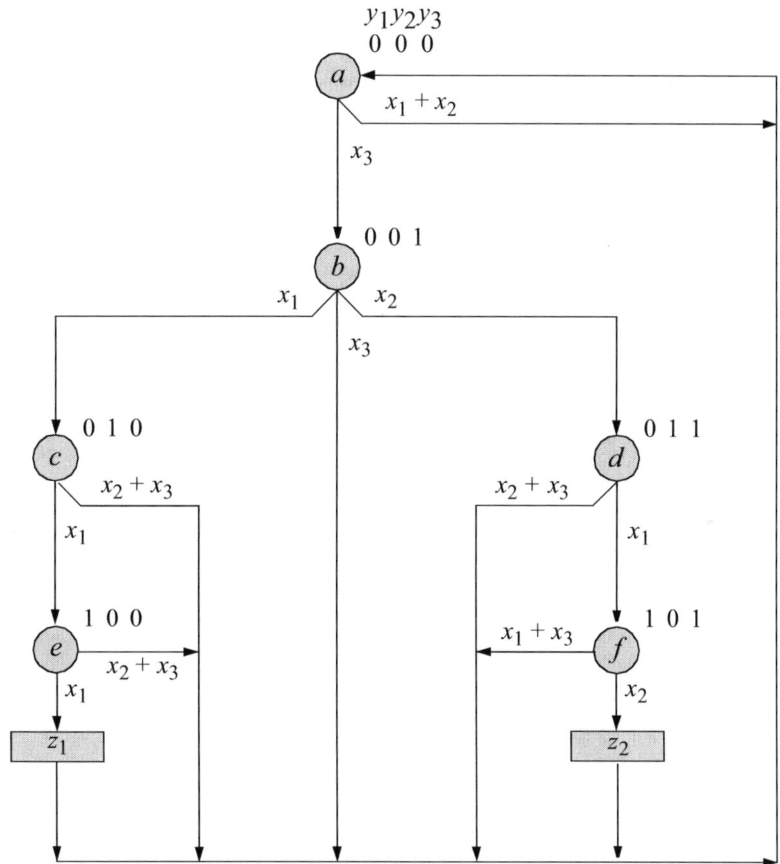

3.21 The state diagram shown below is for a Mealy pulse-mode asynchronous sequential machine. Design the structural module for the machine using built-in primitives and instantiated T flip-flops. Obtain the test bench module and the outputs.

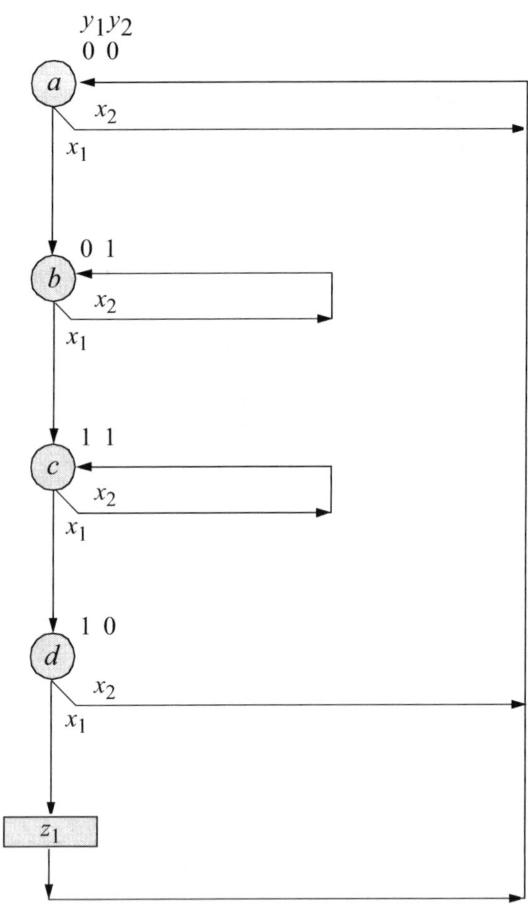

4

Computer Arithmetic Design Using Verilog HDL

4.1 Introduction

This chapter provides techniques for designing different types of adders, subtractors, multipliers, and dividers using Verilog HDL. The number representations that will be used are fixed-point, binary-coded decimal (BCD), and floating-point. For fixed-point addition, the radix point is placed to the immediate right of the number for integers or to the immediate left of the number for fractions. For floating-point addition, the numbers consist of the following three fields: a sign bit s; an exponent e; and a fraction f. These parts represent a number that is obtained by multiplying the fraction f by the radix r, raised to the power of the exponent e, as shown in Equation 4.1 for the number A, where f and e are signed fixed-point numbers, and r is the radix (or base).

$$A = f \times r^e \tag{4.1}$$

4.2 Fixed-Point Addition

Before the actual design process is presented, the addition operation will be illustrated. There are two operands that are added in an addition operation: the *augend* and the *addend*. The addend is added to the augend to produce a *sum*. If there is a *carry-in*, then the carry-in is added to the augend and addend to yield a sum and *carry-out*. The truth table for binary addition is shown in Table 4.1. When adding 1 + 1 = 2, the number 2 in binary is 10_2. When adding 1 + 1 + 1 = 3, the number 3 in binary is 11_2.

The *radix complement* of binary numbers (*2s complement*) is obtained by complementing each bit of the corresponding positive binary number and adding 1 to the low-order bit position. For example, let $A = 0001\ 1100_2 = +28_{10}$ and $A' = 1110\ 0011$ = 1110 0100 = –28. To obtain the value of a negative number count the weight of the 0s and add 1. Examples of addition operations are shown in Table 4.2, which add two 8-bit positive and negative operands.

Table 4.1 Truth Table for a Full Adder for Binary Addition

Augend (*a*)	Addend (*b*)	Carry-in (*cin*)	Carry-out (*cout*)	Sum
0	0	0	0	0
0	0	1	0	1
0	1	0	0	1
0	1	1	1	0
1	0	0	0	1
1	0	1	1	0
1	1	0	1	0
1	1	1	1	1

Table 4.2 Examples of Addition for Two Eight-Bit Signed Operands

	2^7	2^6	2^5	2^4	2^3	2^2	2^1	2^0	Value
Augend =	0	0	0	1	1	0	1	0	+26
+ Addend =	0	0	1	1	1	1	0	1	+61
Sum =	0	1	0	1	0	1	1	1	+87
Augend =	1	1	0	0	1	0	0	0	–56
+ Addend =	1	1	0	1	0	0	0	1	–47
Sum =	1	0	0	1	1	0	0	1	–103

4.2.1 Full Adder

A full adder can be designed from two half adders. A half adder adds two operand bits *a* and *b*, and produces two outputs *sum* and *carry-out*. The truth table for a half adder is shown in Table 4.3 and the equations for a half adder are shown in Equation 4.2.

From Table 4.1, the equations for the sum and carry-out of a full adder are shown in Equation 4.3. The logic diagram for a full adder is shown in Figure 4.1.

Table 4.3 Truth Table for a Half Adder

Augend (a)	Addend (b)	Carry-out ($cout$)	Sum
0	0	0	0
0	1	0	1
1	0	0	1
1	1	1	0

$$sum = a'b + ab' = a \oplus b \qquad cout = ab \tag{4.2}$$

$$\begin{aligned} sum &= a'b'cin + a'bcin' + ab'cin' + abcin \\ &= a \oplus b \oplus cin \\ cout &= a'bcin + ab'cin + abcin' + abcin \\ &= ab + (a \oplus b)cin \end{aligned} \tag{4.3}$$

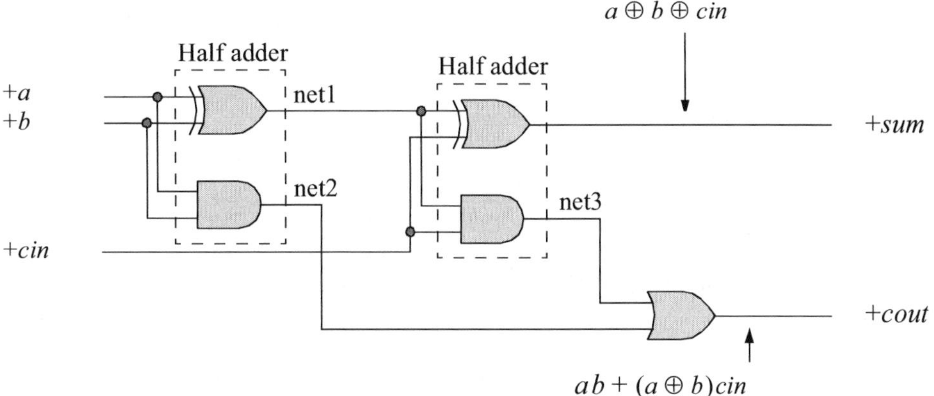

Figure 4.1 Logic diagram for a full adder using two half adders.

The structural design module is shown in Figure 4.2 using built-in primitives. The test bench module and the outputs are shown in Figures 4.3 and 4.4, respectively.

```
//full adder using built-in primitives
module full_adder_bip (a, b, cin, sum, cout);

//define inputs and outputs
input a, b, cin;
output sum, cout;

//design the full adder
//design the sum
xor   inst1 (net1, a, b);
and   inst2 (net2, a, b);
xor   inst3 (sum, net1, cin);

//define the carry-out
and   inst4 (net3, net1, cin);
or    inst5 (cout, net3, net2);

endmodule
```

Figure 4.2 Structural design module for the full adder.

```
//test bench for full adder using built-in primitives
module full_adder_bip_tb;

reg a, b, cin;      //inputs are reg for test bench
wire sum, cout;     //outputs are wire for test bench

//apply input vectors
initial
begin: apply_stimulus
   reg[3:0] invect;      //invect[3] terminates the for loop
   for (invect = 0; invect < 8; invect = invect + 1)
      begin
         {a, b, cin} = invect [3:0];
         #10 $display ("abcin = %b, cout = %b, sum = %b",
                        {a, b, cin}, cout, sum);
   end
end

//instantiate the module into the test bench
full_adder_bip inst1 (a, b, cin, sum, cout);

endmodule
```

Figure 4.3 Test bench module for the full adder.

```
abcin = 000, cout = 0, sum = 0
abcin = 001, cout = 0, sum = 1
abcin = 010, cout = 0, sum = 1
abcin = 011, cout = 1, sum = 0
abcin = 100, cout = 0, sum = 1
abcin = 101, cout = 1, sum = 0
abcin = 110, cout = 1, sum = 0
abcin = 111, cout = 1, sum = 1
```

Figure 4.4 Outputs for the full adder.

4.2.2 Three-Bit Adder

A 3-bit adder will be designed using the continuous assignment statement **assign**. Recall that the continuous assignment statement has the following syntax with optional drive strength and delay:

assign [drive_strength] [delay] left-hand side target = right-hand side expression

The design also uses the concatenation operator **{ }**, which forms a single operand from two or more operands by joining the different operands in sequence separated by commas. The operands to be appended are contained within braces. The size of the operands must be known before concatenation takes place. The design module utilizes the concatenation operator as follows:

```
assign   {cout, sum} = a + b + cin;
```

The dataflow design module is shown in Figure 4.5. The test bench module and the outputs are shown in Figures 4.6 and 4.7, respectively. Since operands *a* and *b* are both 3-bit operands and *cin* is a 1-bit operand, the *cout* and *sum* variables will contain 128 values.

```
//dataflow for a 3-bit adder
module adder3_df (a, b, cin, sum, cout);

input [2:0] a, b;     //define inputs and outputs
input cin;
output [2:0] sum;
output cout;

assign {cout, sum} = a + b + cin;
endmodule
```

Figure 4.5 Dataflow design module for the 3-bit adder.

```
//test bench for 3-bit dataflow adder
module adder3_df_tb;

reg [2:0] a, b;       //inputs are reg for test bench
reg cin;
wire [2:0] sum;       //outputs are wire for test bench
wire cout;

//apply stimulus
initial
begin : apply_stimulus
   reg [7:0] invect;
   for (invect = 0; invect < 128; invect = invect + 1)
      begin
         {a, b, cin} = invect [7:0];
         #10 $display ("a=%b, b=%b, cin=%b,
                  cout=%b, sum=%b", a, b, cin, cout, sum);
      end
end

//instantiate the module into the test bench
adder3_df inst1 (a, b, cin, sum, cout);
endmodule
```

Figure 4.6 Test bench module for the 3-bit adder.

```
a=000, b=000, cin=0, cout=0, sum=000
a=000, b=000, cin=1, cout=0, sum=001
a=000, b=001, cin=0, cout=0, sum=001
a=000, b=001, cin=1, cout=0, sum=010
a=000, b=010, cin=0, cout=0, sum=010

a=000, b=010, cin=1, cout=0, sum=011
a=000, b=011, cin=0, cout=0, sum=011
a=000, b=011, cin=1, cout=0, sum=100
a=000, b=100, cin=0, cout=0, sum=100
a=000, b=100, cin=1, cout=0, sum=101

a=000, b=101, cin=0, cout=0, sum=101
a=000, b=101, cin=1, cout=0, sum=110
a=000, b=110, cin=0, cout=0, sum=110
a=000, b=110, cin=1, cout=0, sum=111
a=000, b=111, cin=0, cout=0, sum=111

                        //continued on next page
```

Figure 4.7 Outputs for the 3-bit adder.

```
a=000, b=111, cin=1, cout=1, sum=000
a=001, b=000, cin=0, cout=0, sum=001
a=001, b=000, cin=1, cout=0, sum=010
a=001, b=001, cin=0, cout=0, sum=010
a=001, b=001, cin=1, cout=0, sum=011

a=001, b=010, cin=0, cout=0, sum=011
a=001, b=010, cin=1, cout=0, sum=100
a=001, b=011, cin=0, cout=0, sum=100
a=001, b=011, cin=1, cout=0, sum=101
a=001, b=100, cin=0, cout=0, sum=101

a=001, b=100, cin=1, cout=0, sum=110
a=001, b=101, cin=0, cout=0, sum=110
a=001, b=101, cin=1, cout=0, sum=111
a=001, b=110, cin=0, cout=0, sum=111
a=001, b=110, cin=1, cout=1, sum=000

a=001, b=111, cin=0, cout=1, sum=000
a=001, b=111, cin=1, cout=1, sum=001
a=010, b=000, cin=0, cout=0, sum=010
a=010, b=000, cin=1, cout=0, sum=011
a=010, b=001, cin=0, cout=0, sum=011

a=010, b=001, cin=1, cout=0, sum=100
a=010, b=010, cin=0, cout=0, sum=100
a=010, b=010, cin=1, cout=0, sum=101
a=010, b=011, cin=0, cout=0, sum=101
a=010, b=011, cin=1, cout=0, sum=110

a=010, b=100, cin=0, cout=0, sum=110
a=010, b=100, cin=1, cout=0, sum=111
a=010, b=101, cin=0, cout=0, sum=111
a=010, b=101, cin=1, cout=1, sum=000
a=010, b=110, cin=0, cout=1, sum=000

a=010, b=110, cin=1, cout=1, sum=001
a=010, b=111, cin=0, cout=1, sum=001
a=010, b=111, cin=1, cout=1, sum=010
a=011, b=000, cin=0, cout=0, sum=011
a=011, b=000, cin=1, cout=0, sum=100

a=011, b=001, cin=0, cout=0, sum=100
a=011, b=001, cin=1, cout=0, sum=101
a=011, b=010, cin=0, cout=0, sum=101
a=011, b=010, cin=1, cout=0, sum=110
a=011, b=011, cin=0, cout=0, sum=110   //continued next page
```

Figure 4.7 (Continued)

```
a=011, b=011, cin=1, cout=0, sum=111
a=011, b=100, cin=0, cout=0, sum=111
a=011, b=100, cin=1, cout=1, sum=000
a=011, b=101, cin=0, cout=1, sum=000
a=011, b=101, cin=1, cout=1, sum=001

a=011, b=110, cin=0, cout=1, sum=001
a=011, b=110, cin=1, cout=1, sum=010
a=011, b=111, cin=0, cout=1, sum=010
a=011, b=111, cin=1, cout=1, sum=011
a=100, b=000, cin=0, cout=0, sum=100

a=100, b=000, cin=1, cout=0, sum=101
a=100, b=001, cin=0, cout=0, sum=101
a=100, b=001, cin=1, cout=0, sum=110
a=100, b=010, cin=0, cout=0, sum=110
a=100, b=010, cin=1, cout=0, sum=111

a=100, b=011, cin=0, cout=0, sum=111
a=100, b=011, cin=1, cout=1, sum=000
a=100, b=100, cin=0, cout=1, sum=000
a=100, b=100, cin=1, cout=1, sum=001
a=100, b=101, cin=0, cout=1, sum=001

a=100, b=101, cin=1, cout=1, sum=010
a=100, b=110, cin=0, cout=1, sum=010
a=100, b=110, cin=1, cout=1, sum=011
a=100, b=111, cin=0, cout=1, sum=011
a=100, b=111, cin=1, cout=1, sum=100

a=101, b=000, cin=0, cout=0, sum=101
a=101, b=000, cin=1, cout=0, sum=110
a=101, b=001, cin=0, cout=0, sum=110
a=101, b=001, cin=1, cout=0, sum=111
a=101, b=010, cin=0, cout=0, sum=111

a=101, b=010, cin=1, cout=1, sum=000
a=101, b=011, cin=0, cout=1, sum=000
a=101, b=011, cin=1, cout=1, sum=001
a=101, b=100, cin=0, cout=1, sum=001
a=101, b=100, cin=1, cout=1, sum=010

a=101, b=101, cin=0, cout=1, sum=010
a=101, b=101, cin=1, cout=1, sum=011
a=101, b=110, cin=0, cout=1, sum=011
a=101, b=110, cin=1, cout=1, sum=100
a=101, b=111, cin=0, cout=1, sum=100   //continued next page
```

Figure 4.7 (Continued)

```
a=101, b=111, cin=1, cout=1, sum=101
a=110, b=000, cin=0, cout=0, sum=110
a=110, b=000, cin=1, cout=0, sum=111
a=110, b=001, cin=0, cout=0, sum=111
a=110, b=001, cin=1, cout=1, sum=000

a=110, b=010, cin=0, cout=1, sum=000
a=110, b=010, cin=1, cout=1, sum=001
a=110, b=011, cin=0, cout=1, sum=001
a=110, b=011, cin=1, cout=1, sum=010
a=110, b=100, cin=0, cout=1, sum=010

a=110, b=100, cin=1, cout=1, sum=011
a=110, b=101, cin=0, cout=1, sum=011
a=110, b=101, cin=1, cout=1, sum=100
a=110, b=110, cin=0, cout=1, sum=100
a=110, b=110, cin=1, cout=1, sum=101

a=110, b=111, cin=0, cout=1, sum=101
a=110, b=111, cin=1, cout=1, sum=110
a=111, b=000, cin=0, cout=0, sum=111
a=111, b=000, cin=1, cout=1, sum=000
a=111, b=001, cin=0, cout=1, sum=000

a=111, b=001, cin=1, cout=1, sum=001
a=111, b=010, cin=0, cout=1, sum=001
a=111, b=010, cin=1, cout=1, sum=010
a=111, b=011, cin=0, cout=1, sum=010
a=111, b=011, cin=1, cout=1, sum=011

a=111, b=100, cin=0, cout=1, sum=011
a=111, b=100, cin=1, cout=1, sum=100
a=111, b=101, cin=0, cout=1, sum=100
a=111, b=101, cin=1, cout=1, sum=101
a=111, b=110, cin=0, cout=1, sum=101

a=111, b=110, cin=1, cout=1, sum=110
a=111, b=111, cin=0, cout=1, sum=110
a=111, b=111, cin=1, cout=1, sum=111
```

Figure 4.7 (Continued)

4.2.3 Four-Bit Ripple-Carry Adder

A ripple-carry adder is not considered a high-speed adder, but requires less logic than a high-speed adder using the carry lookahead technique. The carry lookahead method expresses the carry-out of any stage as a function of a_i and b_i and the carry-in *cin* to the low-order stage. An *n*-stage ripple adder requires *n* full adders. The full adder of Section 4.2.1 will be used in this design. It will be instantiated four times into the structural design module. The logic diagram for a 4-bit ripple-carry adder is shown in Figure 4.8 in which the carries propagate (or ripple) through the adder.

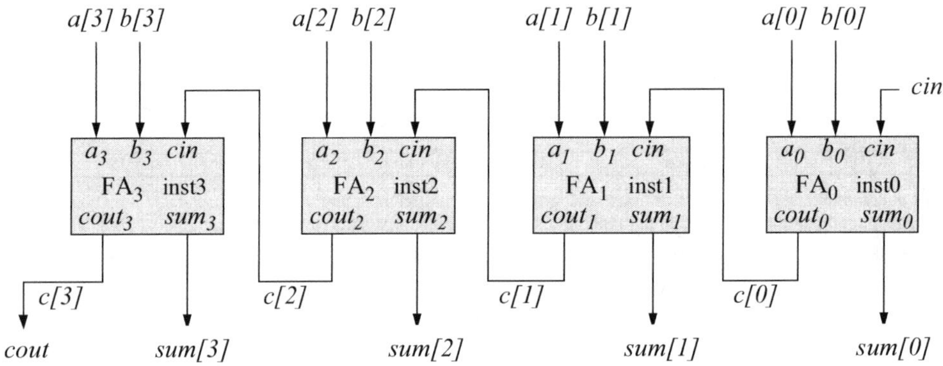

Figure 4.8 Logic diagram for a 4-bit ripple-carry adder.

The structural design module is shown in Figure 4.9. The inputs are two 4-bit vectors, *a[3:0]* and *b[3:0]*, where *a[0]* and *b[0]* are the low-order bits of the augend *A* and the addend *B*, respectively. There is also a scalar input *cin*. The outputs are a 4-bit vector *sum[3:0]* and a scalar output *cout*. The ripple-carries are internal nets represented by a 4-bit vector *c[3:0]*, which connects the carries between the adder stages. The test bench module and outputs are shown in Figures 4.10 and 4.11, respectively.

```
//structural for four-bit ripple-carry adder
module adder4_ripple_carry (a, b, cin, sum, cout);

input [3:0] a, b;     //define inputs and outputs
input cin;
output [3:0] sum;
output cout;

//define internal nets
wire [3:0] c;

//define output
assign cout = c[3];          //continued on next page
```

Figure 4.9 Structural design module for the 4-bit ripple-carry adder.

```
//design the ripple-carry adder
//instantiating the full adders
full_adder_bip inst0 (a[0], b[0], cin, sum[0], c[0]);

full_adder_bip inst1 (a[1], b[1], c[0], sum[1], c[1]);

full_adder_bip inst2 (a[2], b[2], c[1], sum[2], c[2]);

full_adder_bip inst3 (a[3], b[3], c[2], sum[3], c[3]);

endmodule
```

Figure 4.9 (Continued)

```
//test bench for the four-bit ripple-carry adder

module adder4_ripple_carry_tb;

reg [3:0] a, b;      //inputs are reg for test bench
reg cin;

wire [3:0] sum;      //outputs are wire for test bench
wire cout;

//display variables
initial
$monitor ("a = %b, b = %b, cin = %b, cout = %b, sum = %b",
            a, b, cin, cout, sum);

initial     //apply input vectors
begin
   #0    a = 4'b0000;   b = 4'b0001;   cin = 1'b0;
   #10   a = 4'b0010;   b = 4'b0001;   cin = 1'b1;
   #10   a = 4'b0011;   b = 4'b0101;   cin = 1'b1;
   #10   a = 4'b1010;   b = 4'b1001;   cin = 1'b0;
   #10   a = 4'b0111;   b = 4'b0111;   cin = 1'b1;

   #10   a = 4'b1010;   b = 4'b0111;   cin = 1'b0;
   #10   a = 4'b1110;   b = 4'b0111;   cin = 1'b0;
   #10   a = 4'b1100;   b = 4'b1100;   cin = 1'b1;
   #10   a = 4'b1111;   b = 4'b0110;   cin = 1'b1;
   #10   a = 4'b1011;   b = 4'b1000;   cin = 1'b0;
                            //continued on next page
```

Figure 4.10 Test bench module for the 4-bit ripple-carry adder.

```
    #10   a = 4'b1111;   b = 4'b0000;   cin = 1'b1;
    #10   a = 4'b1101;   b = 4'b1100;   cin = 1'b0;
    #10   a = 4'b1000;   b = 4'b0111;   cin = 1'b1;
    #10   a = 4'b0001;   b = 4'b1110;   cin = 1'b0;
    #10   a = 4'b1111;   b = 4'b1111;   cin = 1'b1;

    #10   $stop;
end

//instaniate the module into the test bench
adder4_ripple_carry inst1 (a, b, cin, sum, cout);

endmodule
```

Figure 4.10 (Continued)

```
a = 0000, b = 0001, cin = 0, cout = 0, sum = 0001
a = 0010, b = 0001, cin = 1, cout = 0, sum = 0100
a = 0011, b = 0101, cin = 1, cout = 0, sum = 1001
a = 1010, b = 1001, cin = 0, cout = 1, sum = 0011
a = 0111, b = 0111, cin = 1, cout = 0, sum = 1111

a = 1010, b = 0111, cin = 0, cout = 1, sum = 0001
a = 1110, b = 0111, cin = 0, cout = 1, sum = 0101
a = 1100, b = 1100, cin = 1, cout = 1, sum = 1001
a = 1111, b = 0110, cin = 1, cout = 1, sum = 0110
a = 1011, b = 1000, cin = 0, cout = 1, sum = 0011

a = 1111, b = 0000, cin = 1, cout = 1, sum = 0000
a = 1101, b = 1100, cin = 0, cout = 1, sum = 1001
a = 1000, b = 0111, cin = 1, cout = 1, sum = 0000
a = 0001, b = 1110, cin = 0, cout = 0, sum = 1111
a = 1111, b = 1111, cin = 1, cout = 1, sum = 1111
```

Figure 4.11 Outputs for the 4-bit ripple-carry adder.

4.2.4 Carry Lookahead Adder

This section designs a 4-bit carry lookahead adder using built-in primitives and the **assign** statement. The speed of an add operation can be increased by expressing the carry-out of any stage of the adder as a function of the two operand bits *a* and *b* of that stage and the carry-in to the low-order stage of the adder. Two auxiliary functions can be defined as follows:

Generate $G_i = a_i b_i$
Propagate $P_i = a_i \oplus b_i$

The carry *generate* function G_i specifies where a carry is generated at the *i*th stage. The carry *propagate* function P_i is true when the *i*th stage will pass (or propagate) the incoming carry c_{i-1} to the next higher stage$_{i+1}$. The carry-out c_i of any stage$_i$ can be defined as shown in Equation 4.4.

$$\begin{aligned} c_i &= a_i' b_i c_{i-1} + a_i b_i' c_{i-1} + a_i b_i \\ &= a_i b_i + (a_i \oplus b_i) c_{i-1} \\ &= G_i + P_i c_{i-1} \end{aligned} \tag{4.4}$$

Equation 4.4 indicates that the generate G_i and propagate P_i functions for any carry c_i can be obtained when the operand inputs are applied to the adder. The equation can be applied recursively to obtain the set of carry equations shown in Equation 4.5 in terms of the variables G_i, P_i, and c_{-1}. Equation 4.5 is for a 4-bit adder (3:0), where c_{-1} is the carry-in to the low-stage of the adder.

$$\begin{aligned} c_0 &= G_0 + P_0 c_{-1} \\ c_1 &= G_1 + P_1 c_0 \\ &= G_1 + P_1 (G_0 + P_0 c_{-1}) \\ &= G_1 + P_1 G_0 + P_1 P_0 c_{-1} \\ c_2 &= G_2 + P_2 c_1 \\ &= G_2 + P_2 (G_1 + P_1 G_0 + P_1 P_0 c_{-1}) \\ &= G_2 + P_2 G_1 + P_2 P_1 G_0 + P_2 P_1 P_0 c_{-1} \\ c_3 &= G_3 + P_3 c_2 \\ &= G_3 + P_3 (G_2 + P_2 G_1 + P_2 P_1 G_0 + P_2 P_1 P_0 c_{-1}) \\ &= G_3 + P_3 G_2 + P_3 P_2 G_1 + P_3 P_2 P_1 G_0 + P_3 P_2 P_1 P_0 c_{-1} \end{aligned} \tag{4.5}$$

The 4-bit carry lookahead adder will be designed using Equation 4.5 using built-in primitives and the **assign** statement. The block diagram of the adder is shown in Figure 4.12, where the augend is *a[3:0]*, the addend is *b[3:0]*, and the sum is *s[3:0]*. There are also four internal carries: *c3*, *c2*, *c1*, and *c0*. The structural design module is shown in Figure 4.13. The test bench module and the outputs are shown in Figures 4.14 and 4.15, respectively.

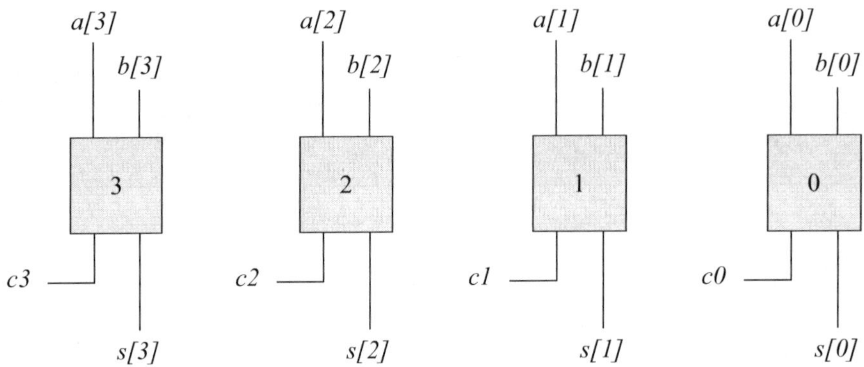

Figure 4.12 Block diagram of a 4-bit adder to be implemented as a carry look-ahead adder using built-in primitives.

```
//structural for four-bit carry lookahead adder using
//built-in primitives and conditional assignment

module adder_4_cla (a, b, cin, sum, cout);

input [3:0] a, b;     //define inputs and outputs
input cin;
output [3:0] sum;
output cout;

//design the logic for the generate functions
and     (g0, a[0], b[0]),
        (g1, a[1], b[1]),
        (g2, a[2], b[2]),
        (g3, a[3], b[3]);

//design the logic for the propagate functions
xor     (p0, a[0], b[0]),
        (p1, a[1], b[1]),
        (p2, a[2], b[2]),
        (p3, a[3], b[3]);

//design the logic for the sum equations
xor     (sum[0], p0, cin),
        (sum[1], p1, c0),
        (sum[2], p2, c1),
        (sum[3], p3, c2);          //continued on next page
```

Figure 4.13 Structural design module for the 4-bit carry lookahead adder.

```
//design the logic for the carry equations
//using the continuous assign statement
assign   c0 = g0 | (p0 & cin),

         c1 = g1 | (p1 & g0) | (p1 & p0 & cin),

         c2 = g2 | (p2 & g1) | (p2 & p1 & g0)
                 | (p2 & p1 & p0 & cin),

         c3 = g3 | (p3 & g2) | (p3 & p2 & g1)
                 | (p3 & p2 & p1 & g0)
                 |(p3 & p2 & p1 & p0 & cin);

//design the logic for cout using assign
assign   cout = c3;

endmodule
```

Figure 4.13 (Continued)

```
//test bench for the four-bit carry lookahead adder
module adder_4_cla_tb;

reg [3:0] a, b;       //inputs are reg for test bench
reg cin;
wire [3:0] sum;       //outputs are wire for test bench
wire cout;

//display variables
initial
$monitor ("a = %b, b = %b, cin = %b, cout = %b, sum = %b",
            a, b, cin, cout, sum);

//define input sequence
initial
begin
   #0    a = 4'b0000;   b = 4'b0000;   cin = 1'b0;
                                    //cout = 0, sum = 0000
   #10   a = 4'b0001;   b = 4'b0010;   cin = 1'b0;
                                    //cout = 0, sum = 0011
   #10   a = 4'b0010;   b = 4'b0110;   cin = 1'b0;
                                    //cout = 0, sum = 1000
   #10   a = 4'b0111;   b = 4'b0111;   cin = 1'b0;
                                    //cout = 0, sum = 1110
                              //continued on next page
```

Figure 4.14 Test bench module for the 4-bit carry lookahead adder.

```
   #10   a = 4'b1001;    b = 4'b0110;    cin = 1'b0;
                              //cout = 0, sum = 1111
   #10   a = 4'b1100;    b = 4'b1100;    cin = 1'b0;
                              //cout = 1, sum = 1000
   #10   a = 4'b1111;    b = 4'b1110;    cin = 1'b0;
                              //cout = 1, sum = 1101
   #10   a = 4'b1110;    b = 4'b1110;    cin = 1'b1;
                              //cout = 1, sum = 1101

   #10   a = 4'b1111;    b = 4'b1111;    cin = 1'b1;
                              //cout = 1, sum = 1111
   #10   a = 4'b1010;    b = 4'b1010;    cin = 1'b1;
                              //cout = 1, sum = 0101
   #10   a = 4'b1000;    b = 4'b1000;    cin = 1'b0;
                              //cout = 1, sum = 0000
   #10   a = 4'b1101;    b = 4'b1000;    cin = 1'b1;
                              //cout = 1, sum = 0110

   #10   a = 4'b1000;    b = 4'b1111;    cin = 1'b0;
                              //cout = 1, sum = 0111
   #10   a = 4'b0011;    b = 4'b1010;    cin = 1'b1;
                              //cout = 0, sum = 1110
   #10   a = 4'b0100;    b = 4'b0100;    cin = 1'b0;
                              //cout = 0, sum = 1000
   #10   a = 4'b1110;    b = 4'b0000;    cin = 1'b1;
                              //cout = 0, sum = 1111

   #10   $stop;
end

//instantiate the module into the test bench
adder_4_cla inst1 (a, b, cin, sum, cout);

endmodule
```

Figure 4.14 (Continued)

```
a = 0000, b = 0000, cin = 0, cout = 0, sum = 0000
a = 0001, b = 0010, cin = 0, cout = 0, sum = 0011
a = 0010, b = 0110, cin = 0, cout = 0, sum = 1000
a = 0111, b = 0111, cin = 0, cout = 0, sum = 1110

a = 1001, b = 0110, cin = 0, cout = 0, sum = 1111
a = 1100, b = 1100, cin = 0, cout = 1, sum = 1000
a = 1111, b = 1110, cin = 0, cout = 1, sum = 1101
a = 1110, b = 1110, cin = 1, cout = 1, sum = 1101   //next pg
```

Figure 4.15 Outputs for the 4-bit carry lookahead adder.

```
a = 1111, b = 1111, cin = 1, cout = 1, sum = 1111
a = 1010, b = 1010, cin = 1, cout = 1, sum = 0101
a = 1000, b = 1000, cin = 0, cout = 1, sum = 0000
a = 1101, b = 1000, cin = 1, cout = 1, sum = 0110

a = 1000, b = 1111, cin = 0, cout = 1, sum = 0111
a = 0011, b = 1010, cin = 1, cout = 0, sum = 1110
a = 0100, b = 0100, cin = 0, cout = 0, sum = 1000
a = 1110, b = 0000, cin = 1, cout = 0, sum = 1111
```

Figure 4.15 (Continued)

4.3 Fixed-Point Subtraction

Fixed-point subtraction is performed by subtracting the subtrahend from the minuend according to the rules shown in Table 4.4. An example is shown in Figure 4.16 using eight bits in which the subtrahend 0010 0101 (+37) is subtracted from the minuend 0011 0110 (+54), resulting in a difference of 0001 0001 (+17).

Table 4.4 Truth Table for Subtraction

0 – 0	=	0	
0 – 1	=	1	with a borrow from the next higher-order minuend
1 – 0	=	1	
1 – 1	=	0	

		2^7	2^6	2^5	2^4	2^3	2^2	2^1	2^0
	A (Minuend) = +54	0	0	1	1	0	1	1	0
–)	*B* (Subtrahend) = +37	0	0	1	0	0	1	0	1
	D (Difference) = +17	0	0	0	1	0	0	0	1

Figure 4.16 Example of subtraction using eight bits.

Computers use an adder to perform the subtract operation by adding the radix complement of the subtrahend to the minuend. The *r*s complement is obtained from the $r - 1$ complement by adding 1. For radix 2, the 2s complement is obtained by adding 1 to the 1s complement. The 1s complement is obtained by inverting all bits in the subtrahend. Thus, let *A* and *B* be two *n*-bit operands, where *A* is the minuend and *B* is the subtrahend as follows:

$$A = a_{n-1}\, a_{n-2} \ldots a_1\, a_0$$

$$B = b_{n-1}\, b_{n-2} \ldots b_1\, b_0$$

Therefore, $A - B = A + (B' + 1)$, where B' is the 1s complement of B and $(B' + 1)$ is the 2s complement of B. Examples of subtraction are shown below for both positive and negative 8-bit operands using the 2s complement method.

	$A =$	0	0	0	0	1	1	1	1	+15
–)	$B =$	0	1	1	0	0	0	0	0	+96
					↓					
		0	0	0	0	1	1	1	1	+15
+)		1	0	1	0	0	0	0	0	–96
		1	0	1	0	1	1	1	1	–81

	$A =$	1	0	1	1	0	0	0	1	–79
–)	$B =$	1	1	1	0	0	1	0	0	–28
					↓					
		1	0	1	1	0	0	0	1	–79
+)		0	0	0	1	1	1	0	0	+28
		1	1	0	0	1	1	0	1	–51

	$A =$	1	0	0	0	0	1	1	1	–121
–)	$B =$	1	1	1	0	0	1	1	0	–26
					↓					
		1	0	0	0	0	1	1	1	–121
+)		0	0	0	1	1	0	1	0	+26
		1	0	1	0	0	0	0	1	–95

```
      A = 0 0 0 1 0 0 1 1     +19
  −)  B = 0 1 0 1 1 1 0 0     +92
  ---------------------------------
                 ↓

          0 0 0 1 0 0 1 1     +19
  +)      1 0 1 0 0 1 0 0     −92
  ---------------------------------
          1 0 1 1 0 1 1 1     −73
```

4.3.1 Four-Bit Ripple Subtractor

A 4-bit subtractor will be designed using built-in primitives and instantiated full adders. A full adder design is shown on page 410 of this chapter using built-in primitives. Since four bits are used in this design, examples of 4-bit subtract operations are shown below.

```
    A = 0 1 1 1  (+7)                     A = 0 1 1 1
 −) B = 0 1 0 0  (+4)    ——→   +) (B' + 1) = 1 1 0 0
    -------------------                  -------------
                 (+3)                 1 ← 0 0 1 1


    A = 0 1 1 0  (+6)                     A = 0 1 1 0
 −) B = 1 1 0 0  (−4)    ——→   +) (B' + 1) = 0 1 0 0
    -------------------                  -------------
                 (+10)                0 ← 1 0 1 0

Result is overflow for four bits      + 10 in 2s complement for five bits


    A = 1 0 0 0  (−8)                     A = 1 0 0 0
 −) B = 0 0 1 0  (+2)    ——→   +) (B' + 1) = 1 1 1 0
    -------------------                  -------------
                 (−10)                1 ← 0 1 1 0

Result is overflow for four bits      − 10 in 2s complement for five bits
```

The logic diagram for the 4-bit subtractor is shown in Figure 4.17 using four full adders (FA) and four inverters. The design module is shown in Figure 4.18 using built-in primitives and instantiated full adders that were designed using built-in primitives. The test bench module and the outputs are shown in Figures 4.19 and 4.20, respectively.

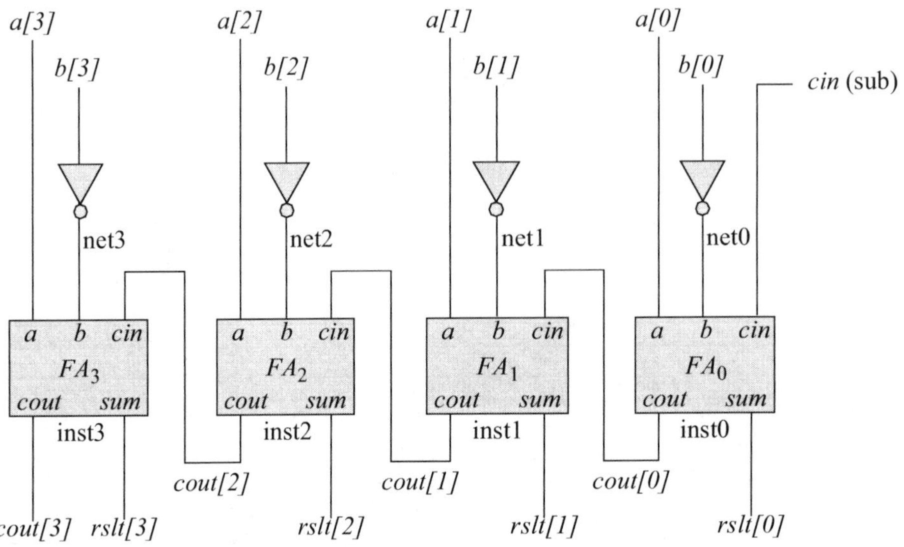

Figure 4.17 Logic diagram for a 4-bit subtractor.

```
//structural for a 4-bit subtractor
//using bip and instantiated full adders

module sub_4bit_bip (a, b, cin, rslt, cout);

//define inputs and outputs
input [3:0] a, b;
input cin;
output [3:0] rslt, cout;

//define internal nets
wire net0, net1, net2, net3;

//design the logic for stage 0
not   (net0, b[0]);
full_adder_bip inst0 (a[0], net0, cin, rslt[0], cout[0]);

//design the logic for stage 1
not   (net1, b[1]);
full_adder_bip inst1 (a[1], net1, cout[0],
               rslt[1], cout[1]);
                                  //continued on next page
```

Figure 4.18 Structural design module for a 4-bit subtractor.

```
//design the logic for stage 2
not    (net2, b[2]);
full_adder_bip inst2 (a[2], net2, cout[1], rslt[2],
cout[2]);

//design the logic for stage 3
not    (net3, b[3]);
full_adder_bip inst3 (a[3], net3, cout[2], rslt[3],
cout[3]);

endmodule
```

Figure 4.18 (Continued)

```
//test bench for 4-bit subtractor

module sub_4bit_bip_tb;

//inputs are reg for test bench
//outputs are wire for test bench
reg [3:0] a, b;
reg cin;
wire [3:0] rslt, cout;

//display variables
initial
$monitor ("a = %b, b = %b, cin = %b, rslt = %b, cout = %b",
            a, b, cin, rslt, cout);

//apply input vectors
initial
begin
   #0    a = 4'b0110;   b = 4'b0010;   cin = 1'b1;
   #10   a = 4'b1100;   b = 4'b0110;   cin = 1'b1;
   #10   a = 4'b1110;   b = 4'b1010;   cin = 1'b1;
   #10   a = 4'b1110;   b = 4'b0011;   cin = 1'b1;

   #10   a = 4'b1111;   b = 4'b0010;   cin = 1'b1;
   #10   a = 4'b1110;   b = 4'b0110;   cin = 1'b1;
   #10   a = 4'b1110;   b = 4'b1111;   cin = 1'b1;
   #10   a = 4'b1111;   b = 4'b0011;   cin = 1'b1;

                                    //continued on next page
```

Figure 4.19 Test bench module for the 4-bit subtractor.

```
    #10    a = 4'b0001;    b = 4'b0010;    cin = 1'b1;
    #10    a = 4'b0001;    b = 4'b0001;    cin = 1'b1;
    #10    a = 4'b1000;    b = 4'b0111;    cin = 1'b1;
    #10    a = 4'b1001;    b = 4'b0011;    cin = 1'b1;

    #10    $stop;
end

//instantiate the module into the test bench
sub_4bit_bip inst1 (a, b, cin, rslt, cout);

endmodule
```

Figure 4.19 (Continued)

```
a = 0110, b = 0010, cin = 1, rslt = 0100, cout = 1111
a = 1100, b = 0110, cin = 1, rslt = 0110, cout = 1001
a = 1110, b = 1010, cin = 1, rslt = 0100, cout = 1111
a = 1110, b = 0011, cin = 1, rslt = 1011, cout = 1100

a = 1111, b = 0010, cin = 1, rslt = 1101, cout = 1111
a = 1110, b = 0110, cin = 1, rslt = 1000, cout = 1111
a = 1110, b = 1111, cin = 1, rslt = 1111, cout = 0000
a = 1111, b = 0011, cin = 1, rslt = 1100, cout = 1111

a = 0001, b = 0010, cin = 1, rslt = 1111, cout = 0001
a = 0001, b = 0001, cin = 1, rslt = 0000, cout = 1111
a = 1000, b = 0111, cin = 1, rslt = 0001, cout = 1000
a = 1001, b = 0011, cin = 1, rslt = 0110, cout = 1001
```

Figure 4.20 Outputs for the 4-bit subtractor.

4.3.2 Eight-Bit Subtractor

This example designs an 8-bit subtractor using behavioral modeling. Designing a module in behavioral modeling is an abstraction of the functional operation of the design. It does not implement the design at the gate level. Behavioral modeling is an algorithmic approach to hardware implementation and represents a higher level of abstraction than other modeling methods.

The behavioral design module is shown in Figure 4.21. The test bench module and the outputs are shown in Figures 4.22 and 4.23, respectively.

```
//behavioral 8-bit subtractor
module sub_8bit_bh (a, b, rslt);

//define inputs and outputs
input [7:0] a, b;
output [7:0] rslt;

//variables used in always are declared as reg
reg [7:0] rslt;

//neg_b is used in the subtract operation
reg [7:0] neg_b = ~b + 1;

always @ (a or b)
begin
   rslt = a + neg_b;
end

endmodule
```

Figure 4.21 Behavioral design module for the 8-bit subtractor.

```
//test bench for the 8-bit subtractor
module sub_8bit_bh_tb;

reg [7:0] a, b;     //inputs are reg for test bench
wire [7:0] rslt;    //outputs are wire for test bench

initial             //display variables
$monitor ("a = %b, b = %b, rslt = %b", a, b, rslt);

//apply input vectors
initial
begin
   #0    a = 8'b0000_0011;    b = 8'b0000_0001; //3-1 = 2
   #10   a = 8'b0000_0100;    b = 8'b0000_0011; //4-3 = 1
   #10   a = 8'b0000_0110;    b = 8'b0000_0011; //6-3 = 3
   #10   a = 8'b0000_1110;    b = 8'b0000_0111; //14-7 = 7

   #10   a = 8'b0000_1100;    b = 8'b0000_0101; //12-5 = 7
   #10   a = 8'b0100_1100;    b = 8'b0001_0101; //76-21 = 55
   #10   a = 8'b0011_0001;    b = 8'b0001_1000; //49-24 = 25
   #10   a = 8'b0111_0001;    b = 8'b0011_1001; //113-57 = 56
                                   //continued on next page
```

Figure 4.22 Test bench module for the 8-bit subtractor.

```
    #10   a = 8'b1000_0001;     b = 8'b1000_0001; //-127+127=0
    #10   a = 8'b0110_0001;     b = 8'b0010_0001; //97-33 = 64
    #10   a = 8'b1100_0110;     b = 8'b1000_0101; //-58+123=65
    #10   a = 8'b0101_0101;     b = 8'b0000_1111; //85-15 = 70

    #10   a = 8'b1111_1000;     b = 8'b0000_0010; //-8-2 = -10

    #10   $stop;
end

//instantiate the module into the test bench
sub_8bit_bh inst1 (a, b, rslt);

endmodule
```

Figure 4.22 (Continued)

```
a = 00000011, b = 00000001, rslt = 00000010
a = 00000100, b = 00000011, rslt = 00000001
a = 00000110, b = 00000011, rslt = 00000011
a = 00001110, b = 00000111, rslt = 00000111

a = 00001100, b = 00000101, rslt = 00000111
a = 01001100, b = 00010101, rslt = 00110111
a = 00110001, b = 00011000, rslt = 00011001
a = 01110001, b = 00111001, rslt = 00111000

a = 10000001, b = 10000001, rslt = 00000000
a = 01100001, b = 00100001, rslt = 01000000
a = 11000110, b = 10000101, rslt = 01000001
a = 01010101, b = 00001111, rslt = 01000110

a = 11111000, b = 00000010, rslt = 11110110
```

Figure 4.23 Outputs for the 8-bit subtractor.

4.3.3 Four-Bit Dataflow Adder/Subtractor

This section presents the dataflow design module of a 4-bit fixed-point ripple adder/ subtractor using the continuous assignment statement **assign** and instantiated full adders. It is desirable to have the adder unit perform both addition and subtraction since there is no advantage to having a separate adder and subtractor. A ripple-carry adder will be modified so that it can also perform subtraction while still maintaining the ability to add. The operands are signed numbers in 2s complement representation. In order to form the 2s complement from the 1s complement, the carry-in to the low-order stage of the adder will be a 1 if subtraction is to be performed. The logic diagram is shown in Figure 4.24.

Overflow is detected if the carries out of bit 2 and bit 3 are different. If the operation is addition, then overflow can be further defined as shown in Equation 4.6. If the operation is subtraction, then overflow can be further defined as shown in Equation 4.7, where the variable *neg_b[7]* is the sign bit of the 2s complement of operand *B*.

$$\text{Overflow} = a[7]\ b[7]\ rslt[7]' + a[7]'\ b[7]'\ rslt[7] \tag{4.6}$$

$$\text{Overflow} = a[7]\ neg\_b[7]\ rslt[7]' + a[7]'\ neg\_b[7]'\ rslt[7] \tag{4.7}$$

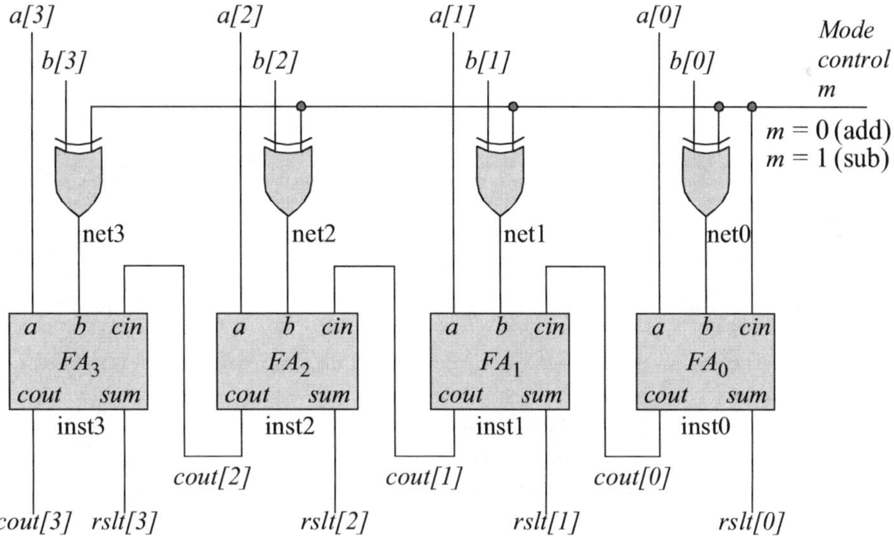

Figure 4.24 Logic diagram for a 4-bit ripple adder/subtractor. If the mode control input $m = 0$, then the operation is addition; if the mode control input $m = 1$, then the operation is subtraction.

The full adder that will be instantiated into the module for the adder/subtractor is shown in Figure 4.25. The dataflow design module for the 4-bit adder/subtractor is shown in Figure 4.26. The test bench module is shown in Figure 4.27, which displays the outputs in decimal notation. The outputs are shown in Figure 4.28.

```
//dataflow full adder
module full_adder (a, b, cin, sum, cout);

//list all inputs and outputs
input a, b, cin;
output sum, cout;

//continuous assign
assign sum = (a ^ b) ^ cin;
assign cout = cin & (a ^ b) | (a & b);

endmodule
```

Figure 4.25 Full adder design module to be instantiated into the 4-bit adder/subtractor.

```
//structural module for an adder/subtractor
module add_sub_4bits_assign (a, b, m, rslt, cout, ovfl);

//define inputs and outputs
input [3:0] a, b;
input m;     m = 0 is add; m = 1 is sub
output [3:0] rslt, cout;
output ovfl;

wire net0, net1, net2, net3;      //define internal nets

//define overflow
assign ovfl = (cout[3] ^ cout[2]);

//-----------------------------------------------
//instantiate the xor and the full adder for FA0
assign net0 = (b[0] ^ m);

full_adder inst0 (a[0], net0, m, rslt[0], cout[0]);
                  //a, b, cin, sum, cout
//-----------------------------------------------
//instantiate the xor and the full adder for FA1
assign net1 = (b[1] ^ m);

full_adder inst1 (a[1], net1, cout[0], rslt[1], cout[1]);
                  //a, b, cin, sum, cout
                                    //continued on next page
```

Figure 4.26 Dataflow design module for the 4-bit adder/subtractor.

```
//--------------------------------------------------
//instantiate the xor and the full adder for FA2
assign net2 = (b[2] ^ m);

full_adder inst2 (a[2], net2, cout[1], rslt[2], cout[2]);
                  //a, b, cin, sum, cout

//--------------------------------------------------
//instantiate the xor and the full adder for FA3
assign net3 = (b[3] ^ m);

full_adder inst3 (a[3], net3, cout[2], rslt[3], cout[3]);
                  //a, b, cin, sum, cout

endmodule
```

Figure 4.26 (Continued)

```
////test bench for structural adder-subtractor
module add_sub_4bits_assign_tb;

//inputs are reg for test bench
//outputs are wire for test bench
reg [3:0] a, b;
reg m;      m = 0 is add; m = 1 is sub
wire [3:0] rslt, cout;
wire  ovfl;

//display variables
initial
$monitor ("a = %d, b = %d, m = %d, rslt = %d,
            cout[3] = %b, cout[2] = %b, ovfl = %b",
               a, b, m, rslt, cout[3], cout[2], ovfl);

//apply input vectors
initial
begin
//addition; m = 0
   #0    a = 4'b0000;   b = 4'b0001;   m = 1'b0;
   #10   a = 4'b0010;   b = 4'b0101;   m = 1'b0;
   #10   a = 4'b0110;   b = 4'b0001;   m = 1'b0;
   #10   a = 4'b0101;   b = 4'b0001;   m = 1'b0;
                                  //continued on next page
```

Figure 4.27 Test bench module for the 4-bit adder/subtractor.

```
//subtraction  m = 1
   #10   a = 4'b0111;   b = 4'b0101;   m = 1'b1;
   #10   a = 4'b0101;   b = 4'b0100;   m = 1'b1;
   #10   a = 4'b0110;   b = 4'b0011;   m = 1'b1;
   #10   a = 4'b0110;   b = 4'b0010;   m = 1'b1;

//overflow
   #10   a = 4'b0111;   b = 4'b0101;   m = 1'b0;   //add
   #10   a = 4'b1000;   b = 4'b1011;   m = 1'b0;   //add
   #10   a = 4'b0110;   b = 4'b1100;   m = 1'b1;   //sub
   #10   a = 4'b1000;   b = 4'b0010;   m = 1'b1;   //sub

   #10   $stop;
end

//instantiate the module into the test bench
add_sub_4bits_assign inst1 (a, b, m, rslt, cout, ovfl);

endmodule
```

Figure 4.27 (Continued)

```
Addition
a = 0, b = 1, m = 0, rslt = 1, cout[3] = 0, cout[2] = 0, ovfl = 0
a = 2, b = 5, m = 0, rslt = 7, cout[3] = 0, cout[2] = 0, ovfl = 0
a = 6, b = 1, m = 0, rslt = 7, cout[3] = 0, cout[2] = 0, ovfl = 0
a = 5, b = 1, m = 0, rslt = 6, cout[3] = 0, cout[2] = 0, ovfl = 0

Subtraction
a = 7, b = 5, m = 1, rslt = 2, cout[3] = 1, cout[2] = 1, ovfl = 0
a = 5, b = 4, m = 1, rslt = 1, cout[3] = 1, cout[2] = 1, ovfl = 0
a = 6, b = 3, m = 1, rslt = 3, cout[3] = 1, cout[2] = 1, ovfl = 0
a = 6, b = 2, m = 1, rslt = 4, cout[3] = 1, cout[2] = 1, ovfl = 0

Overflow for addition
a = 7, b = 5, m = 0, rslt = 12, cout[3] = 0, cout[2] = 1, ovfl = 1
a = 8, b = 11, m = 0, rslt = 3, cout[3] = 1, cout[2] = 0, ovfl = 1

Overflow for subtraction
a = 6, b = 12, m = 1, rslt = 10, cout[3] = 0, cout[2] = 1,
                                                      ovfl = 1
a = 8, b = 2, m = 1, rslt = 6, cout[3] = 1, cout[2] = 0, ovfl = 1
```

Figure 4.28 Outputs for the 4-bit adder/subtractor.

Examine the *rslt* outputs for the addition overflow of Figure 4.28, where $a = 7$ and $b = 5$, with a result = 12. Also, where $a = 8$ and $b = 11$, where the result = 3. Overflow occurs as shown below for both operations. Also, examine the *rslt* outputs for the subtraction overflow of Figure 4.28, where $a = 6$ and $b = 12$, with a result = 10. Also, where $a = 8$ and $b = 2$, where the result = 6. Overflow occurs as shown below for both operations. The maximum value for four signed bits is +7 or –8. Notice that the results agree with Equations 4.6 and 4.7.

```
    0111               1000
+)  0101           +)  1011
    1100 (–4)          0011 (+3)
```

```
    0110          0110
–)  1100  →   +)  0100
                  1010 (–6)
```

```
    1000          1000
–)  0010  →   +)  1110
                  0110 (+6)
```

4.3.4 Eight-Bit Behavioral Adder/Subtractor

This section presents an 8-bit adder/subtractor that is synthesized using behavioral modeling. It is similar to the previous 4-bit adder/subtractor, but does not instantiate a full adder in the implementation. Overflow is defined as in the previous section and the equations are replicated below in Equations 4.8 and 4.9, for convenience.

$$\text{Overflow} = a[7]\ b[7]\ rslt[7]' + a[7]'\ b[7]'\ rslt[7] \tag{4.8}$$

$$\text{Overflow} = a[7]\ neg\_b[7]\ rslt[7]' + a[7]'\ neg\_b[7]'\ rslt[7] \tag{4.9}$$

A logic diagram is not required for behavioral modeling because this method of implementation describes the *behavior* of a digital system and is not concerned with the direct implementation of logic gates, but more with the architecture of the system. This is an algorithmic approach to hardware implementation and represents a higher level of abstraction.

Since behavioral modeling uses the **always** procedural construct statement, the variables used in the **always** statement are declared as type **reg**. The behavioral module is shown in Figure 4.29, which specifies an internal register *neg_b[7:0]* = *b*' + 1 indicating the 2s complement of the subtrahend, to be used in the overflow equation. The test bench module is shown in Figure 4.30 and the outputs are shown in Figure 4.31.

```
//behavioral 8-bit adder/subtractor
module add_subtract_bh (a, b, mode, rslt, ovfl);

input [7:0] a, b;      //define inputs and outputs
input  mode;
output [7:0] rslt;
output ovfl;

//variables rslt and ovfl are left-hand side targets
//in the always block and are declared as type reg
reg [7:0] rslt;
reg ovfl;

wire [7:0] a, b;       //since inputs default to wire
wire mode;             //the type wire is not required

//neg_b = ~b + 1 specifies an internal register
reg [7:0] neg_b = ~b + 1;

always @ (a or b or mode)
begin
   if (mode == 0)      //add
      begin
         rslt = a + b;
         ovfl =(a[7] & b[7] & ~rslt[7]) |
                (~a[7] & ~b[7] & rslt[7]);
      end
   else                //subtract
      begin
         rslt = a + neg_b;
         ovfl =(a[7] & neg_b[7] & ~rslt[7]) |
                (~a[7] & ~neg_b[7] & rslt[7]);
   end
end

endmodule
```

Figure 4.29 Behavioral design module for the 8-bit adder/subtractor.

```
//test bench for the 8-bit adder/subtractor
module add_subtract_bh_tb;

//inputs are reg for test bench
//outputs are wire for test bench
reg [7:0] a, b;
reg mode;
wire [7:0] rslt;
wire ovfl;

initial      //display variables
$monitor ("a=%b, b=%b, mode=%b, result=%b, ovfl=%b",
            a, b, mode, rslt, ovfl);

initial      //apply input vectors
begin
   #0    a = 8'b0000_0000;   b = 8'b0000_0001;   mode = 1'b0;
   #10   a = 8'b0000_0000;   b = 8'b0000_0001;   mode = 1'b1;

   #10   a = 8'b0000_0001;   b = 8'b1111_1001;   mode = 1'b0;
   #10   a = 8'b0000_0001;   b = 8'b1111_1001;   mode = 1'b1;

   #10   a = 8'b0000_0001;   b = 8'b1000_0001;   mode = 1'b0;
   #10   a = 8'b0000_0001;   b = 8'b1000_0001;   mode = 1'b1;
                                                 //ovfl = 1

   #10   a = 8'b1111_0000;   b = 8'b0000_0001;   mode = 1'b0;
   #10   a = 8'b1111_0000;   b = 8'b0000_0001;   mode = 1'b1;

   #10   a = 8'b0110_1101;   b = 8'b0100_0101;   mode = 1'b0;
                                                 //ovfl = 1
   #10   a = 8'b0010_1101;   b = 8'b0000_0101;   mode = 1'b1;

   #10   a = 8'b0000_0110;   b = 8'b0000_0001;   mode = 1'b0;
   #10   a = 8'b0000_0110;   b = 8'b0000_0001;   mode = 1'b1;

   #10   a = 8'b0001_0101;   b = 8'b0011_0001;   mode = 1'b0;
   #10   a = 8'b0001_0101;   b = 8'b0011_0001;   mode = 1'b1;

   #10   a = 8'b1000_0000;   b = 8'b1001_1100;   mode = 1'b0;
                                                 //ovfl = 1
   #10   a = 8'b1000_0000;   b = 8'b1001_1100;   mode = 1'b1;

   #10   a = 8'b1000_0101;   b = 8'b0010_0001;   mode = 1'b0;
   #10   a = 8'b1000_0101;   b = 8'b0010_0001;   mode = 1'b1;
                                                 //ovfl = 1
                                       //continued on next page
```

Figure 4.30 Test bench module for the 8-bit adder/subtractor.

```
   #10   a = 8'b1111_1111;   b = 8'b1111_1111;   mode = 1'b0;
   #10   a = 8'b1111_1111;   b = 8'b1111_1111;   mode = 1'b1;

   #10   $stop;
end

//instantiate the module into the test bench
add_subtract_bh inst1 (a, b, mode, rslt, ovfl);

endmodule
```

Figure 4.30 (Continued)

```
a=00000000, b=00000001, mode=0, result=00000001, ovfl=0
a=00000000, b=00000001, mode=1, result=11111111, ovfl=0
a=00000001, b=11111001, mode=0, result=11111010, ovfl=0
a=00000001, b=11111001, mode=1, result=00001000, ovfl=0
a=00000001, b=10000001, mode=0, result=10000010, ovfl=0
a=00000001, b=10000001, mode=1, result=10000000, ovfl=1

a=11110000, b=00000001, mode=0, result=11110001, ovfl=0
a=11110000, b=00000001, mode=1, result=11101111, ovfl=0
a=01101101, b=01000101, mode=0, result=10110010, ovfl=1

a=00101101, b=00000101, mode=1, result=00101000, ovfl=0
a=00000110, b=00000001, mode=0, result=00000111, ovfl=0
a=00000110, b=00000001, mode=1, result=00000101, ovfl=0
a=00010101, b=00110001, mode=0, result=01000110, ovfl=0
a=00010101, b=00110001, mode=1, result=11100100, ovfl=0
a=10000000, b=10011100, mode=0, result=00011100, ovfl=1

a=10000000, b=10011100, mode=1, result=11100100, ovfl=0
a=10000101, b=00100001, mode=0, result=10100110, ovfl=0
a=10000101, b=00100001, mode=1, result=01100100, ovfl=1

a=11111111, b=11111111, mode=0, result=11111110, ovfl=0
a=11111111, b=11111111, mode=1, result=00000000, ovfl=0
```

Figure 4.31 Outputs for the 8-bit adder/subtractor.

The operands and results for the addition operations of Figure 4.31 that result in an overflow are shown below together with the calculations.

```
a=01101101, b=01000101, mode=0, result=10110010, ovfl=1

a=10000000, b=10011100, mode=0, result=00011100, ovfl=1
```

	0110 1101		1000 0000
+)	0100 0101	+)	1001 1100
	1011 0010		0001 1100

The operands and results for the subtraction operations of Figure 4.31 that result in an overflow are shown below together with the calculations.

```
a=00000001, b=10000001, mode=1, result=10000000, ovfl=1

a=10000101, b=00100001, mode=1, result=01100100, ovfl=1
```

	0000 0001			0000 0001
–)	1000 0001	→	+)	0111 1111
				1000 0000

	1000 0101			1000 0101
–)	0010 0001	→	+)	1101 1111
				0110 0100

4.4 Fixed-Point Multiplication

This section presents the multiplication of two fixed-point binary operands in the 2s complement number representation. An n-bit multiplicand is multiplied by an n-bit multiplier to produce a $2n$-bit product. The multiplication algorithm consists of multiplying the multiplicand by the low-order multiplier bit to obtain a partial product. If the multiplier bit is a 1, then the multiplicand becomes the partial product; if the multiplier bit is a 0, then zeroes become the partial product. The partial product is then shifted left one bit position.

The multiplicand is then multiplied by the next higher-order multiplier bit to obtain a second partial product. The process repeats for all remaining multiplier bits, at which time the partial products are added to obtain the product. If both operands have the same sign, then the sign of the product is positive. If the signs of the operands are different, then the sign of the product is negative. Four examples are shown below containing four variations of the operands:

positive multiplicand and positive multiplier
negative multiplicand and positive multiplier
positive multiplicand and negative multiplier
negative multiplicand and negative multiplier

Multiplicand *A*					0	1	1	1	+7
Multiplier *B*				×)	0	1	1	1	+7
Partial products	0	0	0	0	0	1	1	1	
	0	0	0	0	1	1	1		
	0	0	0	1	1	1			
	0	0	0	0	0				
Product *P*	0	0	1	1	0	0	0	1	+49

Multiplicand *A*					1	0	1	0	–6
Multiplier *B*				×)	0	1	1	1	+7
Partial products	1	1	1	1	1	0	1	0	
	1	1	1	1	0	1	0		
	1	1	1	0	1	0			
	0	0	0	0	0				
Product *P*	1	1	0	1	0	1	1	0	–42

For a positive multiplicand and a negative multiplier, either 2s complement both operands or 2s complement the multiplier, perform the multiplication, then 2s complementing the result. For the problem below, the multiplicand is +5 and the multiplier is –6, which will be 2s complemented to a value of +6.

Multiplicand *A*					0	1	0	1	+5
Multiplier *B*				×)	0	1	1	0	(–6) +6
Partial products	0	0	0	0	0	0	0	0	
	0	0	0	0	1	0	1		
	0	0	0	1	0	1			
	0	0	0	0	0				
	0	0	0	1	1	1	1	0	
Product *P*	1	1	1	0	0	0	1	0	–30

For a negative multiplicand and a negative multiplier, 2s complement both operands, then perform the multiplication to yield the correct product. For the problem below, the multiplicand is –5 and the multiplier is –5. Both operands will be 2s complemented before the multiply operation to obtain the correct product.

Multiplicand *A*					1	0	1	1	–5
Multiplier *B*				×)	1	0	1	1	–5
Partial products	1	1	1	1	1	0	1	1	
	1	1	1	1	0	1	1		
	1	1	1	0	1	1			
	1	1	0	1	1				
Product *P*	1	0	1	1	0	1	0	1	–75

Multiplicand *A*					0	1	0	1	+5
Multiplier *B*				×)	0	1	0	1	+5
Partial products	0	0	0	0	0	1	0	1	
	0	0	0	0	0	0	0		
	0	0	0	1	0	1			
	0	0	0	0	0				
Product *P*	0	0	0	1	1	0	0	1	+25

Examples will now be presented using Verilog HDL to design various multipliers using different design methodologies. The multiplicand and multiplier are both n-bit operands and produce a $2n$-bit result.

4.4.1 Behavioral Four-Bit Multiplier

This section designs a 4-bit multiplier using behavioral modeling. The multiplicand is *a[3:0]*, the multiplier is *b[3:0]*, and the product is *prod[7:0]*. A scalar *start* signal is used to initiate the multiply operation. A count-down sequence counter *count* is initialized to a value of 4 (0100) before the operation begins, because there are four bits in both operands. When the counter reaches a value of 0000, the multiply operation is finished and the 8-bit product is in register *prod[7:0]*. A comparison is made initially to make certain that both operands are nonzero — if either operand has a value of zero, then the operation is terminated. The behavioral design module is shown in Figure 4.32. The test bench module and the outputs are shown in Figures 4.33 and 4.34, respectively.

```
//behavioral add-shift multiply

module mul_add_shift3 (a, b, prod, start);


//define inputs and outputs
input [3:0] a, b;
input start;
output [7:0] prod;


//variables are declared as reg in always
reg [7:0] prod;
reg [3:0] b_reg;
reg [3:0] count;


always @ (posedge start)
begin
   b_reg = b;
   prod = 0;
   count = 4'b0100;

   if ((a!=0) && (b!=0))

      while (count)
         begin
            prod = {((({4{b_reg[0]}} & a) + prod[7:4]),
                        prod[3:1]};
            b_reg = b_reg >> 1;
            count = count - 1;
         end

end

endmodule
```

Figure 4.32 Behavioral design module for a 4-bit multiplier.

```
//test bench for add-shift multiplier

module mul_add_shift3_tb;

//inputs are reg for test bench
//outputs are wire for test bench
reg [3:0] a, b;
reg start;
wire [7:0] prod;

//display variables
initial
$monitor ("a = %b, b = %b, prod = %b", a, b, prod);

//apply input vectors
initial
begin
   #0    start = 1'b0;  a = 4'b0110;   b = 4'b0110;

   #10   start = 1'b1; #10 start = 1'b0;

   #10   a = 4'b0010; b = 4'b0110;
               #10 start = 1'b1; #10 start = 1'b0;

   #10   a = 4'b0111; b = 4'b0101;
               #10 start = 1'b1; #10 start = 1'b0;

   #10   a = 4'b0111; b = 4'b0111;
               #10 start = 1'b1; #10 start = 1'b0;

   #10   a = 4'b0101; b = 4'b0101;
               #10 start = 1'b1; #10 start = 1'b0;

   #10   a = 4'b0111; b = 4'b0011;
               #10 start = 1'b1; #10 start = 1'b0;

   #10   a = 4'b0100; b = 4'b0110;
               #10 start = 1'b1; #10 start = 1'b0;

   #10   $stop;
end

//instantiate the module into the test bench
mul_add_shift3 inst1 (a, b, prod, start);

endmodule
```

Figure 4.33 Test bench for the 4-bit multiplier.

```
a = 0110, b = 0110, prod = 0010_0100
a = 0010, b = 0110, prod = 0010_0100
a = 0010, b = 0110, prod = 0000_1100
a = 0111, b = 0101, prod = 0000_1100
a = 0111, b = 0101, prod = 0010_0011
a = 0111, b = 0111, prod = 0010_0011
a = 0111, b = 0111, prod = 0011_0001
a = 0101, b = 0101, prod = 0011_0001
a = 0101, b = 0101, prod = 0001_1001
a = 0111, b = 0011, prod = 0001_1001
a = 0111, b = 0011, prod = 0001_0101
a = 0100, b = 0110, prod = 0001_0101
a = 0100, b = 0110, prod = 0001_1000
```

Figure 4.34 Outputs for the 4-bit multiplier.

4.4.2 Three-Bit Array Multiplier

This section presents the Verilog design of a high-speed 3-bit array multiplier. Array multipliers are designed using a planar array of full adders. An example of a general array multiply algorithm is shown in Figure 4.35 for two 3-bit operands.

The multiplicand is *A[2:0]* and the multiplier is *B[2:0]*, where *a[0]* and *b[0]* are the low-order bits of *A* and *B*, respectively. The two operands generate a product of *P[5:0]*. Each bit in the multiplicand is multiplied by the low-order bit b_0 of the multiplier. This is equivalent to the AND function and generates the first of three partial products. Each bit in the multiplicand is then multiplied by bit b_1 of the multiplier. The resulting partial product is shifted one bit position to the left. The process is repeated for bit b_2 of the multiplier. The partial products are then added together to form the product. A carry-out of any column is added to the next higher-order column.

Multiplicand *A*				a_2	a_1	a_0
Multiplier *B*			×)	b_2	b_1	b_0
Partial product 1				a_2b_0	a_1b_0	a_0b_0
Partial product 2			a_2b_1	a_1b_1	a_0b_1	
Partial product 3		a_2b_2	a_1b_2	a_0b_2		
Product *P*	2^5	2^4	2^3	2^2	2^1	2^0

Figure 4.35 General array multiply algorithm for two 3-bit operands.

The logic diagram for the 3-bit array multiplier is shown in Figure 4.36 utilizing full adders as the array elements and showing the generated partial products that correspond to those shown in Figure 4.35. The third row of full adders adds the sum and carry-out of the previous columns.

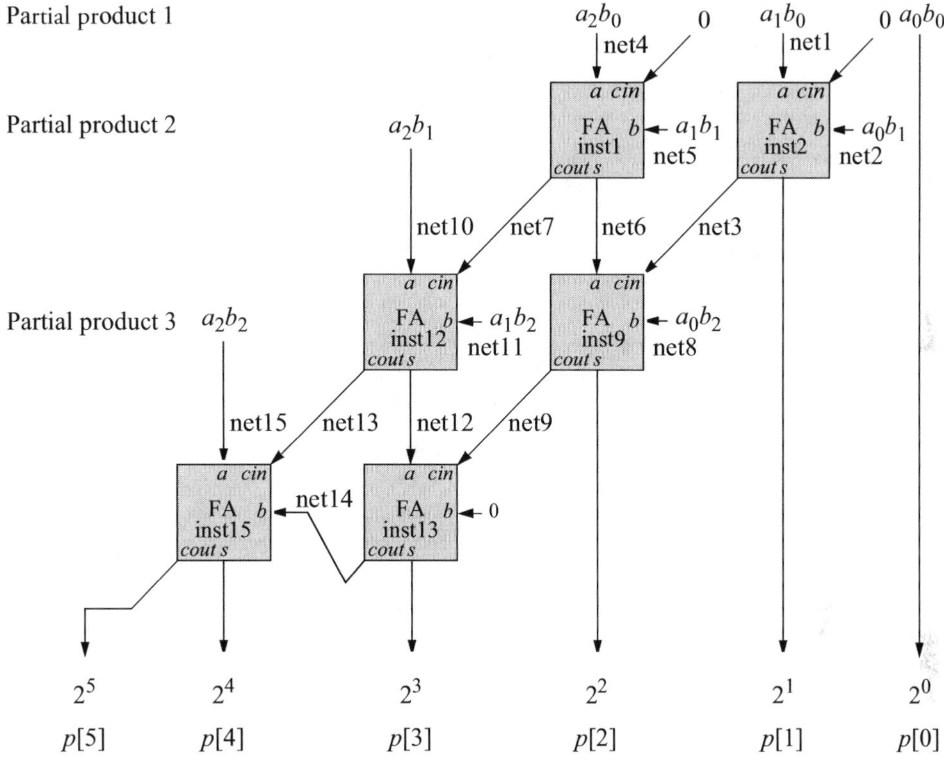

Figure 4.36 Logic diagram for the 3-bit array multiplier.

The structural design module is shown in Figure 4.37 using built-in primitives and instantiated full adders that were designed using dataflow modeling. The dataflow design module for the full adder is shown on page 432, Figure 4.25. The test bench module is shown in Figure 4.38 using all combinations of the three multiplicand bits and the three multiplier bits. The input vectors are treated as unsigned binary numbers. The outputs are shown in Figure 4.39 in decimal notation.

```
//structural for 3-bit array multiplier using bip

module array_mul3_bip (a, b, prod);

//define inputs and output
input [2:0] a, b;
output [5:0] prod;                //continued on next page
```

Figure 4.37 Structural design module for the 3-bit array multiplier.

```
//define internal nets
wire net1, net2, net3, net4, net5, net6, net7, net8, net9,
      net10, net11, net12, net13, net14, net15;

//instantiate the logic for prod[0]
and   (prod[0], a[0], b[0]);

//instantiate the logic for prod[1]
and   (net1, a[1], b[0]);
and   (net2, a[0], b[1]);
full_adder inst1 (net1, net2, 1'b0, prod[1], net3);
//a, b, cin, sum, cout

//instantiate the logic for prod[2]
and   (net4, a[2], b[0]);
and   (net5, a[1], b[1]);
full_adder inst2 (net4, net5, 1'b0, net6, net7);
and   (net8, a[0], b[2]);
full_adder inst3 (net6, net8, net3, prod[2], net9);

//instantiate the logic for prod[3]
and   (net10, a[2], b[1]);
and   (net11, a[1], b[2]);
full_adder inst4 (net10, net11, net7, net12, net13);
full_adder inst5 (net12, 1'b0, net9, prod[3], net14);

//instantiate the logic for prod[4] and prod [5]
and   (net15, a[2], b[2]);
full_adder inst6 (net15, net14, net13, prod[4], prod[5]);

endmodule
```

Figure 4.37 (Continued)

```
//test bench for structural 3-bit array multiplier using bip

module array_mu3_bip_tb;

//inputs are reg for test bench
//outputs are wire for test bench
reg [2:0] a, b;
wire [5:0] prod;

                                     //continued on next page
```

Figure 4.38 Test bench for the 3-bit array multiplier.

```
//apply stimulus and display variables
initial
begin: apply_stimulus
   reg [6:0] invect;
      for (invect = 0; invect < 64; invect = invect + 1)
      begin
         {a, b} = invect [6:0];
         #10 $display ("a = %d, b = %d, prod = %d", a, b,
prod);
      end
end

//instantiate the module into the test bench
array_mul3_bip inst1 (a, b, prod);

endmodule
```

Figure 4.38 (Continued)

```
a = 0, b = 0, prod = 0            a = 2, b = 4, prod = 8
a = 0, b = 1, prod = 0            a = 2, b = 5, prod = 10
a = 0, b = 2, prod = 0            a = 2, b = 6, prod = 12
a = 0, b = 3, prod = 0            a = 2, b = 7, prod = 14
a = 0, b = 4, prod = 0            a = 3, b = 0, prod = 0

a = 0, b = 5, prod = 0            a = 3, b = 1, prod = 3
a = 0, b = 6, prod = 0            a = 3, b = 2, prod = 6
a = 0, b = 7, prod = 0            a = 3, b = 3, prod = 9
a = 1, b = 0, prod = 0            a = 3, b = 4, prod = 12
a = 1, b = 1, prod = 1            a = 3, b = 5, prod = 15

a = 1, b = 2, prod = 2            a = 3, b = 6, prod = 18
a = 1, b = 3, prod = 3            a = 3, b = 7, prod = 21
a = 1, b = 4, prod = 4            a = 4, b = 0, prod = 0
a = 1, b = 5, prod = 5            a = 4, b = 1, prod = 4
a = 1, b = 6, prod = 6            a = 4, b = 2, prod = 8

a = 1, b = 7, prod = 7            a = 4, b = 3, prod = 12
a = 2, b = 0, prod = 0            a = 4, b = 4, prod = 16
a = 2, b = 1, prod = 2            a = 4, b = 5, prod = 20
a = 2, b = 2, prod = 4            a = 4, b = 6, prod = 24
a = 2, b = 3, prod = 6            a = 4, b = 7, prod = 28
                                      //continued on next page
```

Figure 4.39 Outputs for the 3-bit array multiplier.

```
a = 5, b = 0, prod = 0
a = 5, b = 1, prod = 5
a = 5, b = 2, prod = 10
a = 5, b = 3, prod = 15
a = 5, b = 4, prod = 20

a = 5, b = 5, prod = 25
a = 5, b = 6, prod = 30
a = 5, b = 7, prod = 35
a = 6, b = 0, prod = 0
a = 6, b = 1, prod = 6

a = 6, b = 2, prod = 12
a = 6, b = 3, prod = 18
a = 6, b = 4, prod = 24
a = 6, b = 5, prod = 30
a = 6, b = 6, prod = 36

a = 6, b = 7, prod = 42
a = 7, b = 0, prod = 0
a = 7, b = 1, prod = 7
a = 7, b = 2, prod = 14
a = 7, b = 3, prod = 21

a = 7, b = 4, prod = 28
a = 7, b = 5, prod = 35
a = 7, b = 6, prod = 42
a = 7, b = 7, prod = 49
```

Figure 4.39 (Continued)

4.4.3 Four-Bit Dataflow Multiplication Using the Multiply Operator

This multiplication example is relatively simple compared to other design methodologies. It uses the Verilog operator for multiplication (*). Some of the operators are shown in Table 4.5. The dataflow design module is shown in Figure 4.40 using the **assign** statement. The test bench module and the outputs are shown in Figures 4.41 and 4.42, respectively.

Table 4.5 Verilog HDL Operators

Operator Type	Operator Symbol	Operation
Arithmetic	+	Add
	–	Subtract
	*	Multiply
	/	Divide
Logical	&	AND
	|	OR
	^	Exclusive-OR
	~^ or ^~	Exclusive-NOR

```
//dataflow for 4-bit multiplier
module mul_4bits_assign (mpcnd, mplyr, prod);

//define inputs and outputs
input [3:0] mpcnd, mplyr;
output [7:0] prod;

//calculate the product
assign prod = mpcnd * mplyr;

endmodule
```

Figure 4.40 Dataflow design module for the 4-bit multiplier.

```
//test bench for the 4-bit multiplier
module mul_4bits_assign_tb;

//inputs are reg for test bench
//outputs are wire for test bench
reg [3:0] mpcnd, mplyr;
wire [7:0] prod;

initial      //display variables
$monitor ("mpcnd = %b, mplyr = %b, product = %d",
            mpcnd, mplyr, prod);

initial      //apply input vectors
begin
   #0    mpcnd = 4'b0001;   mplyr = 4'b0010;   //prod = 2
   #10   mpcnd = 4'b0101;   mplyr = 4'b0010;   //prod = 10
   #10   mpcnd = 4'b0111;   mplyr = 4'b0011;   //prod = 21
   #10   mpcnd = 4'b0110;   mplyr = 4'b0110;   //prod = 36

   #10   mpcnd = 4'b1000;   mplyr = 4'b0010;   //prod = 16
   #10   mpcnd = 4'b1010;   mplyr = 4'b0010;   //prod = 20
   #10   mpcnd = 4'b1111;   mplyr = 4'b0011;   //prod = 30
   #10   mpcnd = 4'b1100;   mplyr = 4'b0110;   //prod = 72

   #10   $stop;
end

//instantiate the module into the test bench
mul_4bits_assign inst1 (mpcnd, mplyr, prod);

endmodule
```

Figure 4.41 Test bench module for the 4-bit multiplier.

```
mpcnd = 1,  mplyr = 2, product = 2
mpcnd = 5,  mplyr = 2, product = 10
mpcnd = 7,  mplyr = 3, product = 21
mpcnd = 6,  mplyr = 6, product = 36

mpcnd = 8,  mplyr = 2, product = 16
mpcnd = 10, mplyr = 2, product = 20
mpcnd = 15, mplyr = 3, product = 45
mpcnd = 12, mplyr = 6, product = 72
```

Figure 4.42 Outputs for the 4-bit multiplier.

4.5 Fixed-Point Division

Division is usually slower than multiplication and occurs less frequently. The equation that represents the concept of division is shown below and includes the $2n$-bit dividend, the n-bit divisor, the n-bit quotient, and the n-bit remainder.

$$2n\text{-bit dividend} = (n\text{-bit divisor} \times n\text{-bit quotient}) + n\text{-bit remainder}$$

In general, the following equations also apply:

$$\text{Dividend / Divisor} = \text{Quotient}$$
$$\text{Dividend} = \text{Divisor x Quotient} + \text{Remainder}$$

The remainder has the same sign as the dividend. In contrast to multiplication, division is not commutative; that is, $A/B \neq B/A$, except when $A = B$, where A and B are the dividend and divisor, respectively. The process of division is one of successive subtract, shift, and compare operations. An example is shown in Figure 4.43 using *restoring division*, where the dividend = 24 and the divisor = 7, yielding a quotient of 3 and a remainder of 3.

Restoring division examines the state of the carry-out when the dividend is subtracted from the partial remainder. This determines the relative magnitudes of the divisor and partial remainder. If the carry-out = 0, then the partial remainder is restored to its previous value by adding the divisor to the partial remainder. If the carry-out = 1, then there is no restore operation. The partial remainder (high-order half of the dividend) and the low-order half of the dividend are then shifted left one bit position and the process repeats for each bit in the divisor.

The combined behavioral and dataflow design module is shown in Figure 4.44 and uses the **if** and **else** conditional statements together with the continuous assignment statement **assign**. The inputs are an 8-bit dividend, *a[7:0]*; a 4-bit divisor, *b[3:0]*; and a scalar signal, *start*, which initiates the divide operation. The output is an 8-bit register *rslt[7:0]* containing the quotient and remainder. Shifting is accomplished by the left-shift operator (<<) as follows: *rslt* = *rslt* << 1. The test bench module is shown in Figure 4.45 illustrating a variety of dividends and divisors. The outputs are shown in Figure 4.46.

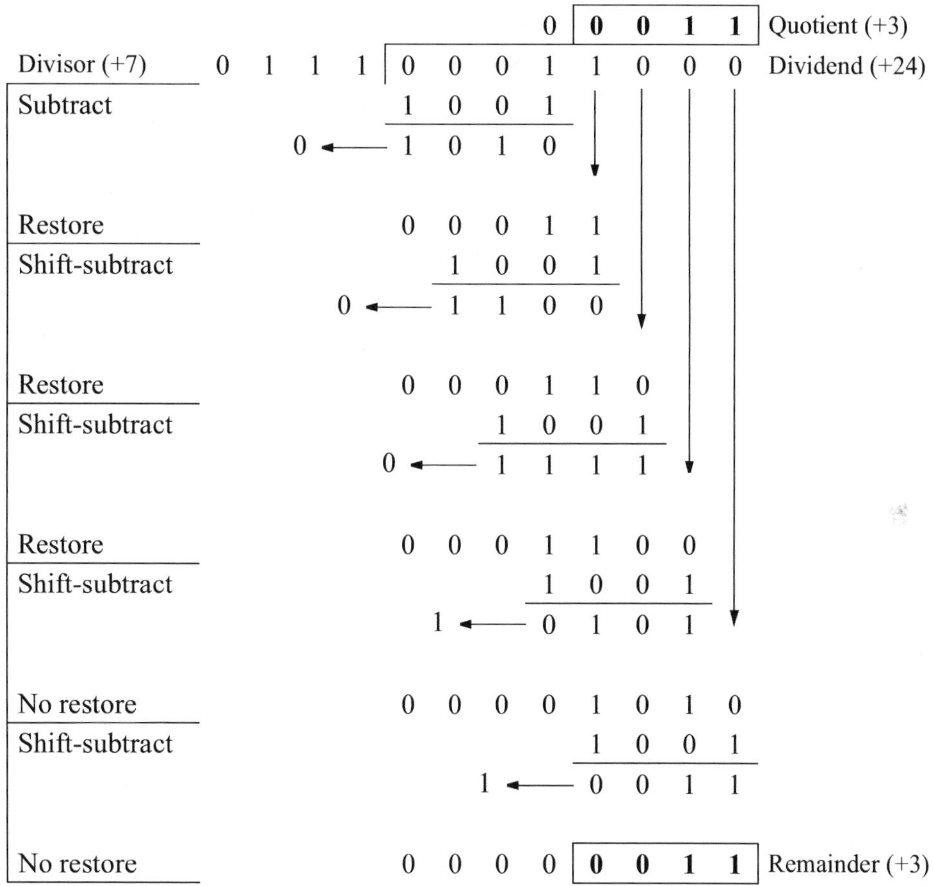

Figure 4.43 Example of fixed-point restoring division.

Subtracting the divisor from the partial remainder is realized by the following statement, which adds the negation of the divisor to the partial product and concatenates the sum with the low-order four bits from the previous partial remainder:

```
rslt = {(rslt[7:4] + b_neg), rslt[3:0]};
```

Then the sign bit (*rslt[7]*) of the sum is tested for a value of 1 or 0. If the sign is 1 (negative), then this indicates that the divisor was greater than the high-order half of the previous partial remainder. Thus, a 0 is placed in the low-order quotient bit. This sequence is executed by the following statements, after which the sequence counter is then decremented by 1:

```
if (rslt[7] == 1)
begin
    rst = {(rslt[7:4] + b), rslt[3:1], 1'b0};
```

If the sign bit (*rslt[7]*) is 0 (positive), then this indicates that the divisor was less than the high-order half of the previous partial remainder. Therefore, no restoration of the partial remainder is required and a 1 is placed in the low-order quotient bit, as shown in the following statement, after which the sequence counter is then decremented by 1:

```
rslt = {rslt[7:1], 1'b1};
```

```
//mixed-design for restoring division

module div_restoring (a, b, start, rslt);

//define inputs and outputs
input [7:0] a;
input [3:0] b;
input start;
output [7:0] rslt;

//define internal net
wire [3:0] b_bar;

//define internal registers
//variables used in always are declared as reg
reg [3:0] b_neg;
reg [7:0] rslt;
reg [3:0] count;

assign b_bar = ~b;

always @ (b_bar)
   b_neg = b_bar + 1;

                                  //continued on next page
```

Figure 4.44 Combined behavioral and dataflow design module for restoring division.

```
//execute the behavioral statements within the always block
always @ (posedge start)
begin
   rslt = a;
   count = 4'b0100;

if ((a!=0) && (b!=0))
   while (count)
      begin
         rslt = rslt << 1;
         rslt = {(rslt[7:4] + b_neg), rslt[3:0]};
            if (rslt[7] == 1)
               begin
                  rslt = {(rslt[7:4] + b), rslt[3:1], 1'b0};
                  count = count - 1;
               end

            else
               begin
                  rslt = {rslt[7:1], 1'b1};
                  count = count - 1;
               end
      end
end

endmodule
```

Figure 4.44 (Continued)

```
//test bench for restoring division
module div_restoring_tb;

//inputs are reg for test bench
//outputs are wire for test bench
reg [7:0] a;
reg [3:0] b;
reg start;
wire [7:0] rslt;

//display variables
initial
$monitor ("a = %d, b = %d, quot = %d, rem = %d",
            a, b, rslt[3:0], rslt[7:4]);
                                    //continued on next page
```

Figure 4.45 Test bench module for restoring division.

```
initial      //apply input vectors
begin
   #0     start = 1'b0;
          a = 8'b0000_1101;     b = 4'b0101;
   #10    start = 1'b1;
   #10    start = 1'b0;

   #10    a = 8'b0011_1100;     b = 4'b0111;
   #10    start = 1'b1;
   #10    start = 1'b0;

   #10    a = 8'b0101_0010;     b = 4'b0110;
   #10    start = 1'b1;
   #10    start = 1'b0;

   #10    a = 8'b0011_1000;     b = 4'b0111;
   #10    start = 1'b1;
   #10    start = 1'b0;

   #10    a = 8'b0110_0100;     b = 4'b0111;
   #10    start = 1'b1;
   #10    start = 1'b0;

   #10    a = 8'b0110_1110;     b = 4'b0111;
   #10    start = 1'b1;
   #10    start = 1'b0;

   #10    a = 8'b0010_0101;     b = 4'b0011;
   #10    start = 1'b1;
   #10    start = 1'b0;

   #10    a = 8'b0100_1000;     b = 4'b0111;
   #10    start = 1'b1;
   #10    start = 1'b0;

   #10    a = 8'b0101_0100;     b = 4'b0110;
   #10    start = 1'b1;
   #10    start = 1'b0;

   #10    a = 8'b0010_1110;     b = 4'b0101;
   #10    start = 1'b1;
   #10    start = 1'b0;
   #10    $stop;
end

div_restoring inst1 (a, b, start, rslt);//instantiate module
endmodule
```

Figure 4.45 (Continued)

```
a = 13,  b = 5, quot = 2,  rem = 3
a = 60,  b = 7, quot = 2,  rem = 3

a = 60,  b = 7, quot = 8,  rem = 4
a = 82,  b = 6, quot = 8,  rem = 4

a = 82,  b = 6, quot = 13, rem = 4
a = 56,  b = 7, quot = 13, rem = 4

a = 56,  b = 7, quot = 8,  rem = 0
a = 100, b = 7, quot = 8,  rem = 0

a = 100, b = 7, quot = 14, rem = 2
a = 110, b = 7, quot = 14, rem = 2

a = 110, b = 7, quot = 15, rem = 5
a = 37,  b = 3, quot = 15, rem = 5

a = 37,  b = 3, quot = 12, rem = 1
a = 72,  b = 7, quot = 12, rem = 1

a = 72,  b = 7, quot = 10, rem = 2
a = 84,  b = 6, quot = 10, rem = 2

a = 84,  b = 6, quot = 14, rem = 0
a = 46,  b = 5, quot = 14, rem = 0

a = 46,  b = 5, quot = 9,  rem = 1
```

Figure 4.46 Outputs for restoring division.

4.6 Arithmetic and Logic Unit

Arithmetic and logic units (ALUs) perform the arithmetic operations of addition, subtraction, multiplication, and division in fixed-point, decimal, and floating-point number representations. They also perform the logical operations of AND, OR, complementation (negation), exclusive-OR, and exclusive-NOR. An ALU is the central part of the computer and, together with the control unit, form the processor.

Most computer operations are executed by the ALU. For example, operands located in memory can be transferred to the ALU, where an operation is performed on the operands, then the result is stored in memory. The ALU performs calculations in different number representations, performs logical operations, and performs comparisons. The ALU also performs shift operations such as shift right algebraic, shift right logical, shift left algebraic, and shift left logical. The algebraic shift operations refer to signed operands in 2s complement representation; the logical shift operations refer to unsigned operands. ALUs also perform operations on packed and unpacked decimal operands.

This section will present the design of a fixed-point eight-function ALU using behavioral modeling with the **parameter** keyword and the **case** statement. The **parameter** keyword will assign values to the operation codes for the following operations: addition, subtraction, multiplication, AND, OR, NOT, exclusive-OR, and exclusive-NOR.

The **case** statement executes one of several different procedural statements depending on the comparison of an expression with a case item (see the format shown below). The expression and the case item are compared bit-by-bit and must match exactly. The statement that is associated with a case item may be a single procedural statement or a block of statements delimited by the keywords **begin** . . . **end.** The **case** statement has the following syntax:

```
case (expression)
    case_item1 : procedural_statement1;
    case_item2 : procedural_statement2;
    case_item3 : procedural_statement3;
                 .
                 .
                 .
    case_itemn : procedural_statementn;
    default : default_statement;
endcase
```

The behavioral design module is shown in Figure 4.47 which designs the following arithmetic operations: addition, subtraction, and multiplication. The module also designs the following logical operations: AND, OR, NOT, exclusive-OR, and exclusive-NOR. The test bench module is shown in Figure 4.48 and the outputs are shown in Figure 4.49.

```
//behavioral 8-function arithmetic and logic unit
module alu_8_bh (a, b, opcode, rslt);

//define inputs and output
input [3:0] a, b;
input [2:0] opcode;
output [7:0] rslt;

//the rslt is left-hand side target in always
//and is declared as type reg
reg [7:0] rslt;                    //continued on next page
```

Figure 4.47 Behavioral design module for the eight-function ALU.

```
//define operation codes
//parameter defines a constant
parameter   add_op = 3'b000,
            sub_op = 3'b001,
            mul_op = 3'b010,
            and_op = 3'b011,
            or_op  = 3'b100,
            not_op = 3'b101,  //negation
            xor_op = 3'b110,
            xnor_op = 3'b111;

//perform the operations
always @ (a or b or opcode)
begin
   case (opcode)
      add_op: rslt = a + b;
      sub_op: rslt = a - b;
      mul_op: rslt = a * b;
      and_op: rslt = a & b;      //also ab
      or_op:  rslt = a | b;
      not_op: rslt = ~a;         //also ~b
      xor_op: rslt = a ^ b;
      xnor_op: rslt = ~(a ^ b);
   endcase
end

endmodule
```

Figure 4.47 (Continued)

```
//test bench for 8-function arithmetic and logic unit
module alu_8_bh_tb;

//inputs are reg for test bench
//outputs are wire for test bench
reg [3:0] a, b;
reg [2:0] opcode;
wire [7:0] rslt;

initial      //display variables
$monitor ("a = %b, b = %b, opcode = %b, result = %b",
            a, b, opcode, rslt);
                                    //continued on next page
```

Figure 4.48 Test bench module for the eight-function ALU.

```
initial      //apply input vectors
begin
//add operation   1 + 2 and 6 + 6
   #0     a = 4'b0001;   b = 4'b0010;   opcode = 3'b000;
   #10    a = 4'b0110;   b = 4'b0110;   opcode = 3'b000;


//subtract operation   12 - 3 and 13 - 10
   #10    a = 4'b1100;   b = 4'b0011;   opcode = 3'b001;
   #10    a = 4'b1101;   b = 4'b1010;   opcode = 3'b001;


//multiply operation   12 x 7 and 15 x 3
   #10    a = 4'b1100;   b = 4'b0111;   opcode = 3'b010;
   #10    a = 4'b1111;   b = 4'b0011;   opcode = 3'b010;


//AND operation
   #10    a = 4'b1100;   b = 4'b0111;   opcode = 3'b011;
   #10    a = 4'b1101;   b = 4'b1011;   opcode = 3'b011;


//OR operation
   #10    a = 4'b0101;   b = 4'b1011;   opcode = 3'b100;
   #10    a = 4'b1001;   b = 4'b1010;   opcode = 3'b100;


//NOT operation
   #10    a = 4'b1001;                  opcode = 3'b101;
   #10    a = 4'b0011;                  opcode = 3'b101;


//exclusive-OR operation
   #10    a = 4'b0111;   b = 4'b1011;   opcode = 3'b110;
   #10    a = 4'b1010;   b = 4'b0101;   opcode = 3'b110;


//exclusive-NOR operation
   #10    a = 4'b0110;   b = 4'b0110;  opcode = 3'b111;
   #10    a = 4'b0011;   b = 4'b1110;  opcode = 3'b111;

   #20    $stop;
end

//instantiate the module into the test bench
alu_8_bh inst1 (a, b, opcode, rslt);

endmodule
```

Figure 4.48 (Continued)

```
add
a = 0001, b = 0010, opcode = 000, result = 00000011
a = 0110, b = 0110, opcode = 000, result = 00001100

subtract
a = 1100, b = 0011, opcode = 001, result = 00001001
a = 1101, b = 1010, opcode = 001, result = 00000011

multiply
a = 1100, b = 0111, opcode = 010, result = 01010100
a = 1111, b = 0011, opcode = 010, result = 00101101

AND
a = 1100, b = 0111, opcode = 011, result = 00000100
a = 1101, b = 1011, opcode = 011, result = 00001001

OR
a = 0101, b = 1011, opcode = 100, result = 00001111
a = 1001, b = 1010, opcode = 100, result = 00001011

NOT
a = 1001, b = 1010, opcode = 101, result = 11110110
a = 0011, b = 1010, opcode = 101, result = 11111100

XOR
a = 0111, b = 1011, opcode = 110, result = 00001100
a = 1010, b = 0101, opcode = 110, result = 00001111

XNOR
a = 0110, b = 0110, opcode = 111, result = 11111111
a = 0011, b = 1110, opcode = 111, result = 11110010
```

Figure 4.49 Outputs for the eight-function ALU.

4.7 Decimal Addition

A single element in a decimal arithmetic processor has nine inputs and five outputs as shown in Figure 4.50. Each operand (*operand A* and *operand B*) is represented by a 4-bit binary-coded decimal (BCD) digit. A carry-in (*cin*) bit is also provided from the previous lower-order addition element. The outputs are a 4-bit BCD digit called the result (*result*) and a carry-out (*cout*).

The most common code for BCD arithmetic is the 8421 code, an example of which is shown below in Figure 4.51. If the sum exceeds nine, then an adjustment is required by adding six (0110) to the result, also shown in Figure 4.51.

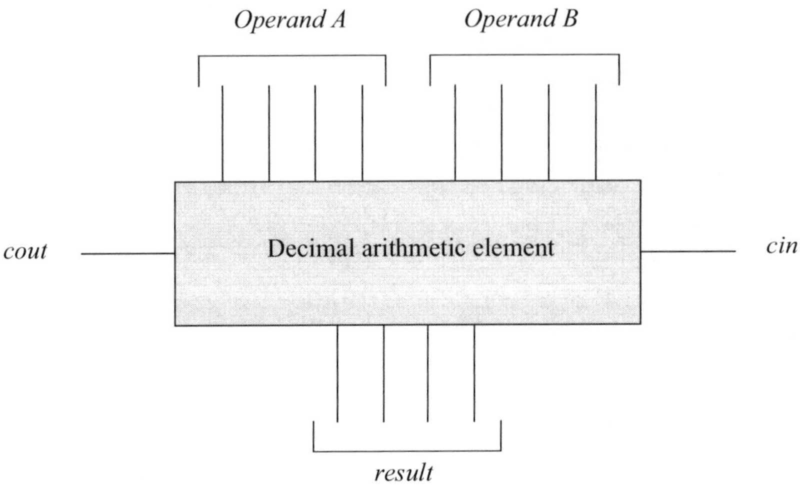

Figure 4.50 A single-digit binary-coded decimal (BCD) element.

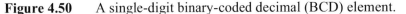

		8421	8421
26		0010	0110
+) 33	+)	0011	0011
59		0101	1001

(a)

		1110
Adjustment	+)	0110
	1	0100

(b)

Figure 4.51 Example of BCD addition; (a) no adjustment required and (b) adjustment required.

Table 4.6 shows the results that require no adjustment and the results that require adjustment. Whenever the unadjusted BCD sum produces a carry-out, the sum must be corrected by adding six. An adjustment is required whenever bit positions 8 and 4 are both 1s or whenever bit positions 8 and 2 are both 1s. Equation 4.10 illustrates this requirement when a carry is generated.

$$\text{Carry} = cout_8 + bit_8\ bit_4 + bit_8\ bit_2 \tag{4.10}$$

Table 4.6 BCD Addition Results for No Adjustment/Adjustment

	BCD Result 8421	Decimal Value		Carry	Valid BCD Result
	0000	0	No adjustment required		0000
	0001	1	↓		0001
	0010	2			0010
	0011	3			0011
	0100	4			0100
	0101	5			0101
	0110	6			0110
	1110	7			0111
	1000	8			1000
	1001	9			1001
	1010	10	Adjustment required	1	0001 0000
	1011	11	↓	1	0001 0001
	1100	12		1	0001 0010
	1101	13		1	0001 0011
	1110	14		1	0001 0100
	1111	15		1	0001 0101
1	0000	16		1	0001 0110
1	0001	17		1	0001 0111
1	0010	18		1	0001 1000
1	0011	19		1	0001 1001

The three examples shown below illustrate both valid and invalid BCD digits with the appropriate sum correction to yield valid BCD results.

Example 4.1 The numbers 58_{10} and 73_{10} will be added in BCD to yield a result of 131_{10} as shown below. Both *intermediate sums* (1011 and 1101) are invalid for BCD; therefore, 0110 must be added to the intermediate sums. Any carry that results from adding six to the intermediate sum is ignored because it provides no new information. The result of the BCD add operation is 0001 0011 0001. The carry produced from the low-order decade is also referred to as the *auxiliary carry*.

58		0101	1000
+) 73		0111	0011
131		1 ←	1011
	1 ←	1101	0110
	↓	0110	0001
	↓	0011	↓
	0001	0011	0001

Example 4.2 Another example of BCD addition is shown below, in which the intermediate sums are valid BCD numbers, but there is a carry-out of the high-order decade. Whenever the unadjusted sum produces a carry-out, the intermediate sum must be corrected by adding six.

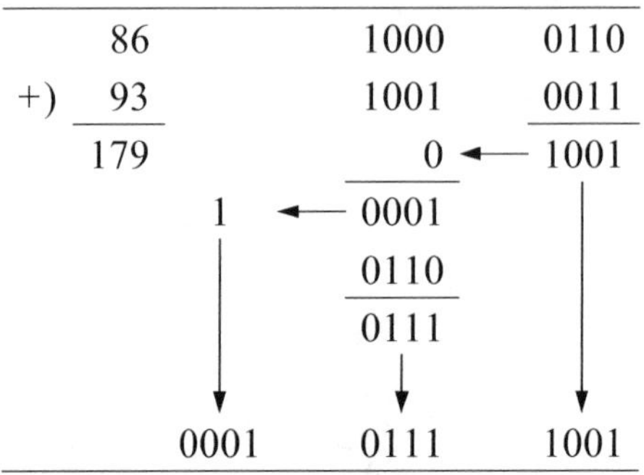

Example 4.3 This example illustrates all three conditions listed in Equation 4.10. The numbers 968 and 762 will be added using BCD arithmetic. The intermediate sums are: 0001 for the hundreds decade, 1101 for the tens decade, and 1010 for the units decade.

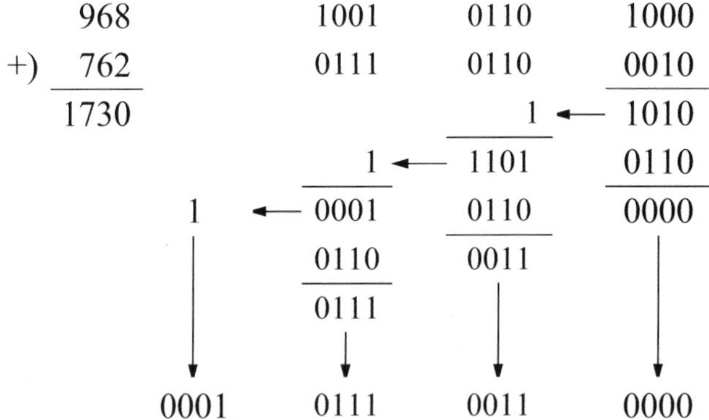

4.7.1 Decimal Addition with Sum Correction

A single stage of a decimal adder is shown in Figure 4.52. The carry-out of the decade corresponds to Equation 4.10, which is reproduced below for convenience. The carry-out of *adder1* — in conjunction with the logic indicated by Equation 4.10 — specifies the carry-out of the decade and is connected to inputs b_4 and b_2 of *adder2* with $b_8\, b_1 = 00$. This corrects an invalid decimal digit. The carry-out of *adder2* can be ignored, because it provides no new information.

$$\text{Carry} = cout_8 + bit_8\, bit_4 + bit_8\, bit_2$$

This decimal adder stage can be used in conjunction with other identical stages to design an n-digit parallel decimal adder. The carry-out of stage$_i$ connects to the carry-in of stage$_{i+1}$; therefore, this is a *ripple adder* for decimal operands. The Verilog behavioral design module of a typical 4-bit adder is shown in Figure 4.53.

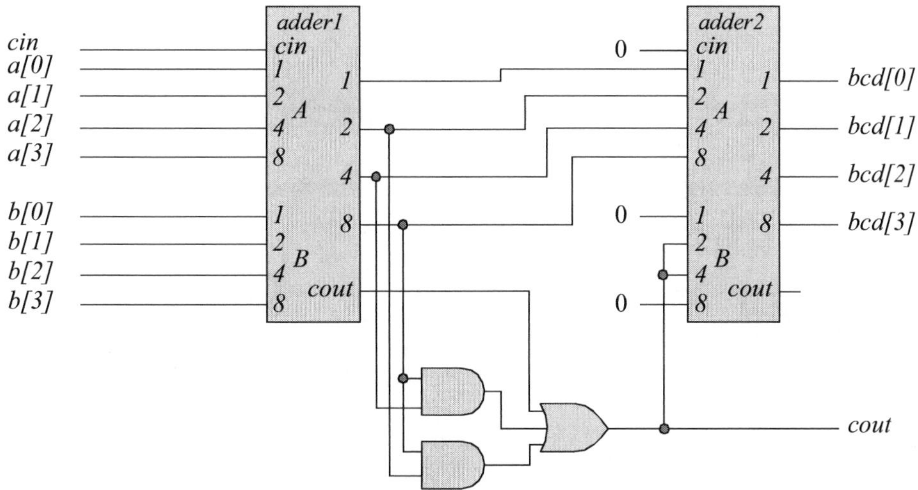

Figure 4.52 Typical stage of an n-digit decimal adder.

```
//behavioral model for a 4-bit adder
module adder4 (a, b, cin, sum, cout);

input [3:0] a, b;     //define inputs and outputs
input cin;
output [3:0] sum;
output cout;

reg [3:0] sum;        //variables in always are register
reg cout;

always @ (a or b or cin)    //perform the add operation
begin
   sum  = a + b + cin;
   cout = (a[3] & b[3]) |
         ((a[3] | b[3]) & (a[2] & b[2])) |
         ((a[3] | b[3]) & (a[2] | b[2]) & (a[1] & b[1])) |
         ((a[3] | b[3]) & (a[2] | b[2]) & (a[1] | b[1]) &
                  (a[0] & b[0])) |
         ((a[3] | b[3]) & (a[2] | b[2]) & (a[1] | b[1]) &
                  (a[0] | b[0]) & cin);
end
endmodule
```

Figure 4.53 Four-bit adder to be used in decimal addition.

The design module for one stage of a decimal adder is shown in Figure 4.54 using built-in primitives and the instantiated 4-bit adder of Figure 4.53. The test bench module and the outputs are shown in Figures 4.55 and 4.56, respectively.

```
//mixed-design bcd adder

module add_bcd (a, b, cin, bcd, cout, invalid_inputs);

//define inputs and outputs
input [3:0] a, b;
input cin;
output [3:0] bcd;
output cout, invalid_inputs;

reg invalid_inputs;      //reg if used in always statement

//define internal nets
wire [3:0] sum;
wire cout3, net3, net4;

//check for invalid inputs
always @ (a or b)
begin
   if ((a > 4'b1001) || (b > 4'b1001))
      invalid_inputs = 1'b1;
   else
      invalid_inputs = 1'b0;
end


//instantiate the logic for adder_1
adder4 inst1 (a[3:0], b[3:0], cin, sum[3:0], cout3);

//instantiate the logic for adder_2
adder4 inst2 (sum[3:0], {1'b0, cout, cout, 1'b0},
               1'b0, bcd[3:0]);


//instantiate the logic for intermediate sum adjustment
and    (net3, sum[3], sum[1]);

and    (net4, sum[3], sum[2]);

or     (cout, cout3, net3, net4);

endmodule
```

Figure 4.54 Mixed-design module for one stage of a decimal adder.

```
//test bench for mixed-design add_bcd
module add_bcd_tb;

//inputs are reg for test bench
//outputs are wire for test bench
reg [3:0] a, b;
reg cin;
wire [3:0] bcd;
wire cout;
wire invalid_inputs;

//display variables
initial
$monitor ("a=%b, b=%b, cin=%b, cout=%b,
            bcd=%b, invalid_inputs=%b",
         a, b, cin, cout, bcd, invalid_inputs);

//apply input vectors
initial
begin
   #0    a = 4'b0011;   b = 4'b0011;   cin = 1'b0;
   #10   a = 4'b0101;   b = 4'b0110;   cin = 1'b0;
   #10   a = 4'b0101;   b = 4'b0100;   cin = 1'b0;
   #10   a = 4'b0111;   b = 4'b1000;   cin = 1'b0;
   #10   a = 4'b0111;   b = 4'b0111;   cin = 1'b0;
   #10   a = 4'b1000;   b = 4'b1001;   cin = 1'b0;
   #10   a = 4'b1001;   b = 4'b1001;   cin = 1'b0;
   #10   a = 4'b0101;   b = 4'b0110;   cin = 1'b1;
   #10   a = 4'b0111;   b = 4'b1000;   cin = 1'b1;
   #10   a = 4'b1001;   b = 4'b1001;   cin = 1'b1;
   #10   a = 4'b1000;   b = 4'b1000;   cin = 1'b0;
   #10   a = 4'b1000;   b = 4'b1000;   cin = 1'b1;
   #10   a = 4'b1001;   b = 4'b0111;   cin = 1'b0;
   #10   a = 4'b0111;   b = 4'b0010;   cin = 1'b1;
   #10   a = 4'b0011;   b = 4'b1000;   cin = 1'b0;

//three invalid inputs
   #10   a = 4'b1010;   b = 4'b0001;   cin = 1'b0;
   #10   a = 4'b0011;   b = 4'b1100;   cin = 1'b0;
   #10   a = 4'b1011;   b = 4'b1110;   cin = 1'b1;
   #10   $stop;
end

//instantiate the module into the test bench
add_bcd inst1 (a, b, cin, bcd, cout, invalid_inputs);

endmodule
```

Figure 4.55 Test bench module for one stage of the decimal adder.

```
a=0101, b=0110, cin=0, cout=1, bcd=0001, invalid_inputs=0
a=0101, b=0100, cin=0, cout=0, bcd=1001, invalid_inputs=0
a=0111, b=1000, cin=0, cout=1, bcd=0101, invalid_inputs=0

a=0111, b=0111, cin=0, cout=1, bcd=0100, invalid_inputs=0
a=1000, b=1001, cin=0, cout=1, bcd=0111, invalid_inputs=0
a=1001, b=1001, cin=0, cout=1, bcd=1000, invalid_inputs=0

a=0101, b=0110, cin=1, cout=1, bcd=0010, invalid_inputs=0
a=0111, b=1000, cin=1, cout=1, bcd=0110, invalid_inputs=0
a=1001, b=1001, cin=1, cout=1, bcd=1001, invalid_inputs=0

a=1000, b=1000, cin=0, cout=1, bcd=0110, invalid_inputs=0
a=1000, b=1000, cin=1, cout=1, bcd=0111, invalid_inputs=0
a=1001, b=0111, cin=0, cout=1, bcd=0110, invalid_inputs=0

a=0111, b=0010, cin=1, cout=1, bcd=0000, invalid_inputs=0
a=0011, b=1000, cin=0, cout=1, bcd=0001, invalid_inputs=0

a=1010, b=0001, cin=0, cout=1, bcd=0001, invalid_inputs=1
a=0011, b=1100, cin=0, cout=1, bcd=0101, invalid_inputs=1
a=1011, b=1110, cin=1, cout=1, bcd=0000, invalid_inputs=1
```

Figure 4.56 Outputs for the BCD adder.

4.7.2 Decimal Addition Using Multiplexers for Sum Correction

An alternative approach to determine whether to add six to correct an invalid decimal number is to use a multiplexer. The two operands are added in *adder1* as before; however, a value of six is always added to this intermediate sum in *adder2*, as shown in Figure 4.57. The sums from *adder1* and *adder2* are then applied to four 2:1 multiplexers. Selection of the *adder1* sum or the *adder2* sum is determined by ORing the carry-out of both adders — *cout1* and *cout2* — to generate a select input – *cout* – to the multiplexers, as shown below.

$$\text{Multiplexer select input} = \begin{cases} \text{Select adder1 sum if } cout \text{ is } 0 \\ \text{Select adder2 sum if } cout \text{ is } 1 \end{cases}$$

The decimal adder module instantiates a dataflow design module for the 2:1 multiplexer shown in Figure 4.58 and a behavioral design module for the 4-bit adder shown in Figure 4.59, which is reproduced from Figure 4.53 for convenience.

```
A =                      0 0 1 1
B =                   +) 0 1 1 0
      cout1 = 0 ◄──      1 0 0 1    adder_1    Intermediate sum
                      +) 0 1 1 0
      cout2 = 0 ◄──      1 1 1 1    adder_2

                    cout    s0 = 0 ──► adder_1 sum = 0  1001
```

```
A =                      1 0 0 0
B =                   +) 1 0 0 1
      cout1 = 1 ◄──      0 0 0 1    adder_1    Intermediate sum
                      +) 0 1 1 0
      cout2 = 0 ◄──      0 1 1 1    adder_2

                    cout    s0 = 1 ──► adder_2 sum = 1  0111
```

```
A =                      0 1 1 1
B =                   +) 1 0 0 0
      cout3 = 0 ◄──      1 1 1 1    adder_1    Intermediate sum
                      +) 0 1 1 0
      cout8 = 1 ◄──      0 1 0 1    adder_2

                    cout    s0 = 1 ──► adder_2 sum = 1  0101
```

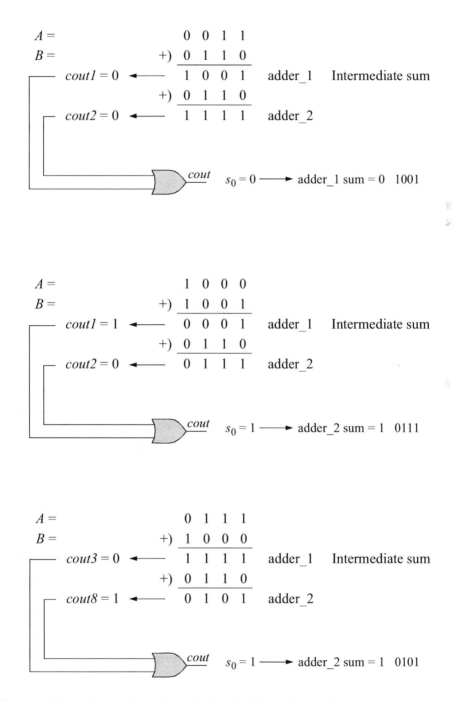

Figure 4.57 Examples using a decimal adder with multiplexers.

```
//dataflow 2:1 multiplexer
module mux2_df (sel, data, z1);

//define inputs and output
input sel;
input [1:0] data;
output z1;

assign z1 = (~sel & data[0]) | (sel & data[1]);

endmodule
```

Figure 4.58 Two-to-one multiplexer used in the BCD adder.

```
//behavioral model for a 4-bit adder
module adder4 (a, b, cin, sum, cout);

input [3:0] a, b;     //define inputs and outputs
input cin;
output [3:0] sum;
output cout;

reg [3:0] sum;        //variables used in always are register
reg cout;

always @ (a or b or cin)    //perform the add operation
begin
   sum  = a + b + cin;
   cout = (a[3] & b[3]) |
         ((a[3] | b[3]) & (a[2] & b[2])) |
         ((a[3] | b[3]) & (a[2] | b[2]) & (a[1] & b[1])) |
         ((a[3] | b[3]) & (a[2] | b[2]) & (a[1] | b[1]) &
                   (a[0] & b[0])) |
         ((a[3] | b[3]) & (a[2] | b[2]) & (a[1] | b[1]) &
                   (a[0] | b[0]) & cin);
end
endmodule
```

Figure 4.59 Four-bit adder to be used in decimal addition using multiplexers.

The logic diagram is shown in Figure 4.60, which shows the instantiation names and net names that will be used in the mixed design module. The mixed-design module is shown in Figure 4.61. The test bench module and the outputs are shown in Figures 4.62 and 4.63, respectively.

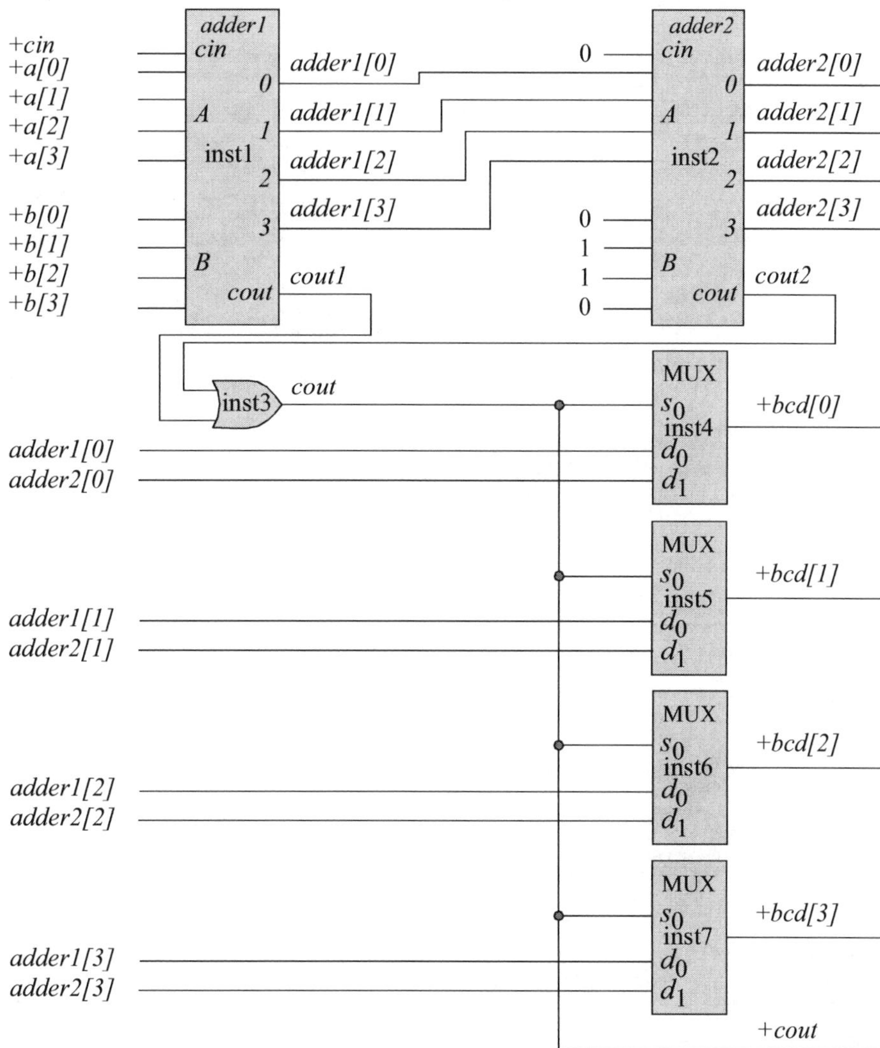

Figure 4.60 Decimal addition using multiplexers to obtain a valid decimal digit.

```
//mixed-design bcd adder using multiplexers
module add_bcd_mux_bip (a, b, cin, bcd, cout,
                         invalid_inputs);

input [3:0] a, b;     //define inputs and outputs
input cin;
output [3:0] bcd;
output cout, invalid_inputs;     //continued on next page
```

Figure 4.61 Mixed-design module for the BCD adder using multiplexers.

```
//reg if used in always statement
reg invalid_inputs;

wire [3:0] adder1, adder2; //define internal nets
wire cout1, cout2;

always @ (a or b)           //check for invalid inputs
begin
   if ((a > 4'b1001) || (b > 4'b1001))
      invalid_inputs = 1'b1;
   else
      invalid_inputs = 1'b0;
end

//instantiate the adder for adder1
adder4 inst1 (a[3:0], b[3:0], cin, adder1, cout1);

//instantiate the adder for adder2
adder4 inst2 (adder1, {1'b0, 1'b1, 1'b1, 1'b0},
               1'b0, adder2, cout2);

//instantiate the multiplexer select logic
or (cout, cout2, cout1);

//instantiate the 2:1 multiplexers
mux2_df inst3 (cout, {adder2[0], adder1[0]}, bcd[0]);

mux2_df inst4 (cout, {adder2[1], adder1[1]}, bcd[1]);

mux2_df inst5 (cout, {adder2[2], adder1[2]}, bcd[2]);

mux2_df inst7 (cout, {adder2[3], adder1[3]}, bcd[3]);

endmodule
```

Figure 4.61 (Continued)

```
//test bench mixed-design bcd adder using multiplexers
module add_bcd_mux_bip_tb;

//inputs are reg for test bench
//outputs are wire for test bench
reg [3:0] a, b;
reg cin;
wire [3:0] bcd;
wire cout, invalid_inputs;         //continued on next page
```

Figure 4.62 Test bench module for the BCD adder using multiplexers.

```
//display variables
initial
$monitor ("a=%b, b=%b, cin=%b, cout=%b, bcd=%b,
               invalid_inputs=%b",
            a, b, cin, cout, bcd, invalid_inputs);


//apply input vectors
initial
begin
   #0    a = 4'b0011;   b = 4'b0011;   cin = 1'b0;
   #10   a = 4'b0101;   b = 4'b0110;   cin = 1'b0;
   #10   a = 4'b0111;   b = 4'b1000;   cin = 1'b0;
   #10   a = 4'b0111;   b = 4'b0111;   cin = 1'b0;

   #10   a = 4'b1000;   b = 4'b1000;   cin = 1'b0;
   #10   a = 4'b1000;   b = 4'b1001;   cin = 1'b0;
   #10   a = 4'b1001;   b = 4'b1001;   cin = 1'b0;
   #10   a = 4'b0101;   b = 4'b0110;   cin = 1'b1;

   #10   a = 4'b0110;   b = 4'b0111;   cin = 1'b0;
   #10   a = 4'b0111;   b = 4'b1000;   cin = 1'b1;
   #10   a = 4'b1001;   b = 4'b1001;   cin = 1'b1;
   #10   a = 4'b1001;   b = 4'b1000;   cin = 1'b0;

   #10   a = 4'b0111;   b = 4'b1011;   cin = 1'b0;
                                          //invalid inputs
   #10   a = 4'b1111;   b = 4'b1000;   cin = 1'b0;
                                          //invalid inputs
   #10   a = 4'b1101;   b = 4'b1010;   cin = 1'b1;
                                          //invalid inputs

   #10   $stop;

end

//instantiate the module into the test bench
add_bcd_mux_bip inst1 (a, b, cin, bcd, cout, invalid_inputs);

endmodule
```

Figure 4.62 (Continued)

```
a=0011, b=0011, cin=0, cout=0, bcd=0110, invalid_inputs=0
a=0101, b=0110, cin=0, cout=1, bcd=0001, invalid_inputs=0
a=0111, b=1000, cin=0, cout=1, bcd=0101, invalid_inputs=0
a=0111, b=0111, cin=0, cout=1, bcd=0100, invalid_inputs=0

a=1000, b=1000, cin=0, cout=1, bcd=0110, invalid_inputs=0
a=1000, b=1001, cin=0, cout=1, bcd=0111, invalid_inputs=0
a=1001, b=1001, cin=0, cout=1, bcd=1000, invalid_inputs=0
a=0101, b=0110, cin=1, cout=1, bcd=0010, invalid_inputs=0

a=0110, b=0111, cin=0, cout=1, bcd=0011, invalid_inputs=0
a=0111, b=1000, cin=1, cout=1, bcd=0110, invalid_inputs=0
a=1001, b=1001, cin=1, cout=1, bcd=1001, invalid_inputs=0
a=1001, b=1000, cin=0, cout=1, bcd=0111, invalid_inputs=0

a=0111, b=1011, cin=0, cout=1, bcd=1000, invalid_inputs=1
a=1111, b=1000, cin=0, cout=1, bcd=1101, invalid_inputs=1
a=1101, b=1010, cin=1, cout=1, bcd=1110, invalid_inputs=1
```

Figure 4.63 Outputs for the BCD adder using multiplexers.

4.8 Decimal Subtraction

Subtraction of BCD numbers is performed by adding the radix (r) complement + 1 of the subtrahend to the minuend. This concept is shown in Equation 4.11, where operand A is the minuend, operand B is the subtrahend, and $(B' + 1)$ is the 10s complement (9s complement + 1) for BCD.

$$A - B = A + (B' + 1) \tag{4.11}$$

Examples are shown below that exemplify the principles of decimal subtraction. The result of each example is true subtraction.

True addition is where the result is the sum of the two numbers, regardless of the sign and corresponds to one of the following conditions:

$$(+A) \quad + \quad (+B)$$
$$(-A) \quad + \quad (-B)$$
$$(+A) \quad - \quad (-B)$$
$$(-A) \quad - \quad (+B)$$

True subtraction is where the result is the difference of the two numbers, regardless of the sign and corresponds to one of the following conditions:

$$\begin{array}{ccc} (+A) & - & (+B) \\ (-A) & - & (-B) \\ (+A) & + & (-B) \\ (-A) & + & (+B) \end{array}$$

Example 4.4 The subtrahend $+57_{10}$ will be subtracted from the minuend $+84_{10}$. This yields a difference of $+27_{10}$. The 10s complement of the subtrahend is obtained as follows using radix 10 numbers: $9 - 5 = 4$; $9 - 7 = 2 + 1 = 3$, where 4 and 2 are the 9s complement of 5 and 7, respectively. A carry-out of the high-order decade indicates that the result is a positive number in BCD.

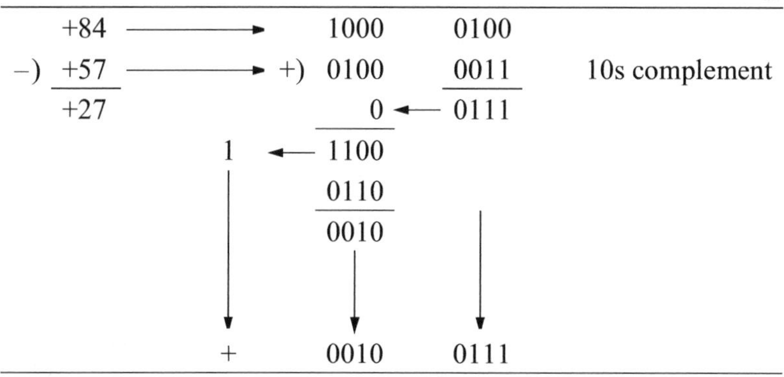

Example 4.5 The subtrahend $+98_{10}$ will be subtracted from the minuend $+73_{10}$. This yields a difference of -25_{10}. The 10s complement of the subtrahend is obtained as follows using radix 10 numbers: $9 - 9 = 0$; $9 - 8 = 1 + 1 = 2$, where 0 and 1 are the 9s complement of 9 and 8, respectively. A carry-out of 0 from the high-order decade indicates that the result is a negative BCD number in 10s complement. To obtain the result in radix 10, form the 10s complement of 75_{10}, which will yield 25_{10}.

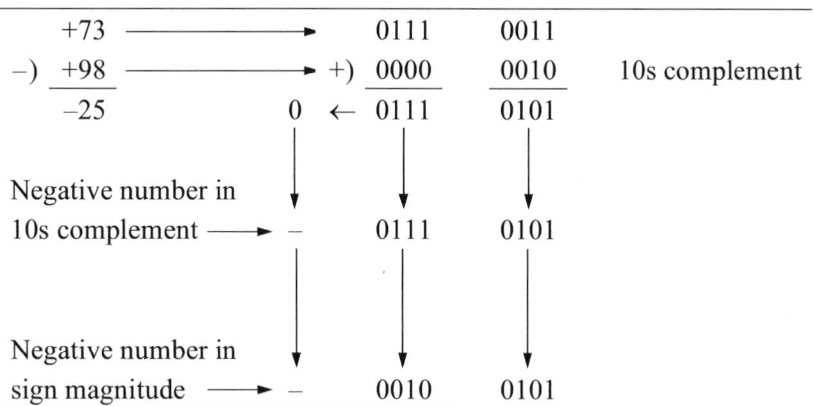

Example 4.6 The following decimal numbers will be added using BCD arithmetic: +54 and –23, as shown below. This can be considered as true subtraction, because the result is the difference of the two numbers, ignoring the signs. A carry of 1 from the high-order decade indicates a positive number.

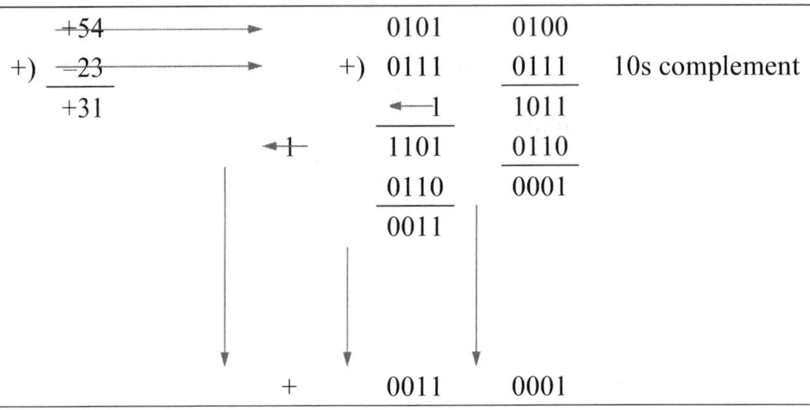

Example 4.7 The following decimal numbers will be subtracted using BCD arithmetic: +617 and +842, resulting in a difference of –225, as shown below. A carry of 0 from the high-order decade indicates a negative number in 10s complement notation.

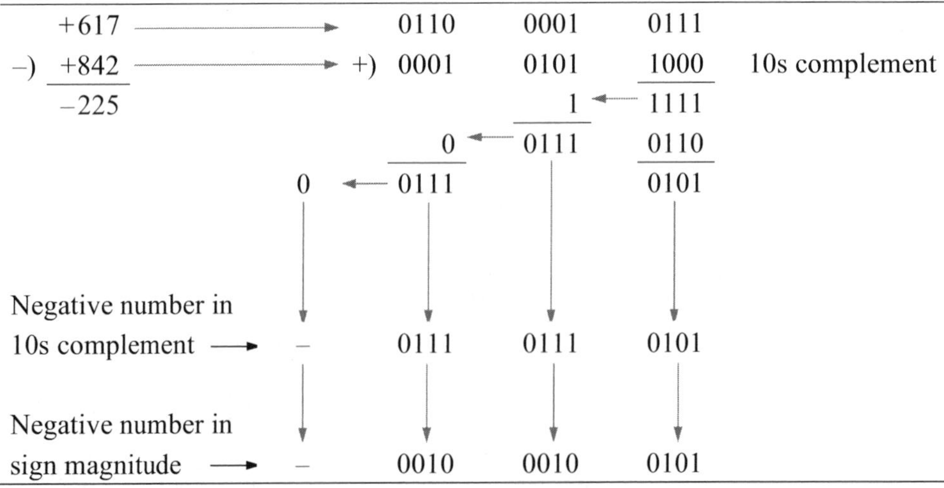

4.8.1 Decimal Subtraction Using Full Adders and Built-In Primitives for Four Bits

A 4-bit binary subtraction unit – including BCD – will be designed using instantiated full adders that were designed using built-in primitives. The Verilog design module for the full adder is reproduced in Figure 4.64 for convenience. The logic diagram for the 4-bit subtraction unit is shown in Figure 4.65.

```
//full adder using built-in primitives
module full_adder_bip (a, b, cin, sum, cout);

//define inputs and outputs
input a, b, cin;
output sum, cout;

//design the full adder
//and design the sum
xor   inst1 (net1, a, b);
and   inst2 (net2, a, b);
xor   inst3 (sum, net1, cin);

//design the carry-out
and   inst4 (net3, net1, cin);
or    inst5 (cout, net3, net2);

endmodule
```

Figure 4.64 Full adder to be used in the design of a 4-bit subtractor.

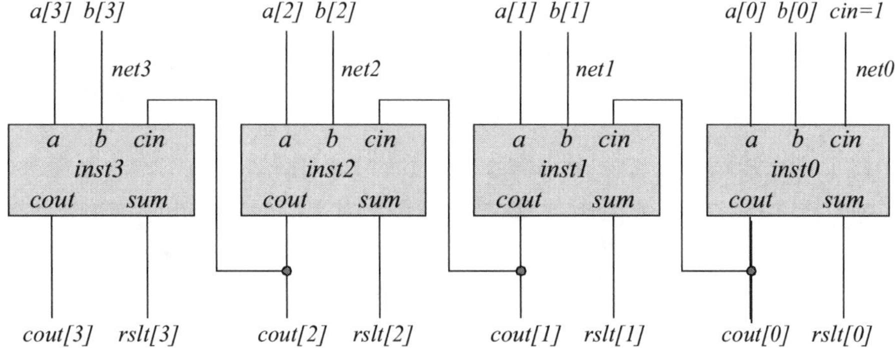

Figure 4.65 Logic diagram for the 4-bit subtraction unit.

The Verilog design module for the 4-bit subtraction unit is shown in Figure 4.66 using built-in primitives and instantiated full adders that were designed using built-in primitives. The test bench module is shown in Figure 4.67 and the outputs are shown in Figure 4.68.

```
//structural for a 4-bit subtractor
//using bip and instantiated full adders

module sub_4bit_bip (a, b, cin, rslt, cout);

//define inputs and outputs
input [3:0] a, b;
input cin;
output [3:0] rslt, cout;

//define internal nets
wire net0, net1, net2, net3;

//design the logic for stage 0
not   (net0, b[0]);
full_adder_bip inst0 (a[0], net0, cin, rslt[0], cout[0]);

//design the logic for stage 1
not   (net1, b[1]);
full_adder_bip inst1 (a[1], net1, cout[0], rslt[1],
                           cout[1]);

//design the logic for stage 2
not   (net2, b[2]);
full_adder_bip inst2 (a[2], net2, cout[1], rslt[2],
                           cout[2]);

//design the logic for stage 3
not   (net3, b[3]);
full_adder_bip inst3 (a[3], net3, cout[2], rslt[3],
                           cout[3]);

endmodule
```

Figure 4.66 Design module for the 4-bit subtractor.

```
//test bench for 4-bit subtractor

module sub_4bit_bip_tb;


//inputs are reg for test bench
//outputs are wire for test bench
reg [3:0] a, b;
reg cin;
wire [3:0] rslt, cout;


//display variables
initial
$monitor ("a = %b, b = %b, cin = %b, rslt = %b, cout = %b",
            a, b, cin, rslt, cout);


//apply input vectors
initial
begin
   #0    a = 4'b0110;   b = 4'b0010;   cin = 1'b1;
   #10   a = 4'b1100;   b = 4'b0110;   cin = 1'b1;
   #10   a = 4'b1110;   b = 4'b1010;   cin = 1'b1;
   #10   a = 4'b1110;   b = 4'b0011;   cin = 1'b1;

   #10   a = 4'b1111;   b = 4'b0010;   cin = 1'b1;
   #10   a = 4'b1110;   b = 4'b0110;   cin = 1'b1;
   #10   a = 4'b1110;   b = 4'b1111;   cin = 1'b1;
   #10   a = 4'b1111;   b = 4'b0011;   cin = 1'b1;

   #10   a = 4'b0001;   b = 4'b0010;   cin = 1'b1;
   #10   a = 4'b0001;   b = 4'b0001;   cin = 1'b1;
   #10   a = 4'b1000;   b = 4'b0111;   cin = 1'b1;
   #10   a = 4'b1001;   b = 4'b0011;   cin = 1'b1;

   #10   $stop;

end

//instantiate the module into the test bench
sub_4bit_bip inst1 (a, b, cin, rslt, cout);

endmodule
```

Figure 4.67 Test bench module for the 4-bit subtractor.

```
a = 0110, b = 0010, cin = 1, rslt = 0100, cout = 1111
a = 1100, b = 0110, cin = 1, rslt = 0110, cout = 1001
a = 1110, b = 1010, cin = 1, rslt = 0100, cout = 1111
a = 1110, b = 0011, cin = 1, rslt = 1011, cout = 1100

a = 1111, b = 0010, cin = 1, rslt = 1101, cout = 1111
a = 1110, b = 0110, cin = 1, rslt = 1000, cout = 1111
a = 1110, b = 1111, cin = 1, rslt = 1111, cout = 0000
a = 1111, b = 0011, cin = 1, rslt = 1100, cout = 1111

a = 0001, b = 0010, cin = 1, rslt = 1111, cout = 0001
a = 0001, b = 0001, cin = 1, rslt = 0000, cout = 1111
a = 1000, b = 0111, cin = 1, rslt = 0001, cout = 1000
a = 1001, b = 0011, cin = 1, rslt = 0110, cout = 1001
```

Figure 4.68 Outputs for the 4-bit subtractor.

Notice the seventh entry in the outputs: 1110 – 1111 (14 – 15). The result is 1111_2 (-1_{10}). This result is obtained as shown below, which is the BCD equivalent of the binary number.

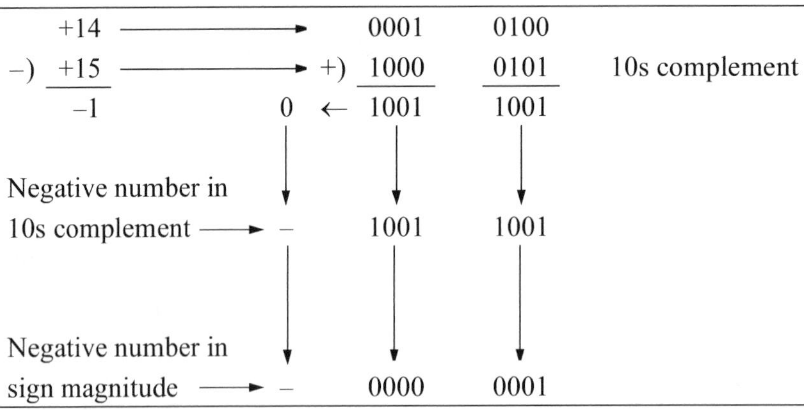

4.8.2 Decimal/Binary Subtraction Using Full Adders and Built-In Primitives for Eight Bits

The design in the previous section will be expanded to design a subtraction unit for eight bits using built-in primitives and instantiated full adders that were designed using built-in primitives. Some subtract operations demonstrate decimal subtraction. The logic diagram is shown in Figure 4.69.

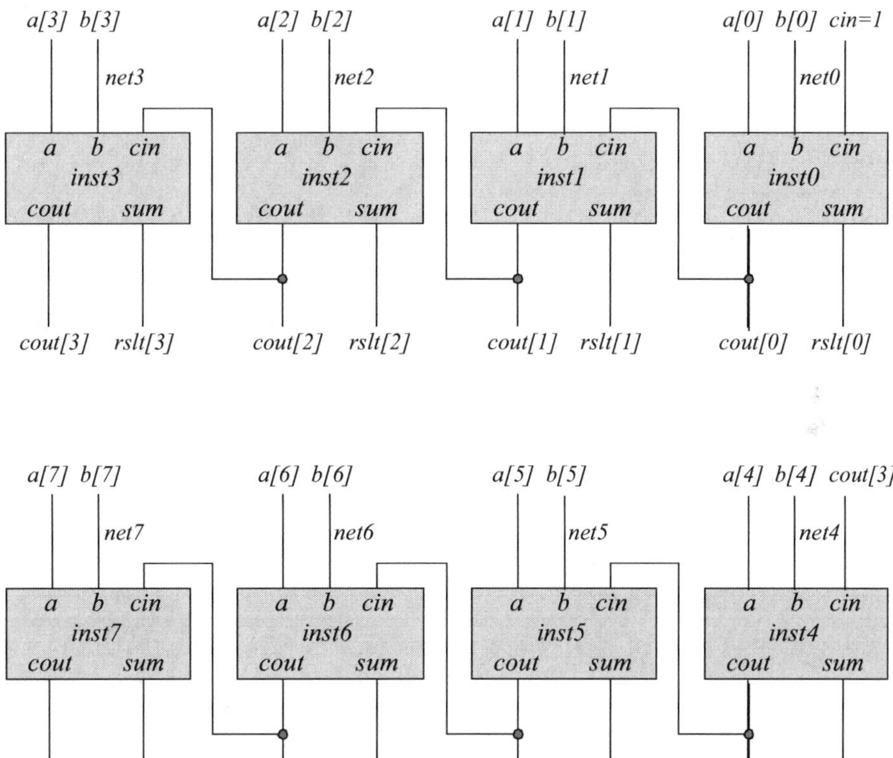

Figure 4.69 Logic diagram for the 8-bit binary subtraction unit.

The design module is shown in Figure 4.70 using built-in primitives and instantiated full adders that were designed using built-in primitives. The test bench module and the outputs are shown in Figures 4.71 and 4.72, respectively.

```
//structural for an 8-bit subtractor
//using bip and instantiated full adders
module sub_8bit_bip (a, b, cin, rslt, cout);

//define inputs and outputs
input [7:0] a, b;
input cin;
output [7:0] rslt, cout;

//define internal nets
wire net0, net1, net2, net3, net4, net5, net6, net7;
                              //continued on next page
```

Figure 4.70 Design module for the 8-bit subtraction unit.

```
//-----------------------------------------------------------
//full_adder_bip: a, b, cin, rslt, cout
//-----------------------------------------------------------

//design the logic for stage 0
not   (net0, b[0]);
full_adder_bip inst0 (a[0], net0, cin, rslt[0], cout[0]);

//design the logic for stage 1
not   (net1, b[1]);
full_adder_bip inst1 (a[1], net1, cout[0], rslt[1],
                                 cout[1]);

//design the logic for stage 2
not   (net2, b[2]);
full_adder_bip inst2 (a[2], net2, cout[1], rslt[2],
                                 cout[2]);

//design the logic for stage 3
not   (net3, b[3]);
full_adder_bip inst3 (a[3], net3, cout[2], rslt[3],
                                 cout[3]);

//-----------------------------------------------------------
//full_adder_bip: a, b, cin, rslt, cout
//-----------------------------------------------------------

//design the logic for stage 4
not   (net4, b[4]);
full_adder_bip inst4 (a[4], net4, cout[3], rslt[4],
                                 cout[4]);

//design the logic for stage 5
not   (net5, b[5]);
full_adder_bip inst5 (a[5], net5, cout[4], rslt[5],
                                 cout[5]);

//design the logic for stage 6
not   (net6, b[6]);
full_adder_bip inst6 (a[6], net6, cout[5], rslt[6],
                                 cout[6]);

//design the logic for stage 7
not   (net7, b[7]);
full_adder_bip inst7 (a[7], net7, cout[6], rslt[7],
                                 cout[7]);

endmodule
```

Figure 4.70 (Continued)

```
//test bench for 8-bit subtractor

module sub_8bit_bip_tb;


//inputs are reg for test bench
//outputs are wire for test bench
reg [7:0] a, b;
reg cin;
wire [7:0] rslt, cout;


//display variables
initial
$monitor ("a = %b, b = %b, cin = %b, rslt = %b, cout = %b",
            a, b, cin, rslt, cout);


//apply input vectors
initial
begin
   #0    cin = 1'b1;
   #0    a = 8'b0001_1001;    b = 8'b0000_0011; //25-03=22
   #10   a = 8'b0010_1100;    b = 8'b0000_0110; //44-06=38
   #10   a = 8'b0000_1110;    b = 8'b0000_1010; //14-10=04
   #10   a = 8'b0000_1110;    b = 8'b0000_0011; //14-03=11

   #10   a = 8'b0100_0000;    b = 8'b0010_0010; //64-18=46
   #10   a = 8'b0000_1110;    b = 8'b0000_0110; //14-06=08
   #10   a = 8'b1001_1110;    b = 8'b0000_1000; //158-08=150
   #10   a = 8'b1000_1111;    b = 8'b1000_1100; //143-140=03

   #10   a = 8'b0011_1001;    b = 8'b0000_1000; //57-08=49
   #10   a = 8'b0000_0001;    b = 8'b0000_0001; //01-01=00
   #10   a = 8'b0110_1000;    b = 8'b0011_0111; //104-55=49
   #10   a = 8'b0000_1001;    b = 8'b0000_0011; //09-03=06

   #10   $stop;
end

//instantiate the module into the test bench
sub_8bit_bip inst1 (a, b, cin, rslt, cout);

endmodule
```

Figure 4.71 Test bench module for the 8-bit subtraction unit.

```
a = 00011001, b = 00000011, cin = 1, rslt = 00010110
a = 00101100, b = 00000110, cin = 1, rslt = 00100110
a = 00001110, b = 00001010, cin = 1, rslt = 00000100
a = 00001110, b = 00000011, cin = 1, rslt = 00001011

a = 01000000, b = 00100010, cin = 1, rslt = 00011110
a = 00001110, b = 00000110, cin = 1, rslt = 00001000
a = 10011110, b = 00001000, cin = 1, rslt = 10010110
a = 10001111, b = 10001100, cin = 1, rslt = 00000011

a = 00111001, b = 00001000, cin = 1, rslt = 00110001
a = 00000001, b = 00000001, cin = 1, rslt = 00000000
a = 01101000, b = 00110111, cin = 1, rslt = 00110001
a = 00001001, b = 00000011, cin = 1, rslt = 00000110
```

Figure 4.72 Outputs for the 8-bit subtraction unit.

4.8.3 Eight-Bit Decimal Subtraction Unit with Built-In Primitives and Full Adders Designed Using Behavioral Modeling

Before presenting the organization for a decimal subtraction unit, a 9s complementer will be designed which will be used in the subtractor module together with a carry-in ($cin = 1$) to form the 10s complement of the subtrahend. The 9s complementer is required because BCD is not a self-complementing code; that is, it cannot form the diminished-radix complement ($r - 1$ complement) by inverting the four bits of each decade. The truth table for the 9s complementer is shown in Table 4.6.

Table 4.6 Nines Complementer

Subtrahend				9s Complement			
b[3]	*b[2]*	*b[1]*	*b[0]*	*f[3]*	*f[2]*	*f[1]*	*f[0]*
0	0	0	0	1	0	0	1
0	0	0	1	1	0	0	0
0	0	1	0	0	1	1	1
0	0	1	1	0	1	1	0
0	1	0	0	0	1	0	1
0	1	0	1	0	1	0	0
0	1	1	0	0	0	1	1
0	1	1	1	0	0	1	0
1	0	0	0	0	0	0	1
1	0	0	1	0	0	0	0

The equations for the 9s complementer are shown in Equation 4.12. The logic diagram for the 9s complementer is shown in Figure 4.73. The structural design module for the 9s complementer using built-in primitives is shown in Figure 4.74. The test bench module and the outputs are shown in Figures 4.75 and 4.76, respectively.

$$
\begin{aligned}
&f[0] = b[0]' \\
&f[1] = b[1] \\
&f[2] = (b[2] \oplus b[1]) \\
&f[3] = b[3]'b[2]'b[1]'
\end{aligned}
\tag{4.12}
$$

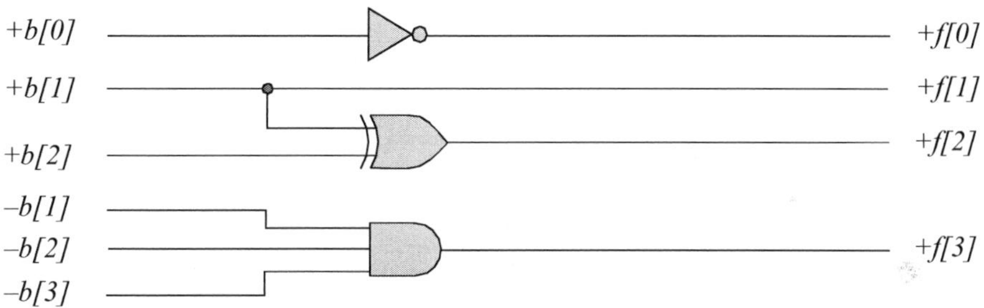

Figure 4.73 Logic diagram for the 9s complementer.

```
//9s complementer using built-in primitives

module nines_comp_sub_bip (b, f);

//define inputs and outputs
input [3:0] b;
output [3:0] f;

//design the logic for the 9s complementer
not       (f[0], b[0]);
assign    f[1] = b[1];
xor       (f[2], b[1], b[2]);
and       (f[3], ~b[1], ~b[2], ~b[3]);

endmodule
```

Figure 4.74 Design module for the 9s complementer.

```
//test bench for the 9s complementer
module nines_comp_sub_bip_tb;

//inputs are reg for test bench
//outputs are wire for test bench
reg [3:0] b;
wire [3:0] f;

//display variables
initial
$monitor ("b = %b, f = %b", b, f);

//apply input vectors
initial
begin
    #0    b = 4'b0000;
    #10   b = 4'b0001;
    #10   b = 4'b0010;
    #10   b = 4'b0011;
    #10   b = 4'b0100;
    #10   b = 4'b0101;
    #10   b = 4'b0110;
    #10   b = 4'b0111;
    #10   b = 4'b1000;
    #10   b = 4'b1001;
    #10   $stop;
end

//instantiate the module into the test bench
nines_comp_sub_bip inst1 (b, f);

endmodule
```

Figure 4.75 Test bench module for the 9s complementer.

```
b = 0000, f = 1001
b = 0001, f = 1000
b = 0010, f = 0111
b = 0011, f = 0110
b = 0100, f = 0101
b = 0101, f = 0100
b = 0110, f = 0011
b = 0111, f = 0010
b = 1000, f = 0001
b = 1001, f = 0000
```

Figure 4.76 Outputs for the 9s complementer.

The logic diagram for the BCD subtractor is shown in Figure 4.77. The minuend *a[3:0]* connects to the *A* inputs of a fixed-point adder for the units decade; the subtrahend *b[3:0]* connects to the inputs of a 9s complementer whose outputs *f[3:0]* connect to the *B* inputs of the adder, which has outputs *sum[3:0]* and *cout3*. The *aux_cy* output adds six to the *B* inputs of the succeeding adder to yield the outputs *bcd[3:0]*.

In a similar manner, the minuend *a[7:4]* connects to the *A* inputs of the adder for the tens decade; the subtrahend *b[7:4]* connects to a 9s complementer whose outputs *f[7:4]* connect to the *B* inputs of the adder, which generates *sum[7:4]* and *cout7*. The *cout* output adds six to the *B* inputs of the succeeding adder to yield the outputs *bcd[7:4]*.

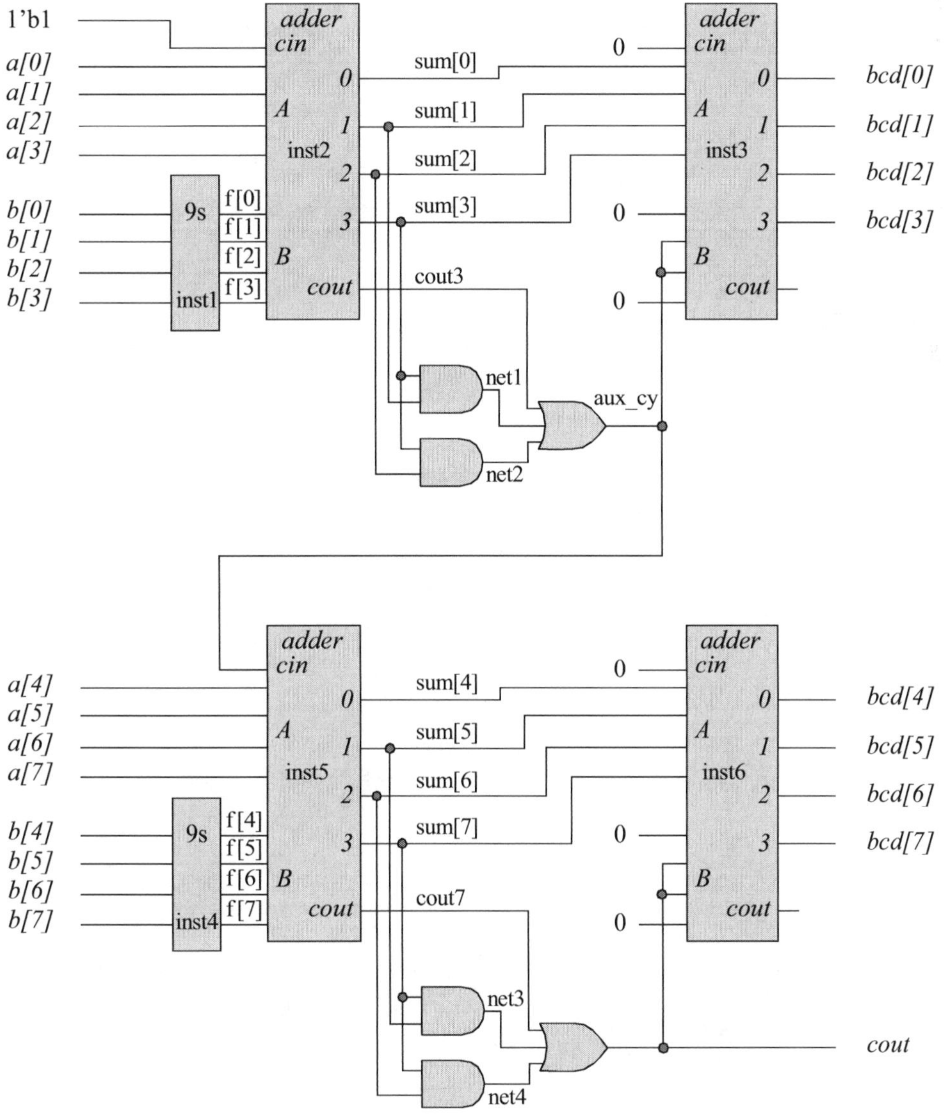

Figure 4.77 Logic diagram for the two-stage BCD subtractor.

The behavioral design module for the 4-bit fixed-point adder is shown in Figure 4.78. The test bench module and the outputs are shown in Figures 4.79 and 4.80, respectively.

The structural design module for the 8-bit subtractor is shown in Figure 4.81. There are two input operands, the minuend *a[7:0]* and the subtrahend *b[7:0]*, and one input mode control (*cin* = 1) to specify a subtract operation. There are two outputs: *bcd[7:0]*, which represents a valid BCD number, and a carry-out, *cout*. The test bench is shown in Figure 4.82 and contains operands for subtraction, including numbers that result in negative differences in BCD. The outputs are shown in Figure 4.83.

```
//behavioral model for a 4-bit adder

module adder4 (a, b, cin, sum, cout);

//define inputs and outputs
input [3:0] a, b;
input cin;
output [3:0] sum;
output cout;

//variables are reg in always
reg [3:0] sum;
reg cout;

//perform the sum and carry-out operations
always @ (a or b or cin)
begin
sum   = a + b + cin;
cout =  (a[3] & b[3]) |
       ((a[3] | b[3]) & (a[2] & b[2])) |
       ((a[3] | b[3]) & (a[2] | b[2]) & (a[1] & b[1])) |
       ((a[3] | b[3]) & (a[2] | b[2]) & (a[1] | b[1])
                      & (a[0] & b[0])) |
       ((a[3] | b[3]) & (a[2] | b[2]) & (a[1] | b[1])
                      & (a[0] | b[0]) & cin);

end

endmodule
```

Figure 4.78 Behavioral design module for 4-bit adder.

```
//test bench for the 4-bit adder

module adder4_tb;


//inputs are reg for test bench
//outputs are wire for test bench
reg [3:0] a, b;
reg cin;
wire [3:0] sum;
wire cout;


//display variables
initial
$monitor ("a=%b, b=%b, cin=%b, cout=%b, sum=%b",
            a, b, cin, cout, sum);


//apply input vectors
initial
begin
   #0    a=4'b0000;  b=4'b0000;  cin=1'b0;
   #10   a=4'b0001;  b=4'b0001;  cin=1'b0;
   #10   a=4'b0001;  b=4'b0011;  cin=1'b0;
   #10   a=4'b0101;  b=4'b0001;  cin=1'b0;
   #10   a=4'b0111;  b=4'b0001;  cin=1'b0;

   #10   a=4'b0101;  b=4'b0101;  cin=1'b0;
   #10   a=4'b1001;  b=4'b0101;  cin=1'b1;
   #10   a=4'b1000;  b=4'b1000;  cin=1'b1;
   #10   a=4'b1011;  b=4'b1110;  cin=1'b1;
   #10   a=4'b1111;  b=4'b1111;  cin=1'b1;

   #10   $stop;

end


//instantiate the module into the test bench
adder4 inst1 adder4 (a, b, cin, sum, cout);

endmodule
```

Figure 4.79 Test bench module for 4-bit adder.

```
a=0000, b=0000, cin=0, cout=0, sum=0000
a=0001, b=0001, cin=0, cout=0, sum=0010
a=0001, b=0011, cin=0, cout=0, sum=0100
a=0101, b=0001, cin=0, cout=0, sum=0110
a=0111, b=0001, cin=0, cout=0, sum=1000

a=0101, b=0101, cin=0, cout=0, sum=1010
a=1001, b=0101, cin=1, cout=0, sum=1111
a=1000, b=1000, cin=1, cout=1, sum=0001
a=1011, b=1110, cin=1, cout=1, sum=1010
a=1111, b=1111, cin=1, cout=1, sum=1111
```

Figure 4.80 Outputs for 4-bit adder.

```
//structural bcd subtractor
module sub_8bit_struc (a, b, bcd, cout);

input [7:0] a, b;     //define inputs and outputs
output [7:0] bcd;
output cout;

wire [7:0] f;         //define internal nets
wire [7:0] sum;
wire cout3, net1, net2, aux_cy;
wire cout7, net3, net4;

//-------------------------------------------------------
//instantiate the logic for the units stage [3:0]
//instantiate the 9s complementer
nines_comp_sub_bip inst1 (b[3:0], f[3:0]);

//instantiate the adder for the intermediate sum for units
adder4 inst2 (a[3:0], f[3:0], 1'b1, sum[3:0], cout3);

//instantiate the logic gates
and   (net1, sum[3], sum[1]);
and   (net2, sum[3], sum[2]);
or    (aux_cy, cout3, net1, net2);

//instantiate the adder for the bcd sum [3:0]
adder4 inst3 (sum[3:0], {1'b0, aux_cy, aux_cy, 1'b0},
                  1'b0, bcd[3:0], 1'b0);

                                    //continued on next page
```

Figure 4.81 Structural design module for the 8-bit decimal subtractor.

```
//------------------------------------------------------------
//instantiate the logic for the tens stage [7:4]
//instantiate the 9s complementer
nines_comp_sub_bip inst4 (b[7:4], f[7:4]);

//instantiate the adder for the intermediate sum for tens
adder4 inst5 (a[7:4], f[7:4], aux_cy, sum[7:4], cout7);

//instantiate the logic gates
and    (net3, sum[7], sum[5]);
and    (net4, sum[7], sum[6]);
or     (cout, cout7, net3, net4);

//instantiate the adder for the bcd sum [7:4]
adder4 inst6 (sum[7:4], {1'b0, cout, cout, 1'b0},
                  1'b0, bcd[7:4], 1'b0);

//------------------------------------------------------------
endmodule
```

Figure 4.81 (Continued)

```
//test bench for the decimal eight-bit subtractor
module sub_8bit_struc_tb;

//inputs are reg for test bench
//outputs are wire for test bench
reg [7:0] a, b;
wire [7:0] bcd;
wire cout;

initial     //display variables
$monitor ("a = %b, b = %b, bcd_tens = %b, bcd_units = %b",
            a, b, bcd[7:4], bcd[3:0]);

//apply input vectors
initial
begin
   #0    a = 8'b0001_1001;    b = 8'b0000_0011;   //19-03=16
   #10   a = 8'b0111_0110;    b = 8'b0100_0010;   //76-42=34
   #10   a = 8'b1001_1001;    b = 8'b0110_0110;   //99-66=33
   #10   a = 8'b1000_0101;    b = 8'b0001_0100;   //85-14=71
                                 //continued on next page
```

Figure 4.82 Test bench module for the 8-bit decimal subtractor.

```
    #10   a = 8'b0101_0101;     b = 8'b0100_0100;   //55-44=11
    #10   a = 8'b0011_0011;     b = 8'b0010_0111;   //33-27=06
    #10   a = 8'b0011_0011;     b = 8'b0110_0110;   //33-66=-33
    #10   a = 8'b0001_0001;     b = 8'b1001_1001;   //11-99=-18
    #10   $stop;
end

//instantiate the module into the test bench
sub_8bit_struc inst1 (a, b, bcd, cout);

endmodule
```

Figure 4.82 (Continued)

```
a = 00011001, b = 00000011, bcd_tens = 0001, bcd_units = 0110
a = 01110110, b = 01000010, bcd_tens = 0011, bcd_units = 0100
a = 10011001, b = 01100110, bcd_tens = 0011, bcd_units = 0011
a = 10000101, b = 00010100, bcd_tens = 0111, bcd_units = 0001

a = 01010101, b = 01000100, bcd_tens = 0001, bcd_units = 0001
a = 00110011, b = 00100111, bcd_tens = 0000, bcd_units = 0110
a = 00110011, b = 01100110, bcd_tens = 0110, bcd_units = 0111
a = 00010001, b = 10011001, bcd_tens = 0001, bcd_units = 0010
```

Figure 4.83 Outputs for the 8-bit decimal subtractor.

Observe the last two entries in the outputs of Figure 4.83. The operations are as follows: 33 – 66 = –33 and 11 – 99 = –88. The two subtractions are obtained as follows:

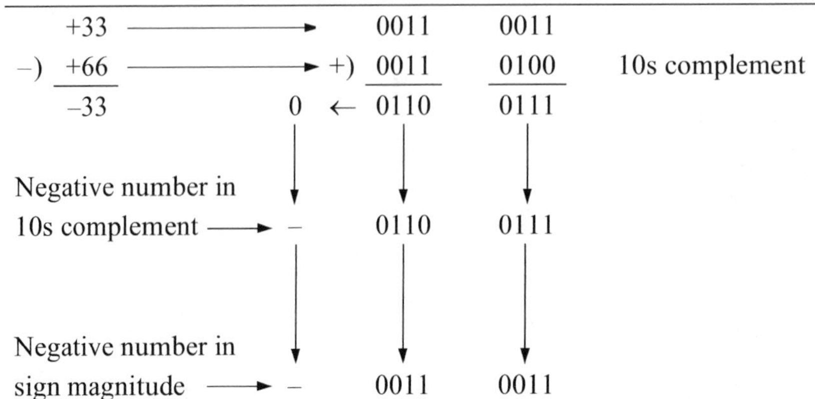

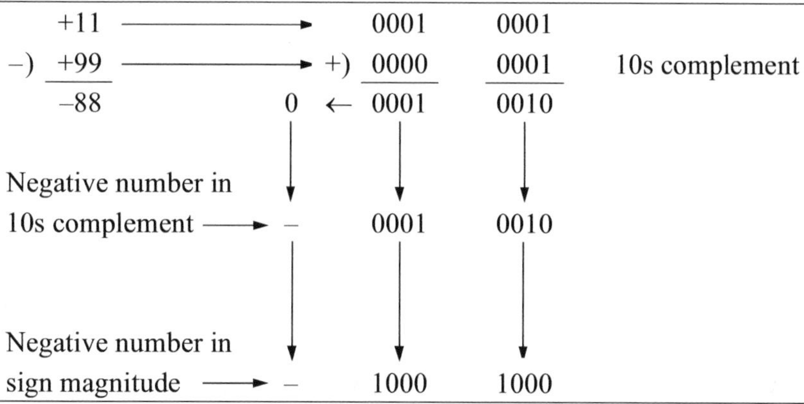

4.9 Decimal Multiplication

There are three operands in multiplication: the *multiplicand* is multiplied by the *multiplier* to produce a *product*. Decimal arithmetic operations can be performed in the fixed-point number representation then converting the result to the binary-coded decimal (BCD) number representation. This method is used in this section for a decimal multiplication operation using behavioral modeling.

Converting from binary to BCD is accomplished by multiplying the BCD number by two repeatedly. Multiplying by two is accomplished by a left shift of one bit position followed by an adjustment, if necessary. For example, a left shift of BCD 1001 (9_{10}) results in 1 0010 which is 18 in binary, but only 12 in BCD. Adding six to the low-order BCD digit results in 1 1000, which is the required value of 18_{10}.

Instead of adding six after the shift, the same result can be achieved by adding three before the shift since a left shift multiplies any number by two. BCD digits in the range 0–4 do not require an adjustment before being shifted left, because the shifted number will be in the range 0–8, which can be contained in a 4-bit BCD digit. However, if the number to be shifted is in the range 5–9, then an adjustment will be required before the left shift, because the shifted number will be in the range 10–18, which requires two BCD digits. Therefore, three is added to the digit prior to the next left shift of 1-bit position.

The multiplication of 9 x 9 is shown below, which yields a product of 81. Table 4.7 shows the procedure for converting from binary 0101 0001_2 (81_{10}) to BCD. Since there are 8 bits in the binary number, 8 left-shift operations are required, yielding the resulting BCD number of 1000 0001_{BCD}. Concatenated registers *A* and *B* are shifted left one bit position during each sequence. During the final left shift operation, no adjustment is performed.

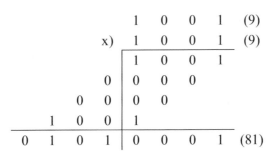

				1	0	0	1	(9)
			x)	1	0	0	1	(9)
				1	0	0	1	
			0	0	0	0		
		0	0	0	0			
	1	0	0	1				
0	1	0	1	0	0	0	1	(81)

Table 4.7 Example of Binary-to-Decimal Conversion

	A Register (BCD)			*B* Register (Binary)	
	15 ... 12	11 ... 8		7 ... 4	3 ... 0
	0 0 0 0	0 0 0 0	←	0 1 0 1	0 0 0 1
Shift left 1	0 0 0 0	0 0 0 0		1 0 1 0	0 0 1 0
Shift left 1	0 0 0 0	0 0 0 1		0 1 0 0	0 1 0 0
Shift left 1	0 0 0 0	0 0 1 0		1 0 0 0	1 0 0 0
Shift left 1	0 0 0 0	0 1 0 1		0 0 0 1	0 0 0 0
Add 3		0 0 1 1			
	0 0 0 0	1 0 0 0			
Shift left 1	0 0 0 1	0 0 0 0		0 0 1 0	0 0 0 0
Shift left 1	0 0 1 0	0 0 0 0		0 1 0 0	0 0 0 0
Shift left 1	0 1 0 0	0 0 0 0		1 0 0 0	0 0 0 0
Shift left 1	**1 0 0 0**	**0 0 0 1**		0 0 0 0	0 0 0 0

The procedure shown in Table 4.7 will be used for BCD multiplication by performing the multiply operation in the fixed-point number representation, and then converting the product to BCD notation. The design will be implemented using behavioral modeling. A 16-bit left-shift register — consisting of two 8-bit registers *A*, *a_reg*, and *B*, *b-reg*, in concatenation — is used for the shifting sequence.

A shift counter is used to determine the number of shift sequences to be executed. Since the final shift sequence is a left-shift operation only (no adjustment), the shift counter is set to a value of the binary length minus one; then a final left shift operation occurs. A **while** loop determines the number of times that the procedural statements within the loop are executed and is a function of the shift counter value.

The **while** loop executes a procedural statement or a block of procedural statements as long as a Boolean expression returns a value of true (≥ 1). When the procedural statements are executed, the Boolean expression is reevaluated. The loop is executed until the expression returns a value of false, in this case a shift counter value of zero. If the evaluation of the expression is false, then the **while** loop is terminated and control is passed to the next statement in the module. If the expression is false before the loop is initially entered, then the **while** loop is never executed.

The behavioral module is shown in Figure 4.84, where register *A* is reset to all zeroes and register *B* contains the product of the multiplicand and the multiplier. The test bench is shown in Figure 4.85, in which several input vectors are applied to the multiplicand and the multiplier, including binary and decimal values. The outputs are shown in Figure 4.86

```
//behavioral bcd multiplier
module mul_bcd_behav2 (a, b, bcd);

input [3:0] a;     //define inputs and outputs
input [3:0] b;
output [7:0] bcd;

//variables are declared as reg in always
reg [7:0] a_reg, b_reg;
reg [15:0] shift_reg;
reg [3:0] shift_ctr;

always @ (a or b)
begin
   shift_ctr = 4'b0111;          //7 shift sequences
   a_reg = 8'b0000_0000;         //reset register a
   b_reg = a * b;                //register b contains product

   shift_reg = {a_reg, b_reg};    //regs a, b are concatenated

   while (shift_ctr)
      begin
         shift_reg = shift_reg << 1;
            if (shift_reg[11:8] > 4'b0100)
               shift_reg[11:8] = shift_reg[11:8] + 4'b0011;

            if (shift_reg[15:12] > 4'b0100)
               shift_reg[15:12] = shift_reg[15:12] + 4'b0011;

         shift_ctr = shift_ctr - 1;
      end
         shift_reg = shift_reg << 1;
end

assign bcd = shift_reg[15:8];

endmodule
```

Figure 4.84 Behavioral design module for the decimal multiplier.

```
//test bench for bcd multiplier

module mul_bcd_behav2_tb;

//inputs are reg for test bench
//outputs are wire for test bench
reg [3:0] a;
reg [3:0] b;
wire [7:0] bcd;

//display variables
initial
$monitor ("a = %b, b = %b, bcd = %b", a, b, bcd);

//apply input vectors
initial
begin
   #0    a = 4'b0110;   b = 4'b1001;      //6 x 9 = 54

   #10   a = 4'b0010;   b = 4'b0110;      //2 x 6 = 12

   #10   a = 4'b0111;   b = 4'b0111;      //7 x 7 = 49

   #10   a = 4'b1001;   b = 4'b1000;      //9 x 8 = 72

   #10   a = 4'b0111;   b = 4'b1001;      //7 x 9 = 63

   #10   a = 4'b0100;   b = 4'b0100;      //4 x 4 = 16

//-----------------------------------------------------------
   #10   a = 4'd9;      b = 4'd9;         //9 x 9 = 81

   #10   a = 4'd7;      b = 4'd6;         //7 x 6 = 42

   #10   a = 4'd8;      b = 4'd8;         //8 x 8 = 64

   #10   a = 4'd5;      b = 4'd7;         //5 x 7 = 35

   #10   $stop;

end

//instantiate the module into the test bench
mul_bcd_behav2 inst1 (a, b, bcd);

endmodule
```

Figure 4.85 Test bench module for the decimal multiplier.

```
a = 0110, b = 1001, bcd = 0101_0100
a = 0010, b = 0110, bcd = 0001_0010
a = 0111, b = 0111, bcd = 0100_1001
a = 1001, b = 1000, bcd = 0111_0010
a = 0111, b = 1001, bcd = 0110_0011
a = 0100, b = 0100, bcd = 0001_0110

//-----------------------------------------------------------
a = 1001, b = 1001, bcd = 1000_0001
a = 0111, b = 0110, bcd = 0100_0010
a = 1000, b = 1000, bcd = 0110_0100
a = 0101, b = 0111, bcd = 0011_0101
```

Figure 4.86 Outputs for the decimal multiplier.

4.10 Decimal Division

Unlike multiplication, division is not commutative; that is, $A/B \neq B/A$, except when $A = B$, where A and B are the dividend and divisor, respectively. In general, the operands are as shown below, where A is the $2n$-bit dividend and B is the n-bit divisor. The quotient is Q and the remainder is R, both of which are n bits.

$$A = a_{2n-1}\ a_{2n-2} \ldots a_n\ a_{n-1} \ldots a_1\ a_0$$
$$B = b_{n-1}\ b_{n-2} \ldots b_1\ b_0$$
$$Q = q_{n-1}\ q_{n-2} \ldots q_1\ q_0$$
$$R = r_{n-1}\ r_{n-2} \ldots r_1\ r_0$$

The sign of the quotient is determined by the following equation:

$$q_{n-1} = a_{2n-1} \oplus b_{n-1}$$

The remainder has the same sign as the dividend. The process of division is one of shift, subtract, and compare operations. A simple method to perform binary-coded decimal (BCD) division is to implement the design using the fixed-point division algorithm and then convert the resulting quotient and remainder to BCD.

A review of the algorithm for fixed-point division is appropriate at this time. The dividend is initially shifted left one bit position. Then the divisor is subtracted from the dividend. Subtraction is accomplished by adding the 2s complement of the divisor. If the high-order bit of the subtract operation is 1, then a 0 is placed in the next lower-order bit position of the quotient; if the high-order bit is 0, then a 1 is placed in the next lower-order bit position of the quotient. The concatenated partial remainder and dividend are then shifted left one bit position. Fixed-point binary restoring division requires one subtraction for each quotient bit. Figure 4.87 illustrates the procedure using a dividend of 0011_1101 (61_{10}) and a divisor of 0111 (7_{10}) to yield a quotient of 8_{10} and a remainder of 5_{10}.

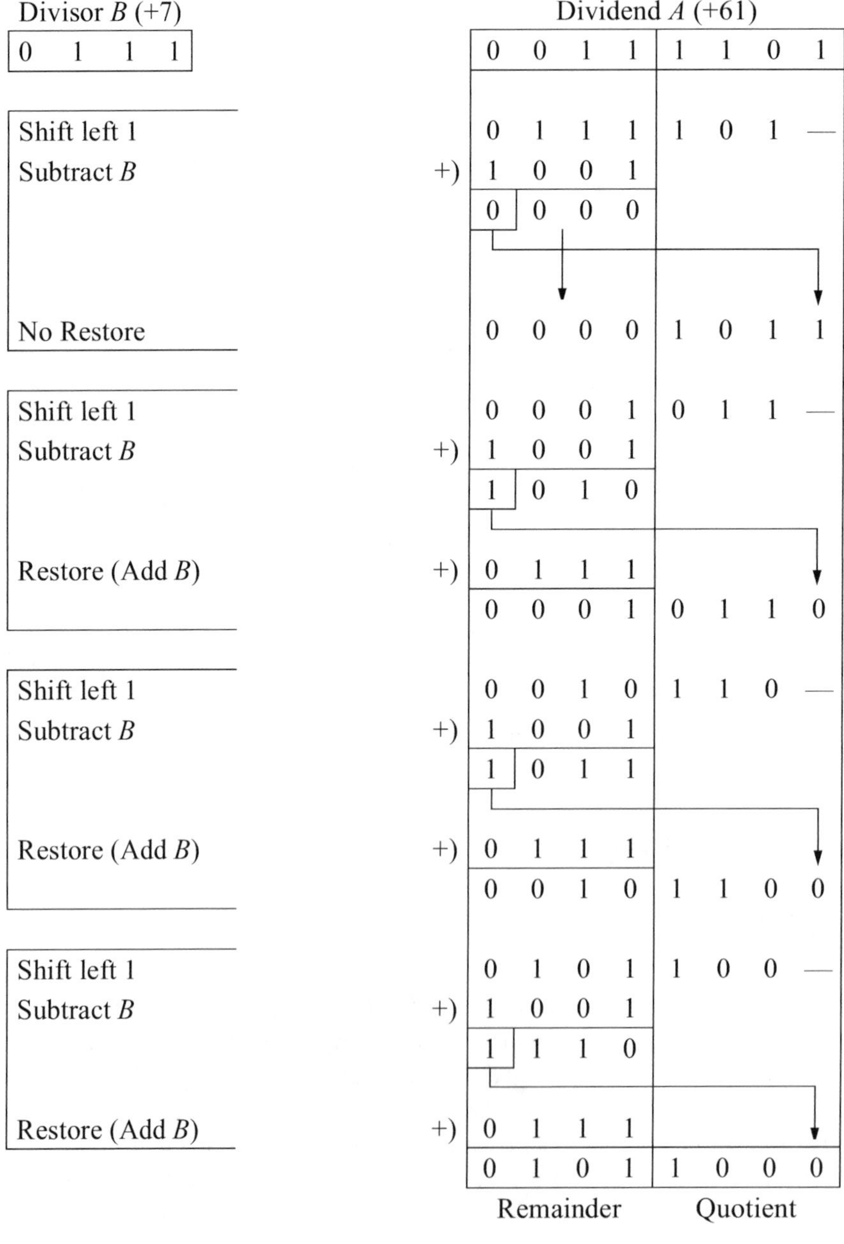

Figure 4.87 Example of fixed-point restoring division.

A second example is shown in Figure 4.88 which illustrates the procedure using a dividend of 0011_0100 (52_{10}) and a divisor of 0101 (5_{10}) to yield a quotient of 10_{10} and a remainder of 2_{10}. Since there are four bits in the divisor, there are four cycles. Like the previous example, the low-order quotient bit is left blank after each left shift operation. Then the divisor is subtracted from the high-order half of the dividend. This sequence repeats for all four bits of the divisor.

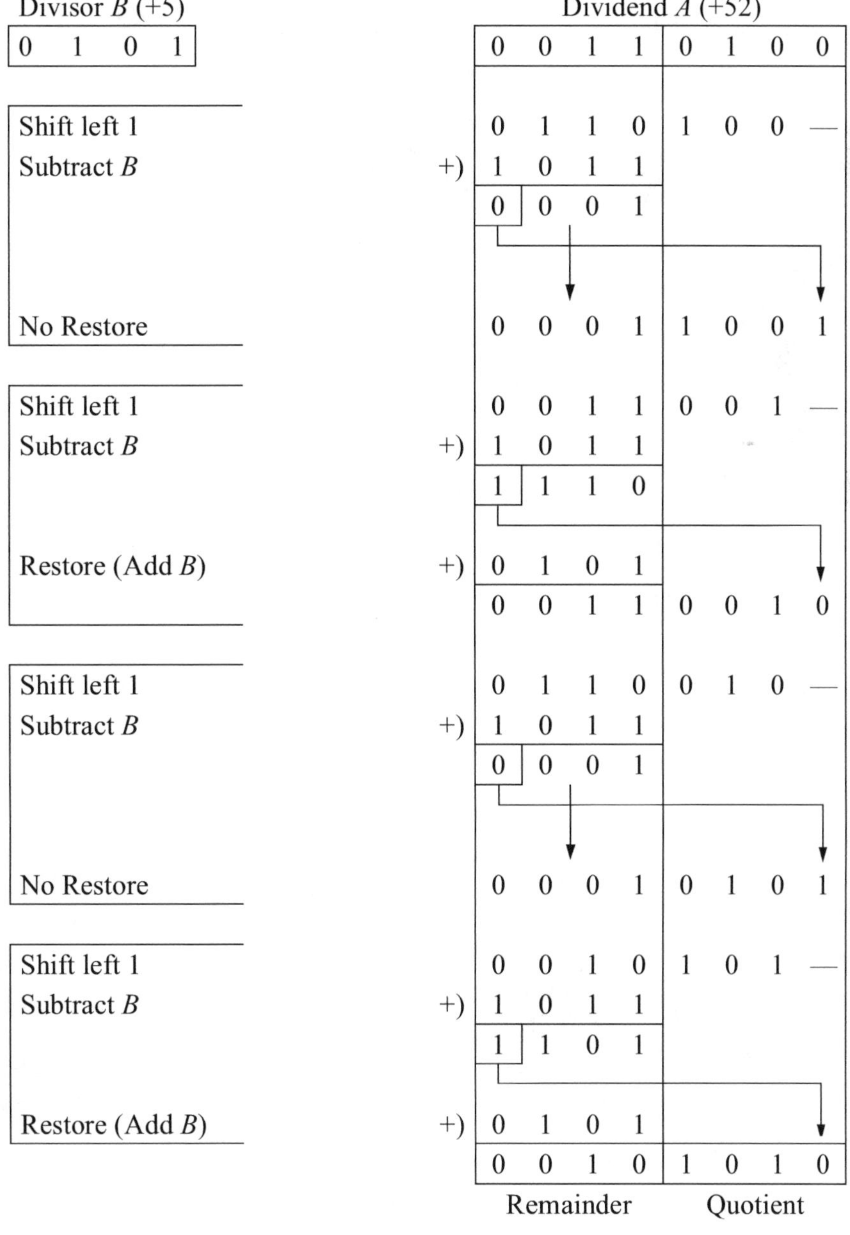

Figure 4.88 Example of fixed-point division.

The behavioral and dataflow module is shown in Figure 4.89. The test bench module is shown in Figure 4.90 and the outputs are shown in Figure 4.91. The dividend is an 8-bit vector, *a[7:0]*; the divisor is a 4-bit vector, *b[3:0]*; and the result is an 8-bit quotient/remainder vector, *rslt[7:0]*, with a carry-out of the high-order quotient bit, *cout_quot*.

The operation begins on the positive assertion of a *start* pulse and follows the algorithm outlined in the previous discussion. The *rslt[7:0]* register is set to the value of the dividend, *a[7:0]*, and a sequence counter, *count*, is set to a value of four (0100), which is the size of the divisor. If the dividend and divisor are both nonzero, then the process continues until the count-down counter, *count*, reaches a value of zero, controlled by the **while** loop.

```
//mixed-design for bcd restoring division

module div_bcd2 (a, b, start, rslt, cout_quot);

//define inputs and outputs
input [7:0] a;
input [3:0] b;
input start;
output [7:0] rslt;
output cout_quot;

//variables are declared as reg in always
wire [3:0] b_bar;

//define internal registers
reg [3:0] b_neg;
reg [7:0] rslt;
reg [3:0] count;
reg [3:0] quot;
reg cout_quot;

assign b_bar = ~b;
always @ (b_bar)
   b_neg = b_bar + 1;

always @ (posedge start)
begin
   rslt = a;
   count = 4'b0100;

                                 //continued on next page
```

Figure 4.89 Design module for the decimal divisor.

```
if ((a!=0) && (b!=0))
      while (count)
         begin
            rslt = rslt << 1;
            rslt = {(rslt[7:4] + b_neg), rslt[3:0]};
               if (rslt[7] == 1)     //restore
                  begin
                     rslt = {(rslt[7:4] + b), rslt[3:1],
                              1'b0};
                     count = count - 1;
                  end

               else              //no restore
                  begin
                     rslt = {rslt[7:1], 1'b1};
                     count = count - 1;
                  end
         end

   if (rslt[3:0] > 4'b1001)    //convert to bcd
      {cout_quot, rslt[3:0]} = rslt[3:0] + 4'b0110;

   else
         cout_quot = 1'b0;
end

endmodule
```

Figure 4.89 (Continued)

```
//test bench for bcd restoring division


module div_bcd2_tb;


//inputs are reg for test bench
//outputs are wire for test bench
reg [7:0] a;
reg [3:0] b;
reg start;
wire [7:0] rslt;
wire cout_quot;
                                            //continued on next page
```

Figure 4.90 Test bench module for decimal division.

```
//display variables
initial
$monitor ("a = %b, b = %b, quot_tens = %b,
            quot_units = %b, rem %b",
         a, b, {{3{1'b0}}, cout_quot},
            rslt[3:0], rslt[7:4]);

//apply input vectors
initial
begin
   #0    start = 1'b0;

         //60 / 7; quot = 8, rem = 4
         a = 8'b0011_1100;     b = 4'b0111;
   #10   start = 1'b1;
   #10   start = 1'b0;

         //13 / 5; quot = 2, rem = 3
   #10   a = 8'b0000_1101;     b = 4'b0101;
   #10   start = 1'b1;
   #10   start = 1'b0;

         //60 / 7; quot = 8, rem = 4
   #10   a = 8'b0011_1100;     b = 4'b0111;
   #10   start = 1'b1;
   #10   start = 1'b0;

         //82 / 6; quot = 13, rem = 4
   #10   a = 8'b0101_0010;     b = 4'b0110;
   #10   start = 1'b1;
   #10   start = 1'b0;

         //56 / 7; quot = 8, rem = 0
   #10   a = 8'b0011_1000;     b = 4'b0111;
   #10   start = 1'b1;
   #10   start = 1'b0;

         //100 / 7; quot = 14, rem = 2
   #10   a = 8'b0110_0100;     b = 4'b0111;
   #10   start = 1'b1;
   #10   start = 1'b0;

         //110 / 7; quot = 15, rem = 5
   #10   a = 8'b0110_1110;     b = 4'b0111;
   #10   start = 1'b1;
   #10   start = 1'b0;
                                       //continued on next page
```

Figure 4.90 (Continued)

```
        //99 / 9; quot = 11, rem = 0
    #10   a = 8'b0110_0011;     b = 4'b1001;
    #10   start = 1'b1;
    #10   start = 1'b0;

        //52 / 5; quot = 10, rem = 2
    #10   a = 8'b0011_0100;     b = 4'b0101;
    #10   start = 1'b1;
    #10   start = 1'b0;

        //88 / 9; quot = 9, rem = 7
    #10   a = 8'b0101_1000;     b = 4'b1001;
    #10   start = 1'b1;
    #10   start = 1'b0;

        //130 / 9; quot = 14, rem = 4
    #10   a = 8'b1000_0010;     b = 4'b1001;
    #10   start = 1'b1;
    #10   start = 1'b0;

    #10   $stop;
end

//instantiate the module into the test bench
div_bcd2 inst1 (a, b, start, rslt, cout_quot);

endmodule
```

Figure 4.90 (Continued)

```
//start = 0
a = 00111100, b = 0111, quot_tens = 000x, quot_units = xxxx,
                        rem xxxx
//start = 1 ------------------------------------------------
a = 00111100, b = 0111, quot_tens = 0000, quot_units = 1000,
                        rem 0100
//start = 0
a = 00001101, b = 0101, quot_tens = 0000, quot_units = 1000,
                        rem 0100
//start = 1 ------------------------------------------------
a = 00001101, b = 0101, quot_tens = 0000, quot_units = 0010,
                        rem 0011
//start = 0
a = 0011110, b = 0111, quot_tens = 0000, quot_units = 0010,
                        rem 0011    //continued on next page
```

Figure 4.91 Outputs for the decimal divisor.

```
//start = 1
//start = 0
//continues in the same sequence

a = 00111100, b = 0111, quot_tens = 0000, quot_units = 1000,
                        rem 0100
a = 01010010, b = 0110, quot_tens = 0000, quot_units = 1000,
                        rem 0100

a = 01010010, b = 0110, quot_tens = 0001, quot_units = 0011,
                        rem 0100
a = 00111000, b = 0111, quot_tens = 0001, quot_units = 0011,
                        rem 0100

a = 00111000, b = 0111, quot_tens = 0000, quot_units = 1000,
                        rem 0000
a = 01100100, b = 0111, quot_tens = 0000, quot_units = 1000,
                        rem 0000

a = 01100100, b = 0111, quot_tens = 0001, quot_units = 0100,
                        rem 0010
a = 01101110, b = 0111, quot_tens = 0001, quot_units = 0100,
                        rem 0010

a = 01101110, b = 0111, quot_tens = 0001, quot_units = 0101,
                        rem 0101
a = 01100011, b = 1001, quot_tens = 0001, quot_units = 0101,
                        rem 0101

a = 01100011, b = 1001, quot_tens = 0001, quot_units = 0001,
                        rem 0000
a = 00110100, b = 0101, quot_tens = 0001, quot_units = 0001,
                        rem 0000

a = 00110100, b = 0101, quot_tens = 0001, quot_units = 0000,
                        rem 0010
a = 01011000, b = 1001, quot_tens = 0001, quot_units = 0000,
                        rem 0010

a = 01011000, b = 1001, quot_tens = 0000, quot_units = 1001,
                        rem 0111
a = 10000010, b = 1001, quot_tens = 0000, quot_units = 1001,
                        rem 0111

a = 10000010, b = 1001, quot_tens = 0001, quot_units = 0100,
                        rem 0100
```

Figure 4.91 (Continued)

4.11 Floating-Point Addition

Fixed-point notation assumes that the radix point is in a fixed location within the number, either at the right end of the number for integers or at the left end of the number for fractions. A floating-point number consists of three parts: a fraction *f*, an exponent *e*, and a sign bit associated with the number. The floating-point number *A* is obtained by multiplying the fraction *f* by a radix *r* that is raised to the power of *e*, as shown below,

$$A = f \times r^e$$

where the fraction and exponent are signed numbers in 2s complement notation. The fraction and exponent are also referred to as the *mantissa* (or *significand*) and *characteristic*, respectively.

By adjusting the magnitude of the exponent *e*, the radix point can be made to *float* around the fraction, thus, the notation $A = f \times r^e$ is referred to as *floating-point* notation. Consider an example of a floating-point number in radix 10, as shown below.

$$A = 0.00025768 \times 10^{+4}$$

The number *A* can also be written as $A = 2.5768 \times 10^0$ or as $A = 25.768 \times 10^{-5}$. When the fraction is shifted *k* positions to the left, the exponent is decreased by *k*; when the fraction is shifted *k* positions to the right, the exponent is increased by *k*. The standard for representing floating-point numbers in a 32-bit single-precision format is shown in Figure 4.92. Double-precision floating-point numbers are represented in a 64-bit format: a 52-bit fraction, an 11-bit exponent, and a sign bit.

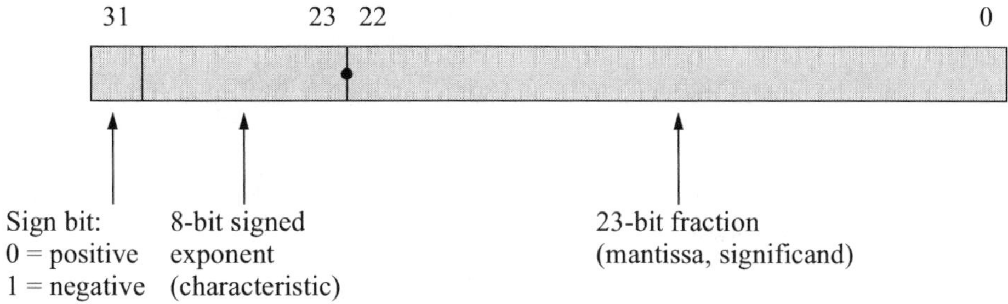

Figure 4.92 32-bit floating-point single-precision format.

Fractions in the IEEE format shown in Figure 4.92 are normalized; that is, the leftmost significant bit is a 1. Figure 4.93 shows unnormalized and normalized numbers in the 32-bit format. Since there will always be a 1 to the immediate right of the radix point, sometimes the 1 bit is not explicitly shown — it is an implied 1.

	S	Exponent	Fraction
Unnormalized	0	0 0 0 0 0 1 • 1 1	0 0 1 1 1 x ... x

$+ \ .0011x \ldots x \times 2^7$

	S	Exponent	Fraction
Normalized	0	0 0 0 0 0 1 • 0 0	1 1 x ... x 0 0 0

$+ \ 1.11x \ldots x000 \times 2^4$

Figure 4.93 Unnormalized and normalized floating-point numbers.

As stated previously, the exponents are signed numbers in 2s complement. However, when adding or subtracting floating-point numbers, the exponents are compared and made equal, which results in a right shift of the fraction with the smaller exponent. A simple comparator can be used for the comparison if the exponents are unsigned.

As the exponents are being formed, a *bias* constant is added to the exponents such that all exponents are positive internally. Since the exponents are eight bits for the single-precision format, the bias constant is +127. Therefore, the biased exponent has a range of

$$0 \leq e_{\text{biased}} \leq 255$$

For example, if the exponents are represented by n bits, then the bias is $2^{n-1} - 1$. For $n = 4$, the most positive number is 0111 (+7). Therefore, all biased exponents are of the form shown in Equation 4.13. When adding two fractions, the exponents must be made equal. An example is shown below using radix 10 numbers. If the exponents are not equal, the result will be incorrect.

$$e_{\text{biased}} = e_{\text{unbiased}} + 2^{n-1} - 1 \tag{4.13}$$

$$\begin{array}{rlcrcl} & .154 \times 10^2 & = & 15.4 & & 0.154 \times 10^2 \\ +) & .430 \times 10^1 & = & 4.3 & & 0.043 \times 10^2 \\ \hline & .573 & & 19.7 & & 0.197 \times 10^2 \end{array}$$

↑ Incorrect result (.573)

↑ Correct result (0.197×10^2)

For a floating-point addition operation, if the signs of the operands are the same ($A_{sign} \oplus B_{sign} = 0$), then this is referred to as *true addition* and the fractions are added. True addition corresponds to one of the following conditions:

$$(+A) + (+B)$$
$$(-A) + (-B)$$
$$(+A) - (-B)$$
$$(-A) - (+B)$$

Examples will now be presented that illustrate floating-point addition for both positive and negative numbers and for conditions requiring no postnormalization and postnormalization. *Postnormalization* occurs when the resulting fraction overflows, requiring a right shift of one bit position with a corresponding increment of the exponent. The bit causing the overflow is shifted right into the high-order fraction bit position.

The examples for conditions requiring no postnormalization are shown in Figures 4.94 and 4.95. The examples for conditions requiring postnormalization are shown in Figures 4.96 and 4.97.

Figure 4.94 shows an example of floating-point addition when adding $A = +14$ and $B = +39$ to yield a sum of +53, in which the 8-bit fractions are not properly aligned initially and there is no postnormalization required.

Before alignment

$A =$	[0]	.	1	1	1	0	0	0	0	0	$\times 2^4$	+14
$B =$	0	.	1	0	0	1	1	1	0	0	$\times 2^6$	+39

After alignment

$A =$	0	.	0	0	1	1	1	0	0	0	$\times 2^6$	+14
$B =$	0	.	1	0	0	1	1	1	0	0	$\times 2^6$	+39
$A + B =$	0	.	1	1	0	1	0	1	0	0	$\times 2^6$	+53

Figure 4.94 Example of floating-point addition in which the fractions are not properly aligned initially and there is no postnormalization.

Figure 4.95 shows an example of floating-point addition when adding $A = -18.50$ and $B = -37.75$ to yield a sum of -56.25, in which the 8-bit fractions are not properly aligned initially and there is no postnormalization required.

```
Before alignment
     A = [1] . 1 0 0 1 0 1 0 0    × 2^5     –18.50
     B =  1  . 1 0 0 1 0 1 1 1    × 2^6     –37.75

After alignment
     A =  1  . 0 1 0 0 1 0 1 0    × 2^6     –18.50
     B =  1  . 1 0 0 1 0 1 1 1    × 2^6     –37.75
         ----------------------
 A + B =  1  . 1 1 1 0 0 0 0 1    × 2^6     –56.25
```

Figure 4.95 Example of floating-point addition in which the fractions are not properly aligned initially and there is no postnormalization.

Figure 4.96 shows an example of floating-point addition when adding $A = +14$ and $B = +23$ to yield a sum of +37, in which the 8-bit fractions are not properly aligned initially and postnormalization is required.

```
Before alignment
      A = [0] . 1 1 1 0 0 0 0 0    × 2^4     +14
  +)  B =  0  . 1 0 1 1 1 0 0 0    × 2^5     +23

After alignment
      A =  0  . 0 1 1 1 0 0 0 0    × 2^5     +14
  +)  B =  0  . 1 0 1 1 1 0 0 0    × 2^5     +23
          ----------------------
       1 <--  . 0 0 1 0 1 0 0 0    × 2^5     +37
        \
         \
          v
           0  . 1 0 0 1 0 1 0 0    × 2^6   Normalize +37
```

Figure 4.96 Example of floating-point addition in which the fractions are not properly aligned initially and postnormalization is required.

Figure 4.97 shows an example of floating-point addition when adding $A = -12$ and $B = -29$ to yield a sum of –41, in which the 8-bit fractions are not properly aligned initially and postnormalization is required.

```
Before alignment
    A = [1] . 1 1 0 0 0 0 0 0    × 2^4          –12
+)  B =  1  . 1 1 1 0 1 0 0 0    × 2^5          –29

After alignment
    A =  1  . 0 1 1 0 0 0 0 0    × 2^5          –12
+)  B =  1  . 1 1 1 0 1 0 0 0    × 2^5          –29
    -------------------------------
      1 ←── . 0 1 0 0 1 0 0 0    × 2^5          –41
        \
         ↘
         1  . 1 0 1 0 0 1 0 0    × 2^6   Normalize –41
```

Figure 4.97 Example of floating-point addition in which the fractions are not properly aligned initially and postnormalization is required.

The design of a floating-point adder using Verilog HDL will be implemented using behavioral modeling for the single-precision format of 32 bits. Figure 4.98 illustrates the behavioral design module. There are two inputs: the augend *flp_a[31:0]* and the addend *flp_b[31:0]*. There are three outputs: the sign of the floating-point number, the exponent, and the sum. The augend consists of a sign bit, *sign_a*; an 8-bit exponent, *exp_a[7:0]*; and a 23-bit fraction, *fract_a[22:0]*. The addend consists of the sign bit, *sign_b*; the 8-bit exponent, *exp_b[7:0]*; and the 23-bit fraction, *fract_b[22:0]*.

The exponents are biased by adding the bias constant of +127 (0111 1111) prior to the addition operation. Then the fractions are aligned by comparing the exponents. A counter, *ctr_align[7:0]*, is set to the difference between the two exponents. The fraction with the smaller exponent is shifted right one bit position, the exponent is incremented by one, and the alignment counter is decremented by one. This process repeats until the alignment counter decrements to a value of zero at which point the fractions are aligned and the addition operation can then be performed. If the exponents are equal initially, then there is no alignment and the addition operation is performed immediately.

If there is a fraction overflow, then the corresponding result is obtained by the following Verilog statement: *{cout, sum} = fract_a + fract_b;*, which allows for a carry-out to be concatenated with the sum. In that case, the carry-out and sum are shifted right one bit position in concatenation to postnormalize the result as follows: *{cout, sum} = {cout, sum} >> 1;* and the sign of the result is set equal to the sign of the augend.

The test bench module is shown in Figure 4.99 illustrating four different augends and addends. The outputs are shown in Figure 4.100.

```
//behavioral floating-point addition
module add_flp5 (flp_a, flp_b, sign, exponent, sum);

//define inputs and outputs
input [31:0] flp_a, flp_b;
output [22:0] sum;
output sign;
output [7:0] exponent;

//variables used in always block
//are declared as registers
reg sign_a, sign_b;
reg [7:0] exp_a, exp_b;
reg [7:0] exp_a_bias, exp_b_bias;
reg [22:0] fract_a, fract_b;
reg [7:0] ctr_align;
reg [22:0] sum;
reg sign;
reg [7:0] exponent;
reg cout;

//define operand signs, exponents, and fractions
always @ (flp_a or flp_b)
begin
   sign_a = flp_a [31];
   sign_b = flp_b [31];

   exp_a = flp_a [30:23];
   exp_b = flp_b [30:23];

   fract_a = flp_a [22:0];
   fract_b = flp_b [22:0];

//shift implied 1 into high-order fraction bit position
   fract_a = fract_a >> 1;
   fract_a[22] = 1'b1;

   fract_b = fract_b >> 1;
   fract_b[22] = 1'b1;

//bias exponents
   exp_a_bias = exp_a + 8'b0111_1111;
   exp_b_bias = exp_b + 8'b0111_1111;

                                    //continued on next page
```

Figure 4.98 Behavioral design module for the floating-point adder.

```
//align fractions
   if (exp_a_bias < exp_b_bias)
      ctr_align = exp_b_bias - exp_a_bias;

      while (ctr_align)
         begin
            fract_a = fract_a >> 1;
            exp_a_bias = exp_a_bias + 1;
            ctr_align = ctr_align - 1;
         end

   if (exp_b_bias < exp_a_bias)
      ctr_align = exp_a_bias - exp_b_bias;

      while (ctr_align)
         begin
            fract_b = fract_b >> 1;
            exp_b_bias = exp_b_bias + 1;
            ctr_align = ctr_align - 1;
         end

//obtain result
{cout, sum} = fract_a + fract_b;

//normalize result
   if (cout == 1)
      {cout, sum} = {cout, sum} >> 1;

   sign = sign_a;
   exponent = exp_b_bias;

end

endmodule
```

Figure 4.98 (Continued)

```
//test bench for floating-point addition
module add_flp5_tb;

reg [31:0] flp_a, flp_b;    //inputs are reg for test bench
wire sign;                  //outputs are wire for test bench
wire [7:0] exponent;
wire [22:0] sum;                   //continued on next page
```

Figure 4.99 Test bench module for the floating-point adder.

```
//display variables
initial
$monitor ("sign = %b, exp_biased = %b, sum = %b",
            sign, exponent, sum);


//apply input vectors
initial
begin
      //+12 + +35 = +47
      //          s ----e---- -------------f-------------
#0    flp_a = 32'b0_0000_0100_1000_0000_0000_0000_0000_000;
      flp_b = 32'b0_0000_0110_0001_1000_0000_0000_0000_000;


      //+26.5 + +4.375 = +30.875
      //          s ----e---- -------------f-------------
#10   flp_a = 32'b0_0000_0101_1010_1000_0000_0000_0000_000;
      flp_b = 32'b0_0000_0011_0001_1000_0000_0000_0000_000;


      //+11 + +34 = +45
      //          s ----e---- -------------f-------------
#10   flp_a = 32'b0_0000_0100_0110_0000_0000_0000_0000_000;
      flp_b = 32'b0_0000_0110_0001_0000_0000_0000_0000_000;


      //+23.75 + +87.125 = +110.875
      //          s ----e---- -------------f-------------
#10   flp_a = 32'b0_0000_0101_0111_1100_0000_0000_0000_000;
      flp_b = 32'b0_0000_0111_0101_1100_1000_0000_0000_000;


#10   $stop;

end

//instantiate the module into the test bench
add_flp5 inst1 (flp_a, flp_b, sign, exponent, sum);

endmodule
```

Figure 4.99 (Continued)

```
sign = 0, exp_biased = 10000101,
      sum = 1011_1100_0000_0000_0000_000

sign = 0, exp_biased = 10000100,
      sum = 1111_0111_0000_0000_0000_000

sign = 0, exp_biased = 10000101,
      sum = 10110100000000000000000

sign = 0, exp_biased = 10000110,
      sum = 1101_1101_1100_0000_0000_000
```

Figure 4.100 Outputs for the floating-point adder.

The biased exponent from output number one of Figure 4.100 is reproduced as shown in Figure 4.101. The unbiased exponent is obtained as follows: Subtract +127 from the biased exponent — which is the larger of the two original exponents — by adding the 2s complement of +127 (–127) to obtain the unbiased exponent.

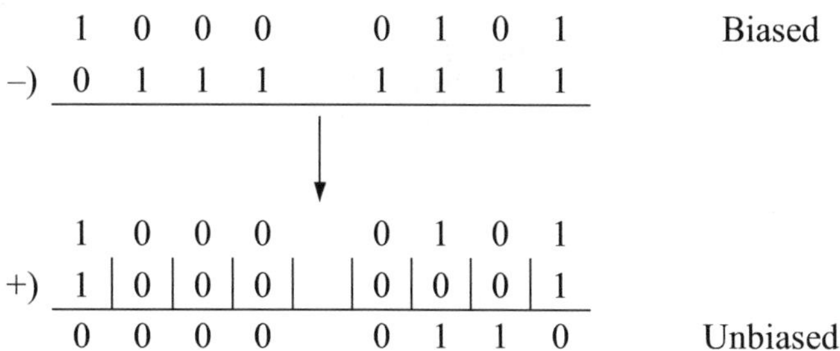

Figure 4.101 Obtain an unbiased exponent from output number one.

In a similar manner, the unbiased exponent from output number two of Figure 4.100 is obtained, as shown in Figure 4.102

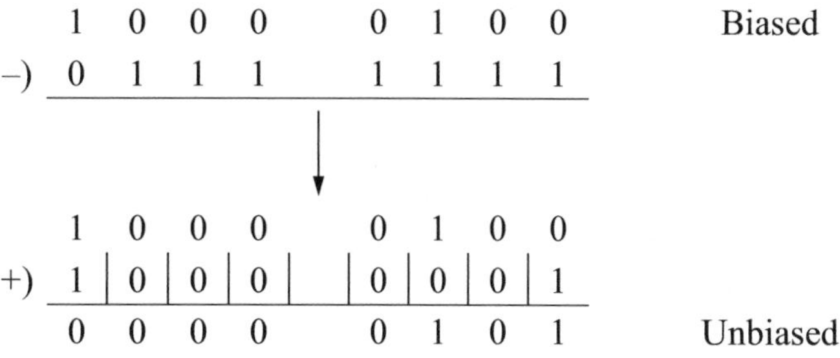

Figure 4.102 Obtain an unbiased exponent from output number three.

Output number two has a sum of +30.875, which results in the following binary number:

```
sign = 0, exp_biased = 10000100,
        sum = 1111_0111_0000_0000_0000_000
```

Since the unbiased exponent is five, the floating-point fraction has the following value: `11110.1110_0000_0000_0000_00`, which has a decimal value of

$$2^4 + 2^3 + 2^2 + 2^1 + 2^0 . 2^{-1} + 2^{-2} + 2^{-3}$$
$$1 \quad 1 \quad 1 \quad 1 \quad 0 \;.\; 0.5 + 0.25 + 0.125 = 30.875$$

4.12 Floating-Point Subtraction

Floating-point subtraction is similar to floating-point addition because subtraction is accomplished by adding the 2s complement of the subtrahend. Therefore, fraction overflow can also occur in subtraction. If the signs of the operands are the same ($A_{sign} \oplus B_{sign} = 0$) and the operation is subtraction, then this is referred to as *true subtraction* and the fractions are subtracted. If the signs of the operands are different ($A_{sign} \oplus B_{sign} = 1$) and the operation is addition, then this is also specified as *true subtraction*. True subtraction corresponds to one of the following conditions:

$$(+A) \; - \; (+B)$$
$$(-A) \; - \; (-B)$$
$$(+A) \; + \; (-B)$$
$$(-A) \; + \; (+B)$$

The minuend and subtrahend are both normalized fractions properly aligned with biased exponents. Subtraction can yield a result that is either true addition or true subtraction. *True addition* produces a result that is the sum of the two operands disregarding the signs; *true subtraction* produces a result that is the difference of the two operands disregarding the signs. There are four cases that yield true addition, as shown below and eight cases that yield true subtraction, as shown below.

True addition

(–Small number) – (+Large number)
(–Large number) – (+Small number)
(+Large number) – (–Small number)
(+Small number) – (–Large number)

True subtraction

(+Large number) – (+Small number)
(+Small number) – (+Large number)
(–Small number) – (–Large number)
(–Large number) – (–Small number)

(+Small number) + (–Large number)
(–Small number) + (+Large number)
(+Large number) + (–Small number)
(–Large number) + (+Small number)

Shown below are six examples that illustrate some of the variations of true addition and true subtraction. For true addition these include the following:

(–Small number) – (+Large number)
(+Large number) – (–Small number)

For true subtraction these include the following:

(+Small number) – (+Large number),
(–Large number) – (–Small number),
(+Small number) + (–Large number)
(+Large number) + (–Small number)

Example 4.8 True addition will be performed by the following subtract operation: (–Small number) – (+Large number) for the decimal numbers (–24) – (+30) to yield a result of –54.

Before alignment													
	$A =$	1	.	1	1	0	0	0	0	0	0	$\times 2^5$	–24
	$B =$	0	.	1	1	1	1	0	0	0	0	$\times 2^5$	+30
After alignment (already aligned)													
	$A =$	1	.	1	1	0	0	0	0	0	0	$\times 2^5$	–24
	+) $B =$	1	.	1	1	1	1	0	0	0	0	$\times 2^5$	–30
	1	←	.	1	0	1	1	0	0	0	0	$\times 2^5$	
Postnormalize		1	.	1	1	0	1	1	0	0	0	$\times 2^6$	–54

Example 4.9 True addition will be performed by the following subtract operation: (+Large number) – (–Small number) for the decimal numbers (+38) – (–15) to yield a result of +53.

Before alignment						
	$A =$	0 .	1 0 0 1	1 0 0 0	$\times 2^6$	+38
	$B =$	1 .	1 1 1 1	0 0 0 0	$\times 2^4$	–15
After alignment	$A =$	0 .	1 0 0 1	1 0 0 0	$\times 2^6$	+38
	$B =$	1 .	0 0 1 1	1 1 0 0	$\times 2^6$	–15
Add fractions						
	$A =$	0 .	1 0 0 1	1 0 0 0	$\times 2^6$	+38
	+) $B =$	1 .	0 0 1 1	1 1 0 0	$\times 2^6$	–15
	0 ←	.	1 1 0 1	0 1 0 0	$\times 2^6$	+53
No postnormalize		0 .	1 1 0 1	0 1 0 0	$\times 2^6$	+53

Example 4.10 True subtraction will be performed by the following subtract operation: (+Small number) – (+Large number) for the decimal numbers (+15) – (+38) to yield a result of –23.

Before alignment						
	$A =$	0 .	1 1 1 1	0 0 0 0	$\times 2^4$	+15
	$B =$	0 .	1 0 0 1	1 0 0 0	$\times 2^6$	+38
After alignment						
	$A =$	0 .	0 0 1 1	1 1 0 0	$\times 2^6$	
	+) $B' + 1 =$	0 .	0 1 1 0	1 0 0 0	$\times 2^6$	
Add fractions	$A =$	0 .	0 0 1 1	1 1 0 0	$\times 2^6$	+15
	+) $B =$	0 .	0 1 1 0	1 0 0 0	$\times 2^6$	+38
	0 ←	.	1 0 1 0	0 1 0 0	$\times 2^6$	
2s complement		1 .	0 1 0 1	1 1 0 0	$\times 2^6$	
Postnormalize		1 .	1 0 1 1	1 0 0 0	$\times 2^5$	–23

Example 4.11 True subtraction will be performed by the following subtract operation: (–Large number) – (–Small number) for the decimal numbers (–52) – (–39) to yield a result of –13.

Before alignment													
	$A =$	1	.	1	1	0	1	0	0	0	0	$\times 2^6$	–52
	$B =$	1	.	1	0	0	1	1	1	0	0	$\times 2^6$	–39
After alignment (already aligned)													
	$A =$	1	.	1	1	0	1	0	0	0	0	$\times 2^6$	
	+) $B' + 1 =$	1	.	0	1	1	0	0	1	0	0	$\times 2^6$	
	1 ←		.	0	0	1	1	0	1	0	0	$\times 2^6$	
		1	.	0	0	1	1	0	1	0	0	$\times 2^6$	–13
Postnormalize		1	.	1	1	0	1	0	0	0	0	$\times 2^4$	–13

Example 4.12 True subtraction will be performed by the following subtract operation: (+Small number) + (–Large number) for the decimal numbers (+30) + (–38) to yield a result of –8.

Before alignment													
	$A =$	0	.	1	1	1	1	0	0	0	0	$\times 2^5$	+30
	$B =$	1	.	1	0	0	1	1	0	0	0	$\times 2^6$	–38
After alignment													
	$A =$	0	.	0	1	1	1	1	0	0	0	$\times 2^6$	+30
	$B =$	1	.	1	0	0	1	1	0	0	0	$\times 2^6$	–38
Add fractions													
	$A =$	0	.	0	1	1	1	1	0	0	0	$\times 2^6$	
	+) $B' + 1 =$	1	.	0	1	1	0	1	0	0	0	$\times 2^6$	
	0 ←		.	1	1	1	0	0	0	0	0	$\times 2^6$	
2s complement		1	.	0	0	1	0	0	0	0	0	$\times 2^6$	–8
Postnormalize		1	.	1	0	0	0	0	0	0	0	$\times 2^4$	–8

Example 4.13 True subtraction will be performed by the following subtract operation: (–Small number) + (+Large number) for the decimal numbers (–30) + (+52) to yield a result of +22.

Before alignment						
$A =$	1	.	0 1 1 1	1 0 0 0	$\times 2^6$	–30
$B =$	0	.	1 1 0 1	0 0 0 0	$\times 2^6$	+52
After alignment (already aligned)						
$A =$	1	.	0 1 1 1	1 0 0 0	$\times 2^6$	
+) $B' + 1 =$	0	.	0 0 1 1	0 0 0 0	$\times 2^6$	
0 ←		.	1 0 1 0	1 0 0 0	$\times 2^6$	
2s complement	0	.	0 1 0 1	1 0 0 0	$\times 2^6$	+22
Postnormalize	0	.	1 0 1 1	0 0 0 0	$\times 2^5$	+22

4.12.1 True Addition and True Subtraction

Using the techniques described in Example 4.8 through Example 4.13, a behavioral design module will be designed that illustrates true addition and different methods of true subtraction. The rules shown below will help to explain the different techniques.

True addition

For true addition, the following rules apply, whether

$fract\_a| < |fract\_b|$ or $|fract\_a| > |fract\_b|$:

(1) Bias the exponents.
(2) Align the fractions
(4) Perform the addition.
(5) The sign of the result is the sign of the minuend.
(6) If carry-out =1, then {cout, rslt} >> 1.
(7) Increment the exponent by 1.

True subtraction

For true subtraction, the following rules apply, depending on whether

$fract\_a| < |fract\_b|$ or $|fract\_a| > |fract\_b|$

and also on the state of a *mode* control input, and on the signs of the operands. If *mode* = 0, then the operation is addition; if *mode* = 1, then the operation is subtraction.

- Bias the exponents.
- Align the fractions

- If | *fract_a* | < | *fract_b* | and *mode* = 0 and sign of *fract_a* ≠ sign of *fract_b*.
  - (1) 2s complement *fract_b*.
  - (2) Perform the addition.
  - (3) Sign of the result = A_{sign}'.

- If | *fract_a* | < | *fract_b* | and *mode* = 1 and sign of *fract_a* = sign of *fract_b*.
  - (1) 2s complement *fract_b*.
  - (2) Perform the addition.
  - (3) Sign of the result = A_{sign}'.

- If | *fract_a* | > | *fract_b* | and *mode* = 0 and sign of *fract_a* ≠ sign of *fract_b*.
  - (1) 2s complement *fract_b*.
  - (2) Perform the addition.
  - (3) Sign of the result = A_{sign}.
  - (4) Postnormalize, if necessary (shift left 1 and decrement the exponent).

- If | *fract_a* | > | *fract_b* | and *mode* = 1 and sign of *fract_a* = sign of *fract_b*.
  - (1) 2s complement *fract_b*.
  - (2) Perform the addition.
  - (3) Sign of the result = A_{sign}.
  - (4) Postnormalize, if necessary (shift left 1 and decrement the exponent).

Example 4.14 Figure 4.103 illustrates the behavioral design module that illustrates a true addition segment, a true subtraction segment in which | *fract_a* | < | *fract_b* | and *mode* = 1, a true subtraction segment in which | *fract_a* | < | *fract_b* | and *mode* = 0, a true subtraction segment in which | *fract_a* | > | *fract_b* | and *mode* = 0, and a true subtraction segment in which | *fract_a* | > | *fract_b* | and *mode* = 1.

The test bench module is shown in Figure 4.104, which applies many floating-point input vectors to the test bench module. The outputs are shown in Figure 4.105 illustrating both the biased exponents, the unbiased exponents, and the result of the operation.

```
//behavioral floating-point addition and subtraction
module sub_flp5 (flp_a, flp_b, mode, sign,
                    exponent, exp_unbiased, rslt);

input [31:0] flp_a, flp_b;          //define inputs and outputs
input mode;                         //continued on next page
```

Figure 4.103 Behavioral design module for true addition and true subtraction.

```
output sign;
output [7:0] exponent, exp_unbiased;
output [22:0] rslt;

//variables used in an always block
//are declared as registers
reg sign_a, sign_b;
reg [7:0] exp_a, exp_b;
reg [7:0] exp_a_bias, exp_b_bias;
reg [22:0] fract_a, fract_b;
reg [7:0] ctr_align;
reg [22:0] rslt;
reg sign;
reg [7:0] exponent, exp_unbiased;
reg cout;

//
==============================================================
//define sign, exponent, and fraction
always @ (flp_a or flp_b)
begin
   sign_a = flp_a[31];
   sign_b = flp_b[31];

   exp_a = flp_a[30:23];
   exp_b = flp_b[30:23];

   fract_a = flp_a[22:0];
   fract_b = flp_b[22:0];

//bias exponents
   exp_a_bias = exp_a + 8'b0111_1111;
   exp_b_bias = exp_b + 8'b0111_1111;

//align fractions
   if (exp_a_bias < exp_b_bias)
      ctr_align = exp_b_bias - exp_a_bias;

      while (ctr_align)
         begin
            fract_a = fract_a >> 1;
            exp_a_bias = exp_a_bias + 1;
            ctr_align = ctr_align - 1;
         end

                                    //continued on next page
```

Figure 4.103 (Continued)

```
   if (exp_b_bias < exp_a_bias)
      ctr_align = exp_a_bias - exp_b_bias;

      while (ctr_align)
         begin
            fract_b = fract_b >> 1;
            exp_b_bias = exp_b_bias + 1;
            ctr_align = ctr_align - 1;
         end

//==========================================================
//true addition
   if ((mode == 1) & (sign_a != sign_b))
      begin
         {cout, rslt} = fract_a + fract_b;
         sign = sign_a;

         //postnormalize
         if (cout == 1)
            begin
               {cout, rslt} = {cout, rslt} >> 1;
               exp_b_bias = exp_b_bias + 1;
            end
      end

//==========================================================
//true subtraction: fract_a < fract_b, mode = 1,
                                         sign_a = sign_b
   if ((fract_a < fract_b) & (mode == 1) & (sign_a == sign_b))
      begin
         fract_b = ~fract_b + 1;
         {cout, rslt} = fract_a + fract_b;
         sign = ~sign_a;

         if (rslt[22] == 1)
            rslt = ~rslt+ 1;

      //postnormalize
         while (rslt[22] == 0)
            begin
               rslt = rslt << 1;
               exp_b_bias = exp_b_bias - 1;
            end
      end

                                   //continued on next page
```

Figure 4.103 (Continued)

```
//==============================================================
//true subtraction: fract_a < fract_b, mode = 0,
                            sign_a != sign_b
   if ((fract_a < fract_b) & (mode == 0) & (sign_a != sign_b))
      begin
         fract_b = ~fract_b + 1;
         {cout, rslt} = fract_a + fract_b;
         sign = ~sign_a;

         if (rslt[22] == 1)
            rslt = ~rslt + 1;

      //postnormalize
         while (rslt[22] == 0)
            begin
               rslt = rslt << 1;
               exp_b_bias = exp_b_bias - 1;
            end
      end

//==============================================================
//true subtraction: fract_a > fract_b, mode = 0,
                            sign_a != sign_b
   if ((fract_a > fract_b) & (mode == 0) & (sign_a != sign_b))
      begin
         fract_b = ~fract_b + 1;
         {cout, rslt} = fract_a + fract_b;
         sign = sign_a;

      //postnormalize
         while (rslt[22] == 0)
            begin
               rslt = rslt << 1;
               exp_b_bias = exp_b_bias - 1;
            end
      end

//==============================================================
//true subtraction: fract_a > fract_b, mode = 1,
                            sign_a = sign_b
   if ((fract_a > fract_b) & (mode == 1) & (sign_a == sign_b))
      begin
         fract_b = ~fract_b + 1;
         {cout, rslt} = fract_a + fract_b;
         sign = sign_a;

                                    //continued on next page
```

Figure 4.103 (Continued)

```
      //postnormalize
         while (rslt[22] == 0)
            begin
               rslt = rslt << 1;
               exp_b_bias = exp_b_bias - 1;
            end
      end

//============================================================
exponent = exp_b_bias;
exp_unbiased = exp_b_bias - 8'b0111_1111;
end

endmodule
```

Figure 4.103 (Continued)

```
//test bench for floating-point subtraction
module sub_flp5_tb;

//inputs are reg in test bench
//outputs are wire in test bench
reg [31:0] flp_a, flp_b;
reg mode;
wire sign;
wire [7:0] exponent, exp_unbiased;
wire [22:0] rslt;

//display variables
initial
$monitor ("sign = %b, exp_biased = %b, exp_unbiased = %b,
         rslt = %b", sign, exponent, exp_unbiased, rslt);

//apply input vectors
initial
begin
//============================================================
//true addition: mode = 1, sign_a != sign_b
      //(-19) - (+25) = -44
      //          s ----e---- -------------f-------------
#0    flp_a = 32'b1_0000_0101_1001_1000_0000_0000_0000_000;
      flp_b = 32'b0_0000_0101_1100_1000_0000_0000_0000_000;
      mode = 1'b1;                //continued on next page
```

Figure 4.104 Test bench module for true addition and true subtraction.

```
        //(-22) - (+28) = -50
        //          s ----e---- --------------f-------------
#10     flp_a = 32'b1_0000_0101_1011_0000_0000_0000_0000_000;
        flp_b = 32'b0_0000_0101_1110_0000_0000_0000_0000_000;

        //(-16) - (+23) = -39
        //          s ----e---- --------------f-------------
#10     flp_a = 32'b1_0000_0101_1000_0000_0000_0000_0000_000;
        flp_b = 32'b0_0000_0101_1011_1000_0000_0000_0000_000;

        //(+34) - (-11) = +45
        //          s ----e---- --------------f-------------
#10     flp_a = 32'b0_0000_0110_1000_1000_0000_0000_0000_000;
        flp_b = 32'b1_0000_0100_1011_0000_0000_0000_0000_000;

        //(-127) - (+76) = -203
        //          s ----e---- --------------f-------------
#10     flp_a = 32'b1_0000_0111_1111_1110_0000_0000_0000_000;
        flp_b = 32'b0_0000_0111_1001_1000_0000_0000_0000_000;

//============================================================
//true subtraction: fract_a < fract_b, mode = 1,
                                        sign_a = sign_b
        //(+11) - (+34) = -23
        //          s ----e---- --------------f-------------
#10     flp_a = 32'b0_0000_0100_1011_0000_0000_0000_0000_000;
        flp_b = 32'b0_0000_0110_1000_1000_0000_0000_0000_000;

        //(-23) - (-36) = +13
        //          s ----e---- --------------f-------------
#10     flp_a = 32'b1_0000_0101_1011_1000_0000_0000_0000_000;
        flp_b = 32'b1_0000_0110_1001_0000_0000_0000_0000_000;

        //(-7) - (-38) = +31
        //          s ----e---- --------------f-------------
#10     flp_a = 32'b1_0000_0011_1110_0000_0000_0000_0000_000;
        flp_b = 32'b1_0000_0110_1001_1000_0000_0000_0000_000;

        //(+47) - (+72) = -25
        //          s ----e---- --------------f-------------
#10     flp_a = 32'b0_0000_0110_1011_1100_0000_0000_0000_000;
        flp_b = 32'b0_0000_0111_1001_0000_0000_0000_0000_000;

//============================================================

                                        //continued on next page
```

Figure 4.104 (Continued)

```
//true subtraction: fract_a < fract_b, mode = 0,
                                   sign_a != sign_b
      //(-36) + (+30) = -6
      //          s ----e---- --------------f-------------
#10   flp_a = 32'b1_0000_0110_1001_0000_0000_0000_0000_000;
      flp_b = 32'b0_0000_0110_0111_1000_0000_0000_0000_000;
      mode = 1'b0;

      //(+22) + (-16) = +6
      //          s ----e---- --------------f-------------
#10   flp_a = 32'b0_0000_0101_1011_0000_0000_0000_0000_000;
      flp_b = 32'b1_0000_0101_1000_0000_0000_0000_0000_000;

      //(+28) + (-35) = -7
      //          s ----e---- --------------f-------------
#10   flp_a = 32'b0_0000_0101_1110_0000_0000_0000_0000_000;
      flp_b = 32'b1_0000_0110_1000_1100_0000_0000_0000_000;

      //(+36) + (-140) = -104
      //          s ----e---- --------------f-------------
#10   flp_a = 32'b0_0000_0110_1001_0000_0000_0000_0000_000;
      flp_b = 32'b1_0000_1000_1000_1100_0000_0000_0000_000;

//============================================================
//true subtraction: fract_a > fract_b, mode = 0,
                                   sign_a != sign_b
      //(+45) + (-13) = +32
      //          s ----e---- --------------f-------------
#10   flp_a = 32'b0_0000_0110_1011_0100_0000_0000_0000_000;
      flp_b = 32'b1_0000_0100_1101_0000_0000_0000_0000_000;

      //(+72) + (-46) = +26
      //          s ----e---- --------------f-------------
#10   flp_a = 32'b0_0000_0111_1001_0000_0000_0000_0000_000;
      flp_b = 32'b1_0000_0110_1011_1000_0000_0000_0000_000;

      //(+172) + (-100) = +72
      //          s ----e---- --------------f-------------
#10   flp_a = 32'b0_0000_1000_1010_1100_0000_0000_0000_000;
      flp_b = 32'b1_0000_0111_1100_1000_0000_0000_0000_000;

      //(-172) + (+100) = -72
      //          s ----e---- --------------f-------------
#10   flp_a = 32'b1_0000_1000_1010_1100_0000_0000_0000_000;
      flp_b = 32'b0_0000_0111_1100_1000_0000_0000_0000_000;

                                   //continued on next page
```

Figure 4.104 (Continued)

```
      //(+85.75) + (-70.50) = +15.25
      //           s ----e---- --------------f-------------
#10   flp_a = 32'b0_0000_0111_1010_1011_1000_0000_0000_000;
      flp_b = 32'b1_0000_0111_1000_1101_0000_0000_0000_000;

      //(-96.50) + (+30.25) = -66.25
      //           s ----e---- --------------f-------------
#10   flp_a = 32'b1_0000_0111_1100_0001_0000_0000_0000_000;
      flp_b = 32'b0_0000_0101_1111_0010_0000_0000_0000_000;

//===========================================================
//true subtraction: fract_a > fract_b, mode = 1,
                                        sign_a = sign_b
      //(-45) - (-35) = -12
      //           s ----e---- --------------f-------------
#10   flp_a = 32'b1_0000_0110_1011_1100_0000_0000_0000_000;
      flp_b = 32'b1_0000_0110_1000_1100_0000_0000_0000_000;
      mode = 1'b1;

      //(-130) - (-25) = -105
      //           s ----e---- --------------f-------------
#10   flp_a = 32'b1_0000_1000_1000_0010_0000_0000_0000_000;
      flp_b = 32'b1_0000_0101_1100_1000_0000_0000_0000_000;

      //(+105) - (+5) = +100
      //           s ----e---- --------------f-------------
#10   flp_a = 32'b0_0000_0111_1101_0010_0000_0000_0000_000;
      flp_b = 32'b0_0000_0011_1010_0000_0000_0000_0000_000;

      //(+36.5) - (+5.75) = +30.75
      //           s ----e---- --------------f-------------
#10   flp_a = 32'b0_0000_0110_1001_0010_0000_0000_0000_000;
      flp_b = 32'b0_0000_0011_1011_1000_0000_0000_0000_000;

      //(+5276) - (+4528) = +748
      //           s ----e---- --------------f-------------
#10   flp_a = 32'b0_0000_1101_1010_0100_1110_0000_0000_000;
      flp_b = 32'b0_0000_1101_1000_1101_1000_0000_0000_000;

      //(+963.50) - (+520.25) = +443.25
      //           s ----e---- --------------f-------------
#10   flp_a = 32'b0_0000_1010_1111_0000_1110_0000_0000_000;
      flp_b = 32'b0_0000_1010_1000_0010_0001_0000_0000_000;

#10   $stop;
end
                                    //continued on next page
```

Figure 4.104 (Continued)

```
//instantiate the module into the test bench
sub_flp5 inst1 (flp_a, flp_b, mode, sign,
                     exponent, exp_unbiased, rslt);

endmodule
```

Figure 4.104 (Continued)

```
True Addition: mode = 1, sign_a != sign_b

sign = 1, exp_biased = 10000101, exp_unbiased = 00000110,
rslt = 1011000000000000000000 -44

sign = 1, exp_biased = 10000101, exp_unbiased = 00000110,
rslt = 1100100000000000000000 -50

sign = 1, exp_biased = 10000101, exp_unbiased = 00000110,
rslt = 1001110000000000000000 -39

sign = 0, exp_biased = 10000101, exp_unbiased = 00000110,
rslt = 1011010000000000000000 +45

sign = 1, exp_biased = 10000111, exp_unbiased = 00001000,
rslt = 1100101100000000000000 -203
----------------------------------------------------------

True Subtraction: fract_a < fract_b, mode = 1,
                     sign_a = sign_b

sign = 1, exp_biased = 10000100, exp_unbiased = 00000101,
rslt = 1011100000000000000000 -23

sign = 0, exp_biased = 10000011, exp_unbiased = 00000100,
rslt = 1101000000000000000000 +13

sign = 0, exp_biased = 10000100, exp_unbiased = 00000101,
rslt = 1111100000000000000000 +31

sign = 1, exp_biased = 10000100, exp_unbiased = 00000101,
rslt = 1100100000000000000000 -25
----------------------------------------------------------

                                  //continued on next page
```

Figure 4.105 Outputs for true addition and true subtraction.

```
True Subtraction: fract_a < fract_b, mode = 0,
                        sign_a != sign_b

sign = 1, exp_biased = 10000010, exp_unbiased = 00000011,
rslt = 1100000000000000000000 -6

sign = 0, exp_biased = 10000010, exp_unbiased = 00000011,
rslt = 1100000000000000000000 +6

sign = 1, exp_biased = 10000010, exp_unbiased = 00000011,
rslt = 1110000000000000000000 -7

sign = 1, exp_biased = 10000110, exp_unbiased = 00000111,
rslt = 1101000000000000000000 -104
-----------------------------------------------------------

True Subtraction: fract_a > fract_b, mode = 0,
                        sign_a != sign_b

sign = 0, exp_biased = 10000101, exp_unbiased = 00000110,
rslt = 1000000000000000000000 +32

sign = 0, exp_biased = 10000100, exp_unbiased = 00000101,
rslt = 1101000000000000000000 +26

sign = 0, exp_biased = 10000110, exp_unbiased = 00000111,
rslt = 1001000000000000000000 +72

sign = 1, exp_biased = 10000110, exp_unbiased = 00000111,
rslt = 1001000000000000000000 -72

sign = 0, exp_biased = 10000011, exp_unbiased = 00000100,
rslt = 1111010000000000000000 +15.25

sign = 1, exp_biased = 10000110, exp_unbiased = 00000111,
rslt = 1000010010000000000000 -66.25
-----------------------------------------------------------

                                  //continued on next page
```

Figure 4.105 (Continued)

```
True Subtraction: fract_a > fract_b, mode = 1,
                     sign_a = sign_b

sign = 1, exp_biased = 10000011, exp_unbiased = 00000100,
rslt = 11000000000000000000000 -12

sign = 1, exp_biased = 10000110, exp_unbiased = 00000111,
rslt = 11010010000000000000000 -105

sign = 0, exp_biased = 10000110, exp_unbiased = 00000111,
rslt = 11001000000000000000000 +100

sign = 0, exp_biased = 10000100, exp_unbiased = 00000101,
rslt = 11110110000000000000000 +30.75

sign = 0, exp_biased = 10001001, exp_unbiased = 00001010,
rslt = 10111011000000000000000 +748

sign = 0, exp_biased = 10001000, exp_unbiased = 00001001,
rslt = 11011101101000000000000 +443.25
------------------------------------------------------------
```

Figure 4.105 (Continued)

4.13 Floating-Point Multiplication

In floating-point multiplication, the fractions are multiplied and the exponents are added. The fractions are multiplied by any of the methods previously used in fixed-point multiplication. The operands are two normalized floating-point operands. Fraction multiplication and exponent addition are two independent operations and can be done in parallel. Floating-point multiplication is defined as shown in Equation 4.14.

$$\begin{aligned} A \times B &= (f_A \times r^{eA}) \times (f_B \times r^{eB}) \\ &= (f_A \times f_B) \times r^{(eA + eB)} \end{aligned} \tag{4.14}$$

The sign of the product is determined by the signs of the operands as shown below.

$$A_{sign} \oplus B_{sign}$$

The single-precision format will be the primary format used in this chapter. The multiplication algorithm is partitioned into five parts:

1. Check for zero operands. If $A = 0$ or $B = 0$, then the product = 0.
2. Determine the sign of the product.
3. Add exponents and subtract the bias.
4. Multiply fractions. Steps 3 and 4 can be done in parallel, but both must be completed before step 5.
5. Normalize the product.

The sequential add-shift multiplication technique will be used in the Verilog design example. Two examples are shown in Example 4.15 and Example 4.16 using the paper-and-pencil method for 4-bit multiplicands and 4-bit multipliers in order to review the multiplication technique. A more detailed example of the add-shift method is shown in Example 4.17 showing the actual add-shift technique as it relates to a Verilog design example.

Example 4.15 Let the multiplicand and multiplier be two positive 4-bit operands as shown below, where *a[3:0]* = 0111 (+7) and *b[3:0]* = 0110 (+6) to produce a product *p[7:0]* = 0010 1010 (+42). A multiplier bit of 0 enters 0s in the partial product; a multiplier bit of 1 copies the multiplicand to the partial product.

Multiplicand *A*						0	1	1	1	+7
Multiplier *B*					×)	0	1	1	0	+6
Partial products	0	0	0	0		0	0	0	0	
	0	0	0	0		1	1	1		
	0	0	0	1		1	1			
	0	0	0	0		0				
Product *P*	0	0	1	0		1	0	1	0	+42

Example 4.16 This example multiplies a negative multiplicand by a positive multiplier. The multiplicand is *a[3:0]* = 1011 (–5); the multiplier is *b[3:0]* = 0011 (+3) to produce a product of *p[7:0]* = 1111 0001 (–15).

Multiplicand *A*						1	0	1	1	–5
Multiplier *B*					×)	0	0	1	1	+3
Partial products	1	1	1	1		1	0	1	1	
	1	1	1	1		0	1	1		
	0	0	0	0		0	0			
	0	0	0	0		0				
Product *P*	1	1	1	1		0	0	0	1	–15

Example 4.17 Each step is an add-shift-right sequence. A multiplicand fraction *fract_a* = 0.1101 1000 × 2^5 (+27) is multiplied by a multiplier *fract_b* = 0.1100 1000 × 2^5 (+25) with a partial product *prod* = 0000 0000 to produce a product of *prod* = 0.1010 1000 1100 0000 x 2^{10} (+675)

fract_a (+27)	Count		*prod*	*fract_b* (+25)	
1101 1000		*prod*	0000 0000	1100 1000	
		+)	0000 0000		Add-shift
		0	0000 0000	1100 1000	
	7		0000 0000	0110 0100	
		+)	0000 0000		Add-shift
		0	0000 0000	0110 0100	
	6		0000 0000	0011 0010	
		+)	0000 0000		Add-shift
		0	0000 0000	0011 0010	
	5		0000 0000	0001 1001	
		+)	1101 1000		Add-shift
		0	1101 1000	0001 1001	
	4		0110 1100	0000 1100	
		+)	0000 0000		Add-shift
		0	0110 0110	0000 1100	
	3		0011 0110	0000 0110	
		+)	0000 0000		Add-shift
		0	0011 0110	0000 0110	
	2		0001 1011	0000 0011	
		+)	1101 1000		Add-shift
		0	1111 0011	0000 0011	
	1		0111 1001	1000 0001	
		+)	1101 1000		Add-shift
		1	0101 0001	1000 0001	
	0		1010 1000	1100 0000	× 2^{10} (+675)

The Verilog behavioral design module for a version similar to Example 4.17 is shown in Figure 4.106 using the multiplication algorithm previously shown. A floating-point format for the design is shown below for 14 bits. Floating-point multiplication using the sequential add-shift method for two operands, *flp_a[13:0]* and *flp_b[13:0]*, will be used. The test bench for several different input vectors, including both positive and negative operands is shown in Figure 4.107. The outputs are shown in Figure 4.108.

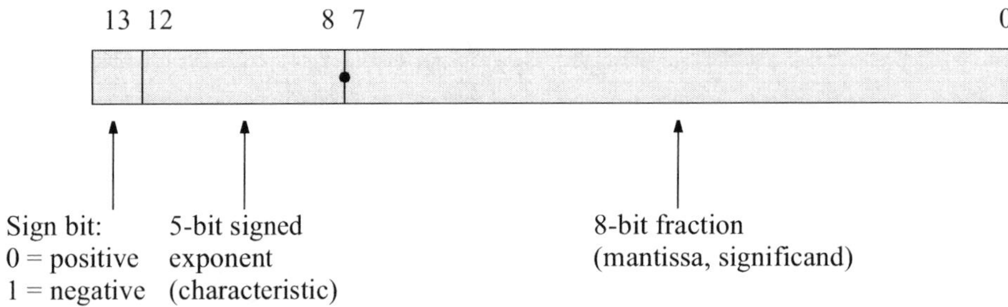

The behavioral module employs two **always** statements — one to define the fields of the single-precision format when the floating-point operands change value, and one to perform the multiplication when a *start* signal is asserted.

Since the fractions contain 8 bits, a count-down counter is set to a value of 1000 (8) to accommodate the add-shift procedure. The fractions are then checked to determine if either fraction is zero. If both fractions are nonzero, then the multiplication begins.

```
//behavioral floating-point multiplication

module mul_flp4 (flp_a , flp_b, start, sign, exponent,
                     exp_unbiased, cout, prod);

//define inputs and outputs
input [13:0] flp_a, flp_b;
input start;
output sign;
output [4:0] exponent, exp_unbiased;
output cout;
output [15:0] prod;

                                   //continued on next page
```

Figure 4.106 Behavioral design module for floating-point multiplication using the sequential add-shift method.

```
//variables used in an always block are declared as registers
reg sign_a, sign_b;
reg [4:0] exp_a, exp_b;
reg [4:0] exp_a_bias, exp_b_bias;
reg [4:0] exp_sum;
reg [7:0] fract_a, fract_b;
reg [7:0] fract_b_reg;
reg sign;
reg [4:0] exponent, exp_unbiased;
reg cout;
reg [15:0] prod;
reg [3:0] count;


//define sign, exponent, and fraction
always @ (flp_a or flp_b)
begin
   sign_a = flp_a[13];
   sign_b = flp_b[13];

   exp_a = flp_a[12:8];
   exp_b = flp_b[12:8];

   fract_a = flp_a[7:0];
   fract_b = flp_b[7:0];


//bias exponents
   exp_a_bias = exp_a + 5'b01111;
   exp_b_bias = exp_b + 5'b01111;


//add exponents
   exp_sum = exp_a_bias + exp_b_bias;

//remove one bias
   exponent = exp_sum - 5'b01111;


   exp_unbiased = exponent - 5'b01111;
end

//multiply fractions
always @ (posedge start)
begin
   fract_b_reg = fract_b;
   prod = 0;

                                        //continued on next page
```

Figure 4.106 (Continued)

```
      count = 4'b1000;
         if ((fract_a != 0) && (fract_b != 0))
            while (count)
               begin
                  {cout, prod[15:8]} = (({8{fract_b_reg[0]}}
                     & fract_a) + prod[15:8]);
                  prod = {cout, prod[15:8], prod[7:1]};
                  fract_b_reg = fract_b_reg >> 1;
                  count = count - 1;
               end

//postnormalize result
   while (prod[15] == 0)
      begin
         prod = prod << 1;
         exp_unbiased = exp_unbiased - 1;
      end

   sign = sign_a ^ sign_b;

end

endmodule
```

Figure 4.106 (Continued)

```
//test bench for floating-point multiplication
module mul_flp4_tb;

//inputs are reg for test bench
//outputs are wire for test bench
reg    [13:0] flp_a, flp_b;
reg    start;
wire   sign;
wire   [4:0] exponent, exp_unbiased;
wire   [4:0] exp_sum;
wire   [15:0] prod;

//display variables
initial
$monitor ("sign = %b, exp_unbiased = %b, prod = %b",
           sign, exp_unbiased, prod); //continued on next page
```

Figure 4.107 Test bench module for floating-point multiplication using the sequential add-shift method.

```
//apply input vectors
initial
begin
   #0     start = 1'b0;
          //+5 x +3 = +15
          //           s    e       f
          flp_a = 14'b0_00011_1010_0000;
          flp_b = 14'b0_00010_1100_0000;
   #10    start = 1'b1;
   #10    start = 1'b0;

          //+7 x -5 = -35
          //           s    e       f
   #0     flp_a = 14'b0_00011_1110_0000;
          flp_b = 14'b1_00011_1010_0000;
   #10    start = 1'b1;
   #10    start = 1'b0;

          //+25 x +25 = +625
          //           s    e       f
   #0     flp_a = 14'b0_00101_1100_1000;
          flp_b = 14'b0_00101_1100_1000;
   #10    start = 1'b1;
   #10    start = 1'b0;

          //-7 x -15 = +105
          //           s    e       f
   #0     flp_a = 14'b1_00011_1110_0000;
          flp_b = 14'b1_00100_1111_0000;
   #10    start = 1'b1;
   #10    start = 1'b0;

          //-35 x +72 = -2520
          //           s    e       f
   #0     flp_a = 14'b1_00110_1000_1100;
          flp_b = 14'b0_00111_1001_0000;
   #10    start = 1'b1;
   #10    start = 1'b0;

          //+80 x +37 = +2960
          //           s    e       f
   #0     flp_a = 14'b0_00111_1010_0000;
          flp_b = 14'b0_00110_1001_0100;
   #10    start = 1'b1;
   #10    start = 1'b0;
                                    //continued on next page
```

Figure 4.107 (Continued)

```
         //+34 x -68 = -2312
         //          s   e      f
   #0    flp_a = 14'b0_00110_1000_1000;
         flp_b = 14'b1_00111_1000_1000;
   #10   start = 1'b1;
   #10   start = 1'b0;

         //+27 x +25 = +675
         //          s   e      f
   #0    flp_a = 14'b0_00101_1101_1000;
         flp_b = 14'b1_00101_1100_1000;
   #10   start = 1'b1;
   #10   start = 1'b0;

   #10   $stop;
end

//instantiate the module into the test bench
mul_flp4 inst1 (flp_a , flp_b, start, sign, exponent,
                  exp_unbiased, cout, prod);

endmodule
```

Figure 4.107 (Continued)

```
+5 x +3 = +15
sign = 0, exp_unbiased = 00100, prod = 1111000000000000

+7 x -5 = -35
sign = 1, exp_unbiased = 00110, prod = 1000110000000000

+25 x +25 = +625
sign = 0, exp_unbiased = 01010, prod = 1001110001000000

-7 x -15 = +105
sign = 0, exp_unbiased = 00111, prod = 1101001000000000

-35 x +72 = -2520
sign = 1, exp_unbiased = 01100, prod = 1001110110000000

+80 x +37 = +2960
sign = 0, exp_unbiased = 01100, prod = 1011100100000000
                                      //continued on next page
```

Figure 4.108 Outputs for floating-point multiplication using the sequential add-shift method.

```
+34 x -68 = -2312
sign = 1, exp_unbiased = 01100, prod = 1001000010000000

+27 x +25 = +675
sign = 1, exp_unbiased = 01010, prod = 1010100011000000
```

Figure 4.108 (Continued)

4.14 Floating-Point Division

The division of two floating-point numbers is accomplished by dividing the fractions and subtracting the exponents. The fractions are divided by any of the methods presented in the section on fixed-point division and overflow is checked in the same manner. Fraction division and exponent subtraction are two independent operations and can be done in parallel. Floating-point division is defined as shown in Equation 4.15.

$$\begin{aligned} A / B &= (f_A \times r^{eA}) / (f_B \times r^{eB}) \\ &= (f_A / f_B) \times r^{(eA - eB)} \end{aligned} \tag{4.15}$$

Both operands are checked for a value of zero. If the dividend is zero, then the exponent, quotient, and remainder are set to zero. If the divisor is zero, then the operation is terminated. The sign of the quotient is determined by the signs of the operands as follows: $A_{sign} \oplus B_{sign}$. The division algorithm is defined as shown below.

1. Normalize the operands.
2. Check for zero operands.
3. Determine the sign of the quotient.
4. Align the dividend, if necessary.
5. Subtract the exponents.
6. Add the bias.
7. Divide the fractions.
8. Normalize the result, if necessary.

Restoring division will be used in the behavioral design of the floating-point division example in this section. As stated previously, restoring division examines the state of the carry-out when the dividend is subtracted from the partial remainder. This determines the relative magnitudes of the divisor and partial remainder. If the carry-out = 0, then the partial remainder is restored to its previous value by adding the divisor to the partial remainder. If the carry-out = 1, then there is no restore operation. The partial remainder (high-order half of the dividend) and the low-order half of the dividend are then shifted left one bit position and the process repeats for each bit in the divisor.

Two examples will now be presented to illustrate the technique for floating-point division using the shift-subtract/add restoring division method.

Example 4.18 A dividend fraction *fract_a* = 0.1001 1000 × 2^5 (+19) is divided by a divisor fraction *fract_b* = 0.0101 × 2^4 (+5) to yield a quotient of 0011 × 2^4 (+3) and a remainder of 0100 × 2^4 (+4).

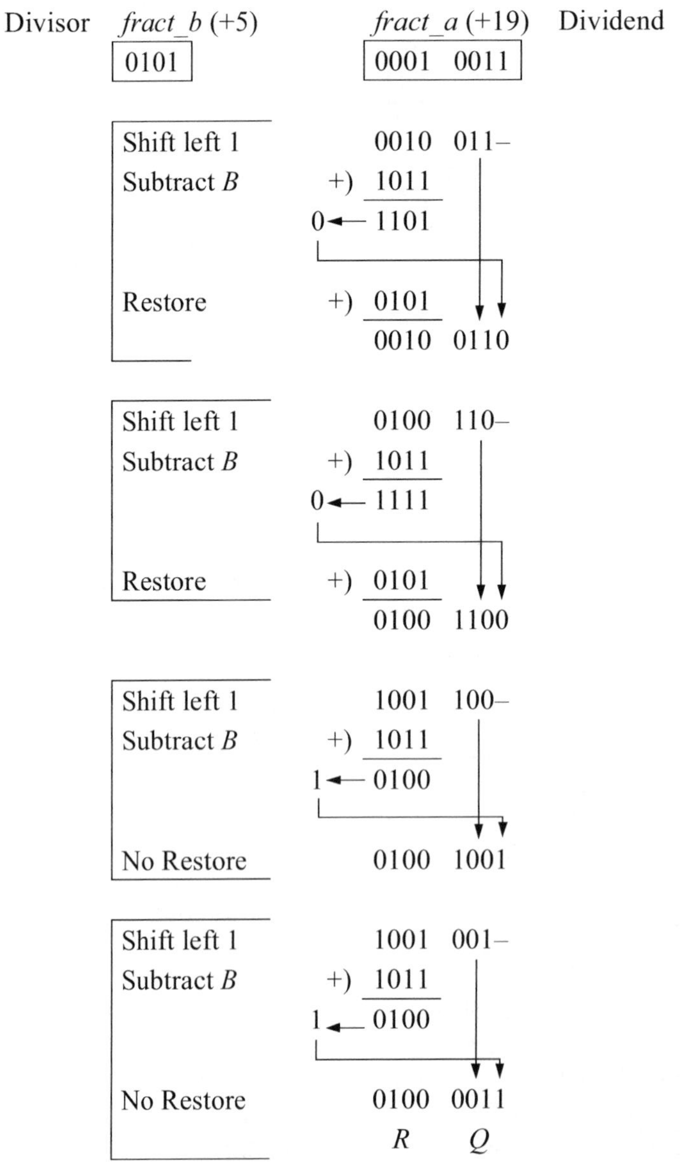

Example 4.19 A dividend fraction *fract_a* = 0.1000 0110 × 2^7 (+67) is divided by a divisor fraction *fract_b* = 0.1000 × 2^4 (+8) to yield a quotient of 1000 × 2^4 (+8) and a remainder of 0011 × 2^4 (+3).

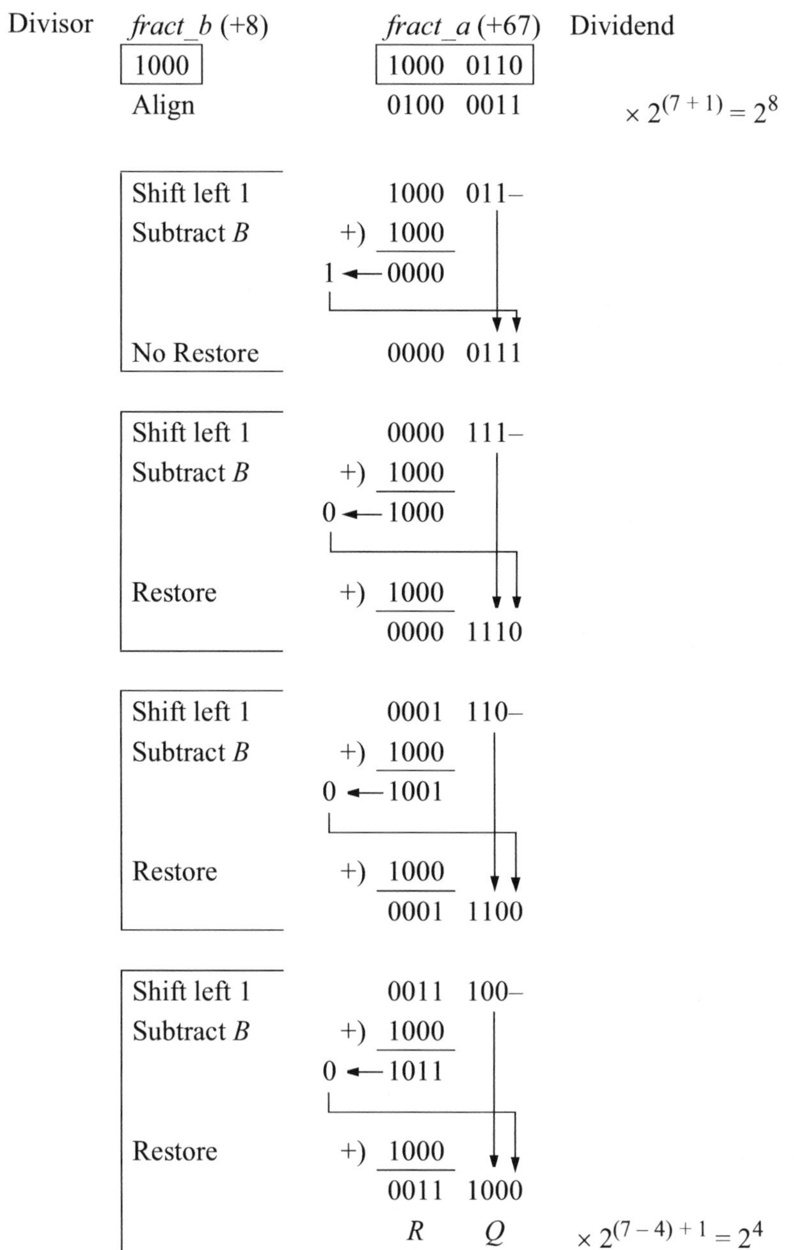

The Verilog behavioral design module for a version similar to Examples 4.18 and 4.19 is shown in Figure 4.109 using the restoring division technique. The module has two operands, a 16-bit dividend *a[15:0]* and an 8-bit divisor *b[7:0]*. The test bench for several different input vectors is shown in Figure 4.110. The outputs are shown in Figure 4.111.

```
//behavioral design for restoring division

module div_restoring_vers6 (a, b, start, rslt);

//define inputs and outputs
input [15:0] a;
input [7:0] b;
input start;
output [15:0] rslt;

//variables in always are declared as variables
wire [7:0] b_bar;

//define internal registers
reg [7:0] b_neg;
reg [15:0] rslt;
reg [3:0] count;

assign b_bar = ~b;
always @ (b_bar)
   b_neg = b_bar + 1;

//execute the division
always @ (posedge start)
begin
   rslt = a;
   count = 5'b1000;

   if ((a!=0) && (b!=0))
      while (count)
         begin
            rslt = rslt << 1;
            rslt = {(rslt[15:8] + b_neg), rslt[7:0]};
               if (rslt[15] == 1)
                  begin
                     rslt = {(rslt[15:8] + b),
                                    rslt[7:1], 1'b0};
                     count = count - 1;
                  end
                                    //continued on next page
```

Figure 4.109 Mixed design module for floating-point restoring division.

```
                    else
                       begin
                          rslt = {rslt[15:1], 1'b1};
                       count = count - 1;
                       end
            end
end
endmodule
```

Figure 4.109 (Continued)

```
//test bench for restoring division
module div_restoring_vers6_tb;

reg [15:0] a;           //inputs are reg for test bench
reg [7:0] b;
reg start;
wire [15:0] rslt;       //outputs are wire for test bench

initial      //display variables
$monitor ("a = %b, b = %b, quot = %b, rem = %b",
            a, b, rslt[7:0], rslt[15:8]);

initial      //apply input vectors
begin
   #0      //(19) / (5) = q3, r4
           start = 1'b0;
           a = 16'b0000_0000_0001_0011;  b = 8'b0000_0101;
   #10     start = 1'b1;
   #10     start = 1'b0;

   #10     //(37) / (10) = q3, r7
           a = 16'b0000_0000_0010_0101;  b = 8'b0000_1010;
   #10     start = 1'b1;
   #10     start = 1'b0;

           //(60) / (7) = q8, r4
   #10     a = 16'b0000_0000_0011_1100;  b = 8'b000_0111;
   #10     start = 1'b1;
   #10     start = 1'b0;

                                   //continued on next page
```

Figure 4.110 Test bench module for floating-point restoring division.

```
         //(4,044) / (127) = q31, r107
   #10   a =16'b0000_1111_1100_1100;   b = 8'b0111_1111;
   #10   start = 1'b1;
   #10   start = 1'b0;

         //(2046) / (126) = q16, r30
   #10   a = 16'b0000_0111_1111_1110;  b = 8'b0111_1110;
   #10   start = 1'b1;
   #10   start = 1'b0;

         //(90) / (15) = q6, r0
   #10   a = 16'b0000_0000_0101_1010;  b = 8'b000_1111;
   #10   start = 1'b1;
   #10   start = 1'b0;

      //(260) / (120) = q2, r20
   #10   a = 16'b0000_0001_0000_0100;  b = 8'b111_1000;
   #10   start = 1'b1;
   #10   start = 1'b0;

      //(204) / (120) = q1, r84
   #10   a = 16'b0000_0000_1100_1100;  b = 8'b111_1000;
   #10   start = 1'b1;
   #10   start = 1'b0;

      //(127) / (127) = q1, r0
   #10   a = 16'b0000_0000_0111_1111;  b = 8'b111_1111;
   #10   start = 1'b1;
   #10   start = 1'b0;

      //(508) / (127) = q4, r0
   #10   a = 16'b0000_0001_1111_1100;  b = 8'b111_1111;
   #10   start = 1'b1;
   #10   start = 1'b0;

      //(0) / (127) = q0, r0
   #10   a = 16'b0000_0000_0000_0000;  b = 8'b111_1111;
   #10   start = 1'b1;
   #10   start = 1'b0;

   #10   $stop;
end

//instantiate the module into the test bench
div_restoring_vers6 inst1 (a, b, start, rslt);

endmodule
```

Figure 4.110 (Continued)

```
19 / 5 = q3, r4
a = 0000_0000_0001_0011, b = 0000_0101,
      quot = 0000_0011, rem = 0000_0100

37 / 10 = q3, r7
a = 0000_0000_0010_0101, b = 0000_1010,
      quot = 0000_0011, rem = 0000_0111

60 / 7 = q8, r4
a = 0000_0000_0011_1100, b = 0000_0111,
      quot = 0000_1000, rem = 0000_0100

4,044 / 127 = q31, r107
a = 0000_1111_1100_1100, b = 0111_1111,
      quot = 0001_1111, rem = 0110_1011

2046 / 126 = q16, r30
a = 0000_0111_1111_1110, b = 0111_1110,
      quot = 0001_0000, rem = 0001_1110

90 / 15 = q6, r0
a = 0000_0000_0101_1010, b = 0000_1111,
      quot = 0000_0110, rem = 0000_0000

260 / 120 = q2, r20
a = 0000_0001_0000_0100, b = 0111_1000,
      quot = 0000_0010, rem = 0001_0100

204 / 120 = q1, r84
a = 0000_0000_1100_1100, b = 0111_1000,
      quot = 0000_0001, rem = 0101_0100

127 / 127 = q1, r0
a = 0000_0000_0111_1111, b = 0111_1111,
      quot = 0000_0001, rem = 0000_0000

508 / 127 = q4, r0
a = 0000_0001_1111_1100, b = 0111_1111,
      quot = 0000_0100, rem = 0000_0000

0 / 127 = q0, r0
a = 0000_0000_0000_0000, b = 0111_1111,
      quot = 0000_0000, rem = 0000_0000
```

Figure 4.111 Outputs for floating-point restoring division.

4.15 Problems

Fixed-Point Addition

4.1 Use structural modeling with built-in primitives to design a single-bit full adder using fixed-point addition. Recall that a full adder has three scalar inputs: augend *a*, addend *b*, and a carry-in *cin* from the previous lower-order stage. There are two scalar outputs: *sum* and carry-out *cout*. The truth table for a full adder is shown below. Note that the *sum* is 1 for an odd number of 1s; the carry-out *cout* is a 1 for two or more 1s.

Truth Table for a Full Adder

		Carry-In	Sum	Carry-Out
a	*b*	*cin*	*sum*	*cout*
0	0	0	0	0
0	0	1	1	0
0	1	0	1	0
0	1	1	0	1
1	0	0	1	0
1	0	1	0	1
1	1	0	0	1
1	1	1	1	1

4.2 A considerable increase in speed can be realized by using a carry lookahead adder rather than using a ripple adder. The increase in speed is achieved by expressing the carry-out $cout_i$ of any stage$_i$ as a function of the two operand bits, a_i and b_i, and the carry-in cin_{-1} to the low-order stage$_0$ of the adder, where the adder is an *n*-bit adder $n_{-1}\ n_{-2} \ldots n_1\ n_0$.

The equation for the carry-out of any stage$_i$ of a carry lookahead adder is $cout_i = a_i b_i + (a_i \oplus b_i)cin_{i-1}$. The carries entering all the bit positions of the adder can be generated simultaneously by a *carry lookahead* generator. This results in a constant addition time that is independent of the length of the adder. A carry will be generated for $a_i\, b_i$. A carry will be propagated for $a_i \oplus b_i$. Therefore, the equation for the carry-out of any stage$_i$ can be defined as $cout_i = G_i + P_i\, cin_{i-1}$. Design a 4-bit carry lookahead adder using built-in primitives and the continuous assignment statement **assign**. Obtain the test bench that adds several combinations of the augend and addend and obtain the outputs.

4.3 Use dataflow modeling with the continuous assignment statement **assign** to design a single-bit full adder. Obtain the design module, the test bench module for all combinations of the inputs, and the outputs.

4.4 Use the full adder designed in Problem 4.3 to design a 4-bit ripple adder using structural modeling. Obtain the structural design module, the test bench module for several input vectors, and the outputs.

4.5 As a final problem in fixed-point addition, design a 4-bit dataflow adder that would be used in an arithmetic and logic unit (ALU) that has three inputs: an *augend*, an *addend*, and a *carry-in*. There is only one output: *sum*. There is no *carry-out* from the 4-bit adder. This is a simple adder that uses the Verilog HDL add operator + for the add operation. Obtain the dataflow design module, the test bench module that applies several input vectors, and the outputs.

Fixed-Point Subtraction

4.6 Design a 4-bit subtractor using structural modeling with dataflow full adders that were implemented with the continuous assignment statement **assign**. Also use built-in primitives in the design of the subtractor. The minuend is *a* (3:0), the subtrahend is *b* (3:0), and the result (difference) is *rslt* (3:0).

Subtraction is performed by adding the 2s complement of the subtrahend to the minuend, where the 2s complement is the 1s complement (negation) plus one. Obtain the test bench module for several input vectors of the minuend and subtrahend and include inputs that produce negative results. Also obtain the outputs.

4.7 Design a 4-bit ripple-carry fixed-point adder/subtractor using built-in primitives and instantiated full adders that were designed using behavioral modeling. There are three inputs: *a*[3:0], *b*[3:0], and a mode control *m*, which is used to determine whether the operation is addition or subtraction. If $m = 0$, then the operation is addition; if $m = 1$, then the operation is subtraction. There are two outputs: *rslt*[3:0] and *cout*[3:0].

Obtain the logic diagram, the design module using structural modeling, the test bench module with combinations of the inputs for both addition and subtraction including overflow, and the outputs.

4.8 Design a 4-bit behavioral adder/subtractor unit with the following three inputs: augend/minuend *a[3:0]*, addend/subtrahend *b[3:0]*, and a *mode* control input to determine the operation to be performed, where addition is defined as *mode* = 0 and subtraction is defined as *mode* = 1. There are two outputs: the result of the operation, *rslt[3:0]*, and an overflow indication, *ovfl*. The operands are signed numbers in 2s complement representation. Obtain the behavioral design module, the test bench module, and the outputs.

Fixed-Point Multiplication

4.9 Design a 3-bit array multiplier using structural modeling. Instantiate full adders that were designed using dataflow modeling with the continuous assignment statement **assign** and 2-input AND gates that were also designed using dataflow modeling. An example of a general array multiply algorithm is shown below for two 3-bit operands.

				a_2	a_1	a_0
			×)	b_2	b_1	b_0
Partial product 1				a_2b_0	a_1b_0	a_0b_0
Partial product 2			a_2b_1	a_1b_1	a_0b_1	
Partial product 3		a_2b_2	a_1b_2	a_0b_2		
	2^5	2^4	2^3	2^2	2^1	2^0

Obtain the structural design module, the test bench module, and the outputs. Obtain all combinations of the multiplicand and the multiplier to yield the 64 combinations of the product. Display the values of the variables in decimal notation.

4.10 Design an add-shift multiplier for two 4-bit operands using behavioral modeling. The multiplicand and multiplier are both unsigned binary operands. Use a scalar *start* signal to initiate the multiply operation for each pair of operands. Obtain the behavioral design module, the test bench module, and the outputs in decimal notation.

Fixed-Point Division

4.11 The sequential shift add/subtract restoring division technique will be utilized in this problem. Obtain the mixed-design (behavioral/dataflow) module for 16-bit dividends and 8-bit divisors. Obtain the test bench module using several different values of the dividend and divisor. Use a scalar *start* signal in the test bench to initiate the divide operation for each pair of operands. Obtain the outputs in both binary (%b) and decimal (%d) notation.

4.12 This problem implements nonrestoring division for 16-bit dividends and 8-bit divisors. The results are also 16 bits; that is 8-bit quotients and 8-bit remainders. Nonrestoring division allows both a positive partial remainder and a negative partial remainder to be utilized in the division process. The final partial remainder may require restoration if the sign is 1 (negative). This is required in order to have a final positive remainder.

Obtain the behavioral design module, the test bench module, and the outputs in both binary notation (%b) and decimal notation (%d) for all operands: dividend *a*, divisor *b*, quotient *q*, and remainder *r*.

Arithmetic and Logic Unit

4.13 Arithmetic and logic units perform the arithmetic operations of addition, subtraction, multiplication, and division. They also perform the logical operations of AND, NAND, OR, NOR, exclusive-OR, and exclusive-NOR. This problem is to design a behavioral module to implement the four operations of add, subtract, multiply, and divide. The operands are eight bits, the operation code is three bits, and the result of the operation is eight bits. The two 8-bit inputs are operands *a[7:0]* and *b[7:0]*. The 3-bit operation code is *opcode[2:0]* and the 8-bit result is *rslt[7:0]*.

Obtain the behavioral design module using the **case** statement for the four arithmetic operations, the test bench module displaying all variables in decimal (%d) notation, and the outputs.

4.14 This problem implements the behavioral design of the logical functions of AND, NAND, OR, NOR, XOR, and XNOR. The operands are four bits, the operation code is three bits, and the result of the operation is four bits. The two 4-bit inputs are operands *a[3:0]* and *b[3:0]*. The 3-bit operation code is *opcode[2:0]* and the 4-bit result is *rslt[3:0]*.

Obtain the behavioral design module using the **case** statement for the six logical operations, the test bench module displaying all variables in binary (%b) notation, and the outputs.

Decimal Addition

4.15 Design a single-digit binary-coded decimal (BCD) adder using behavioral modeling. The design uses two 4-bit adders. Obtain the test bench module for several input variables and the outputs. Assume that all input vectors are valid BCD digits.

4.16 This problem repeats Problem 4.15, but implements the BCD adder design using the equations that represent the outputs of Adder_1, *cout*, and Adder_2. The equation for the carry-out (*cout*) of the BCD adder is

$$cout = cout_8 + bit_8\, bit_4 + bit_8\, bit_2$$

Obtain the behavioral design module, the test bench module displaying the variables in decimal notation, and the outputs.

4.17 Instantiate the design of Problem 4.16 to implement the design of a 2-digit BCD adder. In order to use the same single-digit BCD adder two times — not modified — in this problem, do not let the sum of the two low-order digits exceed nine. Obtain the behavioral design module, the test bench module with the variables displayed in decimal notation, and the outputs.

4.18 Use structural modeling with built-in primitives to design the two digit BCD adder of Problem 4.17. Obtain the structural design module and the test bench module with several input combinations of two operands. Enter augends and addends that produce sums in the units, tens, and hundreds representations. Display all of the outputs in decimal notation.

Decimal Subtraction

4.19 Design a structural module that performs both BCD addition and subtraction in which the result of an operation does not exceed a value of 99. Since the device performs both addition and subtraction, the nines complementer requires an additional input to specify the mode (m) of the operation — either addition ($m = 0$) or subtraction ($m = 1$). The equations for the nines complementer are shown below.

$$f[0] = b[0] \oplus m$$

$$f[1] = b[1]$$

$$f[2] = b[2]m' + (b[2] \oplus b[1])m$$

$$f[3] = b[3]m' + b[3]'b[2]'b[1]'m$$

Obtain the logic diagram using nines complementers, 4-bit adders, required logic gates, and 2-to-1 multiplexers. Then obtain the structural design module using the 4-bit adders, 2-to-1 multiplexers to generate the BCD result, and built-in primitives. Obtain the test bench module using operands for both addition and subtraction. Obtain the outputs.

4.20 This is a relatively simpler problem than Problem 4.19 for BCD subtraction. Design a behavioral module for BCD subtraction using the **always** statement. Obtain the design module, the test bench module, and the outputs in decimal notation.

Decimal Multiplication

4.21 Using the method shown in Section 4.9, design a behavioral module to multiply an 8-bit multiplicand by a 4-bit multiplier to produce a 12-bit binary-coded decimal product. Recall that decimal multiplication can be performed first in fixed-point multiplication then converting the product to binary-coded decimal. Obtain the behavioral design module, the test bench module for several different operands, and the outputs.

Decimal Division

4.22 Design a binary-coded decimal (BCD) divisor module using behavioral modeling. Use the restoring division technique described in Section 4.5. Obtain the fixed-point quotient and remainder, then convert the quotient and remainder to BCD. The dividend is an 8-bit input, the divisor is a 4-bit input, and the quotient and remainder are 4-bit outputs. Obtain the behavioral design module, the test bench module applying several inputs, and the outputs displayed in decimal notation.

Floating-Point Addition

4.23 Design a behavioral module for a floating-point adder that adds two 32-bit fractions in the single-precision floating-point format that results in true addition; that is, $(+A) + (+B)$. The fractions are defined as: *flp_a[31:0]* and *flp_b[31:0]*. The exponents are eight bits defined as: *exp_a[7:0]* and *exp_b[7:0]*. Obtain the behavioral design module, the test bench module, and the outputs.

Floating-Point Subtraction

4.24 Design a behavioral module that performs true subtraction on two 32-bit operands. True subtraction can be defined as follows: $(+A) - (+B)$ or $(-A) - (-B)$, which has the following attributes: *fract_a* > *fract_b* and *sign_a* = *sign_b*. The exponents are eight bits. Obtain the behavioral design module, the test bench module, and the outputs.

4.25 Design a behavioral module that is similar to Problem 4.24 that performs true subtraction on two 32-bit operands. In this version, true subtraction is defined as follows: $(+A) + (-B)$ or $(-A) + (+B)$, which has the following attributes: *fract_a* > *fract_b* and *sign_a* ≠ *sign_b*. The exponents are eight bits. Obtain the behavioral design module, the test bench module, and the outputs.

Floating-Point Multiplication

4.26 Design a behavioral module to perform multiplication on two 32-bit floating-point operands: multiplicand *flp_a[31:0]* and multiplier *flp_b[31:0]*. Use the sequential add-shift technique and a *start* signal to perform the multiplication. Obtain the behavioral module using the **always** statement, the test bench module for ten multiply operations, and the outputs.

4.27 Design a behavioral module for a 32-bit single-precision floating-point multiplication operation for these two operands: multiplicand *flp_a[31:0]* and multiplier *flp_b[31:0]*. The single-precision format is shown below. Use the multiply arithmetic operator (*) to perform the multiply operation. Obtain the behavioral module, the test bench module, and the outputs showing the product as a 23-bit result.

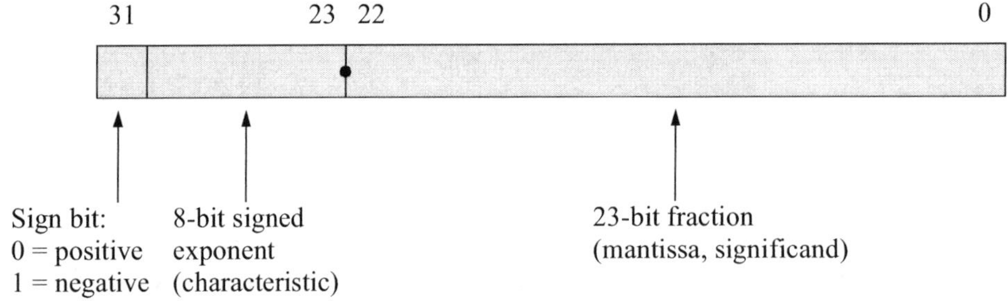

Floating-Point Division

4.28 Using the floating-point formats shown below, design a behavioral module to divide a 14-bit dividend by a 10-bit divisor. Also shown are the operands for the first set of input vectors: 82/9. Obtain the test bench and the outputs.

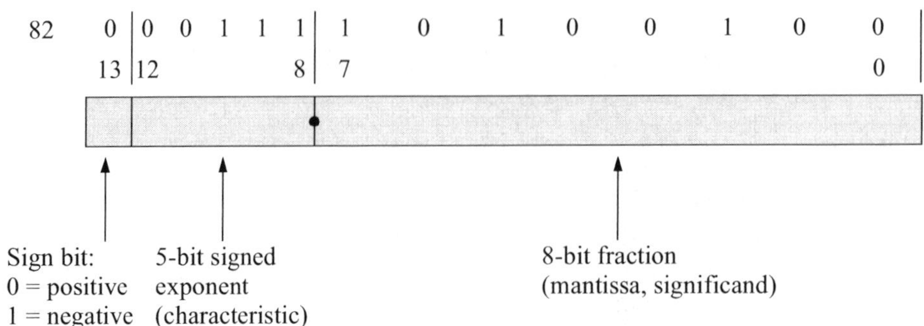

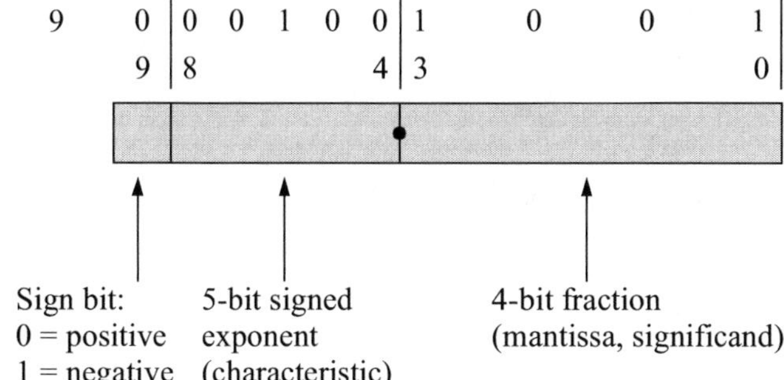
9
0 0 0 1 0 0 1 0 0 1
9 8 4 3 0
Sign bit:
0 = positive
1 = negative
5-bit signed
exponent
(characteristic)
4-bit fraction
(mantissa, significand)

Appendix A

Event Queue

Event management in Verilog hardware description language (HDL) is controlled by an event queue. Verilog modules generate events in the test bench, which provide stimulus to the module under test. These events can then produce new events by the modules under test. Since the Verilog HDL Language Reference Manual (LRM) does not specify a method of handling events, the simulator must provide a way to arrange and schedule these events in order to accurately model delays and obtain the correct order of execution. The manner of implementing the event queue is vendor-dependent.

Time in the event queue advances when every event that is scheduled in that time step is executed. Simulation is finished when all event queues are empty. An event at time t may schedule another event at time t or at time $t + n$.

A.1 Event Handling for Dataflow Assignments

Dataflow constructs consist of continuous assignments using the **assign** statement. The assignment occurs whenever simulation causes a change to the right-hand side expression. Unlike procedural assignments, continuous assignments are order independent — they can be placed anywhere in the module.

Consider the logic diagram shown in Figure A.1 which is represented by the two dataflow modules of Figures A.2 and A.3. The test bench for both modules is shown in Figure A.4. The only difference between the two dataflow modules is the reversal of the two **assign** statements. The order in which the two statements execute is not defined by the Verilog HDL LRM; therefore, the order of execution is indeterminate.

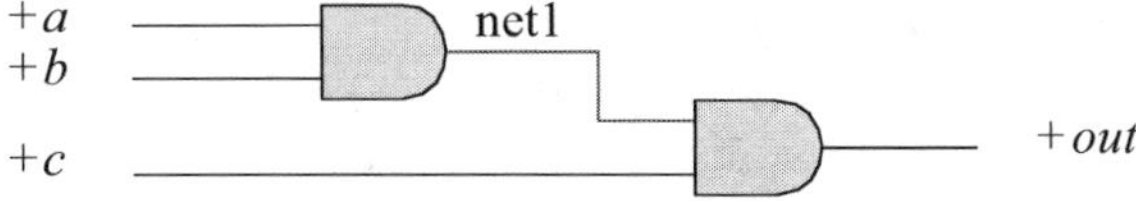

Figure A.1 Logic diagram to demonstrate event handling.

```
module dataflow (a, b, c, out);

input a, b, c;
output out;

wire a, b, c;
wire out;

//define internal net
wire net1;

assign net1 = a & b;
assign out = net1 & c;

endmodule
```

Figure A.2 Dataflow module 1.

```
module dataflow (a, b, c, out)

input a, b, c;
output out;

wire a, b, c;
wire out;

//define internal net
wire net1;

assign out = net1 & c;
assign net1 = a & b;

endmodule
```

Figure A.3 Dataflow module 2.

```
module dataflow_tb;

reg test_a, test_b, test_c;
wire test_out;

initial
begin
   test_a = 1'b1;
   test_b = 1'b0;
   test_c = 1'b0;

   #10    test_b = 1'b1;
          test_c = 1'b1;
   #10    $stop;
end
//instantiate the module
dataflow inst1
   .a(test_a),
   .b(test_b),
   .c(test_c),
   .out(test_out)
   );

endmodule
```

Figure A.4 Test bench for Figures A.2 and A.3.

Assume that the simulator executes the assignment order shown in Figure A.2 first. When the simulator reaches time unit #10 in the test bench, it will evaluate the right-hand side of *test_b = 1'b1;* and place its value in the event queue for an immediate scheduled assignment. Since this is a blocking statement, the next statement will not execute until the assignment has been made. Figure A.5 represents the event queue after the evaluation. The input signal *b* will assume the value of *test_b* through instantiation.

Event queue					
Scheduled event 5	Scheduled event 4	Scheduled event 3	Scheduled event 2	Scheduled event 1	Time unit
				test_b ← 1'b1 b ← 1'b1	$t = \#10$
←				Order of execution	

Figure A.5 Event queue after execution of *test_b = 1'b1;*.

After the assignment has been made, the simulator will execute the *test_c = 1'b1;* statement by evaluating the right-hand side, and then placing its value in the event queue for immediate assignment. The new event queue is shown in Figure A.6. The entry that is not shaded represents an executed assignment.

Event queue					
Scheduled event 5	Scheduled event 4	Scheduled event 3	Scheduled event 2	Scheduled event 1	Time unit
			test_c ← 1'b1 c ← 1'b1	test_b ← 1'b1 b ← 1'b1	$t = \#10$
←				Order of execution	

Figure A.6 Event queue after execution of *test_c = 1'b1;*.

When the two assignments have been made, time unit #10 will have ended in the test bench, which is the top-level module in the hierarchy. The simulator will then enter the instantiated dataflow module during this same time unit and determine that events have occurred on input signals *b* and *c* and execute the two continuous assignments. At this point, inputs *a*, *b*, and *c* will be at a logic 1 level. However, *net1* will still contain a logic 0 level as a result of the first three assignments that executed at time #0 in the test bench. Thus, the statement *assign out = net1 & c;* will evaluate to a logic 0, which will be placed in the event queue and immediately assigned to *out*, as shown in Figure A.7.

Event queue					
Scheduled event 5	Scheduled event 4	Scheduled event 3	Scheduled event 2	Scheduled event 1	Time unit
		out ← 1'b0 test_out ← 1'b0	test_c ← 1'b1 c ← 1'b1	test_b ← 1'b1 b ← 1'b1	$t = \#10$
←				Order of execution	

Figure A.7 Event queue after execution of *assign out = net1 & c;*.

The simulator will then execute the *assign net1 = a & b;* statement in which the right-hand side evaluates to a logic 1 level. This will be placed on the queue and immediately assigned to *net1* as shown in Figure A.8.

Event queue					
Scheduled event 5	Scheduled event 4	Scheduled event 3	Scheduled event 2	Scheduled event 1	Time unit
	net1 ← 1'b1	out ← 1'b0 test_out ← 1'b0	test_c ← 1'b1 c ← 1'b1	test_b ← 1'b1 b ← 1'b1	$t = \#10$
←				Order of execution	

Figure A.8 Event queue after execution of *assign net1 = a & b;*.

When the assignment has been made to *net1*, the simulator will recognize this as an event on *net1*, which will cause all statements that use *net1* to be reevaluated. The only statement to be reevaluated is *assign out = net1 & c;*. Since both *net1* and *c* equal a logic 1 level, the right-hand side will evaluate to a logic 1, resulting in the event queue shown in Figure A.9.

Event queue					
Scheduled event 5	Scheduled event 4	Scheduled event 3	Scheduled event 2	Scheduled event 1	Time unit
out ← 1'b1 test_out ← 1'b1	net1 ← 1'b1	out ← 1'b0 test_out ← 1'b0	test_c ← 1'b1 c ← 1'b1	test_b ← 1'b1 b ← 1'b1	t = #10
←				Order of execution	

Figure A.9 Event queue after execution of *assign out = net1 & c;*.

The test bench signal *test_out* must now be updated because it is connected to *out* through instantiation. Because the signal *out* is not associated with any other statements within the module, the output from the module will now reflect the correct output. Since all statements within the dataflow module have been processed, the simulator will exit the module and return to the test bench. All events have now been processed; therefore, time unit #10 is complete and the simulator will advance the simulation time.

Since the order of executing the **assign** statements is irrelevant, processing of the dataflow events will now begin with the *assign net1 = a & b;* statement to show that the result is the same. The event queue is shown in Figure A.10.

Event queue					
Scheduled event 5	Scheduled event 4	Scheduled event 3	Scheduled event 2	Scheduled event 1	Time unit
		net1 ← 1'b1	test_c ← 1'b1 c ← 1'b1	test_b ← 1'b1 b ← 1'b1	t = #10
←				Order of execution	

Figure A.10 Event queue beginning with the statement *assign net1 = a & b;*.

Once the assignment to *net1* has been made, the simulator recognizes this as a new event on *net1*. The existing event on input *c* requires the evaluation of statement *assign out = net1 & c;*. The right-hand side of the statement will evaluate to a logic 1, and will be placed on the event queue for immediate assignment, as shown in Figure A.11.

Event queue					
Sched-uled event 5	Scheduled event 4	Scheduled event 3	Scheduled event 2	Scheduled event 1	Time unit
	out ← 1'b1 test_out ← 1'b1	net1 ← 1'b1	test_c ← 1'b1 c ← 1'b1	test_b ← 1'b1 b ← 1'b1	t = #10
←				Order of execution	

Figure A.11 Event queue after execution of *assign out = net1 & c;*.

A.2 Event Handling for Blocking Assignments

The blocking assignment operator is the equal (=) symbol. A blocking assignment evaluates the right-hand side arguments and completes the assignment to the left-hand side before executing the next statement; that is, the assignment *blocks* other assignments until the current assignment has been executed.

Example A.1 Consider the code segment shown in Figure A.12 using blocking assignments in conjunction with the event queue of Figure A.13. There are no interstatement delays and no intrastatement delays associated with this code segment. In the first blocking assignment, the right-hand side is evaluated and the assignment is scheduled in the event queue. Program flow is blocked until the assignment is executed. This is true for all blocking assignment statements in this code segment. The assignments all occur in the same simulation time step t.

```
always @ (x2 or x3 or x5 or x7)
begin
   x1 = x2 | x3;
   x4 = x5;
   x6 = x7;
end
```

Figure A.12 Code segment with blocking assignments.

Event queue					
Scheduled event 5	Scheduled event 4	Scheduled event 3	Scheduled event 2	Scheduled event 1	Time unit
		x6 ← x7 (*t*)	x4 ← x5 (*t*)	x1 ← x2 | x3 (*t*)	*t*
←				Order of execution	

Figure A.13 Event queue for Figure A.12.

Example A.2 The code segment shown in Figure A.14 contains an interstatement delay. Both the evaluation and the assignment are delayed by two time units. When the delay has taken place, the right-hand side is evaluated and the assignment is scheduled in the event queue as shown in Figure A.15. The program flow is blocked until the assignment is executed.

```
always @ (x2)
begin
   #2 x1 = x2;
end
```

Figure A.14 Blocking statement with interstatement delay.

Event queue					
Scheduled event 5	Scheduled event 4	Scheduled event 3	Scheduled event 2	Scheduled event 1	Time unit
					t
				x1 ← x2 (*t* + 2)	t + 2
←				Order of execution	

Figure A.15 Event queue for Figure A.14.

Example A.3 The code segment of Figure A.16 shows three statements with interstatement delays of $t + 2$ time units. The first statement does not execute until simulation time $t + 2$ as shown in Figure A.17. The right-hand side $(x_2 \mid x_3)$ is evaluated at the current simulation time which is $t + 2$ time units, and then assigned to the left-hand side. At $t + 2$, x_1 receives the output of $x_2 \mid x_3$.

```
always @ (x2 or x3 or x5 or x7)
begin
   #2 x1 = x2 | x3;
   #2 x4 = x5;
   #2 x6 = x7;
end
```

Figure A.16 Code segment for delayed blocking assignment with interstatement delays.

Event queue					
Scheduled event 5	Scheduled event 4	Scheduled event 3	Scheduled event 2	Scheduled event 1	Time unit
					t
				x1 ← x2 | x3 ($t + 2$)	$t + 2$
				x4 ← x5 ($t + 4$)	$t + 4$
				x6 ← x7 ($t + 6$)	$t + 6$
←				Order of execution	

Figure A.17 Event queue for Figure A.16.

Example A.4 The code segment in Figure A.18 shows three statements using blocking assignments with intrastatement delays. Evaluation of $x_3 = \#2\ x_4$ and $x_5 = \#2\ x_6$ is blocked until x_2 has been assigned to x_1, which occurs at $t + 2$ time units. When the second statement is reached, it is scheduled in the event queue at time $t + 2$, but the assignment to x_3 will not occur until $t + 4$ time units. The evaluation in the third statement is blocked until the assignment is made to x_3. Figure A.19 shows the event queue.

```
always @ (x2 or x4 or x6)
begin
   x1 = #2 x2;     //first statement
   x3 = #2 x4;     //second statement
   x5 = #2 x6;     //third statement
end
```

Figure A.18 Code segment using blocking assignments with interstatement delays.

Event queue					
Scheduled event 5	Scheduled event 4	Scheduled event 3	Scheduled event 2	Scheduled event 1	Time unit
					t
				x1 ← x2 (t)	$t + 2$
				x3 ← x4 ($t + 2$)	$t + 4$
				x5 ← x6 ($t + 4$)	$t + 6$
←				Order of execution	

Figure A.19 Event queue for the code segment of Figure A.18.

A.3 Event Handling for Nonblocking Assignments

Whereas blocking assignments block the sequential execution of an **always** block until the simulator performs the assignment, nonblocking statements evaluate each statement in succession and place the result in the event queue. Assignment occurs when all of the **always** blocks in the module have been processed for the current time unit. The assignment may cause new events that require further processing by the simulator for the current time unit.

Example A.5 For nonblocking statements, the right-hand side is evaluated and the assignment is scheduled at the end of the queue. The program flow continues and the assignment occurs at the end of the time step. This is shown in the code segment of Figure A.20 and the event queue of Figure A.21.

```
always @ (posedge clk)
begin
   x1 <= x2;
end
```

Figure A.20 Code segment for a nonblocking assignment.

Event queue					
Scheduled event 5	Scheduled event 4	Scheduled event 3	Scheduled event 2	Scheduled event 1	Time unit
x1 ← x2 (t)					t
←				Order of execution	

Figure A.21 Event queue for Figure A.20.

Example A.6 The code segment of Figure A.22 shows a nonblocking statement with an interstatement delay. The evaluation is delayed by the timing control, and then the right-hand side expression is evaluated and assignment is scheduled at the end of the queue. Program flow continues and assignment is made at the end of the current time step as shown in the event queue of Figure A.23.

```
always @ (posedge clk)
begin
   #2 x1 <= x2;
end
```

Figure A.22 Nonblocking assignment with interstatement delay.

Event queue					
Scheduled event 5	Scheduled event 4	Scheduled event 3	Scheduled event 2	Scheduled event 1	Time unit
					t
x1 ← x2 ($t + 2$)					$t + 2$
←				Order of execution	

Figure A.23 Event queue for Figure A.22.

Example A.7 The code segment of Figure A.24 shows a nonblocking statement with an intrastatement delay. The right-hand side expression is evaluated and assignment is delayed by the timing control and is scheduled at the end of the queue. Program flow continues and assignment is made at the end of the current time step as shown in the event queue of Figure A.25.

```
always @ (posedge clk)
begin
   x1 <= #2 x2;
end
```

Figure A.24 Nonblocking assignment with intrastatement delay.

Event queue					
Scheduled event 5	Scheduled event 4	Scheduled event 3	Scheduled event 2	Scheduled event 1	Time unit
					t
x1 ← x2 (t)					$t + 2$
←				Order of execution	

Figure A.25 Event queue for Figure A.24.

Example A.8 The code segment of Figure A.26 shows nonblocking statements with intrastatement delays. The right-hand side expressions are evaluated and assignment is delayed by the timing control and is scheduled at the end of the queue. Program flow continues and assignment is made at the end of the current time step as shown in the event queue of Figure A.27.

```
always @ (posedge clk)
begin
   x1 <= #2 x2;
   x3 <= #2 x4;
   x5 <= #2 x6;
end
```

Figure A.26 Nonblocking assignments with intrastatement delays.

Event queue					
Scheduled event 5	Scheduled event 4	Scheduled event 3	Scheduled event 2	Scheduled event 1	Time unit
					t
x5 ← x6 (t)	x3 ← x4 (t)	x1 ← x2 (t)			$t + 2$
←				Order of execution	

Figure A.27 Event queue for Figure A.26.

Example A.9 Figure A.28 shows a code segment using nonblocking assignment with an intrastatement delay. The right-hand expression is evaluated at the current time. The assignment is scheduled, but delayed by the timing control #2. This method allows for propagation delay through a logic element; for example, a *D* flip-flop. The event queue is shown in Figure A.29.

```
always @ (posedge clk)
begin
   q <= #2 d;
end
```


Figure A.28 Code segment using intrastatement delay with blocking assignment.

Event queue					
Scheduled event 5	Scheduled event 4	Scheduled event 3	Scheduled event 2	Scheduled event 1	Time unit
					t
				q ← d (t)	t + 2
←				Order of execution	

Figure A.29 Event queue for the code segment of Figure A.28.

A.4 Event Handling for Mixed Blocking and Nonblocking Assignments

All nonblocking assignments are placed at the end of the queue while all blocking assignments are placed at the beginning of the queue in their respective order of evaluation. Thus, for any given simulation time *t*, all blocking statements are evaluated and assigned first, then all nonblocking statements are evaluated.

This is the reason why combinational logic requires the use of blocking assignments while sequential logic, such as flip-flops, requires the use of nonblocking assignments. In this way, Verilog events can model real hardware in which combinational logic at the input to a flip-flop can stabilize before the clock sets the flip-flop to the state of the input logic. Therefore, blocking assignments are placed at the top of the queue to allow the input data to be stable, whereas nonblocking assignments are placed at the bottom of the queue to be executed after the input data has stabilized.

The logic diagram of Figure A.30 illustrates this concept for two multiplexers connected to the *D* inputs of their respective flip-flops. The multiplexers represent combinational logic; the *D* flip-flops represent sequential logic. The behavioral module is shown in Figure A.31 and the event queue is shown in Figure A.32.

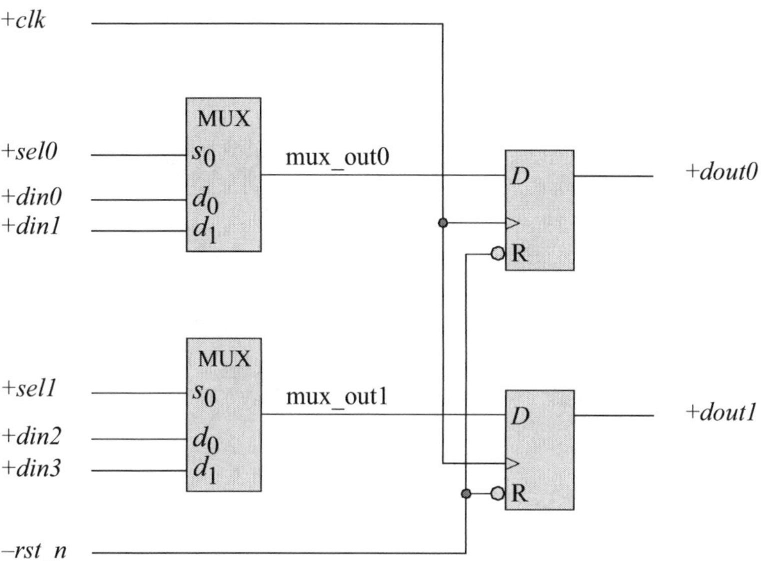

Figure A.30 Combinational logic connected to sequential logic to illustrate the use of blocking and nonblocking assignments.

Because multiplexers are combinational logic, the outputs *mux_out0* and *mux_out1* are placed at the beginning of the queue, as shown in Figure A.32. Nets *mux_out0* and *mux_out1* are in separate **always** blocks; therefore, the order in which they are placed in the queue is arbitrary and can differ with each simulator. The result, however, is the same. If *mux_out0* and *mux_out1* were placed in the same **always** block, then the order in which they are placed in the queue must be the same order as they appear in the **always** block.

Because *dout0* and *dout1* are sequential, they are placed at the end of the queue. Since they appear in separate **always** blocks, the order of their placement in the queue is irrelevant. Once the values of *mux_out0* and *mux_out1* are assigned in the queue, their values will then be used in the assignment of *dout0* and *dout1*; that is, the state of *mux_out0* and *mux_out1* will be set into the *D* flip-flops at the next positive clock transition and assigned to *dout0* and *dout1*.

```
//behavioral module with combinational and sequential logic
//to illustrate their placement in the event queue

module mux_plus_flop (clk, rst_n,
      din0, din1, sel0, dout0,
      din2, din3, sel1, dout1);

input clk, rst_n;
input din0, din1, sel0;
input din2, din3, sel1;
output dout0, dout1;

reg mux_out0, mux_out1;
reg dout0, dout1;

//combinational logic for multiplexers
always @ (din0 or din1 or sel0)
begin
   if (sel0)
      mux_out0 = din1;
   else
      mux_out0 = din0;
end

always @ (din2 or din3 or sel1)
begin
   if (sel1)
      mux_out1 = din3;
   else
      mux_out1 = din2;
end
//continued on next page
```

Figure A.31 Mixed blocking and nonblocking assignments that represent combinational and sequential logic.

```
//sequential logic for D flip-flops
always @ (posedge clk or negedge rst_n)
begin
   if (~rst_n)
      dout0 <= 1'b0;
   else
      dout0 <= mux_out0;
end

always @ (posedge clk or negedge rst_n)
begin
   if (~rst_n)
      dout1 <= 1'b0;
   else
      dout1 <= mux_out1;
end

endmodule
```

Figure A.31 (Continued)

Event queue					
Scheduled event 4	Scheduled event 3	N/A	Scheduled event 2	Scheduled event 1	Time unit
dout1 ← mux_out1 (t)	dout0 ← mux_out0 (t)		mux_out1 ← din3 (t)	mux_out0 ← din1 (t)	*t*
← Order of execution					

Figure A.32 Event queue for Figure A.31.

Appendix B

Verilog Project Procedure

- **Create a folder** (Do only once)
  Local disk (C:) > New Folder <Verilog> > Enter > Exit local disk C.

- **Create a project** (Do for each project)
  Bring up Silos Simulation Environment.
  File > Close Project. Minimize Silos.
  Local disk (C:) > Verilog > File > New Folder <new folder name> Enter.
  Exit Local disk (C:). Maximize Silos.
  File > New Project.
  Create New Project. Save In: Verilog folder.
  Click new folder name. Open.
  Create New Project. Filename: Give project name — usually same name as the folder name. Save
  Project Properties > Cancel.

- **File > New**

  .
  . Design module code goes here
  .

- **File > Save As > File name: <filename.v> > Save**

- **Compile code**
  Edit > Project Properties > Add. Select one or more files to add.
  Click on the file > Open.
  Project Properties. The selected files are shown > OK.
  Load/Reload Input Files. This compiles the code.
  Check screen output for errors. "Simulation stopped at the end of time 0" indicates no compilation errors.

- **Test bench**

  File > New

  .
  . Test bench module code goes here
  .

- **File > Save As > File name: < filename.v> > Save.**

- **Compile test bench**

  Edit > Project Properties > Add. Select one or more files to add.
  Click on the file > Open
  Project Properties. The selected files are shown > OK.
  Load/Reload Input Files. This compiles the code.
  Check screen output for errors. "Simulation stopped at end of time 0" indicates no compilation errors.

- **Binary Output and Waveforms**

  For binary output: click on the GO icon.
  For waveforms: click on the Analyzer icon.
  Click on the Explorer icon. The signals are listed in Silos Explorer.
  Click on the desired signal names.
  Right click. Add Signals to Analyzer.
  Waveforms are displayed.
  Exit Silos Explorer.

- **Change Time Scale**

  With the waveforms displayed, click on Analyzer > X-Axis > Timescale
  Enter Time / div > OK

- **Exit the project**

  Close the waveforms, module, and test bench.
  File > Close Project.

Appendix C

Answers to Select Problems

Chapter 1 Introduction to Logic Design Using Verilog HDL

1.1 Given the equation shown below, obtain the minimized equation for z_1 in a product-of-sums notation and implement the equation using NAND gate built-in primitives. Obtain the design module, the test bench module, and the outputs. Output z_1 is asserted high.

$$z_1(x_1, x_2, x_3, x_4) = \Sigma_m(1, 4, 7, 9, 11, 13) + \Sigma_d(5, 14, 15)$$

x_1x_2 \ x_3x_4	0 0	0 1	1 1	1 0
0 0	0 (0)	1 (1)	0 (3)	0 (2)
0 1	1 (4)	– (5)	1 (7)	0 (6)
1 1	0 (12)	1 (13)	– (15)	– (14)
1 0	0 (8)	1 (9)	1 (11)	0 (10)

z_1

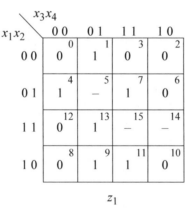

Product of sums: $z_1 = (x_2 + x_4)(x_1' + x_4)(x_1 + x_2 + x_3')(x_3' + x_4)$

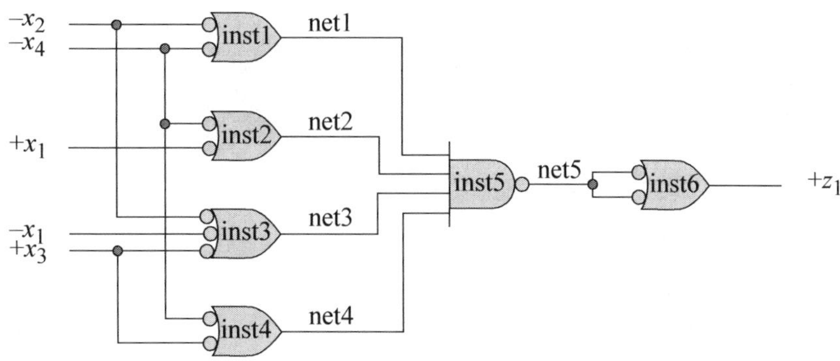

```
//built-in primitives for logic equation as a pos
module log_eqn_pos_nand2 (x1, x2, x3, x4, z1);

input x1, x2, x3, x4;
output z1;

nand  inst1 (net1, ~x2, ~x4),
      inst2 (net2, x1, ~x4),
      inst3 (net3, ~x1, ~x2, x3),
      inst4 (net4, x3, ~x4),
      inst5 (net5, net1, net2, net3, net4),
      inst6 (z1, net5, net5);

endmodule
```

```
//test bench for logic equation as a pos
module log_eqn_pos_nand2_tb;

reg x1, x2, x3, x4;
wire z1;

//apply input vectors and display variables
initial
begin: apply_stimulus
   reg [4:0] invect;
   for (invect=0; invect<16; invect=invect+1)
      begin
         {x1, x2, x3, x4} = invect [4:0];
         #10 $display ("x1 x2 x3 x4 = %b, z1 = %b",
                        {x1, x2, x3, x4}, z1);
      end
end

//instantiate the module into the test bench
log_eqn_pos_nand2 inst1 (x1, x2, x3, x4, z1);

endmodule
```

```
x1 x2 x3 x4 = 0000, z1 = 0
x1 x2 x3 x4 = 0001, z1 = 1
x1 x2 x3 x4 = 0010, z1 = 0
x1 x2 x3 x4 = 0011, z1 = 0
x1 x2 x3 x4 = 0100, z1 = 1
x1 x2 x3 x4 = 0101, z1 = 1
x1 x2 x3 x4 = 0110, z1 = 0
x1 x2 x3 x4 = 0111, z1 = 1
x1 x2 x3 x4 = 1000, z1 = 0
x1 x2 x3 x4 = 1001, z1 = 1
x1 x2 x3 x4 = 1010, z1 = 0
x1 x2 x3 x4 = 1011, z1 = 1
x1 x2 x3 x4 = 1100, z1 = 0
x1 x2 x3 x4 = 1101, z1 = 1
x1 x2 x3 x4 = 1110, z1 = 0
x1 x2 x3 x4 = 1111, z1 = 1
```

1.3 Use AND gate and OR gate built-in primitives to implement a circuit in a sum-of-products form that will generate an output z_1 if an input is greater than or equal to 2 and less than 5; and also greater than or equal to 12 and less than 15. Then obtain the design module, test bench module, and outputs.

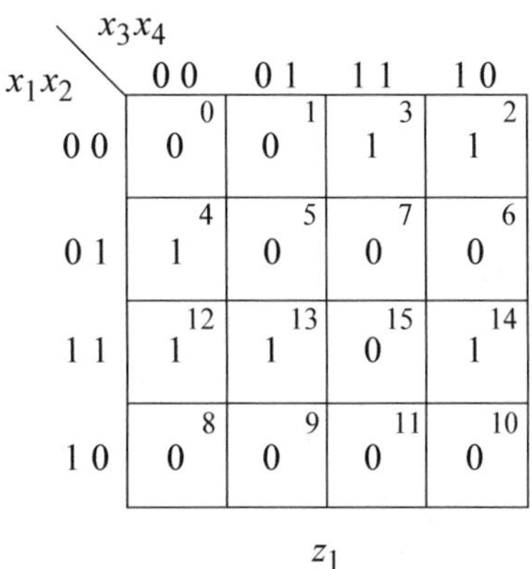

$$z_1 = x_2x_3'x_4' + x_1x_2x_3' + x_1x_2x_4' + x_1'x_2'x_3$$

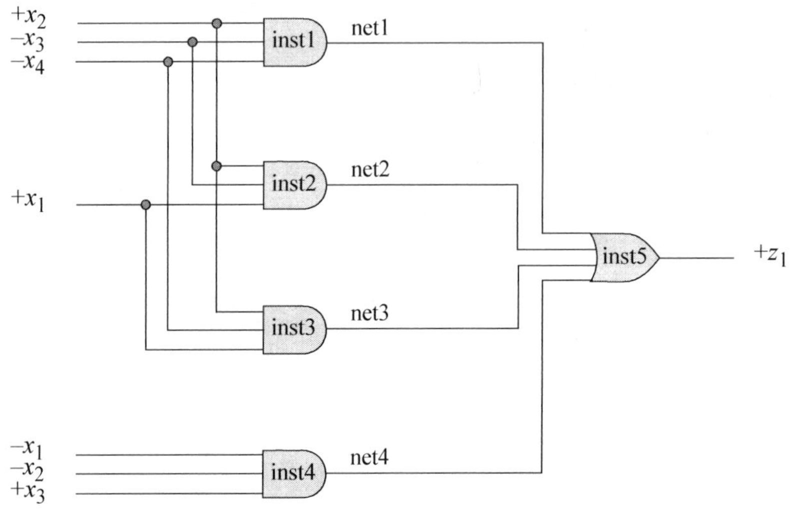


```
//built-in primitives to generate sop

module built_in_sop3 (x1, x2, x3, x4, z1);

//define inputs and output
input x1, x2, x3, x4;
output z1;

//design the logic
and   inst1 (net1, x2, ~x3, ~x4),
      inst2 (net2, x1, x2, ~x3),
      inst3 (net3, x1, x2, ~x4),
      inst4 (net4, ~x1, ~x2, x3);

or    inst5 (z1, net1, net2, net3, net4);

endmodule
```

```
//test bench for module
module built_in_sop3_tb;

reg x1, x2, x3, x4;  //inputs are reg for test bench
wire z1;             //outputs are wire for test bench

//apply input vectors and display variables
initial
begin: apply_stimulus
   reg [4:0] invect;
   for (invect = 0; invect < 16; invect = invect + 1)
      begin
         {x1, x2, x3, x4}= invect [4:0];
         #10$display ("x1 x2 x3 x4 = %b, z1 = %b",
                     {x1, x2, x3, x4}, z1);
      end
end

//instantiate the module into the test bench
built_in_sop3 inst1 (x1, x2, x3, x4, z1);

endmodule
```

```
x1 x2 x3 x4 = 0000, z1 = 0
x1 x2 x3 x4 = 0001, z1 = 0
x1 x2 x3 x4 = 0010, z1 = 1
x1 x2 x3 x4 = 0011, z1 = 1
x1 x2 x3 x4 = 0100, z1 = 1

x1 x2 x3 x4 = 0101, z1 = 0
x1 x2 x3 x4 = 0110, z1 = 0
x1 x2 x3 x4 = 0111, z1 = 0
x1 x2 x3 x4 = 1000, z1 = 0
x1 x2 x3 x4 = 1001, z1 = 0

x1 x2 x3 x4 = 1010, z1 = 0
x1 x2 x3 x4 = 1011, z1 = 0
x1 x2 x3 x4 = 1100, z1 = 1
x1 x2 x3 x4 = 1101, z1 = 1
x1 x2 x3 x4 = 1110, z1 = 1
x1 x2 x3 x4 = 1111, z1 = 0
```

1.6 Design a modulo-8 counter using the D flip-flop that was designed in the *edge-sensitive user-defined primitives* section of this chapter. Use additional logic gate UDPs as necessary. Obtain the design module, the test bench module, and the outputs.

y_1	y_2	y_3
0	0	0
0	0	1
0	1	0
0	1	1
1	0	0
1	0	1
1	1	0
1	1	1
0	0	0

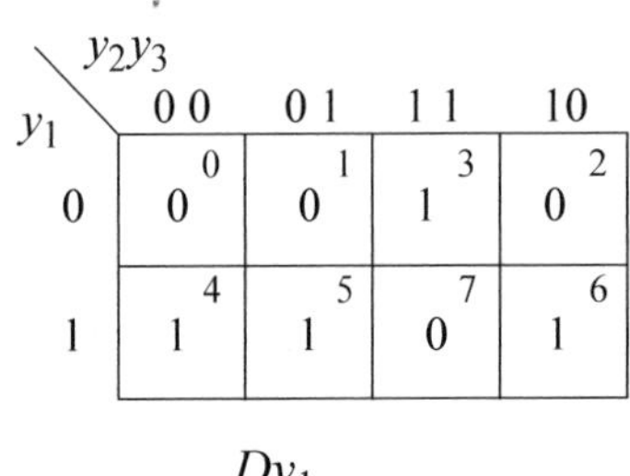

$y_1 \backslash y_2y_3$	00	01	11	10
0	0 (0)	0 (1)	1 (3)	0 (2)
1	1 (4)	1 (5)	0 (7)	1 (6)

Dy_1

$y_1 \backslash y_2y_3$	00	01	11	10
0	0 (0)	1 (1)	0 (3)	1 (2)
1	0 (4)	1 (5)	0 (7)	1 (6)

Dy_2

$y_1 \backslash y_2y_3$	00	01	11	10
0	1 (0)	0 (1)	0 (3)	1 (2)
1	1 (4)	0 (5)	0 (7)	1 (6)

Dy_3

$$Dy_1 = y_1'y_2y_3 + y_1y_2' + y_1y_3'$$

$$Dy_2 = y_2'y_3 + y_2y_3' = y_2 \oplus y_3$$

$$Dy_3 = y_3'$$

```
//modulo-8 counter using udps

module ctr_mod8_udp1 (rst_n, clk, y1, y2, y3);

//define inputs and outputs
input clk, rst_n;
output y1, y2, y3;

//-----------------------------------------------
//instantiate the udp for flip-flop y1
udp_and3 inst1  (net1, ~y1, y2, y3);
udp_and2 inst2  (net2, y1, ~y2);
udp_and2 inst3  (net3, y1, ~y3);
udp_or3  inst4  (net4, net1, net2, net3);

udp_dff_edge1 inst5 (y1, net4, clk, rst_n);

//-----------------------------------------------
//instantiate the udp for flip-flop y2
udp_xor2 inst6  (net5, y2, y3);

udp_dff_edge1 inst7 (y2, net5, clk, rst_n);

//-----------------------------------------------
//instantiate the udp for flip-flop y3
udp_dff_edge1 inst8 (y3, ~y3, clk, rst_n);

endmodule
```

```
//test bench for the modulo-8 counter using udps

module ctr_mod8_udp1_tb;

//inputs are reg for test bench
//outputs are wire for test bench
reg clk, rst_n;
wire y1, y2, y3;

//display variables
initial
$monitor ("{y1 y2 y3} = %b", {y1, y2, y3});

//generate reset
initial
begin
   #0 rst_n = 1'b1;
   #2 rst_n = 1'b0;
   #5 rst_n = 1'b1;
end

//generate clock
initial
begin
   clk = 1'b0;
   forever
      #10clk = ~clk;
end

//determine length of simulation
initial
begin
   repeat (9) @ (posedge clk);
   $stop;
end

//instantiate the module into the test bench
ctr_mod8_udp1 inst1 (rst_n, clk, y1, y2, y3);

endmodule
```

```
{y1 y2 y3} = 000
{y1 y2 y3} = 001
{y1 y2 y3} = 010
{y1 y2 y3} = 011
{y1 y2 y3} = 100
{y1 y2 y3} = 101
{y1 y2 y3} = 110
{y1 y2 y3} = 111
{y1 y2 y3} = 000
```

1.14 Design a binary-to-excess-3 code converter using user-defined primitives of the following types: *udp_and2*, *udp_and3*, *udp_or3*, and *udp_or4*. The binary code is labelled *a*, *b*, *c*, *d*, and the excess-3 code is labelled *w*, *x*, *y*, *z*, where *d* and *z* are the low-order bits of their respective codes. Binary codes above 1100 produce an excess-3 code of 0000. Obtain the design module, the test bench module, and outputs.

Binary code				Excess-3 code			
a	*b*	*c*	*d*	*w*	*x*	*y*	*z*
0	0	0	0	0	0	1	1
0	0	0	1	0	1	0	0
0	0	1	0	0	1	0	1
0	0	1	1	0	1	1	0
0	1	0	0	0	1	1	1
0	1	0	1	1	0	0	0
0	1	1	0	1	0	0	1
0	1	1	1	1	0	1	0
1	0	0	0	1	0	1	1
1	0	0	1	1	1	0	0
1	0	1	0	1	1	0	1
1	0	1	1	1	1	1	0
1	1	0	0	1	1	1	1
1	1	0	1	0	0	0	0
1	1	1	0	0	0	0	0
1	1	1	1	0	0	0	0

$$\overset{\text{net1}}{w = a'bd} + \overset{\text{net2}}{a'bc} + \overset{\text{net3}}{ab'} + \overset{\text{net4}}{ac'd'}$$

$$\overset{\text{net5}}{x = bc'd'} + \overset{\text{net6}}{b'd} + \overset{\text{net7}}{b'c}$$

$$\overset{\text{net8}}{y = c'd'} + \overset{\text{net9}}{a'cd} + \overset{\text{net10}}{b'cd}$$

$$\overset{\text{net11}}{z = c'd'} + \overset{\text{net12}}{a'd'} + \overset{\text{net13}}{b'd'}$$

```
//binary-to-excess-3 using udps

module bin_to_ex3_udp (a, b, c, d, w, x, y, z);

//define inputs and outputs
input a, b, c, d;
output w, x, y, z;

//define internal nets
wire  net1, net2, net3, net4,
      net5, net6, net7,
      net8, net9, net10,
      net11, net12, net13;

//instantiate the udps for high-order excess_3 w
udp_and3 (net1, ~a, b, d);
udp_and3 (net2, ~a, b, c);
udp_and2 (net3, a, ~b);
udp_and3 (net4, a, ~c, ~d);
udp_or4  (w, net1, net2, net3, net4);

//instantiate the udps for excess_3 x
udp_and3 (net5, b, ~c, ~d);
udp_and2 (net6, ~b, d);
udp_and2 (net7, ~b, c);
udp_or3  (x, net5, net6, net7);

//instantiate the udps for excess_3 y
udp_and2 (net8, ~c, ~d);
udp_and3 (net9, ~a, c, d);
udp_and3 (net10, ~b, c, d);
udp_or3  (y, net8, net9, net10);

//instantiate the udps for excess_3 z
udp_and2 (net11, ~c, ~d);
udp_and2 (net12, ~a, ~d);
udp_and2 (net13, ~b, ~d);
udp_or3  (z, net11, net12, net13);

endmodule
```

```
//test bench for binary-to-excess-3 udp module

module bin_to_ex3_udp_tb;

//inputs are reg for test bench
//outputs are wire for test bench
reg a, b, c, d;
wire w, x, y, z;

//apply stimulus and display variables
initial
begin: apply_stimulus
   reg [4:0] invect;
   for (invect = 0; invect < 16; invect = invect + 1)
      begin
         {a, b, c, d} = invect [4:0];
         #10 $display ("abcd = %b, wxyz = %b",
                          {a, b, c, d}, {w, x, y, z});
      end
end

//instantiate the module into the test bench
bin_to_ex3_udp inst1 (a, b, c, d, w, x, y, z);

endmodule
```

```
abcd = 0000, wxyz = 0011
abcd = 0001, wxyz = 0100
abcd = 0010, wxyz = 0101
abcd = 0011, wxyz = 0110

abcd = 0100, wxyz = 0111
abcd = 0101, wxyz = 1000
abcd = 0110, wxyz = 1001
abcd = 0111, wxyz = 1010

abcd = 1000, wxyz = 1011
abcd = 1001, wxyz = 1100
abcd = 1010, wxyz = 1101
abcd = 1011, wxyz = 1110

abcd = 1100, wxyz = 1111
abcd = 1101, wxyz = 0000
abcd = 1110, wxyz = 0000
abcd = 1111, wxyz = 0000
```

1.19 Design a module to execute the four shift operations of *shift left logical* (SLL), *shift left algebraic* (SLA), *shift right logical* (SRL), and *shift right algebraic* (SRA) using the **case** statement. The operands to be shifted are 8-bit operands. Obtain the test bench providing four shift amounts for each shift operation and obtain the outputs.

```
//behavioral logical and algebraic shifter
module four_fctn_shift (a, shift_code, shift_amt,
                           shift_rslt);

//define inputs and outputs
input [7:0] a;
input [1:0] shift_code;
input [2:0] shift_amt;
output [7:0] shift_rslt;

//variables used in always are declared as registers
reg [7:0] reg_a;
reg [7:0] shift_rslt;
reg [15:0] sra_reg;

//define shift codes
//parameter is used to define constants
parameter   sll = 2'b00,
            sla = 2'b01,
            srl = 2'b10,
            sra = 2'b11;

//perform the shift operations
always @ (a or shift_code)
begin
   case (shift_code)
      sll:
         begin
            reg_a = a << shift_amt;
            shift_rslt = reg_a;
         end

   sla:
         begin
            reg_a = a;
            reg_a = reg_a << shift_amt;
            reg_a[7] = a[7];
            shift_rslt = reg_a;
         end

                           //continued on next page
```

```
        sra:
        begin
           sra_reg[15:8] = {8{a[7]}};
           sra_reg[7:0] = a;
           sra_reg = sra_reg >> shift_amt;
           shift_rslt = sra_reg[7:0];
        end

   endcase
end

endmodule
```

```
//test bench for four_fctn_shift module
module four_fctn_shift_tb;

reg [7:0] a;            //inputs are reg for test bench
reg [1:0] shift_code;
reg [2:0] shift_amt;

wire [7:0] shift_rslt;  //outputs are wire

//display variables
initial
$monitor ("a=%b, shift_code=%b, shift_amt=%b,
            shift_rslt=%b",
            a, shift_code, shift_amt, shift_rslt);

//apply input vectors
initial
begin
//-------------------------------------------------------
//sll
   #0    a = 8'b0001_1110;
         shift_code = 2'b00;  shift_amt = 3'b011;

   #10   a = 8'b0111_1101;
         shift_code = 2'b00;  shift_amt = 3'b111;

   #10   a = 8'b0110_0101;
         shift_code = 2'b00;  shift_amt = 3'b101;

   #10   a = 8'b1111_1111;
         shift_code = 2'b00;  shift_amt = 3'b111;
//-------------------------------------------------------
//sla
   #10   a = 8'b0001_1110;
         shift_code = 2'b01;  shift_amt = 3'b011;

   #10   a = 8'b0111_1101;
         shift_code = 2'b01;  shift_amt = 3'b111;

   #10   a = 8'b0110_0101;
         shift_code = 2'b01;  shift_amt = 3'b101;

   #10   a = 8'b1111_1111;
         shift_code = 2'b01;  shift_amt = 3'b111;
//-------------------------------------------------------
                                //continued on next page
```

```
//srl
   #10    a = 8'b0001_1110;
          shift_code = 2'b10;  shift_amt = 3'b011;

   #10    a = 8'b0111_1101;
          shift_code = 2'b10;  shift_amt = 3'b111;

   #10    a = 8'b0110_0101;
          shift_code = 2'b10;  shift_amt = 3'b101;

   #10    a = 8'b1111_1111;
          shift_code = 2'b10;  shift_amt = 3'b111;
//------------------------------------------------------
//sra
   #10    a = 8'b0001_1110;
          shift_code = 2'b11;  shift_amt = 3'b011;

   #10    a = 8'b0111_1101;
          shift_code = 2'b11;  shift_amt = 3'b111;

   #10    a = 8'b0110_0101;
          shift_code = 2'b11;  shift_amt = 3'b101;

   #10    a = 8'b1111_1111;
          shift_code = 2'b11;  shift_amt = 3'b111;
//------------------------------------------------------

   #10    $stop;
end

//instantiate the module into the test bench
four_fctn_shift inst1 (a, shift_code, shift_amt,
                        shift_rslt);

endmodule
```

```
sll = 00,   sla = 01,   srl = 10,   sra = 11

------------------------------------------------------
shift left logical

a = 00011110,  shift_code = 00,
shift_amt = 011,  shift_rslt = 11110000

a = 01111101, shift_code = 00,
shift_amt = 111,  shift_rslt = 10000000

a = 01100101, shift_code = 00,
shift_amt = 101,  shift_rslt = 10100000

a = 11111111, shift_code = 00,
shift_amt = 111,  shift_rslt = 10000000

------------------------------------------------------
shift left algebraic

a = 00011110, shift_code = 01,
shift_amt = 011,  shift_rslt = 01110000

a = 01111101, shift_code = 01,
shift_amt = 111,  shift_rslt = 00000000

a = 01100101, shift_code = 01,
shift_amt = 101,  shift_rslt = 00100000

a = 11111111, shift_code = 01,
shift_amt = 111,  shift_rslt = 10000000

------------------------------------------------------
```

//continued on next page

```
sll = 00,   sla = 01,   srl = 10,   sra = 11

------------------------------------------------------
shift right logical
a = 00011110, shift_code = 10,
shift_amt = 011,  shift_rslt = 00000011

a = 01111101, shift_code = 10,
shift_amt = 111,  shift_rslt = 00000000

a = 01100101, shift_code = 10,
shift_amt = 101,  shift_rslt = 00000011

a = 11111111, shift_code = 10,
shift_amt = 111,  shift_rslt = 00000001

------------------------------------------------------
shift right algebraic
a = 00011110, shift_code = 11,
shift_amt = 011,  shift_rslt = 00000011

a = 01111101, shift_code = 11,
shift_amt = 111,  shift_rslt = 00000000

a = 01100101, shift_code = 11,
shift_amt = 101,  shift_rslt = 00000011

a = 11111111, shift_code = 11,
shift_amt = 111,  shift_rslt = 11111111
------------------------------------------------------
```

1.23 Use structural modeling to design a logic circuit to generate an output z_1 whenever a 4-bit variable — x_1, x_2, x_3, x_4 — has three or more 1s. Implement the module using AND gates and OR gates that were designed using dataflow modeling. Obtain the test bench and the outputs.

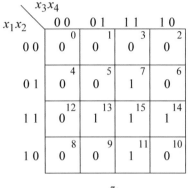

x_1x_2 \ x_3x_4	0 0	0 1	1 1	1 0
0 0	0 0	1 0	3 0	2 0
0 1	4 0	5 0	7 1	6 0
1 1	12 0	13 1	15 1	14 1
1 0	8 0	9 0	11 1	10 0

z_1

$$z_1 = x_1 x_2 x_4 + x_1 x_2 x_3 + x_2 x_3 x_4 + x_1 x_3 x_4$$

```
//structural for the following expression
//z1 = x1 x2 x4 + x1 x2 x3 + x2 x3 x4 + x1 x3 x4

module majority4 (x1, x2, x3, x4, z1);

//define inputs and output
input x1, x2, x3, x4;
output z1;

//define internal nets
wire net1, net2, net3, net4, net5;

//instantiate the logic gates
and3_df  inst1 (x1, x2, x4, net1);
and3_df  inst2 (x1, x2, x3, net2);
and3_df  inst3 (x2, x3, x4, net3);
and3_df  inst4 (x1, x3, x4, net4);
or4_df   inst5 (net1, net2, net3, net4, z1);

endmodule
```

```
//test bench for majority4
module majority4_tb;

//inputs are reg for test bench
//outputs are wire for test bench
reg x1, x2, x3, x4;
wire z1;

//apply input vectors and display outputs
initial
begin: apply_stimulus
   reg [4:0] invect;
   for (invect = 0; invect < 16; invect = invect + 1)
      begin
         {x1, x2, x3, x4} = invect [4:0];
         #10 $display ("x1, x2, x3, x4 = %b, z1 = %b",
                  {x1, x2, x3, x4}, z1);
      end
end

//instantiate the module into the test bench
majority4 inst1 (x1, x2, x3, x4, z1);

endmodule
```

```
x1, x2, x3, x4 = 0000, z1 = 0
x1, x2, x3, x4 = 0001, z1 = 0
x1, x2, x3, x4 = 0010, z1 = 0
x1, x2, x3, x4 = 0011, z1 = 0

x1, x2, x3, x4 = 0100, z1 = 0
x1, x2, x3, x4 = 0101, z1 = 0
x1, x2, x3, x4 = 0110, z1 = 0
x1, x2, x3, x4 = 0111, z1 = 1

x1, x2, x3, x4 = 1000, z1 = 0
x1, x2, x3, x4 = 1001, z1 = 0
x1, x2, x3, x4 = 1010, z1 = 0
x1, x2, x3, x4 = 1011, z1 = 1

x1, x2, x3, x4 = 1100, z1 = 0
x1, x2, x3, x4 = 1101, z1 = 1
x1, x2, x3, x4 = 1110, z1 = 1
x1, x2, x3, x4 = 1111, z1 = 1
```

Chapter 2 Combinational Logic Design Using Verilog HDL

2.3 Design a structural module that will generate a high output z_1 if a 4-bit binary number $x_1x_2x_3x_4$ has a value less than or equal to four or greater than eleven. Generate a Karnaugh map and obtain the equation for z_1 in a sum-of-products form and for z_2 in a product-of-sums form. Instantiate dataflow modules for the logic gates into the structural module. Obtain the design module, the test bench module for all combinations of the inputs, and the outputs.

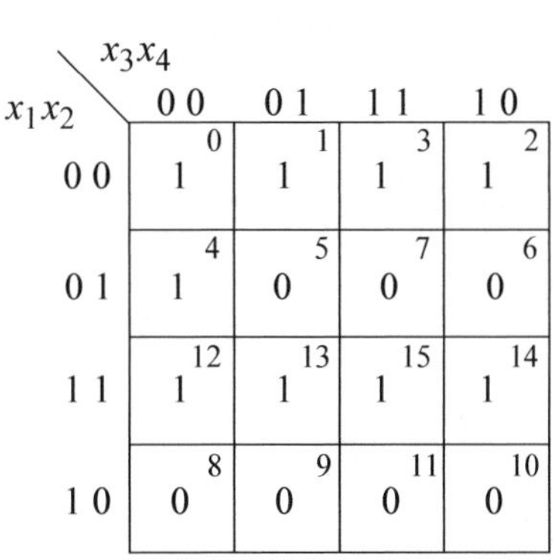

x_1x_2 \ x_3x_4	0 0	0 1	1 1	1 0
0 0	1 (0)	1 (1)	1 (3)	1 (2)
0 1	1 (4)	0 (5)	0 (7)	0 (6)
1 1	1 (12)	1 (13)	1 (15)	1 (14)
1 0	0 (8)	0 (9)	0 (11)	0 (10)

z_1 (z_2)

$$z_1 = x_1'x_2' + x_1x_2 + x_2x_3'x_4'$$
$$= (x_1 \oplus x_2)' + x_2x_3'x_4'$$

$$z_2 = (x_1 + x_2' + x_4')\,(x_1 + x_2' + x_3')\,(x_1' + x_2)$$

```
//structural dataflow number <=4 or >11
module number_range5 (x1, x2, x3, x4, z1, z2);

input x1, x2, x3, x4;   //define inputs and output
output z1, z2;

//define internal nets
wire  net1, net2, net3, net4, net5;

//design the logic for the sum-of-products z1
xnor2_df     inst1 (x1, x2, net1);
and3_df      inst2 (x2, ~x3, ~x4, net2);
or2_df       inst3 (net1, net2, z1);

//design the logic for the product-of-sums z2
or3_df       inst4 (x1, ~x2, ~x4, net3),
             inst5 (x1, ~x2, ~x3, net4);
or2_df       inst6 (~x1, x2, net5);
and3_df      imst7 (net3, net4, net5, z2);
endmodule
```

```
//test bench for number <=4 or >11
module number_range5_tb;

reg x1, x2, x3, x4;  //inputs are reg for test bench
wire z1, z2;         //outputs are wire

//apply input vectors and display variables
initial
begin: apply_stimulus
reg [4:0] invect;
for (invect = 0; invect < 16; invect = invect + 1)
   begin
      {x1, x2, x3, x4} = invect [4:0];
      #10 $display ("x1 x2 x3 x4) = %b, z1 =%b, z2 =%b",
                  {x1, x2, x3, x4}, z1, z2);
   end
end

//instantiate the module into the test bench
number_range5 inst1 (x1, x2, x3, x4, z1, z2);

endmodule
```

```
x1 x2 x3 x4) = 0000, z1 =1, z2 =1
x1 x2 x3 x4) = 0001, z1 =1, z2 =1
x1 x2 x3 x4) = 0010, z1 =1, z2 =1
x1 x2 x3 x4) = 0011, z1 =1, z2 =1

x1 x2 x3 x4) = 0100, z1 =1, z2 =1
x1 x2 x3 x4) = 0101, z1 =0, z2 =0
x1 x2 x3 x4) = 0110, z1 =0, z2 =0
x1 x2 x3 x4) = 0111, z1 =0, z2 =0

x1 x2 x3 x4) = 1000, z1 =0, z2 =0
x1 x2 x3 x4) = 1001, z1 =0, z2 =0
x1 x2 x3 x4) = 1010, z1 =0, z2 =0
x1 x2 x3 x4) = 1011, z1 =0, z2 =0

x1 x2 x3 x4) = 1100, z1 =1, z2 =1
x1 x2 x3 x4) = 1101, z1 =1, z2 =1
x1 x2 x3 x4) = 1110, z1 =1, z2 =1
x1 x2 x3 x4) = 1111, z1 =1, z2 =1
```

2.9 Design a dataflow module for a full adder using logic gates that were designed using dataflow modeling. Recall that a full adder is a combinational circuit that adds two operand bits: the augend *a* and the addend *b* plus a carry-in bit *cin*. The carry-in bit represents the carry-out of the previous lower-order stage. A full adder produces two outputs: a sum bit *sum* and carry-out bit *cout*. The truth table for a full adder is shown below. Obtain the test bench module and the outputs.

a	*b*	*cin*	*cout*	*sum*
0	0	0	0	0
0	0	1	0	1
0	1	0	0	1
0	1	1	1	0
1	0	0	0	1
1	0	1	1	0
1	1	0	1	0
1	1	1	1	1

$$sum = a \oplus b \oplus cin$$

$$cout = cin\,(a \oplus b) + ab$$

```
//dataflow for a full adder

module full_adder_df (a, b, cin, sum, cout);

//define inputs and outputs
input a, b, cin;
output sum, cout;

//define internal nets
wire net1, net2, net3;

//design the sum for the full adder
xor3_df  inst1 (a, b, cin, sum);

//design the carry-out for the full adder
xor2_df  inst2 (a, b, net1);
and2_df  inst3 (cin, net1, net2);
and2_df  inst4 (a, b, net3);
or2_df   inst5 (net2, net3, cout);

endmodule
```

```
//test bench for full adder
module full_adder_df_tb;

//inputs are reg for test bench
//outputs are wire for test bench
reg a, b, cin;
wire sum, cout;

//apply input vectors and display outputs
initial
begin: apply_stimulus
   reg [3:0] invect;
   for (invect = 0; invect < 8; invect = invect + 1)
      begin
         {a, b, cin} = invect [3:0];
         #10 $display ("a, b, cin) = %b,
                        sum = %b, cout = %b",
                        {a, b, cin}, sum, cout);
      end
end

//instantiate the module into the test bench
full_adder_df inst1 (a, b, cin, sum, cout);

endmodule
```

```
a, b, cin) = 000, sum = 0, cout = 0
a, b, cin) = 001, sum = 1, cout = 0
a, b, cin) = 010, sum = 1, cout = 0
a, b, cin) = 011, sum = 0, cout = 1

a, b, cin) = 100, sum = 1, cout = 0
a, b, cin) = 101, sum = 0, cout = 1
a, b, cin) = 110, sum = 0, cout = 1
a, b, cin) = 111, sum = 1, cout = 1
```

2.14 Use structural modeling to design a 4:1 multiplexer using logic gates that were designed using dataflow modeling. Obtain the test bench module and the outputs for 16 combinations of the inputs.

```
//structural for a 4to1 multiplexer
//using dataflow logic gates
module mux_4to1_df (sel, data, z1);

//define inputs and output
input [1:0] sel;
input [3:0] data;
output z1;

//define internal nets
wire net1, net2, net3, net4;

//design the 4to1 multiplexer
and3_df  inst1 (data[0], ~sel[1], ~sel[0], net1),
         inst2 (data[1], ~sel[1], sel[0], net2),
         inst3 (data[2], sel[1], ~sel[0], net3),
         inst4 (data[3], sel[1], sel[0], net4);

or4_df   inst5 (net1, net2, net3, net4, z1);

endmodule
```

```
//test bench for the 4:1 multiplexer structural module

module mux_4to1_df_tb;

//inputs are reg for test bench
//outputs are wire for test bench
reg [1:0] sel;
reg [3:0] data;
wire z1;

initial
$monitor ("sel = %b, data = %b, z1 = %b",
            sel, data, z1);

//apply stimulus
initial
begin
   #0    sel = 2'b00;   data = 4'b0001;    //z1 = 1
   #10   sel = 2'b01;   data = 4'b1001;    //z1 = 0
   #10   sel = 2'b10;   data = 4'b1000;    //z1 = 0
   #10   sel = 2'b11;   data = 4'b1001;    //z1 = 1

   #10   sel = 2'b00;   data = 4'b0100;    //z1 = 0
   #10   sel = 2'b01;   data = 4'b1000;    //z1 = 0
   #10   sel = 2'b10;   data = 4'b1100;    //z1 = 1
   #10   sel = 2'b11;   data = 4'b1101;    //z1 = 1

   #10   sel = 2'b00;   data = 4'b0110;    //z1 = 0
   #10   sel = 2'b01;   data = 4'b0000;    //z1 = 0
   #10   sel = 2'b10;   data = 4'b1001;    //z1 = 0
   #10   sel = 2'b11;   data = 4'b0100;    //z1 = 0

   #10   sel = 2'b00;   data = 4'b0111;    //z1 = 1
   #10   sel = 2'b01;   data = 4'b0010;    //z1 = 1
   #10   sel = 2'b10;   data = 4'b1101;    //z1 = 1
   #10   sel = 2'b11;   data = 4'b1100;    //z1 = 1

   #10   $stop;
end

//instantiate the module into the test bench
mux_4to1_df inst1 (sel, data, z1);

endmodule
```

```
sel[1:0]  data[3:0]

------------------------------
sel = 00, data = 0001, z1 = 1
sel = 01, data = 1001, z1 = 0
sel = 10, data = 1000, z1 = 0
sel = 11, data = 1001, z1 = 1

sel = 00, data = 0100, z1 = 0
sel = 01, data = 1000, z1 = 0
sel = 10, data = 1100, z1 = 1
sel = 11, data = 1101, z1 = 1

sel = 00, data = 0110, z1 = 0
sel = 01, data = 0000, z1 = 0
sel = 10, data = 1001, z1 = 0
sel = 11, data = 0100, z1 = 0

sel = 00, data = 0111, z1 = 1
sel = 01, data = 0010, z1 = 1
sel = 10, data = 1101, z1 = 1
sel = 11, data = 1100, z1 = 1
```

2.18 Obtain a minimized equation for z_1 in a sum-of-products representation and for z_2 in a product-of-sums representation for the Karnaugh map shown below, where the outputs are $12 \leq z_1(z_2) < 3$. The obtain the design module using built-in primitives, the test bench module, and the outputs.

x_1x_2 \ x_3x_4	0 0	0 1	1 1	1 0
0 0	1 (0)	1 (1)	0 (3)	1 (2)
0 1	0 (4)	0 (5)	0 (7)	0 (6)
1 1	1 (12)	1 (13)	1 (15)	1 (14)
1 0	0 (8)	0 (9)	0 (11)	0 (10)

$$z_1 = x_1x_2 + x_1'x_2'x_3' + x_1'x_2'x_4'$$

$$z_2 = (x_1 + x_2')\,(x_1' + x_2)\,(x_1 + x_3' + x_4')$$

```
//built-in primitives for number 12 <= z1 < 3
module sop_pos_bip2 (x1, x2, x3, x4, z1, z2);

//define inputs and outputs
input x1, x2, x3, x4;
output z1, z2;

//design the logic using bips for z1 in a sum-of-products
and    inst1 (net1, x1, x2),
       inst2 (net2, ~x1, ~x2, ~x3),
       inst3 (net3, ~x1, ~x2, ~x4);
or     inst4 (z1, net1, net2, net3);

//design the logic using bips for z1 in a product-of-sums
or     inst5 (net5, x1, ~x2),
       inst6 (net6, ~x1, x2),
       inst7 (net7, x1, ~x3, ~x4);
and    inst8 (z2, net5, net6, net7);

endmodule
```

```
//test bench for number 12 <= z1 < 3
module sop_pos_bip2_tb;

//inputs are reg for test bench
//outputs are wire for test bench
reg x1, x2, x3, x4;
wire z1, z2;

//apply input vectors and display variables
initial
begin: apply_stimulus
   reg [4:0] invect;
   for (invect = 0; invect < 16; invect = invect + 1)
      begin
         {x1, x2, x3, x4} = invect [4:0];
         #10 $display ("{x1 x2 x3 x4} = %b,
                        z1 = %b, z2 = %b",
                        {x1, x2, x3, x4}, z1, z2);
      end
end

//instantiate the module into the test bench
sop_pos_bip2 inst1 (x1, x2, x3, x4, z1, z2);

endmodule
```

```
{x1 x2 x3 x4} = 0000, z1 = 1, z2 = 1
{x1 x2 x3 x4} = 0001, z1 = 1, z2 = 1
{x1 x2 x3 x4} = 0010, z1 = 1, z2 = 1
{x1 x2 x3 x4} = 0011, z1 = 0, z2 = 0

{x1 x2 x3 x4} = 0100, z1 = 0, z2 = 0
{x1 x2 x3 x4} = 0101, z1 = 0, z2 = 0
{x1 x2 x3 x4} = 0110, z1 = 0, z2 = 0
{x1 x2 x3 x4} = 0111, z1 = 0, z2 = 0

{x1 x2 x3 x4} = 1000, z1 = 0, z2 = 0
{x1 x2 x3 x4} = 1001, z1 = 0, z2 = 0
{x1 x2 x3 x4} = 1010, z1 = 0, z2 = 0
{x1 x2 x3 x4} = 1011, z1 = 0, z2 = 0

{x1 x2 x3 x4} = 1100, z1 = 1, z2 = 1
{x1 x2 x3 x4} = 1101, z1 = 1, z2 = 1
{x1 x2 x3 x4} = 1110, z1 = 1, z2 = 1
{x1 x2 x3 x4} = 1111, z1 = 1, z2 = 1
```

Chapter 3 Sequential Logic Design Using Verilog HDL

3.4 The state diagram for a Moore synchronous sequential machine is shown below with three inputs, x_1, x_2, and x_3. There are two outputs, z_1 and z_2. Obtain the structural design module using built-in primitives and instantiated *D* flip-flops that were designed using behavioral modeling. Obtain the test bench module and the outputs. Use the **$random** system task for the test bench module to generate a random value for certain inputs when their value can be considered a "don't care" — either 0 or 1. Use *clk'* to gate the outputs to avoid possible glitches.

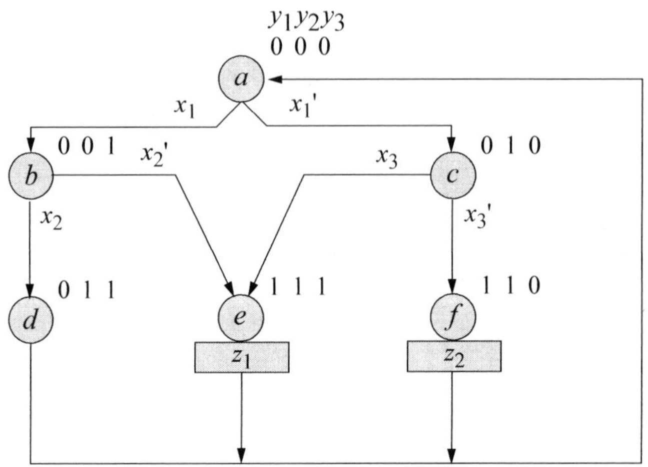

$x_1 \backslash x_2x_3$	0 0	0 1	1 1	10
0	$0\ ^{0}$	$x_2'\ ^{1}$	$0\ ^{3}$	$1\ ^{2}$
1	$-\ ^{4}$	$-\ ^{5}$	$0\ ^{7}$	$0\ ^{6}$

Dy_1

$x_1 \backslash x_2x_3$	0 0	0 1	1 1	10
0	$x_1'\ ^{0}$	$1\ ^{1}$	$0\ ^{3}$	$1\ ^{2}$
1	$-\ ^{4}$	$-\ ^{5}$	$0\ ^{7}$	$0\ ^{6}$

Dy_2

$x_1 \backslash x_2x_3$	0 0	0 1	1 1	10
0	$x_1\ ^{0}$	$1\ ^{1}$	$0\ ^{3}$	$x_3\ ^{2}$
1	$-\ ^{4}$	$-\ ^{5}$	$0\ ^{7}$	$0\ ^{6}$

Dy_3

---

net1 net2

$$Dy_1 = y_2'y_3x_2' + y_1'y_2y_3'$$

net3

---

net4 net5 net6

$$Dy_2 = y_2'x_1' + y_2'y_3 + y_1'y_2y_3'$$

net7

---

net8 net9 net10

$$Dy_3 = y_2'x_1 + y_2'y_3 + y_1'y_2y_3'x_3$$

net11

```
//structural for moore ssm using bip

module moore_ssm_bip (rst_n, clk, x1, x2, x3,
                        y1, y2, y3, z1, z2);

//define inputs and outputs
input rst_n, clk, x1, x2, x3;
output y1, y2, y3, z1, z2;

//define internal nets
wire net1, net2, net3, net4, net5, net6,
      net7, net8, net9, net10, net11;

//-----------------------------------------------
//instantiate the logic for D flip-flop y1
and   (net1, ~y2, y3, ~x2),
      (net2, ~y1, y2, ~y3);

or    (net3, net1, net2);

//instantiate the D flip-flop for y1
d_ff_bh  inst1 (rst_n, clk, net3, y1);
               //reset, clock, D, Q

                        //continued on next page
```

```
//--------------------------------------------------------
//instantiate the logic for D flip-flop y2
and    (net4, ~y2, ~x1),
       (net5, ~y2, y3),
       (net6, ~y1, y2, ~y3);

or     (net7, net4, net5, net6);

//instantiate the D flip-flop for y2
d_ff_bh  inst2 (rst_n, clk, net7, y2);
               //reset, clock, D, Q

//--------------------------------------------------------
//instantiate the logic for D flip-flop y3
and    (net8, ~y2, x1),
       (net9, ~y2, y3),
       (net10, ~y1, y2, ~y3, x3);

or     (net11, net8, net9, net10);

//instantiate the D flip-flop for y3
d_ff_bh  inst3 (rst_n, clk, net11, y3);
               //reset, clock, D, Q

//--------------------------------------------------------
//instantiate the logic for outputs z1 and z2
and    (z1, y1, y2, y3, ~clk),
       (z2, y1, y2, ~y3, ~clk);

endmodule
```

```
//test bench for the moore ssm using bip
module moore_ssm_bip_tb;

//inputs are reg for test bench
//outputs are wire for test bench
reg rst_n, clk, x1, x2, x3;
wire y1, y2, y3, z1, z2;

//display variables
initial
$monitor ("x1 x2 x3 = %b, state = %b, z1 z2 = %b",
            {x1, x2, x3}, {y1, y2, y3}, {z1, z2});
                                //continued on next page
```

```
//test bench for the moore ssm using bip
module moore_ssm_bip_tb;

reg rst_n, clk, x1, x2, x3;        //inputs are reg
wire y1, y2, y3, z1, z2;           //outputs are wire

initial     //display variables
$monitor ("x1 x2 x3 = %b, state = %b, z1 z2 = %b",
            {x1, x2, x3}, {y1, y2, y3}, {z1, z2});

//define clock
initial
begin
   clk = 1'b0;
      forever
      #10   clk = ~clk;
end

//define input sequence
initial
begin
   #0    rst_n = 1'b0;      //reset to state_a
         x1 = 1'b1;  x2 =1'b0;   x3 = 1'b0;
   #5    rst_n = 1'b1;

//------------------------------------------------------
   x1 = 1'b1;  x2 = 1'b0;  x3 = 1'b0;
      @ (posedge clk)

   x1 = 1'b0;  x2 = 1'b0;  x3 = 1'b0;
      @ (posedge clk)   //go to state_e, assert z1

   x1 = $random;  x2 = $random;  x3 = $random;
      @ (posedge clk)   //go to state_a

//------------------------------------------------------
   x1 = 1'b0;  x2 = 1'b0;  x3 = 1'b0;
      @ (posedge clk)

   x1 = 1'b0;  x2 = $random;  x3 = 1'b0;
      @ (posedge clk)   //go to state_f, assert z2

   x1 = $random;  x2 = $random;  x3 = $random;
      @ (posedge clk)   //go to state_a
                          //continued on next page
```

```
//--------------------------------------------------------
   x1 = 1'b0;   x2 = 1'b0;   x3 = 1'b0;
      @ (posedge clk)

   x1 = 1'b0;   x2 = $random;   x3 = 1'b1;
      @ (posedge clk)    //go to state_e, assert z1

   x1 = $random;   x2 = $random;   x3 = $random;
      @ (posedge clk)    //go to state_a

   #10   $stop;

end

//instantiate the module into the test bench
moore_ssm_bip inst1 (rst_n, clk, x1, x2, x3,
                        y1, y2, y3, z1, z2);

endmodule
```

```
x1 x2 x3 = 100, state = 000, z1 z2 = 00
x1 x2 x3 = 100, state = 001, z1 z2 = 00
x1 x2 x3 = 000, state = 111, z1 z2 = 10
x1 x2 x3 = 011, state = 000, z1 z2 = 00

x1 x2 x3 = 000, state = 010, z1 z2 = 00
x1 x2 x3 = 010, state = 110, z1 z2 = 01
x1 x2 x3 = 000, state = 000, z1 z2 = 00

x1 x2 x3 = 001, state = 010, z1 z2 = 00
x1 x2 x3 = 011, state = 111, z1 z2 = 10
x1 x2 x3 = 110, state = 000, z1 z2 = 00
x1 x2 x3 = 110, state = 001, z1 z2 = 00
```

3.10 The timing diagram for a Mealy asynchronous sequential machine is shown below. Design the machine using instantiated logic gates that were designed using dataflow modeling. Obtain the structural design module, the test bench module, and the outputs.

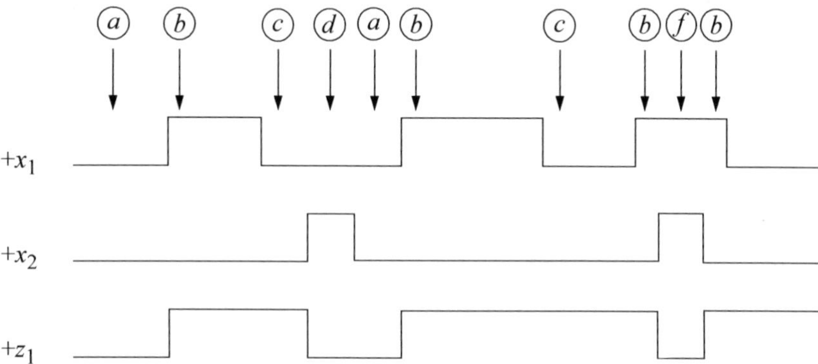

Primitive flow table

x_1x_2	00	01	11	10	z_1
	(a)	d	–	b	0
	c	–	f	(b)	1
	(c)	d	–	b	1
	a	(d)	e	–	0
	–	d	(e)	b	1
	–	d	(f)	b	0

Merger diagram

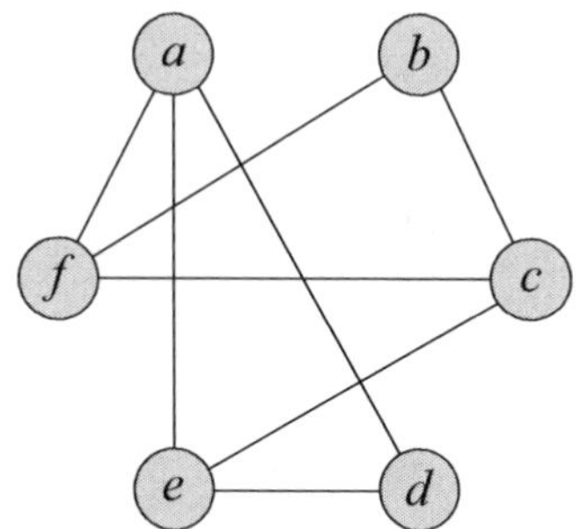

Merged flow table

	00	01	11	10
(a) (d) (e)	(a)	(d)	(e)	b
(b) (c) (f)	(c)	d	(f)	(b)

Excitation map

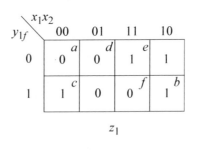

y_{1f} \ x_1x_2	00	01	11	10
0	0 (a)	0 (d)	0 (e)	1
1	1 (c)	0	1 (f)	1 (b)

Y_{1e}

$$Y_{1e} = \underbrace{x_1x_2'}_{\text{net1}} + \underbrace{y_{1f}x_1}_{\text{net2}} + \underbrace{y_{1f}x_2'}_{\text{net3}}$$

Output map

y_{1f} \ x_1x_2	00	01	11	10
0	0 (a)	0 (d)	1 (e)	1
1	1 (c)	0	0 (f)	1 (b)

z_1

$$z_1 = \underbrace{x_1x_2'}_{\text{net1}} + \underbrace{y_{1f}x_2'}_{\text{net3}} + \underbrace{y_{1f}'x_1}_{\text{net4}}$$

```
//structural for sop asynchronous sequential machine
module asm_sop_df (rst_n, x1, x2, y1e, z1);

//define inputs and outputs
input rst_n, x1, x2;
output y1e, z1;

//define internal nets
wire net1, net2, net3, net4;

//-----------------------------------------------
//design the logic for excitation variable Y1e
and2_df  inst1 (x1, ~x2, net1);
and3_df  inst2 (x1, y1e, rst_n, net2),
         inst3 (y1e, ~x2, rst_n, net3);

or3_df   inst4 (net1, net2, net3, y1e);

//-----------------------------------------------
//design the logic for output z1
and2_df  inst5 (~y1e, x1, net4);

or3_df   inst6 (net1, net3, net4, z1);

endmodule
```

```
//test bench for the sop asynchronous sequential
machine

module asm_sop_df_tb;

//inputs are reg for test bench
//outputs are wire for test bench
reg rst_n, x1, x2;
wire y1e, z1;

//display variables
initial
$monitor ("x1 x2 = %b, state = %b, z1 = %b",
            {x1, x2}, y1e, z1);

                         //continued on next page
```

```
//define input vectors
initial
begin
   #0     rst_n = 1'b0;      //reset to state_a
          x1 = 1'b0;  x2 = 1'b0;

   #5     rst_n = 1'b1;

//-------------------------------------------------------
   #10    x1 = 1'b1;  x2 = 1'b0;  //go to state_b;
                                  //assert z1
   #10    x1 = 1'b0;  x2 = 1'b0;  //go to state_c;
                                  //assert z1
   #10    x1 = 1'b0;  x2 = 1'b1;  //go to state_d

   #10    x1 = 1'b0;  x2 = 1'b0;  //go to state_a
   #10    x1 = 1'b1;  x2 = 1'b0;  //go to state_b;
                                  //assert z1
   #20    x1 = 1'b0;  x2 = 1'b0;  //go to state_c;
                                  //assert z1
   #10    x1 = 1'b1;  x2 = 1'b0;  //go to state_b;
                                  //assert z1
   #10    x1 = 1'b1;  x2 = 1'b1;  //go to state_f
   #10    x1 = 1'b1;  x2 = 1'b0;  //go to state_b;
                                  //assert z1
   #10    x1 = 1'b1;  x2 = 1'b1;  //go to state_f
   #10    x1 = 1'b1;  x2 = 1'b0;  //go to state_b;
                                  //assert z1
   #10    x1 = 1'b0;  x2 = 1'b0;  //go to state_c;
                                  //assert z1
   #10    x1 = 1'b1;  x2 = 1'b1;  //go to state_f

   #10    $stop;
end

//instantiate the module into the test bench
asm_sop_df inst1 (rst_n, x1, x2, y1e, z1);

endmodule
```

```
x1 x2 = 00, state = 0, z1 = 0
x1 x2 = 10, state = 1, z1 = 1
x1 x2 = 00, state = 1, z1 = 1
x1 x2 = 01, state = 0, z1 = 0

x1 x2 = 00, state = 0, z1 = 0
x1 x2 = 10, state = 1, z1 = 1
x1 x2 = 00, state = 1, z1 = 1
x1 x2 = 10, state = 1, z1 = 1
x1 x2 = 11, state = 1, z1 = 0
x1 x2 = 10, state = 1, z1 = 1
x1 x2 = 11, state = 1, z1 = 0
x1 x2 = 10, state = 1, z1 = 1
x1 x2 = 00, state = 1, z1 = 1
x1 x2 = 11, state = 1, z1 = 0
```

3.15 The state diagram for a Mealy pulse-mode asynchronous sequential machine is shown below. Synthesize the machine using logic gates that were designed using dataflow modeling and *D* flip-flops that were designed using behavioral modeling. Obtain the structural design module, the test bench module, and the outputs.

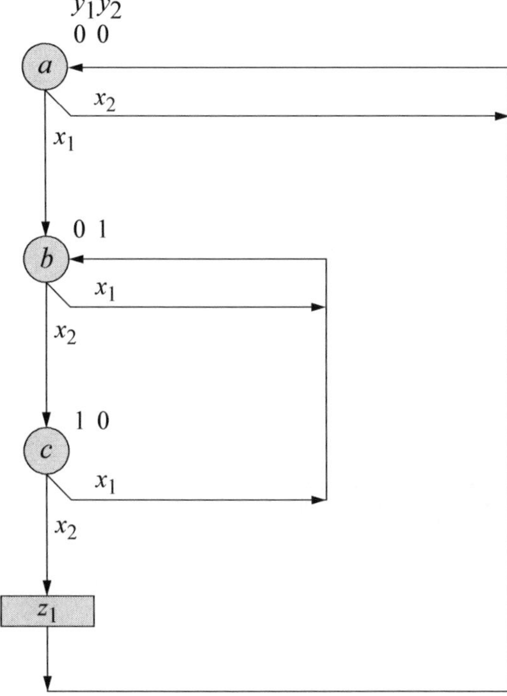

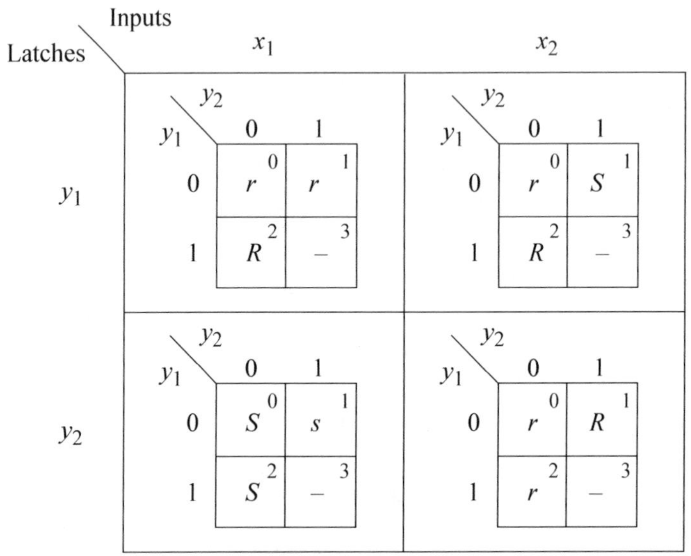

$S\,Ly_1 = y_2x_2$ $\qquad$ $R\,Ly_1 = x_1 + y_1x_2$

$S\,Ly_2 = x_1$ $\qquad$ $R\,Ly_2 = x_2$

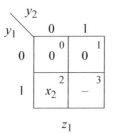

Output map

$y_1 \backslash y_2$	0	1
0	0 [0]	0 [1]
1	x_2 [2]	– [3]

z_1

$$z_1 = y_1y_2'x_2$$
$$= y_1x_2 \text{ (Minimized)}$$

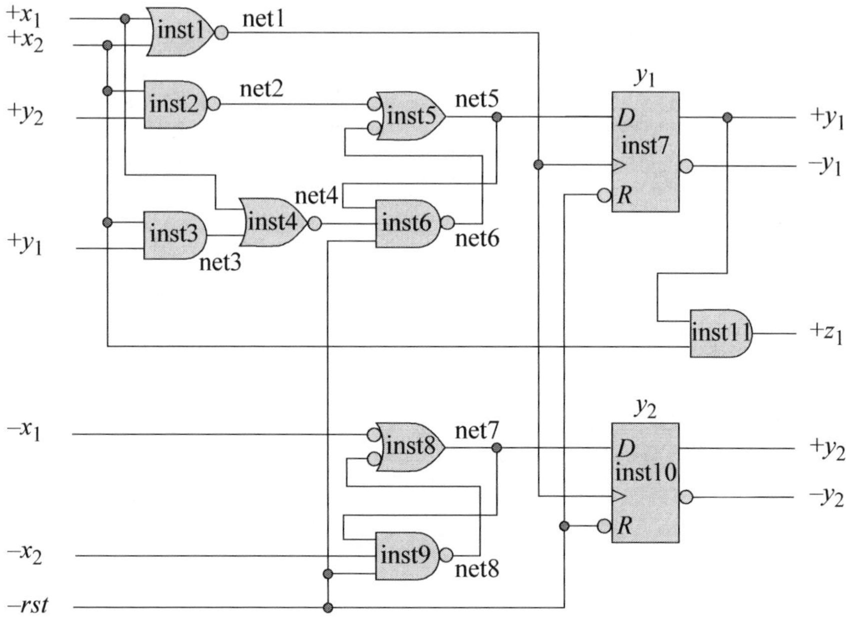


```
//mealy pulse-mode asm using dataflow logic and D ff
module pm_asm_mealy2_dff (rst_n, x1, x2, y1, y2, z1);

input rst_n, x1, x2;     //define inputs and outputs
output y1, y2, z1;

//define internal nets
wire net1, net2, net3, net4, net5, net6, net7, net8;

//define the clock for the D flip-flops
nor2_df  inst1 (x1, x2, net1);

//------------------------------------------------------
//define the logic for latch Ly1 and D flip-flop y1
nand2_df inst2 (y2, x2, net2);
and2_df  inst3 (x2, y1, net3);
nor2_df  inst4 (x1, net3, net4);
nand2_df inst5 (net2, net6, net5);
nand3_df inst6 (net5, net4, rst_n, net6);

//instantiate the D flip-flop for y1
d_ff_bh  inst7 (rst_n, net1, net5, y1);
                //rst, clk, D, Q
                           //continued on next page
```

```
//-----------------------------------------------------------
//define the logic for latch Ly2 and D flip-flop y2
nand2_df inst8 (~x1, net8, net7);
nand3_df inst9 (net7, ~x2, rst_n, net8);

//instantiate the D flip-flop for y2
d_ff_bh  inst10 (rst_n, net1, net7, y2);
                  //rst, clk, D, Q

//-----------------------------------------------------------
//define the logic for output z1
and2_df  inst11 (y1, x2, z1);

endmodule
```

```
//test bench for the mealy pulse-mode asm
module pm_asm_mealy2_dff_tb;

reg rst_n, x1, x2;    //inputs are reg for test bench
wire y1, y2, z1;      //outputs are wire for test bench

//display variables
initial
$monitor ("x1 x2 = %b, state = %b, z1 = %b",
            {x1, x2}, {y1, y2}, z1);

//apply input sequence
initial
begin
   #0    rst_n = 1'b0;
         x1 = 1'b0;  x2 = 1'b0;
   #5    rst_n = 1'b1;

//-----------------------------------------------------------
   #10   x1 = 1'b1;  x2 = 1'b0;  //state_b
   #10   x1 = 1'b0;  x2 = 1'b0;

   #10   x1 = 1'b0;  x2 = 1'b1;  //state_c
   #10   x1 = 1'b0;  x2 = 1'b0;

   #10   x1 = 1'b0;  x2 = 1'b1;  //state_a, assert z1
   #10   x1 = 1'b0;  x2 = 1'b0;

                             //continued on next page
```

```
//-------------------------------------------------------
   #10   x1 = 1'b0;  x2 = 1'b1;  //state_a
   #10   x1 = 1'b0;  x2 = 1'b0;

   #10   x1 = 1'b1;  x2 = 1'b0;  //state_b
   #10   x1 = 1'b0;  x2 = 1'b0;

   #10   x1 = 1'b1;  x2 = 1'b0;  //state_b
   #10   x1 = 1'b0;  x2 = 1'b0;

   #10   x1 = 1'b0;  x2 = 1'b1;  //state_c
   #10   x1 = 1'b0;  x2 = 1'b0;

   #10   x1 = 1'b0;  x2 = 1'b1;  //state_a, assert z1
   #10   x1 = 1'b0;  x2 = 1'b0;

   #10   $stop;
end

//instantiate the module into the test bench
pm_asm_mealy2_dff inst1 (rst_n, x1, x2, y1, y2, z1);

endmodule
```

```
x1 x2 = 00, state = 00, z1 = 0
x1 x2 = 10, state = 00, z1 = 0
x1 x2 = 00, state = 01, z1 = 0
x1 x2 = 01, state = 01, z1 = 0
x1 x2 = 00, state = 10, z1 = 0
x1 x2 = 01, state = 10, z1 = 1
x1 x2 = 00, state = 00, z1 = 0

x1 x2 = 01, state = 00, z1 = 0
x1 x2 = 00, state = 00, z1 = 0
x1 x2 = 10, state = 00, z1 = 0
x1 x2 = 00, state = 01, z1 = 0
x1 x2 = 10, state = 01, z1 = 0
x1 x2 = 00, state = 01, z1 = 0
x1 x2 = 01, state = 01, z1 = 0
x1 x2 = 00, state = 10, z1 = 0
x1 x2 = 01, state = 10, z1 = 1
x1 x2 = 00, state = 00, z1 = 0
```

3.21 The state diagram shown below is for a Mealy pulse-mode asynchronous sequential machine. Design the structural module for the machine using built-in primitives and instantiated T flip-flops. Obtain the test bench module and the outputs.

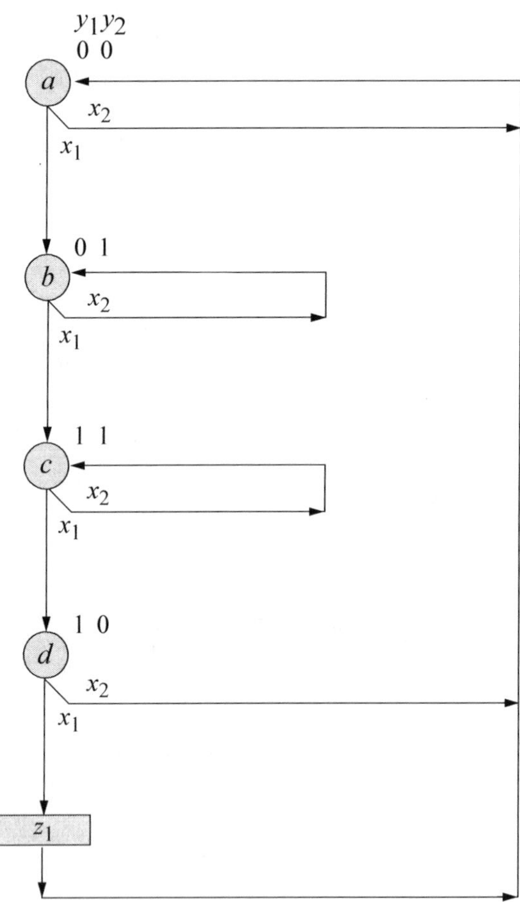

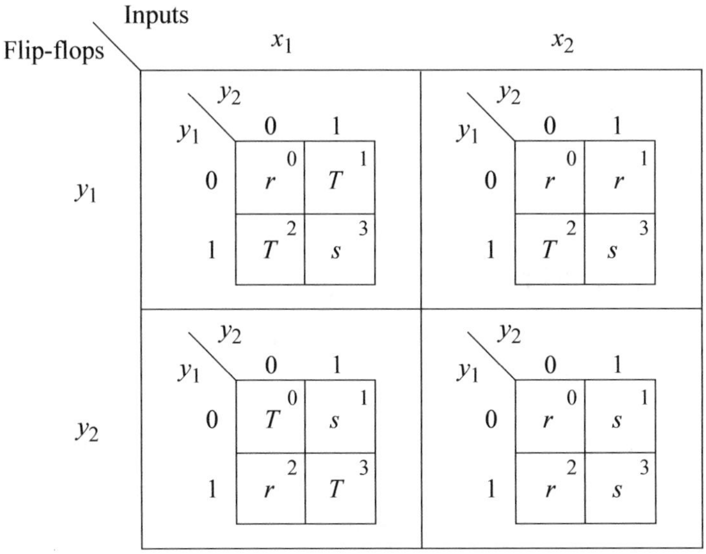

Flip-flops \ Inputs: x_1, x_2

Flip-flop y_1, input x_1:

y_1 \ y_2	0	1
0	(0) r	(1) T
1	(2) T	(3) s

Flip-flop y_1, input x_2:

y_1 \ y_2	0	1
0	(0) r	(1) r
1	(2) T	(3) s

Flip-flop y_2, input x_1:

y_1 \ y_2	0	1
0	(0) T	(1) s
1	(2) r	(3) T

Flip-flop y_2, input x_2:

y_1 \ y_2	0	1
0	(0) r	(1) s
1	(2) r	(3) s

net1 net2 net3

$$Ty_1 = y_1'y_2x_1 + y_1y_2'x_1 + y_1y_2'x_2$$

net4

net5 net6

$$Ty_2 = y_1'y_2'x_1 + y_1y_2x_1$$

net7

$$z_1 = y_1y_2'x_1$$

```
//structural mealy pulse-mode asm using bip and tff

module pm_asm_mealy_bip_tff (rst_n, x1, x2, y, z1);

//define inputs and outputs
input rst_n, x1, x2;
output [1:2] y;
output z1;

//define internal nets
wire net1, net2, net3, net4, net5, net6, net7;

//---------------------------------------------------
//design the logic for T flip-flop y[1]
and    (net1, ~y[1], y[2], x1),
       (net2, y[1], ~y[2], x1),
       (net3, y[1], ~y[2], x2);

or     (net4, net1, net2, net3);

//instantiate the T flip-flop
t_ff_da inst1 (rst_n, net4, nety1);//rst, T, Q

buf    #12    (y[1], nety1);
//nety1 is the output of the T flip-flop.
//y[1] is the output delayed by 12 time units

//---------------------------------------------------
//design the logic for T flip-flop y[2]
and    (net5, ~y[1], ~y[2], x1),
       (net6, y[1], y[2], x1);

or     (net7, net5, net6);

//instantiate the T flip-flop
t_ff_da inst2 (rst_n, net7, nety2);//rst, T, Q

buf    #12    (y[2], nety2);
//nety2 is the output of the T flip-flop.
//y[2] is the output delayed by 12 time units

//---------------------------------------------------
//design the logic for output z1
assign   z1 = y[1] & ~y[2] & x1;

endmodule
```

```
//test bench mealy pulse-mode asm using bip and tff
module pm_asm_mealy_bip_tff_tb;

reg rst_n, x1, x2;      //inputs are reg for test bench
wire [1:2] y;           //outputs are wire for test bench
wire z1;

initial      //display variables
$monitor ("x1 x2 = %b, state = %b, z1 = %b",
            {x1, x2}, y, z1);

//define input sequence
initial
begin
   #0    rst_n = 1'b0;  //reset to state_a
         x1 = 1'b0;  x2 = 1'b0;

   #5    rst_n = 1'b1;

//-------------------------------------------------------
   #10   x1 = 1'b1;  x2 = 1'b0;  //state_b
   #10   x1 = 1'b0;  x2 = 1'b0;

   #10   x1 = 1'b1;  x2 = 1'b0;  //state_c
   #10   x1 = 1'b0;  x2 = 1'b0;

   #10   x1 = 1'b1;  x2 = 1'b0;  //state_d
   #10   x1 = 1'b0;  x2 = 1'b0;

   #10   x1 = 1'b1;  x2 = 1'b0;  //state_a
   #10   x1 = 1'b0;  x2 = 1'b0;  //assert z1

//-------------------------------------------------------
   #10   x1 = 1'b1;  x2 = 1'b0;  //state_b
   #10   x1 = 1'b0;  x2 = 1'b0;

   #10   x1 = 1'b0;  x2 = 1'b1;  //state_b
   #10   x1 = 1'b0;  x2 = 1'b0;

   #10   x1 = 1'b1;  x2 = 1'b0;  //state_c
   #10   x1 = 1'b0;  x2 = 1'b0;

   #10   x1 = 1'b0;  x2 = 1'b1;  //state_c
   #10   x1 = 1'b0;  x2 = 1'b0;

                              //continued on next page
```

```
   #10   x1 = 1'b1;  x2 = 1'b0;  //state_d
   #10   x1 = 1'b0;  x2 = 1'b0;  //assert z1

   #10   x1 = 1'b1;  x2 = 1'b0;  //state_a
   #10   x1 = 1'b0;  x2 = 1'b0;  //assert z1

//-----------------------------------------------------
   #12   $stop;
end

//instantiate the module into the test bench
pm_asm_mealy_bip_tff inst1 (rst_n, x1, x2, y, z1);

endmodule
```

```
x1 x2 = 00, state = 00, z1 = 0
x1 x2 = 10, state = 00, z1 = 0
x1 x2 = 00, state = 00, z1 = 0
x1 x2 = 00, state = 01, z1 = 0
x1 x2 = 10, state = 01, z1 = 0
x1 x2 = 00, state = 01, z1 = 0
x1 x2 = 00, state = 11, z1 = 0
x1 x2 = 10, state = 11, z1 = 0
x1 x2 = 00, state = 11, z1 = 0
x1 x2 = 00, state = 10, z1 = 0
x1 x2 = 10, state = 10, z1 = 1
x1 x2 = 00, state = 10, z1 = 0

x1 x2 = 00, state = 00, z1 = 0
x1 x2 = 10, state = 00, z1 = 0
x1 x2 = 00, state = 00, z1 = 0
x1 x2 = 00, state = 01, z1 = 0
x1 x2 = 01, state = 01, z1 = 0
x1 x2 = 00, state = 01, z1 = 0
x1 x2 = 10, state = 01, z1 = 0
x1 x2 = 00, state = 01, z1 = 0
x1 x2 = 00, state = 11, z1 = 0
x1 x2 = 01, state = 11, z1 = 0
x1 x2 = 00, state = 11, z1 = 0
x1 x2 = 10, state = 11, z1 = 0
x1 x2 = 00, state = 11, z1 = 0
x1 x2 = 00, state = 10, z1 = 0
x1 x2 = 10, state = 10, z1 = 1
x1 x2 = 00, state = 10, z1 = 0
x1 x2 = 00, state = 00, z1 = 0
```

Chapter 4 Computer Arithmetic Design Using Verilog HDL

4.3 Use dataflow modeling with the continuous assignment statement **assign** to design a single-bit full adder. Obtain the design module, the test bench module for all combinations of the inputs, and the outputs.

```
//dataflow full adder
module full_adder (a, b, cin, sum, cout);

input a, b, cin;      //list inputs and outputs
output sum, cout;

//define wires (wire are not required; optional)
wire a, b, cin;
wire sum, cout;

//continuous assignment
assign sum = (a ^ b) ^ cin;
assign cout = cin & (a ^ b) | (a & b);

endmodule
```

```
//test bench for full adder
module full_adder_tb;

//inputs are reg for test bench
//outputs are wire for test bench
reg a, b, cin;
wire sum, cout;

//apply input vectors and display variables
initial
begin: apply_stimulus
   reg [3:0] invect;
   for (invect = 0; invect < 8; invect = invect + 1)
     begin
       {a, b, cin} = invect [3:0];
       #10 $display ("a b cin = %b, sum = %b, cout = %d",
                     {a, b, cin}, sum, cout);
     end
end

//instantiate the module into the test bench
full_adder inst1 (a, b, cin, sum, cout);
endmodule
```

```
a b cin = 000, cout = 0, sum = 0
a b cin = 001, cout = 0, sum = 1

a b cin = 010, cout = 0, sum = 1
a b cin = 011, cout = 1, sum = 0

a b cin = 100, cout = 0, sum = 1
a b cin = 101, cout = 1, sum = 0

a b cin = 110, cout = 1, sum = 0
a b cin = 111, cout = 1, sum = 1
```

4.7 Design a 4-bit ripple-carry fixed-point adder/subtractor using built-in primitives and instantiated full adders that were designed using behavioral modeling. There are three inputs: *a*[3:0], *b*[3:0], and a mode control *m*, which is used to determine whether the operation is addition or subtraction. If $m = 0$, then the operation is addition; if $m = 1$, then the operation is subtraction. There are two outputs: *rslt*[3:0] and *cout*[3:0].

For *n*-bit operands, the range for numbers in 2s complement representation is

$$-2^{n-1} \text{ to } +2^{n-1} - 1$$

where *n* is the number of bits in the operands. Thus, operands *a* and *b* have the following syntax:

$$a_{n-1}\, a_{n-2} \ldots a_1\, a_0$$
$$b_{n-1}\, b_{n-2} \ldots b_1\, b_0$$

In the design module include a method to detect overflow, as follows:

$$\text{Overflow} = cout_{n-1} \oplus cout_{n-2}$$

Obtain the logic diagram, the design module using structural modeling, the test bench module with combinations of the inputs for both addition and subtraction including overflow, and the outputs.

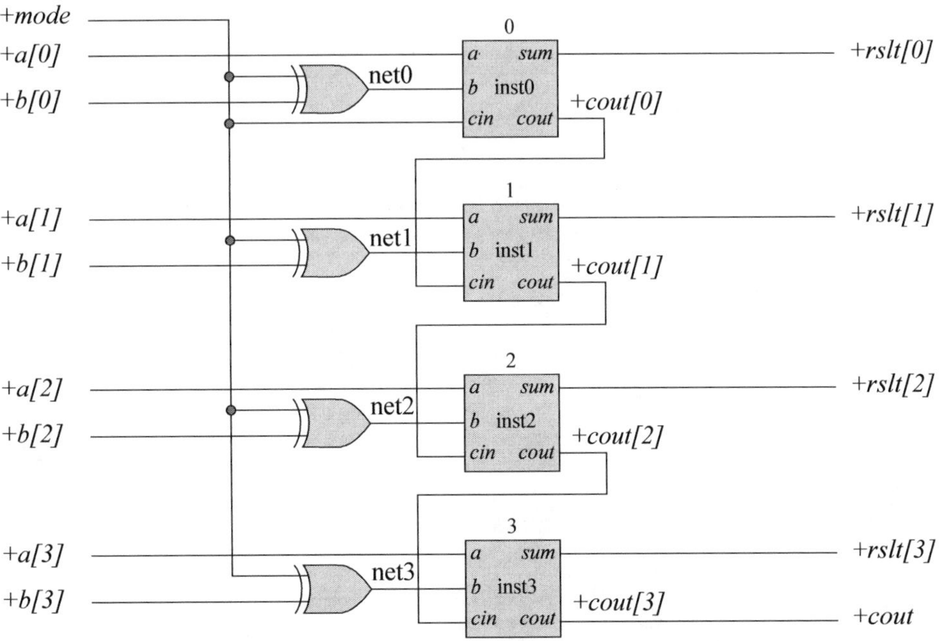


```
//behavioral full adder

module full_adder_bh (a, b, cin, sum, cout);

//define inputs and outputs
input a, b, cin;
output sum, cout;

//variables are reg in always
reg sum, cout;

always @ (a or b or cin)
begin
   sum = a ^ b ^ cin;
   cout = (a & b) | (a & cin) | (b & cin);
end

endmodule
```

```
//structural for a 4-bit adder/subtractor
//using instantiated full adders and bips

module add_sub_4bit_struc (a, b, mode,
                    rslt, cout, ovfl);

//define inputs and outputs
input [3:0] a, b;
input mode;
output [3:0] rslt, cout;
output ovfl;

//define internal nets
wire net0, net1, net2, net3;

//check for overflow
xor (ovfl, cout[3], cout[2]);

//instantiate the logic for rslt[0]
xor (net0, mode, b[0]);
full_adder_bh inst0 (a[0], net0, mode,
         rslt[0], cout[0]);

//instantiate the logic for rslt[1]
xor (net1, mode, b[1]);
full_adder_bh inst1 (a[1], net1, cout[0],
         rslt[1], cout[1]);

//instantiate the logic for rslt[2]
xor (net2, mode, b[2]);
full_adder_bh inst2 (a[2], net2, cout[1],
         rslt[2], cout[2]);

//instantiate the logic for rslt[3]
xor (net3, mode, b[3]);
full_adder_bh inst3 (a[3], net3, cout[2],
         rslt[3], cout[3]);

endmodule
```

```
//test bench for structural 4-bit adder/subtractor
module add_sub_4bit_struc_tb;

//inputs are reg for test bench
//outputs are wire for test bench
reg [3:0] a, b;
reg mode;
wire [3:0] rslt, cout;
wire ovfl;

initial      //display variables
$monitor ("a=%b, b=%b, mode=%b, rslt=%b,
            cout[3]=%b, cout[2]=%b, ovfl=%b",
            a, b, mode, rslt, cout[3], cout[2], ovfl);

//apply input vectors
initial
begin
//addition ------------------------------------------
   #0    a = 4'b0000;   b = 4'b0001;   mode = 1'b0;
   #10   a = 4'b0010;   b = 4'b0101;   mode = 1'b0;
   #10   a = 4'b0110;   b = 4'b0001;   mode = 1'b0;
   #10   a = 4'b1000;   b = 4'b0001;   mode = 1'b0;

//subtraction ---------------------------------------
   #10   a = 4'b1110;   b = 4'b0100;   mode = 1'b1;
   #10   a = 4'b0110;   b = 4'b0011;   mode = 1'b1;
   #10   a = 4'b0111;   b = 4'b0010;   mode = 1'b1;
   #10   a = 4'b0111;   b = 4'b0001;   mode = 1'b1;

//overflow ------------------------------------------
//addition
   #10   a = 4'b0111;   b = 4'b0101;   mode = 1'b0;
   #10   a = 4'b1000;   b = 4'b1011;   mode = 1'b0;

//subtraction
   #10   a = 4'b0110;   b = 4'b1100;   mode = 1'b1;
   #10   a = 4'b1000;   b = 4'b0010;   mode = 1'b1;

   #10   $stop;
end

//instantiate the module into the test bench
add_sub_4bit_struc inst1 (a, b, mode, rslt, cout,
                              ovfl);

endmodule
```

```
Addition -------------------------------------------

a=0000, b=0001, mode=0, rslt=0001,
        cout[3]=0,cout[2]=0, ovfl=0
a=0010, b=0101, mode=0, rslt=0111, cout[3]=0,
        cout[2]=0, ovfl=0
a=0110, b=0001, mode=0, rslt=0111, cout[3]=0,
        cout[2]=0, ovfl=0
a=1000, b=0001, mode=0, rslt=1001, cout[3]=0,
        cout[2]=0, ovfl=0

Subtraction ----------------------------------------

a=1110, b=0100, mode=1, rslt=1010, cout[3]=1,
        cout[2]=1, ovfl=0
a=0110, b=0011, mode=1, rslt=0011, cout[3]=1,
        cout[2]=1, ovfl=0
a=0111, b=0010, mode=1, rslt=0101, cout[3]=1,
        cout[2]=1, ovfl=0
a=0111, b=0001, mode=1, rslt=0110, cout[3]=1,
        cout[2]=1, ovfl=0

Overflow Addition ----------------------------------

a=0111, b=0101, mode=0, rslt=1100, cout[3]=0,
        cout[2]=1, ovfl=1
a=1000, b=1011, mode=0, rslt=0011, cout[3]=1,
        cout[2]=0, ovfl=1

Overflow Subtraction -------------------------------

a=0110, b=1100, mode=1, rslt=1010, cout[3]=0,
        cout[2]=1, ovfl=1
a=1000, b=0010, mode=1, rslt=0110, cout[3]=1,
        cout[2]=0, ovfl=1
```

4.13 Arithmetic and logic units perform the arithmetic operations of addition, subtraction, multiplication, and division. They also perform the logical operations of AND, NAND, OR, NOR, exclusive-OR, and exclusive-NOR. This problem is to design a behavioral module to implement the four operations of add, subtract, multiply, and divide. The operands are eight bits, the operation code is three bits, and the result of the operation is eight bits. The two 8-bit inputs are operands *a[7:0]* and *b[7:0]*. The 3-bit operation code is *opcode[2:0]* and the 8-bit result is *rslt[7:0]*.

Obtain the behavioral design module using the **case** statement for the four arithmetic operations, the test bench module displaying all variables in decimal (%d) notation, and the outputs.

```
//behavioral 4-function arithmetic unit
//add, sub, div, mul

module add_sub_div_mul_bh (a, b, opcode, rslt);

//define inputs and outputs
input [7:0] a, b;
input [2:0] opcode;
output [7:0] rslt;

//variables are reg in always
reg [7:0] rslt;
reg [15:0] rslt_mul;

//define the opcodes
parameter   add_op = 3'b000,
            sub_op = 3'b001,
            div_op = 3'b011,
            mul_op = 3'b100;

//perform the arithmetic operations
always @ (a or b or opcode)
begin
   case (opcode)
      add_op : rslt = a + b;
      sub_op : rslt = a - b;
      div_op : rslt = a / b;
      mul_op : rslt = a * b;

      default : rslt = 0;
   endcase
end

endmodule
```

```
//4fctn arithmetic unit test bench

module add_sub_div_mul_bh_tb;

//inputs are reg for test bench
//outputs are wire for test bench
reg [7:0] a, b;
reg [2:0] opcode;
wire [7:0] rslt;
```

//continued on next page

```
//display variables
initial
$monitor ("a = %d, b = %d, opcode = %d, rslt = %d",
            a, b, opcode, rslt);

initial
begin
//add_op ---------------------------------------------
         //a = 10, b = 20,  rslt - 30
   #10   a = 8'b0000_1010;    b = 8'b0001_0100;
         opcode = 3'b000;

         // a = 98, b = 28, rslt = 126
   #10   a = 8'b0110_0010;    b = 8'b0001_1100;
         opcode = 3'b000;

         // a = 67, b = 60, rslt = 127
   #10   a = 8'b0100_0011;    b = 8'b0011_1100;
         opcode = 3'b000;

         //a = 250, b = 5, rslt = 255
   #10   a = 8'b1111_1010;    b = 8'b0000_0101;
         opcode = 3'b000;

//sub_op ---------------------------------------------
         //a = 128, b = 99, rslt = 29
   #10   a = 8'b1000_0000;    b = 8'b0110_0011;
         opcode = 3'b001;

         //a = 255, b = 15, rslt = 240
   #10   a = 8'b1111_1111;    b = 8'b0000_1111;
         opcode = 3'b001;

         //a = 20, b = 16, rslt = 4
   #10   a = 8'b0001_0100;    b = 8'b0001_0000;
         opcode = 3'b001;

         //a = 255, b = 250, rslt = 5
   #10   a = 8'b1111_1111;    b = 8'b1111_1010;
         opcode = 3'b001;

                                   //continued on next page
```

```
//div_op -------------------------------------------------
        //a = 240, b = 15, rslt = 16
   #10  a = 8'b1111_0000;    b = 8'b0000_1111;
        opcode = 3'b011;

        //a = 16, b = 8, rslt = 2
   #10  a = 8'b0001_0000;    b = 8'b0000_1000;
        opcode = 3'b011;

//mul_op -------------------------------------------------
        //a= 4, b = 4, rslt = 16
   #10  a = 8'b0000_0100;    b = 8'b0000_0100;
        opcode = 3'b100;

        //a= 10, b = 20, rslt = 200
   #10  a = 8'b0000_1010;    b = 8'b0001_0100;
        opcode = 3'b100;

   #10  $stop;
end

//instantiate the module into the test bench
add_sub_div_mul_bh inst1 (a, b, opcode, rslt);

endmodule
```

```
Add
a = 10,  b = 20,  opcode = 0, rslt = 30
a = 98,  b = 28,  opcode = 0, rslt = 126
a = 67,  b = 60,  opcode = 0, rslt = 127
a = 250, b = 5,   opcode = 0, rslt = 255

Subtract
a = 128, b = 99,  opcode = 1, rslt = 29
a = 255, b = 15,  opcode = 1, rslt = 240
a = 20,  b = 16,  opcode = 1, rslt = 4
a = 255, b = 250, opcode = 1, rslt = 5

Divide
a = 240, b = 15,  opcode = 3, rslt = 16
a = 16,  b = 8,   opcode = 3, rslt = 2

Multiply
a = 4,   b = 4,   opcode = 4, rslt = 16
a = 10,  b = 20,  opcode = 4, rslt = 200
```

4.18 Use structural modeling with built-in primitives to design the two-digit BCD adder shown below. Obtain the structural design module and the test bench module with several input combinations of the two operands. Enter augends and addends that produce sums in the units, tens, and hundreds representations. Display all of the outputs in decimal notation.

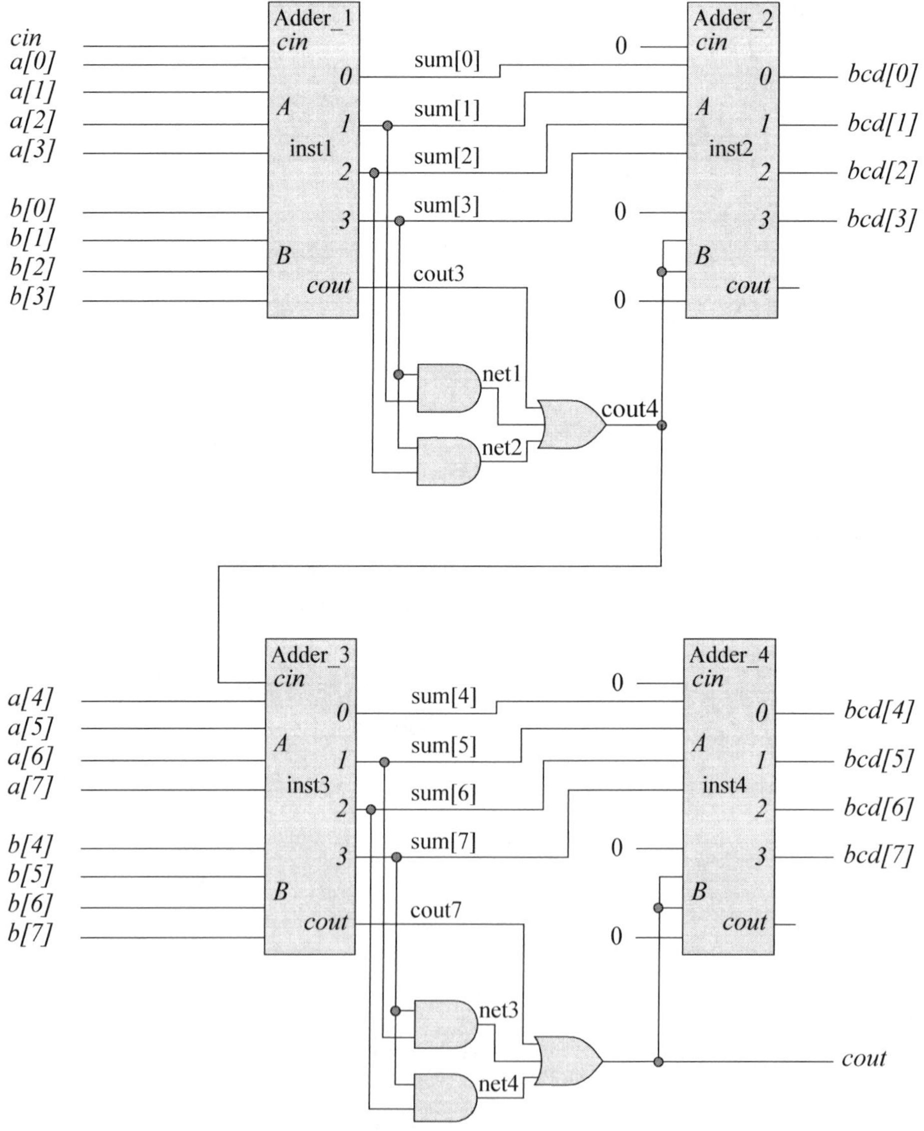

```
//structural design for 2-digit BCD adder

module add_bcd_struc (a, b, cin, bcd, cout);

//define inputs and outputs
input [7:0] a, b;
input cin;
output [7:0] bcd;
output cout;

//define internal nets
wire [7:0] sum;

//-----------------------------------------------------
//instantiate the logic for adder_1 and adder_2
adder4 inst1 (a[3:0], b[3:0], cin, sum[3:0], cout3);

and (net1, sum[3], sum[1]);
and (net2, sum[3], sum[2]);
or  (cout4, cout3, net1, net2);

adder4 inst2 (sum[3:0], {1'b0, cout4, cout4, 1'b0},
              1'b0, bcd[3:0], 1'b0);

//-----------------------------------------------------
//instantiate the logic for adder_3 and adder_4
adder4 inst3 (a[7:4], b[7:4], cout4, sum[7:4], cout7);

and (net3, sum[7], sum[5]);
and (net4, sum[7], sum[6]);
or  (cout, cout7, net3, net4);

adder4 inst4 (sum[7:4], {1'b0, cout, cout, 1'b0},
              1'b0, bcd[7:4], 1'b0);

//-----------------------------------------------------
endmodule
```

```
//test bench for 2-digit BCD adder
module add_bcd_struc_tb;

//inputs are reg for test bench
//outputs are wire for test bench
reg [7:0] a, b;
reg cin;
wire [7:0] bcd;
wire cout;

//display variables
initial
$monitor ("a_ten=%d, a_unit=%d, b_ten=%d, b_unit=%d,
            cin=%d,
            bcd_hund=%d, bcd_ten=%d, bcd_unit=%d",

            a[7:4], a[3:0], b[7:4], b[3:0], cin,
            {{3{1'b0}}, cout}, bcd[7:4], bcd[3:0]);

//apply input vectors
initial
begin
         //03 + 06 = 9
   #0    a = 8'b0000_0011; b = 8'b0000_0110;
               cin = 1'b0;

         //97 + 82 = 179
   #10   a = 8'b1001_0111; b = 8'b1000_0010;
               cin = 1'b0;

         //58 + 24 = 82
   #10   a = 8'b0101_1000; b = 8'b0010_0100;
               cin = 1'b0;

         //25 + 25 + 1 = 51
   #10   a = 8'b0010_0101; b = 8'b0010_0101;
               cin = 1'b1;

         //97 + 99 + 1 = 197
   #10   a = 8'b1001_0111; b = 8'b1001_1001;
               cin = 1'b1;

         //88 + 02 = 90
   #10   a = 8'b1000_1000; b = 8'b0000_0010;
               cin = 1'b0;
                             //continued on next page
```

```
        //99 + 99 = 198
   #10  a = 8'b1001_1001; b = 8'b1001_1001;
             cin = 1'b0;

        //86 + 72 + 1 = 159
   #10  a = 8'b1000_0110; b = 8'b0111_0010;
             cin = 1'b1;

        //1 + 1 = 2
   #10  a = 8'b0000_0001; b = 8'b0000_0001;
             cin = 1'b0;

        //70 + 30 = 100
   #10  a = 8'b0111_0000; b = 8'b0011_0000;
             cin = 1'b0;

        //33 = 66 + 1 = 100
   #10  a = 8'b0011_0011; b = 8'b0110_0110;
             cin = 1'b1;

        //47 + 17 = 64
   #10  a = 8'b0100_0111; b = 8'b0001_0111;
             cin = 1'b0;

        //0 + 0 + 1 = 1
   #10  a = 8'b0000_0000; b = 8'b0000_0000;
             cin = 1'b1;

        //99 + 11 + 1 = 111
   #10  a = 8'b1001_1001; b = 8'b0001_0001;
             cin = 1'b1;

        //0 + 0 + 0 = 0
   #10  a = 8'b0000_0000; b = 8'b0000_0000;
             cin = 1'b0;

   #10  $stop;
end

//instantiate the module into the test bench
add_bcd_struc inst1 (a, b, cin, bcd, cout);

endmodule
```

```
a_ten = 0, a_unit = 3, b_ten = 0, b_unit = 6, cin = 0,
   bcd_hund = 0, bcd_ten = 0, bcd_unit = 9

a_ten = 9, a_unit = 7, b_ten = 8, b_unit = 2, cin = 0,
   bcd_hund = 1, bcd_ten = 7, bcd_unit = 9

a_ten = 5, a_unit = 8, b_ten = 2, b_unit = 4, cin = 0,
   bcd_hund = 0, bcd_ten = 8, bcd_unit = 2

a_ten = 2, a_unit = 5, b_ten = 2, b_unit = 5, cin = 1,
   bcd_hund = 0, bcd_ten = 5, bcd_unit = 1

a_ten = 9, a_unit = 7, b_ten = 9, b_unit = 9, cin = 1,
   bcd_hund = 1, bcd_ten = 9, bcd_unit = 7

a_ten = 8, a_unit = 8, b_ten = 0, b_unit = 2, cin = 0,
   bcd_hund = 0, bcd_ten = 9, bcd_unit = 0

a_ten = 9, a_unit = 9, b_ten = 9, b_unit = 9, cin = 0,
   bcd_hund = 1, bcd_ten = 9, bcd_unit = 8

a_ten = 8, a_unit = 6, b_ten = 7, b_unit = 2, cin = 1,
   bcd_hund = 1, bcd_ten = 5, bcd_unit = 9

a_ten = 0, a_unit = 1, b_ten = 0, b_unit = 1, cin = 0,
   bcd_hund = 0, bcd_ten = 0, bcd_unit = 2

a_ten = 7, a_unit = 0, b_ten = 3, b_unit = 0, cin = 0,
   bcd_hund = 1, bcd_ten = 0, bcd_unit = 0

a_ten = 3, a_unit = 3, b_ten = 6, b_unit = 6, cin = 1,
   bcd_hund = 1, bcd_ten = 0, bcd_unit = 0

a_ten = 4, a_unit = 7, b_ten = 1, b_unit = 7, cin = 0,
   bcd_hund = 0, bcd_ten = 6, bcd_unit = 4

a_ten = 0, a_unit = 0, b_ten = 0, b_unit = 0, cin = 1,
   bcd_hund = 0, bcd_ten = 0, bcd_unit = 1

a_ten = 9, a_unit = 9, b_ten = 1, b_unit = 1, cin = 1,
   bcd_hund = 1, bcd_ten = 1, bcd_unit = 1

a_ten = 0, a_unit = 0, b_ten = 0, b_unit = 0, cin = 0,
   bcd_hund = 0, bcd_ten = 0, bcd_unit = 0
```

4.24 Design a behavioral module that performs true subtraction on two 32-bit operands. True subtraction can be defined as follows: $(+A) - (+B)$ or $(-A) - (-B)$, which has the following attributes: *fract_a* > *fract_b* and *sign_a* = *sign_b*. The exponents are eight bits. Obtain the behavioral design module, the test bench module, and the outputs.

```
//behavioral floating-point subtraction
//true subtraction: fract_a > fract_b, sign_a = sign_b
module sub_flp_bh (flp_a, flp_b, sign,
            exponent, exp_unbiased, rslt);

input [31:0] flp_a, flp_b; //define inputs and outputs
output sign;
output [7:0] exponent, exp_unbiased;
output [22:0] rslt;

//variables in always block are declared as registers
reg sign_a, sign_b;
reg [7:0] exp_a, exp_b;
reg [7:0] exp_a_bias, exp_b_bias;
reg [22:0] fract_a, fract_b;
reg [7:0] ctr_align;
reg [22:0] rslt;
reg sign;
reg [7:0] exponent, exp_unbiased;
reg cout;
                              //continued on next page
```

```
//define the sign, exponent, and fraction
always @ (flp_a or flp_b)
begin
   sign_a = flp_a[31];
   sign_b = flp_b[31];

   exp_a = flp_a[30:23];
   exp_b = flp_b[30:23];

   fract_a = flp_a[22:0];
   fract_b = flp_b[22:0];

//bias the exponents
exp_a_bias = exp_a + 8'b0111_1111;
exp_b_bias = exp_b + 8'b0111_1111;

//align the fractions
   if (exp_a_bias < exp_b_bias)
      ctr_align = exp_b_bias - exp_a_bias;

      while (ctr_align)
         begin
            fract_a = fract_a >> 1;
            exp_a_bias = exp_a_bias + 1;
            ctr_align = ctr_align - 1;
         end

   if (exp_b_bias < exp_a_bias)
      ctr_align = exp_a_bias - exp_b_bias;

      while (ctr_align)
         begin
            fract_b = fract_b >> 1;
            exp_b_bias = exp_b_bias + 1;
            ctr_align = ctr_align - 1;
         end

//------------------------------------------------------
//obtain the rslt
   if (fract_a > fract_b)
      begin
         fract_b = ~fract_b + 1;
         {cout, rslt} = fract_a + fract_b;
         sign = sign_a;
      end
                                    //continued on next page
```

```
//postnormalize
   while (rslt[22] == 0)
      begin
         rslt = rslt << 1;
         exp_b_bias = exp_b_bias - 1;
      end

   exponent = exp_b_bias;
   exp_unbiased = exp_b_bias - 8'b0111_1111;

end

endmodule
```

```
//test bench for floating-point subtraction

module sub_flp_bh_tb;

//inputs are reg for test bench
//outputs are wire for test bench
reg [31:0] flp_a, flp_b;
wire sign;
wire [7:0] exponent, exp_unbiased;
wire [22:0] rslt;

//display variables
initial
$monitor ("sign = %b, exp_unbiased = %b, rslt = %b",
            sign, exp_unbiased, rslt);

//apply input vectors
initial
begin
      //(+32) - (+30) = +2
      //          s ----e---- --------------f-------------
#0    flp_a = 32'b0_0000_1000_0010_0000_0000_0000_0000_000;
      flp_b = 32'b0_0000_1000_0001_1110_0000_0000_0000_000;

      //(-130) - (-25) = -105
      //          s ----e---- --------------f-------------
#10   flp_a = 32'b1_0000_1000_1000_0010_0000_0000_0000_000;
      flp_b = 32'b1_0000_0101_1100_1000_0000_0000_0000_000;

                                        //continued on next page
```

```
      //(+50) - (+40) = +10
      //          s ----e---- -------------f------------
#10   flp_a = 32'b0_0000_1000_0011_0010_0000_0000_0000_000;
      flp_b = 32'b0_0000_1000_0010_1000_0000_0000_0000_000;

      //(+105) - (+5) = +100
      //          s ----e---- -------------f------------
#10   flp_a = 32'b0_0000_0111_1101_0010_0000_0000_0000_000;
      flp_b = 32'b0_0000_0011_1010_0000_0000_0000_0000_000;

      //(+72) - (+47) = +25
      //          s ----e---- -------------f------------
#10   flp_a = 32'b0_0000_0111_1001_0000_0000_0000_0000_000;
      flp_b = 32'b0_0000_0110_1011_1100_0000_0000_0000_000;

      //(-127) - (-60) = -67
      //          s ----e---- -------------f------------
#10   flp_a = 32'b1_0000_0111_1111_1110_0000_0000_0000_000;
      flp_b = 32'b1_0000_0110_1111_0000_0000_0000_0000_000;

      //(+36.5) - (+5.75) = +30.75
      //          s ----e---- -------------f------------
#10   flp_a = 32'b0_0000_0110_1001_0010_0000_0000_0000_000;
      flp_b = 32'b0_0000_0011_1011_1000_0000_0000_0000_000;

      //(-720.75) - (-700.25) = -20.50
      //          s ----e---- -------------f------------
#10   flp_a = 32'b1_0000_1010_1011_0100_0011_0000_0000_000;
      flp_b = 32'b1_0000_1010_1010_1111_0001_0000_0000_000;

      //(+963.50) - (+520.25) = +443.25
      //          s ----e---- -------------f------------
#10   flp_a = 32'b0_0000_1010_1111_0000_1110_0000_0000_000;
      flp_b = 32'b0_0000_1010_1000_0010_0001_0000_0000_000;

      //(+5276) - (+4528) = +748
      //          s ----e---- -------------f------------
#10   flp_a = 32'b0_0000_1101_1010_0100_1110_0000_0000_000;
      flp_b = 32'b0_0000_1101_1000_1101_1000_0000_0000_000;
#10   $stop;
end

//instantiate the module into the test bench
sub_flp_bh inst1 (flp_a, flp_b, sign,
                     exponent, exp_unbiased, rslt);
endmodule
```

```
(+32) - (+30) = +2
sign = 0, exp_unbiased = 0000_0010,
   rslt = 1000_0000_0000_0000_0000_000

(-130) - (-25) = -105
sign = 1, exp_unbiased = 0000_0111,
   rslt = 1101_0010_0000_0000_0000_000

(+50) - (+40) = +10
sign = 0, exp_unbiased = 0000_0100,
   rslt = 1010_0000_0000_0000_0000_000

(+105) - (+5) = +100
sign = 0, exp_unbiased = 0000_0111,
   rslt = 1100_1000_0000_0000_0000_000

(+72) - (+47) = +25
sign = 0, exp_unbiased = 0000_0101,
   rslt = 1100_1000_0000_0000_0000_000

(-127) - (-60) = -67
sign = 1, exp_unbiased = 0000_0111,
   rslt = 1000_0110_0000_0000_0000_000

(+36.5) - (+5.75) = +30.75
sign = 0, exp_unbiased = 0000_0101,
   rslt = 1111_0110_0000_0000_0000_000

(-720.75) - (-700.25) = -20.50
sign = 1, exp_unbiased = 0000_0101,
   rslt = 1010_0100_0000_0000_0000_000

(+963.50) - (+520.25) = +443.25
sign = 0, exp_unbiased = 0000_1001,
   rslt = 1101_1101_1010_0000_0000_000

(+5276) - (+4528) = +748
sign = 0, exp_unbiased = 0000_1010,
   rslt = 1011_1011_0000_0000_0000_000
```

4.27 Design a behavioral module for a 32-bit single-precision floating-point multiplication operation for two operands: multiplicand *flp_a[31:0]* and multiplier *flp_b[31:0]*. The single-precision format is shown below. Use the multiply arithmetic operator (*) to perform the multiply operation. Obtain the behavioral module, the test bench module, and the outputs showing the product as a 23-bit result.

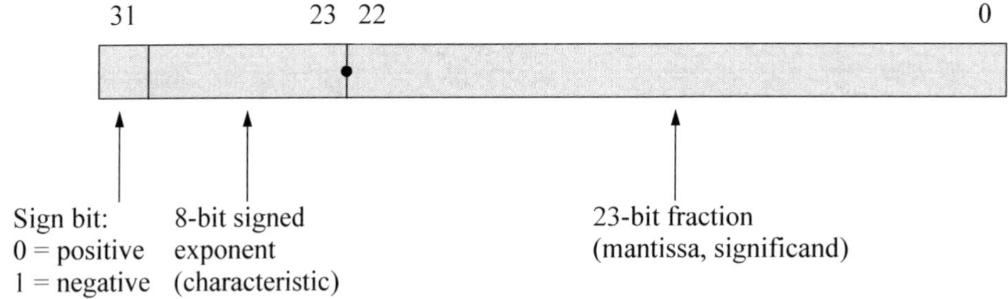

```
//behavioral floating-point multiplication

module mul_flp6 (flp_a, flp_b, sign,
                 exponent, exp_unbiased, exp_sum, prod);

//define inputs and outputs
input [31:0] flp_a, flp_b;
output sign;
output [7:0] exponent, exp_unbiased;
output [8:0] exp_sum;
output [22:0] prod;

//variables in always are declared as registers
reg sign_a, sign_b;
reg [7:0] exp_a, exp_b;
reg [7:0] exp_a_bias, exp_b_bias;
reg [8:0] exp_sum;
reg [22:0] fract_a, fract_b;
reg [45:0] prod_dbl;
reg [22:0] prod;
reg sign;
reg [7:0] exponent, exp_unbiased;
reg cout;
reg zero_opnd;

//define sign, exponent, and fraction
always @ (flp_a or flp_b)
begin
   if ((flp_a != 0) && (flp_b != 0))
      begin
         zero_opnd = 1'b0;

         sign_a = flp_a[31];
         sign_b = flp_b[31];

         exp_a = flp_a[30:23];
         exp_b = flp_b[30:23];

         fract_a = flp_a[22:0];
         fract_b = flp_b[22:0];

//bias exponents
      exp_a_bias = exp_a + 8'b0111_1111;
      exp_b_bias = exp_b + 8'b0111_1111;

                                    //continued on next page
```

```
//add exponents
      exp_sum = exp_a_bias + exp_b_bias;

//remove one bias
      exponent = exp_sum - 8'b0111_1111;
      exp_unbiased = exponent - 8'b0111_1111;

//multiply fractions
      prod_dbl = fract_a * fract_b;
      prod = prod_dbl[45:23];

//postnormalize product
      while (prod[22] == 0)
         begin
            prod = prod << 1;
            exp_unbiased = exp_unbiased - 1;
         end

      sign = sign_a ^ sign_b;

   end

   else
      zero_opnd = 1'b1;
end
endmodule
```

```
//test bench for floating-point multiplication
module mul_flp6_tb;

//inputs are reg in test bench
//outputs are wire in test bench
reg [31:0] flp_a, flp_b;
wire exp_ovfl;
wire sign;
wire zero_opnd;
wire [7:0] exponent, exp_unbiased;
wire [8:0] exp_sum;
wire [22:0] prod;

//display variables
initial
$monitor ("sign=%b, exp_unbiased=%b, prod=%b",
            sign, exp_unbiased, prod);
                              //continued on next page
```

```
//apply input vectors
initial
begin
      //+5 x +3 = +15
      //          s ----e---- -------------f-------------
#0    flp_a = 32'b0_0000_0011_1010_0000_0000_0000_0000_000;
      flp_b = 32'b0_0000_0010_1100_0000_0000_0000_0000_000;

      //+6 x +4 = +24
      //          s ----e---- -------------f-------------
#10   flp_a = 32'b0_0000_0011_1100_0000_0000_0000_0000_000;
      flp_b = 32'b0_0000_0011_1000_0000_0000_0000_0000_000;

      //-5 x +5 = -25
      //          s ----e---- -------------f-------------
#10   flp_a = 32'b1_0000_0011_1010_0000_0000_0000_0000_000;
      flp_b = 32'b0_0000_0011_1010_0000_0000_0000_0000_000;

      //+7 x -5 = -35
      //          s ----e---- -------------f-------------
#10   flp_a = 32'b0_0000_0011_1110_0000_0000_0000_0000_000;
      flp_b = 32'b1_0000_0011_1010_0000_0000_0000_0000_000;

      //+25 x +25 = +625
      //          s ----e---- -------------f-------------
#10   flp_a = 32'b0_0000_0101_1100_1000_0000_0000_0000_000;
      flp_b = 32'b0_0000_0101_1100_1000_0000_0000_0000_000;

      //+76 x +55 = +4180
      //          s ----e---- -------------f-------------
#10   flp_a = 32'b0_0000_0111_1001_1000_0000_0000_0000_000;
      flp_b = 32'b0_0000_0110_1101_1100_0000_0000_0000_000;

      //-48 x -17 = +816
      //          s ----e---- -------------f-------------
#10   flp_a = 32'b1_0000_0110_1100_0000_0000_0000_0000_000;
      flp_b = 32'b1_0000_0101_1000_1000_0000_0000_0000_000;

   //-20 x -20 = +400
      //          s ----e---- -------------f-------------
#10   flp_a = 32'b1_0000_0101_1010_0000_0000_0000_0000_000;
      flp_b = 32'b1_0000_0101_1010_0000_0000_0000_0000_000;

                                   //continued on next page
```

```
   //+64 x +128 = +8192
      //          s ----e---- --------------f-------------
#10   flp_a = 32'b0_0000_0111_1000_0000_0000_0000_0000_000;
      flp_b = 32'b0_0000_1000_1000_0000_0000_0000_0000_000;

#10   $stop;

end

//instantiate the module into the test bench
mul_flp6 inst1 (flp_a, flp_b, sign, exponent,
                     exp_unbiased, exp_sum, prod);

endmodule
```

```
+5 x +3 = +15
sign = 0, exp_unbiased = 0000_0100,
   prod = 1111_0000_0000_0000_0000_000

+6 x +4 = +24
sign = 0, exp_unbiased = 0000_0101,
   prod = 1100_0000_0000_0000_0000_000

-5 x +5 = -25
sign = 1, exp_unbiased = 0000_0101,
   prod = 1100_1000_0000_0000_0000_000

+7 x -5 = -35
sign = 1, exp_unbiased = 0000_0110,
   prod = 1000_1100_0000_0000_0000_000

+25 x +25 = +625
sign = 0, exp_unbiased = 0000_1010,
   prod = 1001_1100_0100_0000_0000_000

+76 x +55 = +4180
sign = 0, exp_unbiased = 0000_1101,
   prod = 1000_0010_1010_0000_0000_000

-48 x -17 = +816
sign = 0, exp_unbiased = 0000_1010,
   prod = 1100_1100_0000_0000_0000_000

                              //continued on next page
```

```
-20 x -20 = +400
sign = 0, exp_unbiased = 0000_1001,
   prod = 1100_1000_0000_0000_0000_000

+20 x -20 = -400
sign = 1, exp_unbiased = 0000_1001,
   prod = 1100_1000_0000_0000_0000_000

+64 x +128 = +8192
sign = 0, exp_unbiased = 0000_1110,
   prod = 1000_0000_0000_0000_0000_000
```

INDEX

Symbols

Numbers

A

D

E

P